A MATHEMATICAL JOURNEY

A Mathematical Journey

SECOND EDITION

Stanley Gudder

Department of Mathematics and Computer Science
University of Denver
Denver, Colorado

McGRAW-HILL, INC.

New York   St. Louis   San Francisco   Auckland   Bogotá   Caracas
Lisbon   London   Madrid   Mexico City   Milan   Montreal   New Delhi
San Juan   Singapore   Sydney   Tokyo   Toronto

A MATHEMATICAL JOURNEY

This book is printed on acid-free paper.

1 2 3 4 5 6 7 8 9 0 VNH VNH 9 0 9 8 7 6 5 4 3

ISBN 0-07-025130-4

This book was set in Times Roman by Paula Gudder and
Electronic Technical Publishing Services.
The editors were Michael Johnson, Karen M. Minette, and Margery Luhrs;
the designer was Joan Greenfield;
the production supervisor was Phil Galea.
The photo researcher was Debra Hershkowitz;
the photo editor was Kathy Bendo.
Von Hoffmann Press, Inc., was printer and binder.

Backcover print: M. C. Escher's "Relativity 1953" was the cover for the first edition of *A Mathematical Journey* by Stanley Gudder. This print can also be found in this new edition in the special section on M. C. Escher in Chapter Two: *Mathematics and Art.*

Library of Congress Cataloging-in-Publication Data

Gudder, Stanley.
    A mathematical journey / Stanley Gudder. — 2nd ed.
      p.   cm.
    Includes bibliographical references and index.
    ISBN 0-07-025130-4
    1. Mathematics.   I. Title.
QA39.2.G83   1994
  510—dc20   93-1359

CONTENTS

TO THE STUDENT

The idea that aptitude for mathematics is rarer than aptitude for other subjects is merely an illusion which is caused by belated or neglected beginners.
— FRIEDRICH JOHANN HERBART (1776–1841), German Educator

There can be no mystery in a result you have discovered for yourself.
— W. W. SAWYER, Twentieth-Century Mathematician

You are about to embark on a mathematical journey, and this will serve as your guidebook. You will see many sights; some will be beautiful and elegant, some have inspired the awe of thinking people for centuries, some will help you in daily life, and others will be merely interesting curiosities. As you can see from the table of contents, mathematics encompasses a huge spectrum of topics. During a tour of limited length, you cannot possibly see everything, so various important topics have been omitted. The ones that are present have been chosen for three reasons. First, they can be understood with a minimal background, requiring at most a grasp of high school algebra. Second, they are interesting and exciting even if you are not scientifically inclined (and there is nothing wrong with that). Third, they are useful. Some of the topics, such as consumer mathematics, computer science, matrices, and statistics, have direct applications to daily life and business. Although they have less direct applications, other topics, such as logic, number theory, geometry, probability, and graph theory, provide you with new ways of thinking and reasoning.

Each mathematical field you encounter will be like a large country which today is very much alive and vibrant yet which in the past has experienced times of despair as well as greatness. Like countries, these fields have common boundaries across which ideas and methods are traded, and the boundaries sometimes shift so that one field becomes part of another. Except for Chapter 1, each chapter of this book is essentially independent of the others. However, you will notice common threads and ideas that tie them all together. As with most adventures, you will encounter some surprises. You will see various fascinating connections between mathematics, art, and psychology. You will learn how to use logic to understand income tax forms. You will see how the Richter scale for measuring the strength of earthquakes and the decibel scale for measuring the loudness of sounds are constructed. You will learn how geometry is used to improve the range and accuracy of golf balls. You will view the beautiful world of fractals. You will find out what the law of supply and demand really means. You will see how probability theory can be used to interpret medical tests. This list goes on and on, and if you are patient you will see them all. Each of these topics may not be covered in the particular course you are taking, but there is nothing to prevent you from reading and understanding them on your own. For additional reading, you may want to consult the annotated bibliography at the end of this book. In the text, references from the bibliography are given in the form (author, date).

As your journey progresses you will view some striking mathematical landmarks called results or theorems. A theorem is a mathematical statement that has been proved to be true. The proof of a theorem is a rigorous demonstration of its validity. The proof is like the foundation of a great building: it is necessary to support the building and keep it stable, but it is of little interest to those admiring the building's external architectural qualities. We are interested in mathematical results because many of them have important applications in the real world. To apply these results you do not have to know their proofs. However, as a thinking, curious person, you probably would like to have an intuitive feeling for why a result works. For example, you can drive a car without understanding how the engine works. But if you do not want to be deceived by car advertisements and auto mechanics, you should have some idea about the construction and principles of car engines. Thus, although we shall not stress proofs in this book, we shall motivate results and explain to some extent why they are true.

To derive maximum benefit from any journey, first go prepared and then gain skills along the way to develop a deeper appreciation for later experiences. To review and reinforce the necessary skills for your mathematical journey, this book begins with an orientation chapter designed to prepare you for mathematical thinking. As you move from one mathematical field to another, the three major themes of concepts, results, and applications will recur. The concepts are usually given in the form of definitions and examples. It is important for you to understand their meaning and intuitive motivation. The applications will give uses for the concepts and results both inside and outside of mathematics.

The mission of this journey is to find an answer to the questions: What is mathematics and what is it good for? Mathematics is both an art and a science, and defining it is as difficult as trying to define music. You can listen to music, study music history and structure, play an instrument, and possibly even compose music, and only then do you really know what music is. You can study what great mathematicians have done and look at mathematical history and structure. But you also need to do mathematics and use mathematics before you have a true feeling for what it is and how it is applied.

This is both a business and a pleasure trip. The business end will require time, concentration, and work. You will need to read this book with pencil and paper in hand. You will need to work many problems and examples. Don't be afraid to experiment and use trial and error. If your first try fails, try again. Let your imagination take wing and you will fly. The pleasure will come from the sense of confidence and accomplishment that you will gain during your travels and from the joy in viewing beautiful and interesting vistas. The great adventure begins...

Bon Voyage!

TO THE
TEACHER

To educate means to influence the whole personality of the student.
— Geоrge Polya, Twentieth-Century Mathematician

The essence of mathematics is not to make simple things complicated but to make complicated things simple.
— Stanley Gudder, Author

Poor teaching leads to the inevitable idea that the subject [mathematics] is only adapted to peculiar minds, when it is the one universal science and the one whose four ground-rules are taught us almost in infancy and reappear in the motions of the universe.
— H. J. S. Smith, Twentieth-Century Mathematician

The objectives of this book are to present modern mathematics from an elementary point of view so as to be understandable to nonmathematics majors and nonscience majors and to foster an appreciation for the art, history, beauty, and applications of mathematics.

There are areas of modern mathematics that can be appreciated and understood at a reasonable level by liberal arts students. The topics chosen for this book include many of these areas. It is not the intention of this book to talk about doing mathematics but actually to *do* mathematics. A subject will not be touched upon and then quickly dropped but will be delved into fairly deeply. Of course, the student readers are not mathematical experts, and so tedious, esoteric details will not be considered. On the other hand, students will not be pampered with "fun and games" mathematics. They will have to stay alert and extend themselves but will be rewarded with the revelation that mathematics can be interesting and even exciting.

As the instructor you have a very important role. For many students the spoken word generates more excitement and enthusiasm than the written word. This book is intended to help you transmit this excitement and enthusiasm to your students. I have found that the material is best presented at a fairly leisurely pace. I have tried to draw the students out, to involve them in the material, to encourage them to question and ask questions, to understand the material completely before moving on. The exercises are an important part of this book and a considerable amount of time should be spent on them. The answers to selected odd-numbered exercises appear in the back of the book, and a teacher's manual is available containing the remaining answers. Starred exercises are more challenging and double-starred exercises are considerably more challenging than the others.

This book has been designed to be used at two different levels. Most of the chapters have optional sections that are at a higher level than the others. For a more rigorous course entire chapters can be covered, while for a more leisurely course optional sections and starred exercises can be omitted. For a cultural survey course you could just cover the first few sections of most of the chapters. Chapter 1, except for the last three sections, presents basic material used in later chapters and should not be omitted. Chapter 10 depends upon Chapter 9, Section 8.5 depends on Section 5.5, and Section 8.7 depends on Section 5.2. However, if you present Chapter 8 and not Chapter 5, the necessary concepts can be covered in one lecture. Otherwise, the chapters are essentially independent and you can pick and choose at will. In fact, one of the pleasures of teaching liberal arts mathematics is the flexibility of topics that is frequently granted to the instructor.

I think that one of the main reasons students have difficulty with a course like this is that the instructor goes too fast. This text contains much more material than one can cover in a year course. In Table 1, I have listed the approximate number of chapters that I recommend for coverage under various hour constraints. Table 2 lists recommended chapters for different types of emphasis.

Table 1

System	Hours/week	Number of chapters
per quarter	3	3–4
per quarter	4	4
per quarter	5	4–5
per semester	3	4–6
per semester	4	6
per semester	5	6–7

Table 2

Emphasis	Chapters
Applications	1, 6, 7, 8, 11
Probability and Statistics	1, 5, 8, 9, 10
Analytical Thought	1, 3, 4, 5, 9, 12
Pure Mathematics	1, 3, 4, 5, 11, 12
Math Appreciation	1, 2, 4, 9, 10
Cultural Survey	First few sections of 1, 2, 3, 4, 5, 8, 9

This is the second edition of a text that originally appeared in 1976. Since the appearance of the text it has become clear that the emphasis of this course has shifted toward more practical applications of mathematics. Because of this trend, the text has been substantially revised. Chapter 6 (Directed Graphs) and Chapter 7 (Group Theory) of the first edition have been deleted since they had a more theoretical character. The remaining chapters have been simplified and updated. They contain less theory and more explanation. Moreover, five new chapters have been added to the second edition. These new chapters include material dealing with practical, every-day applications. The new chapters are Chapter 3 (Logic), Chapter 5 (Geometry), Chapter 6 (Consumer Mathematics), Chapter 8 (Statistics), and Chapter 11 (Matrices). These changes are summarized in Table 3. I believe that this second edition presents a balanced survey of the beauty, culture, and real-world applications of mathematics.

Table 3

Chapters deleted from first edition	New chapters in second edition
Directed Graphs Group Theory	Logic Geometry Consumer Mathematics Statistics Matrices

At the end of the text is an annotated bibliography that you may find useful. You might use it as a source of further reading for your students or for supplementary material. Items in the bibliography are referred to in parentheses, such as (Agostini, 1983).

There are a number of supplements both students and instructors will find helpful. The Student's Solutions Manual contains solutions to selected odd-numbered exercises in the text. The Instructor's Resource Manual contains solutions to those problems that do not appear at the end of the text, tests for each chapter, midterms and finals, as well as tranparency masters that highlight important features in the text. The Professor's Assistant is a computerized test generator that allows the instructor to create tests using questions generated from a standard testbank. This testing system enables the instructor to choose questions either manually or randomly by section, question type, difficulty level, and other criteria. This system is available for IBM, IBM compatible, and Macintosh computers. The Print Test Bank is a printed and bound copy of the questions found in the standard testbank.

I have found that teaching liberal arts mathematics can be a very enjoyable and rewarding experience. This book has evolved slowly and is the outgrowth of years of trial and error and experimentation. I hope it will enhance your teaching endeavors.

PROLOGUE

Mathematicians feel they discover the laws that God himself could not avoid having to follow.
— DAVID MUMFORD, Twentieth-Century Mathematician

Mathematics is a jungle, the Jungle of the Infinite. Most of us see only broad paths driven through it by pioneers, trampled flat by generations of fellow students. A few go out and hack new paths of their own.
— IAN STEWART, Twentieth-Century Mathematician

Fundamentals stay the same.
— DIZZY GILLESPIE (1917–1993), Jazz Musician

One of the main themes of *A Mathematical Journey* is that mathematics abounds in the world around us. Moreover, mathematics obtains its inspiration and motivation from three sources: nature and the betterment of mankind, science and technology, and beauty and intellectual stimulation in its own right.

The patterns, the designs, the structures, the rhythms, the variations of nature are closely intertwined with mathematics. The ancient Greek mathematician Euclid once said: "The laws of nature are but the mathematical thoughts of God."

Birds, insects, fish, and animals seem to have a natural sense of distance, time, and direction. Birds migrate thousands of miles and frequently return to their summer home on the same date every year. Salmon that hatch from eggs in a tiny stream will travel thousands of miles and after three years return to their place of birth. How are these miracles accomplished? Although we are not absolutely certain, these creatures most likely have developed an instinctive means of calculating distance, time, and direction. Aided by their senses of smell, feel, sight, and sound, they are able to process information at a speed and accuracy that would rival a modern computer.

These creatures also frequently have an innate sense of geometry. Bees build their cells in hexagonal form, spiders build spiral webs, birds build nests of curved surfaces, animals stake out territorial boundaries.

Furthermore, mineral crystals are formed in the shape of polyhedra and snowflakes form lacy designs, each one different and yet each one a symmetrical, hexagonal pattern. Flowers exhibit beautiful symmetries: the three-petalled trillium, the four-petalled dogwood, the pentagonal blossom of the mountain laurel, the hexagonal blossoms of the lily, iris, and jonquil. Mathematical progressions frequently appear in nature: the spiral seeds of sunflowers and daisies, the protrusions in pinecones and pineapples.

The laws of science are almost always written in mathematical form. These laws range from those describing the elementary particles and basic forces of nature to ones describing the formation of galaxies and the cosmology of the early universe. Mathematics is used in biology, chemistry, sociology, medicine, and psychology. It is frequently employed in economics, business, and finance. In many ways, it has become indispensable for improving the quality of our lives.

The modern technology of our civilization is based upon mathematics. We have used mathematics and the laws of nature to construct buildings, bridges, automobiles, airplanes, ships, rockets, and satellites. We have designed computers, telephones, television, telescopes, radar, and lasers using the tools and techniques of mathematics. We have gained a better understanding of geology, earthquakes, and weather. We have sent people to the moon, explored our solar system, and developed vast sources of energy including nuclear, solar, and thermal energy. Humans have discovered better ways of planting, growing, and harvesting crops to feed our expanding population. None of this would have been possible without mathematics. Some of this technology has caused the pollution of our environment and mathematics has again been called upon to help control and correct this situation.

Mathematics is even prevalent in the world of art. For example, op art is based almost exclusively on mathematical principles. The famous Dutch artist M. C. Escher used mathematical ideas in much of his work. There are many concepts that appear in both art and mathematics. These include order, proportion, perspective, symmetry, progression, pattern, geometry, convexity, similarity, closure, relativity, transformation, convergence, limit, tessellations, and randomness.

Finally, mathematics has a beauty, excitement, and stimulation in its own right. It is a powerful language with qualities of conciseness, precision, and abstraction. It provides us with a means for thinking logically, precisely, and analytically.

This journey into the world of mathematics will discuss some of the topics that were previously mentioned and many more. It is hoped that you will gain a greater appreciation and understanding for this fascinating and useful intellectual world.

Stanley Gudder

ACKNOWLEDGMENTS

Many times a day I realize how much my own outer and inner life is built upon the labors of my fellow man, both living and dead, and how earnestly I must exert myself in order to give in return as much as I have received and am still receiving.
— ALBERT EINSTEIN (1879–1955), Physicist

I would like to express my appreciation to everyone at McGraw-Hill who worked to bring this book into print. In particular, I thank Michael Johnson, not only for suggesting that I write this second edition, but for his encouragement and enthusiasm about the project. I also thank Margery Luhrs for her advice, constant help, and attention to every detail of the production process, and Karen Minette for all her efforts in coordinating the supplements package. Debra Hershkowitz did an excellent job of photo research. Linda B. Farish of North Lake College checked the accuracy of the problems in great detail. The following reviewers as well as the proofreader and copy editors, gave constructive criticism and were extremely helpful: Ernest Berman, Harry S Truman College; George Bradley, Duquesne University; David Duncan, University of Northern Iowa; Barbara Dunn, St. Louis Community College, Meramec; Anne Flanigan, Hawaii Loa College; Melinda Gougeon, Greenfield Community College; Laurie McManus, St. Louis Community College, Meramec; Maurice Monohan, South Dakota State University; Joseph Moser, West Chester University of Pennsylvania; Christine Robinson, Florida Community College at Jacksonville; Tom Walsh, City College of San Francisco; Sue Welsch, Sierra Nevada College; Bruce Williamson, University of Wisconsin, River Falls; and Donald Zlotek, Luzerne County Community College.

A special thanks to the people at ETP Services Company; it was a pleasure to work with all of them. Their skills, professionalism, cooperation, and efficiency were invaluable for the completion of this project.

Finally, I thank my wife, Paula, for her total dedication to every stage of the production of this book. I could not have carried out this project without her.

Paula and I acknowledge the amazing support and encouragement from all our collegues at the University of Denver and from our wonderful extended family.

With great joy, we dedicate this volume to our loving family
Dolly, Gail, Michael, Walter, Joey, and Hannah
who have created with us a rewarding journey through life.

ORIENTATION

1

He that gives a portion of his time and talent to the investigation of mathematical truth will come to all other questions with a decided advantage.
— WALTER COLTON (1797–1851), American Clergyman and Writer

One of the purposes of this chapter is to give you practice in analytical and logical thinking. The best way to sharpen your skills is to work problems. You may surprise yourself and solve problems you did not think you could possibly do. Give yourself plenty of time; use trial and error; experiment. Some of the problems in this chapter will reappear later, where systematic methods for their solution will be developed.

Another purpose of this chapter is to set the stage for our later studies. The two fundamental concepts upon which all of mathematics is based are sets and functions. Much of this material may be review, but it is important to ensure that we all begin with the necessary common background. Later chapters will rely on a thorough understanding of these two concepts.

1.1 Try Some Puzzles

Use your head to learn the subject—use the subject to learn to use your head.
— G. POLYA, Twentieth-Century Mathematician

Mathematics is a valuable tool for solving quantitative problems. This application is obvious in the physical sciences, which are usually phrased in mathematical language. Moreover, now that the social, biological, and economic sciences are becoming more quantitative, mathematics is important in these areas as well. As this book progresses, you will see applications of mathematics in the physical, biological, social, and economic sciences, but the present section is devoted to the more mundane use of mathematics in solving puzzles.

You may have been at a party where the conversation degenerated into solving little puzzles. These party-type puzzles illustrate the problem-solving capability of

Photo 1.1

Auguste Rodin, *The Thinker*, 1880. Bronze, 79 in. $\times$ $51\frac{1}{4}$ in. $\times$ $55\frac{1}{4}$ in. (The Rodin Museum, Philadelphia, Gift of Jules E. Masbaum.)

mathematics. Although they are usually only interesting amusements, similar puzzles in the past have led to deeper and more important results. Try your hand at solving the following puzzles before you look at the solutions.

PUZZLE 1   A monk leaves at 6:00 a.m., climbs a trail to the top of a mountain at some unknown variable speed (he may stop to meditate, he may jog for a while), and arrives at the summit at 6:00 p.m. He sleeps until 6:00 a.m. and begins his descent along the same trail, arriving at the bottom at 6:00 p.m. Show that there is at least one place on the trail that he crosses at the same time of day going up and coming down.

Solution 1.   First consider a nonmathematical solution. Take a movie of the monk going up the mountain. It is unlikely that this will be a very interesting 12-hour movie deserving of an Oscar, but it can be done. Also take a movie of the monk going down the mountain. Now show these two movies simultaneously on two screens placed right next to each other. Initially you will see the monk at the bottom of the mountain on one screen and at the top on the other, but as the movie progresses, the monks on the screens will approach each other and eventually they will be at the same spot. At this time the monk was at the same spot at the same time on the two successive days.

Solution 2.   Solution 1 is admittedly clumsy and impractical. The mathematical solution is much more satisfying. Draw two curves plotting the monk's distance along the trail versus the time of day, one for his ascent and one for his descent, using the same axes (Figure 1.1). Regardless of what the curves look like, there will always be a point at which they cross, and this demonstrates what was to be shown.   □

This solution illustrates the great power of mathematics. You are given very little information: you do not know how fast the monk is traveling, how high the mountain

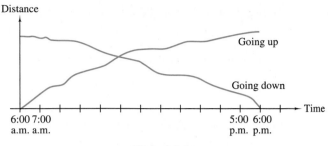

Figure 1.1

is, how steep the trail is, etc. However, you are not asked for much, either. There is no way to determine at what time the monk is at the same spot on the two days or how far he is along the trail at that time. Mathematics is able to give you a limited answer for a given limited amount of information.

PUZZLE 2   A rope is tied tight around the equator of the earth. A second rope is placed 1 foot directly above the first (Figure 1.2). About how much longer is the second rope than the first: (a) 6 ft, (b) 600 ft, (c) 60,000 ft, or (d) 6,000,000 ft?

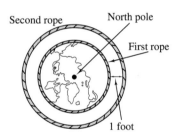

Figure 1.2

Solution.   Certainly the two ropes are very long, and since the second is 1 foot above the first all the way around the earth, you might expect the second to be much longer than the first. But this is not the case. Let l_1 and l_2 be the lengths of the first and second ropes, respectively, and let d be the diameter of the earth. Remembering that the circumference of a circle is π times its diameter, we have

$$l_1 = \pi d \quad \text{and} \quad l_2 = \pi(d+2)$$

so

$$l_2 - l_1 = \pi d + 2\pi - \pi d = 2\pi \approx 6 \text{ ft} \qquad \square$$

PUZZLE 3   This is a problem similar to one given to Friedrich Gauss when he was a small boy in elementary school. Evidently the teacher wanted some time for himself and gave the students a challenging problem to occupy them for a considerable period. Young Gauss solved the problem almost immediately with the pronouncement "there it lies," and the other students turned in their answers an hour later. Only Gauss had found the correct solution. The puzzle was: What is the sum of the first 100 natural numbers? (Recall that a natural number is a counting number or a whole number.) That is, $1 + 2 + 3 + \cdots + 100 = ?$

Solution.   If you add the numbers in order, the solution is very time-consuming, but if you add the first to the last, the second to the second to last, etc., you obtain fifty 101s:

$$(1 + 100) + (2 + 99) + (3 + 98) + \cdots + (50 + 51) = (50)(101) = 5050 \qquad \square$$

This illustrates a basic law of arithmetic, that addition is commutative. Regardless of the order in which numbers are added, you will still get the same answer. This property of addition is sometimes used by accountants to check sums.

PUZZLE 4   Mr. Dithers is paid a salary of \$1000 a week for 22 weeks. Mr. Bumstead earns 1¢ the first week, 2¢ the second week, 4¢ the third, 8¢ the fourth, and 2^{n-1}¢ the nth week for 22 weeks. Which 22-week salary is better?

Solution.   Let S be Bumstead's total 22-week salary. Then

$$S = 1 + 2 + 2^2 + 2^3 + \cdots + 2^{21}$$

Instead of adding all of these numbers, there is a simpler way to find S. Multiply both sides of this equation by 2 to obtain

$$2S = 2 + 2^2 + 2^3 + \cdots + 2^{22}$$

Subtract the original equation from the new equation:

$$S = 2S - S = 2^{22} - 1 = \$41{,}943.03$$

Thus Bumstead's salary is higher. The moral is that starting with a low salary and getting periodic raises is better than having a fixed high salary (if you can live on 1 cent the first week). $\qquad \square$

PUZZLE 5   There are five teams in a football league. Assuming there are no identical records at the end of the season, how many possible rankings are there for the teams of this league?

Solution.   Five different teams can come in first. For each team that comes in first there are four teams that can come in second, giving $(5)(4) = 20$ first- and second-place teams. For each set of first- and second-place teams there are three teams left that can come in third, giving $(5)(4)(3) = 60$ first-, second-, and third-place teams. Continuing in this way gives $(5)(4)(3)(2) = 120$ possible rankings. This number is usually denoted by 5!, called five factorial. $\qquad \square$

PUZZLE 6   *The paradox of the unexpected hanging.* A prisoner stood before the judge on Sunday. The judge pronounced the following sentence: *"You will be hanged at noon on one of the seven days of next week. But you will not know which day it is until you are so informed on the morning of the day of the hanging.*"

The judge was known to be a man who always kept his word. However, the prisoner reasoned as follows: "*I can't be hanged next Sunday. Sunday is the last day of the week. On Saturday afternoon I would know before I was told that the hanging would be on Sunday. So Sunday is positively ruled out. But now they can't hang me on Saturday since by Friday afternoon I would know the hanging would take place Saturday. So Saturday is out. In the same way I can rule out Friday, Thursday, Wednesday, Tuesday, and Monday. Thus the judge's decree cannot be carried out.*"

Nevertheless, the judge's decree was performed on Thursday. How is this possible?

Solution. This paradox was first circulated by word of mouth in the early 1940s and has been a topic of discussion ever since. More than 20 articles about it, written by philosophers, mathematicians, and physicists, have appeared in learned journals. The philosophical ramifications cannot be discussed here, but notice that since the prisoner did not expect to be hanged at all, the Thursday hanging was unexpected and therefore the judge's decree was carried out. What is your interpretation? ☐

PUZZLE 7 A concession stand sells hamburgers for $5.00, hot dogs for $2.00, and mints for 10¢. During a certain period they sold a total of 100 of these items and brought in exactly $100. How many of each item was sold?

Solution. Letting H, D, and M be the number of hamburgers, hot dogs, and mints sold, we have the following two equations:

$$H + D + M = 100$$

$$5H + 2D + \frac{1}{10}M = 100$$

Solving the first equation for D gives

$$D = 100 - H - M$$

Substituting our expression for D into the second equation gives

$$5H + 200 - 2H - 2M + 0.1M = 100$$

Simplifying, we have

$$3H + 100 = 1.9M$$

We thus have one equation and two unknowns, H and M. What do we do now? Well, we have one more piece of information. Since the left side of the last equation is a natural number, $1.9M$ must also be a natural number. This implies that M must be a multiple of 10. Hence, $M = 10M'$ for some natural number M'. We then have

$$3H + 100 = 19M'$$

What values can M' have? Since $19M'$ must be at least 100, M' cannot be 1, 2, 3, 4, or 5. Thus M' is 6, 7, 8, 9, or 10. Let's try $M' = 6$. We then obtain

$$3H + 100 = 114$$

which gives $H = 14/3$. But this does not work since H must be a natural number. Let's try $M' = 7$. We then have

$$3H + 100 = 133$$

which gives $H = 11$. Now solving for M' gives

$$133 = 19M'$$

so $M' = 7$ and $M = 70$. We conclude that $D = 19$. Thus, the stand sold 11 hamburgers, 19 hot dogs, and 70 mints. You can check that M' cannot be 8, 9, or 10, so this is the only possible answer.   □

PUZZLE 8   *To switch or not to switch.* Your friend Jane invites you to play the following game. She has three envelopes numbered 1, 2, and 3. She secretly places a dollar bill in one envelope and leaves the other two empty. You now choose one of the envelopes; say you choose number 1. Of the remaining two envelopes, she points out an empty one, say number 3. You now have the option of switching your choice to number 2. If your final choice contains the dollar, you get to keep it. If not, you must give Jane a dollar. Suppose you play this game many times. What is your best strategy? Should you always remain with your first choice, should you always switch, or should you sometimes switch according to your mood?

   Solution.   You should always switch. If you do, you win two times out of three. But, you may protest, what good does that do? In the above example, the dollar is just as likely to be in envelope 1 as in envelope 2. Why should I switch my choice from 1 to 2? But that's not the point. If you always keep your first choice, then there is only one chance in three that this choice is right, and this does not change when you learn the number of an empty envelope among the other two. Your first choice is either right or wrong. Since it is right one-third of the time, it's wrong two-thirds of the time. But then switching is right two-thirds of the time. This is the best strategy. If you try a mixed strategy of sometimes switching, your chance of winning is something between one-third and two-thirds.

   If you do not believe this, try playing the game, say 50 times, and see for yourself. If you are still not convinced, look at it this way. Suppose there were 100 envelopes, one containing a dollar and the others empty. You choose one, say envelope 1. Now there is only 1 chance out of 100 that this choice is right. You are now told that envelopes 3 through 100 are empty. Should you change your choice to envelope 2? Absolutely; there are 99 chances out of 100 that envelope 2 contains the dollar.   □

EXERCISES

Answers to selected odd-numbered exercises are given in the back of the book. Starred exercises are more challenging and double-starred exercises are very challenging.

   The area of a square whose sides have length 1 in. is, by definition, 1 sq in. Use this fact in Exercises 1–3 to find the areas of other squares.

   1. A square S has sides of length 4 in. Divide S into 1-in. squares to find the area of S. What is this area?

2. A square S has sides of length $1/2$ in. Divide a 1-in. square into smaller squares to find the area of S. What is this area?

3. A square S has sides of length 4.5 in. Divide S into smaller squares and rectangles to find the area of S.

4. The formula for the area of a square whose sides have length a is a^2. If the lengths of all the sides of a square are doubled, what happens to the area?

5. The formula for the area of a rectangle whose sides have lengths a and b is $a \times b$. If the lengths of two opposite sides of a rectangle are doubled, what happens to the area?

6. Suppose the lengths of the two legs (sides at right angles to each other) of a right triangle are a and b. Use the formula in Exercise 5 to find the area of the triangle.

7. If the monk in Puzzle 1 never stops to rest or turn back, show that there is precisely one time at which he was at the same spot the previous day at that time.

8. What is the sum of the first thousand natural numbers? The first million?

* 9. If n is a natural number, find $1 + 2 + 3 + \cdots + n$.

10. Suppose Mr. Dithers is paid \$10,000 a week for 31 weeks and Mr. Bumstead is paid as in Puzzle 4 for 31 weeks. Which 31-week salary is better? How much must Mr. Dithers be paid each week at a fixed salary to equal Bumstead's 31-week salary?

11. What is $1 + 3 + 3^2 + 3^3 + \cdots + 3^{10}$?

*12. If r is any number, find $1 + r + r^2 + r^3 + \cdots + r^n$.

13. Evaluate $4!$, $6!$, $100!/99!$, $20!/18!$.

14. If n is a natural number, show that the following equations hold:

$$n! = n(n-1)! \qquad n \geq 2$$
$$n(n-1)(n-2)(n-3)! = n! \qquad n \geq 4$$
$$\frac{n!}{(n-1)!} = n \qquad n \geq 2$$

15. With a pencil, start at any one of the points in Figure 1.3 and draw exactly four straight lines that include all nine points without lifting the pencil from the paper.

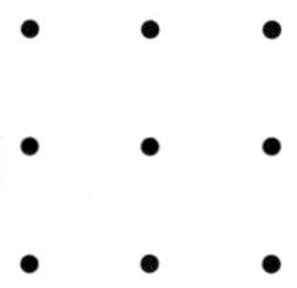

16. Write 100 using only four 9s and any of the arithmetic operations (addition, subtraction, multiplication, and division).

17. A man buys a shirt and a tie for \$9.50. The shirt costs \$5.50 more than the tie. What does each cost?

18. Two doctors are walking down the street. If one doctor is the father of the other doctor's son, how are the two doctors related?

19. A half is a third of it. What is it?

20. How can two American coins equal 30¢ if one is not a nickle?

21. An item is made from lead blanks in a lathe shop. Each blank is enough for one item. Lead shavings accumulated from making six items can be melted and made into a blank. How many items can be made from 36 blanks?

*22. How can you plant 10 trees in 5 rows with 4 trees in each row?

*23. What are the next three logical letters in the sequence OTTFFSSEN...?

24. A bus leaves New York for Boston at noon. An hour later a cyclist leaves Boston for New York, moving, of course, slower than the bus. When bus and cyclist meet, which of the two will be farther from New York?

Figure 1.3

25. What arithmetic symbol can be placed between 2 and 3 to make a number greater than 2 but less than 3?

**26. Three business associates check into a hotel. The clerk charges them $30 for a room, and they each give him a $10 bill. The clerk later discovers he gave them a $25 room, so he gives a bellhop $5 to return to the men. Since the bellhop doesn't know how to divide the $5 equally among the three men, he gives each man $1 and pockets the extra $2. Now each man has paid $9, making a total of $27, and the $2 for the bellhop adds up to $29. What happened to the other dollar?

27. Suppose in Puzzle 8 you are given a first choice but you are not told the number of an empty envelope among the remaining two. You are then given the option to switch. How would this change your strategy?

28. Suppose there were four envelopes, one containing a dollar bill, in Puzzle 8. You again have a first choice and are told the numbers of two empty envelopes among the other three. You are now given the option to switch. What is your best strategy, and what are your chances of winning?

29. If Puzzle 7 were exactly the same except that hamburgers sold for $4.00, how many of each item was sold? Is your answer the only one possible?

30. If Puzzle 7 were exactly the same except that hamburgers sold for $3.00, how many of each item was sold? Is your answer the only one possible?

31. Why would a hair stylist rather cut the hair of two blondes than of one brunette?

32. Why are manholes and their covers circular instead of square?

**33. Diophantus was an ancient Greek mathematician who is often called the father of algebra. Although we know he lived between A.D. 100 and 400, we do not know the exact period. However, his age at death is known because of the following ancient riddle. How long did Diophantus live? *"Diophantus' youth lasted 1/6 of his life. He grew a beard after 1/12 more of his life. After 1/7 more of his life, Diophantus married. Five years later he had a son. The son lived 1/2 as long as his father, and Diophantus died four years after his son's death. All this totals the years Diophantus lived."*

1.2 Why Think Abstractly?

Strange as it may sound, the power of mathematics rests on its evasion of all unnecessary thought and on its wonderful saving of mental operations.   — ERNST MACH (1838–1916), Austrian Physicist

One of the great powers of mathematics is its universal applicability. Its abstractness allows us to apply mathematical results to many special and practical situations. For example, Puzzle 3 showed that $1 + 2 + 3 + \cdots + 100 = (50)(101)$. In Exercise 8 the same reasoning could be applied to show that $1 + 2 + 3 + \cdots + 1000 = (500)(1001)$ and that $1 + 2 + 3 + \cdots + 1,000,000 = (500,000)(1,000,001)$. Is it possible to derive a general formula that would cover all these specific cases? Indeed, this was Exercise 9. If we let

$$S = 1 + 2 + \cdots + n$$

then

$$2S = (1 + 2 + \cdots + n) + (1 + 2 + \cdots + n)$$

Let's reverse the order of the sum in the second set of parentheses and write the terms under those of the first parentheses before adding the two:

$$1 \; + \; 2 \; + \; 3 \; + \cdots + \; n$$
$$\underline{+n \; + (n-1)+(n-2)+\cdots + \; 1}$$
$$2S = (n+1)+(n+1)+(n+1)+\cdots+(n+1) = n(n+1)$$

We have thus found the formula

$$S = \frac{n(n+1)}{2}$$

This formula is really an infinite collection of formulas giving the sum of any string of the first n natural numbers.

As another example, Puzzle 4 showed that $1+2+2^2+2^3+\cdots+2^{21} = 2^{22}-1$. Using similar reasoning, you showed in Exercise 11 that $1+3+3^2+3^3+\cdots+3^{10} = (3^{11}-1)/2$. In Exercise 12 you were asked to derive a general formula that covers all such specific cases. Again let $S = 1 + r + r^2 + \cdots + r^n$ (this is called a **geometric series**). Multiply both sides of this equation by r to obtain

$$rS = r + r^2 + r^3 + \cdots + r^{n+1}$$

Then

$$(r-1)S = rS - S = r^{n+1} - 1$$

and so

$$S = \frac{r^{n+1} - 1}{r - 1}$$

(This equation holds only for $r \neq 1$. If $r = 1$, then $S = n+1$.) Again this gives an abstract formula that can be used in many specific situations. If you let $r = 2$ and $n = 21$, you obtain the answer to Puzzle 4; if you let $r = 3$ and $n = 10$, you have the answer to Exercise 11; each set of values of r and n gives a different valid formula.

In grade school you were taught many facts about fractions. You were probably either very mystified by these facts or very bored. This is because you had to memorize a bunch of facts and were not told why they were true. You were not given these reasons because it required some abstraction, but now you are in a position to understand.

One of the things you were told was that if the denominators are equal you just add the numerators. For example,

$$\frac{2}{7} + \frac{3}{7} = \frac{5}{7}$$

Why is this always true? This statement is true for the present example because of the following rules of arithmetic:

$$\frac{2}{7} + \frac{3}{7} = 2\left(\frac{1}{7}\right) + 3\left(\frac{1}{7}\right) = (2+3)\left(\frac{1}{7}\right) = 5\left(\frac{1}{7}\right) = \frac{5}{7}$$

Be sure that you understand each of the previous steps. Now that you know that you can just add the numerators in this example, how can you be sure that this method works for all examples in which the denominators are equal? Let's replace the natural numbers in the fractions by letters. Since the letters represent natural numbers, the same rules of arithmetic hold for these letters. We thus have

$$\frac{a}{b} + \frac{c}{b} = a\left(\frac{1}{b}\right) + c\left(\frac{1}{b}\right) = (a+c)\left(\frac{1}{b}\right) = \frac{a+c}{b}$$

Now the letters a, b, and c can represent any natural numbers, so we know this result holds for any natural numbers.

If the denominators are not the same, you were told to "cross multiply." For example,

$$\frac{2}{3} + \frac{4}{5} = \frac{2(5) + 3(4)}{3(5)} = \frac{22}{15} = 1\frac{7}{15}$$

Why is this? Applying the rules of arithmetic for this example, we obtain

$$\frac{2}{3} + \frac{4}{5} = \frac{5}{5} \cdot \frac{2}{3} + \frac{3}{3} \cdot \frac{4}{5} = \frac{5(2) + 3(4)}{3(5)} = \frac{22}{15}$$

Again, using letters, we have

$$\frac{a}{b} + \frac{c}{d} = \frac{d}{d} \cdot \frac{a}{b} + \frac{b}{b} \cdot \frac{c}{d} = \frac{ad}{bd} + \frac{bc}{bd} = \frac{ad + bc}{bd}$$

Such reasoning can be used to explain all the facts you were told about fractions.

Now consider another example that illustrates the power of abstract mathematics. Suppose you want to solve the following three problems.

PROBLEM 1   A rowboat race is set up so that the boats start at point A, are rowed to any point on the opposite shore, and end at point B (see Figure 1.4). Naturally, a rower would want to row to a point C so that his total rowing distance is a minimum. Where is C?

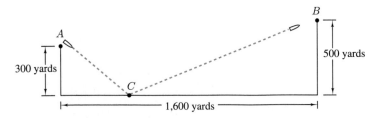

Figure 1.4

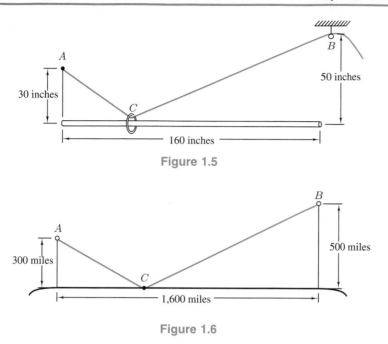

Figure 1.5

Figure 1.6

PROBLEM 2 A rope attached to point A is drawn through a ring on a rod, placed over a peg at B, and pulled (see Figure 1.5). The ring will settle at a position C, for which the total length of rope from A to B is a minimum. Where is C?

PROBLEM 3 An experiment is carried out using a laser emitter in an orbiting earth satellite. If a laser beam is emitted from satellite A and received by satellite B, at what point C of the earth is it reflected, assuming the earth is approximately a flat plane at the point of reflection (see Figure 1.6)? Light travels in a path that takes it from one point to another in the shortest time. Since light travels at a constant speed, the point C must be so located that the total distance traveled is a minimum.

How do we solve these three problems? We first note that although the three problems are superficially different, they have important similarities. If we strip away the nonessentials, we have the abstract mathematical problem diagrammed in Figure 1.7.

We denote the distance from A to C by $\overline{AC}$, etc. Then the problem becomes: Where is C so that $\overline{AC} + \overline{CB}$ is a minimum? In order to solve this problem we must use our imagination. One of the first things to pop into our minds is that a straight line is the shortest distance between two points. But this doesn't seem to help, since a route from A to the line PQ and then to B cannot be straight.

Let us try a technique that is frequently useful in mathematics (and in life as well). Let us try to solve a simpler problem and see if we can use it to solve our original problem. Suppose A' is a point the same distance as A from P but on the other side of the line PQ (Figure 1.8). Now we want C so that $\overline{A'C} + \overline{CB}$ is a minimum. But this is easy. Draw the straight line $A'B$ from A' to B, and C is the intersection of this line with the line PQ. So we have found the right point C for A'. Can we use

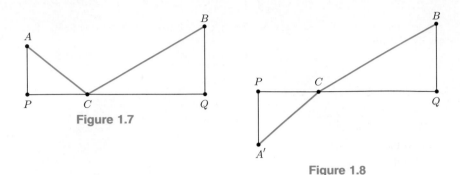

Figure 1.7

Figure 1.8

this to find the C for A? We now use a little plane geometry. Since APC and $A'PC$ are right triangles with equal legs, they have the same length hypotenuse $\overline{AC} = \overline{A'C}$. Since $\overline{A'C} + \overline{CB}$ is a minimum, $\overline{AC} + \overline{CB}$ is a minimum. Thus the point C that solved our simpler problem also turns out to solve our original problem.

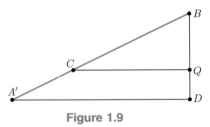

Figure 1.9

Now suppose we know $\overline{AP}$, $\overline{BQ}$, and $\overline{PQ}$. How do we find $\overline{PC}$ and $\overline{CQ}$? Since $A'DB$ and CQB have parallel sides, their sides are proportional (see Figure 1.9). Thus

$$\frac{\overline{CQ}}{\overline{A'D}} = \frac{\overline{BQ}}{\overline{BD}} \quad \text{and} \quad \overline{CQ} = \frac{\overline{A'D} \times \overline{BQ}}{\overline{BD}}$$

Although there seems to be no relationship between rowboats, ropes and rings, and satellites, we have used abstract mathematics to solve problems concerning them as well as many other problems of a similar nature. In order to solve Problems 1 to 3 numerically, we need only substitute the particular numbers of these problems. For example, in Problem 1 we have

$$\overline{CQ} = (1600) \left(\frac{500}{800} \right) = 1000 \text{ yd}$$

Then

$$\overline{PC} = 1600 - 1000 = 600 \text{ yd}$$

EXERCISES

If a is any number and n is a natural number (a positive whole number), we define a^n to be a times itself n times. That is,

$$a^n = \underbrace{a \times a \times \cdots \times a}_{n \text{ times}}$$

Use this definition to solve Exercises 1–8.

1. Find 2^3, 3^2, 4^2, 5^3.

2. Show that $a^3 = aa^2$.

3. Show that $a^4 = a^2a^2$.

4. Show that $1^n = 1$.

5. If n and m are natural numbers, show that $a^{n+m} = a^n a^m$.

6. If the formula $a^{n+m} = a^n a^m$ is to hold for $n = 0$, show that $a^0 = 1$.

7. If the formula $a^{n+m} = a^n a^m$ is to hold for negatives of natural numbers, show that $a^{-n} = 1/a^n$.

8. If n and m are natural numbers, show that $(a^n)^m = a^{nm}$.

* 9. The formula $(a^n)^m = a^{nm}$ need not hold if n and m are not natural numbers. Show this by letting $a = -1$, $n = 2$, and $m = 1/2$.

10. You were told in grade school that

$$\frac{\dfrac{1}{3}}{\dfrac{2}{}} = \frac{2}{3}$$

Using letters instead of numbers, show that this result holds for all fractions of natural numbers.

11. You were told in grade school that

$$\frac{8}{11} \leq \frac{7}{9}$$

since $8(9) \leq 7(11)$. Using letters instead of numbers, show that this result holds for all fractions of natural numbers.

12. What is the solution of Problem 1 if $\overline{AP} = 100$ yd, $\overline{BQ} = 600$ yd, and $\overline{PQ} = 1400$ yd?

13. If in Problem 3 we have $\overline{AP} = 200$ mi, $\overline{PQ} = 1000$ mi, and $\overline{PC} = 200$ mi, how high is satellite B?

14. The electric company wishes to place a pole at the street and connect wires from the pole to the two houses in Figure 1.10 so that the smallest amount of wire is needed. Where should the pole be placed?

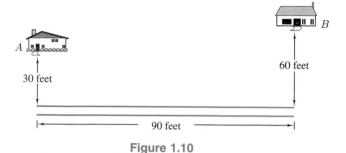

A

30 feet

B

60 feet

90 feet

Figure 1.10

*15. When a light beam is reflected from a plane surface, the angle of incidence equals the angle of reflection. Use this fact to solve Problem 3.

*16. Tom, Dick, and Harry are sitting alongside each other in a movie. Tom always tells the truth, Dick sometimes tells the truth, and Harry never tells the truth. The man on the left says, "The guy in the middle is Tom." The man in the middle says, "I'm Dick," and the man on the right says, "The guy in the middle is Harry." What are the correct names of the three men?

*17. Two guards lead a prisoner into a room containing two doors. The prisoner must select one of the doors. One leads to freedom, and the other leads to certain death. He is allowed to ask one of the guards one question. One guard always lies and the other always tells the truth. What question should he ask to guarantee his freedom?

18. A detachment of soldiers must cross a river. The bridge is broken, and the river is deep. Suddenly the officer in charge spots two boys playing in a rowboat. The rowboat is so small, however, that it can only hold two boys or one soldier. Still, all the soldiers succeed in crossing the river in the boat. How?

19. You are given eight identical-looking coins and a balance scale. Seven of the coins weigh the same, but the eighth is lighter. You may use the balance three times to determine which is lighter. Can you do it?

*20. There are 10 stacks each containing 10 silver dollars. You are given the weight of a real silver dollar and are told that each counterfeit coin weighs 1 gram more than a real silver dollar. You also know that precisely one of the stacks is completely counterfeit, and you can use a scale that weighs by grams. What is the minimum number of weighings needed to determine the counterfeit stack?

21. (This problem can be found in eighth-century writings.) A man has to take a wolf, a goat, and some cabbage across a river. His rowboat has enough room for the man plus either the wolf or the goat or the cabbage. If he takes the cabbage with him, the wolf will eat the goat. If he takes the wolf, the goat will eat the cabbage. Only when the man is present are the goat and cabbage safe from their enemies. All the same, the man carries wolf, goat, and cabbage across the river. How?

22. A blacksmith wants to join five short pieces of chain into a long chain (see Figure 1.11). He could open ring 1 (first operation), link it to ring 2 (second operation), then open ring 3 and link it to ring 4, and so on; but this would take eight operations. How can he do it in six?

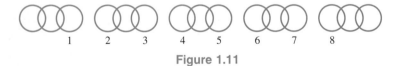

$$\quad 1 \quad\quad 2 \quad 3 \quad\quad 4 \quad 5 \quad\quad 6 \quad 7 \quad\quad 8$$

Figure 1.11

23. A boy has as many sisters as brothers, but each sister has only half as many sisters as brothers. How many brothers and sisters are in the family?

24. A lone goose was flying in the opposite direction from a flock of geese. He cried, "Hello, 100 geese!" The leader of the flock answered, "We aren't 100. If you take twice our number and add half our number and add a quarter our number and finally add you, the result is 100, but—well, you figure it out." How many geese were in the flock?

*25. There is an interesting five-digit number. With a 1 after it, it is 3 times as large as with a 1 before it. What is it?

*26. The devil said to the fool, "Walk across the bridge and I will double the money you have now. In fact, each time you cross I will double your money. But there is one small thing. Since I am so generous you must give me $24 after each crossing." The fool agreed. He crossed the bridge and stopped to count his money—A miracle! it had doubled. He threw $24 to the devil and crossed again. His money doubled, he paid another $24 and crossed a third time. Again his money doubled. But now he had only $24 and he had to give it all to the devil. (Moral?) How much money did he start with?

*27. John is 32 years old. When he was as old as his wife is now, her age was half his age now. How old is his wife now? What is the answer if John is 40 instead of 32? Show that no matter how old John is now, his wife's age is three-fourths his own age.

*28. Farmer Jones plans to go into the fowl business. He plans to spend $100 for a total of 100 roosters, hens, and baby chicks. How many of each kind must he purchase if roosters cost $5 each, hens $1 each, and baby chicks 5¢ each? (He buys at least one rooster.)

29. Three married couples want to cross a river in a rowboat that carries two people at a time. Each husband is very jealous and will not allow his wife to be in the boat or ashore with one or two of the other men unless he is present. How do they get across?

30. A six-story house (not counting the basement) has stairs of the same length from floor to floor. How many times as high is a climb from the first to the sixth floor as a climb from the first to the third floor?

*31. Using only a 3-quart container and a 5-quart container, any whole number of quarts from 1 through 8 can be measured out and transported. How is this possible?

32. "How much does one cost?" "Fifty cents," replied the hardware store clerk. "And how much does 18 cost?" "One dollar." "Okay, I will take 706." "That will be a dollar and a half." What was the customer buying?

33. A triangle has sides of length 13, 18, and 31 in. What is its area?

**34. Fill in the blanks. The number of occurrences in this sentence of 0s is _____ , of 1s is _____ , of 2s is _____ , of 3s is _____ , of 4s is _____ , of 5s is _____ , of 6s is _____ , of 7s is _____ , of 8s is _____ , of 9s is _____ .

1.3 Sets

We don't do things like that in our set.   —OLD SAYING

One of the basic concepts in mathematics is that of a set. Similar concepts come up in everyday life all the time. We have herds of cattle, flocks of birds, schools of fish, football teams, stamp collections, and so forth.

> We shall consider a **set** to be a collection of objects or elements.

This naive definition can lead to trouble, and mathematicians have given more precise definitions for sets, but for what we wish to do here, this intuitive definition will suffice.

What is wrong with our naive definition of a set? For one thing, our definition is circular. If a set is a collection of objects, then what is a collection? A collection is an

Table 1.1

Set	Subsets	Number of subsets
$\emptyset$	$\emptyset$	$1 = 2^0$
$\{a\}$	$\emptyset, \{a\}$	$2 = 2^1$
$\{a, b\}$	$\emptyset, \{a\}, \{b\}, \{a, b\}$	$4 = 2^2$
$\{a, b, c\}$	$\emptyset, \{a\}, \{b\}, \{c\}, \{a, b\},$ $\{a, c\}, \{b, c\}, \{a, b, c\}$	$8 = 2^3$

aggregate of objects. But what is an aggregate? An aggregate is a bunch of objects. What is a bunch? A bunch is a set of objects. Also, what is an object? For another thing, our definition can actually lead to contradictions. We shall consider some of these a little later. However, in this book we will consider only very simple sets, such as sets of numbers and finite sets of points. For such simple sets we will be on safe ground.

A set is frequently described by enclosing a list of its elements within braces. Thus, for example, the set of all natural numbers can be denoted by $\{1, 2, 3, \ldots\}$, the set consisting of the first four natural numbers by $\{1, 2, 3, 4\}$, the set of letters of the alphabet by $\{a, b, c, \ldots, x, y, z\}$. A set need not be a collection of numbers or letters; it can be any collection of objects as long as the elements can be unambiguously identified. Thus the collection of all mountains over 14,000 ft high or the collection of all past presidents of the United States is a set. The collection of all nice people is not a set; since "nice" is a subjective term depending on the individual's interpretation, the elements cannot be unambiguously identified.

Subsets and Supersets

If an object a is an element (or member) of a set A, we write $a \in A$. If a is not an element of a set A, we write $a \notin A$. For example, $2 \in \{1, 2, 3, 4\}$, $5 \notin \{1, 2, 3, 4\}$, and $d \in \{a, b, c, \ldots, x, y, z\}$.

> If every element of a set A is also an element of a set B, we write $A \subseteq B$ and say that A is a **subset** of B and B is a **superset** of A.

Thus, for example, $\{1, 2\} \subseteq \{1, 2, 3, 4\}$, and $\{1, 2, 3, 4\} \subseteq \{1, 2, 3, \ldots\}$. There is a special set called the **empty set**, denoted by $\emptyset$, which contains no elements. Notice that $\emptyset \subseteq A$ for any set A.

How many subsets does a set have? In Table 1.1 we list some sets, their subsets, and the number of subsets. Continuing this pattern, we might make the following conjecture.

> If a set has n elements, then it has 2^n subsets.

This conjecture is true. How can we prove it? Let us take a set $A = \{a_1, a_2, a_3, a_4, a_5\}$ with five elements and see if we can prove that is has $2^5 = 32$ subsets. We could just list all the subsets of A and then count them, but this would be very cumbersome. What we need is a systematic way of describing the subsets of A. The following is such a way. We can make each subset of A correspond to precisely one 5-tuple of 0s and 1s. For example, if $B = \{a_2, a_4, a_5\}$, then B corresponds to the 5-tuple $(0, 1, 0, 1, 1)$. We obtain this correspondence as follows. If $a_j \in B$, then we put a 1 in the jth position of the 5-tuple, and if $a_j \notin B$, then we put a 0 in the jth position. Thus, since $a_1 \notin B$, we have a 0 in the first position, since $a_2 \in B$ we have a 1 in the second position, etc. In this way we see that there are just as many subsets of A as there are 5-tuples of 0s and 1s. How many 5-tuples of 0s and 1s are there? We have two possibilities for the first position (either a 0 or a 1), two for the second, two for the third, etc. It follows that there are $(2)(2) = 2^2$ possibilities for the first two positions, $(2)(2)(2) = 2^3$ possibilities for the first three positions, 2^4 possibilities for the first four positions, and 2^5 possibilities for the first five positions. Hence there are $2^5 = 32$ different 5-tuples of 0s and 1s and this same number of subsets A. By repeating this same argument we can show that if $C = \{a_1, a_2, \ldots, a_n\}$ has n elements then there are as many subsets of C as there are n-tuples of 0s and 1s. Since there are 2^n different n-tuples of 0s and 1s, C has 2^n subsets.

Subsets with a Property

If S is a set and P is a property of certain elements of S, then the subset of S consisting of those elements satisfying property P is denoted by $\{s \in S: s \text{ has property } P\}$. For example, if the natural numbers are denoted by $\mathbb{N} = \{1, 2, 3, \ldots\}$, the set of natural numbers from 1 to 10 may be written $\{n \in \mathbb{N}: 1 \leq n \leq 10\}$, the even natural numbers may be written $\{n \in \mathbb{N}: n = 2k, k \in \mathbb{N}\}$, the odd natural numbers may be written $\{n \in \mathbb{N}: n = 2k - 1, k \in \mathbb{N}\}$, and the natural numbers that are squares may be written $\{n \in \mathbb{N}: n = k^2, k \in \mathbb{N}\}$. Besides the natural numbers $\mathbb{N}$, there are other sets of numbers that will be useful. The **positive rational numbers** $Q = \{m/n: m, n \in \mathbb{N}\}$ are the quotients of natural numbers. The **integers** $I = \{0, 1, -1, 2, -2, \ldots\}$ are the natural numbers together with their negatives and zero. The **real numbers** $\mathbb{R}$ are all possible decimal expansions. Thus 5, 2.376, π, $\sqrt{2}$, $3/4$, -67.1276 are examples of real numbers. Notice that $\mathbb{N} \subseteq Q \subseteq \mathbb{R}$ and $\mathbb{N} \subseteq I \subseteq \mathbb{R}$.

Paradoxes

It is now shown that if we do not use a more precise and restricted definition of sets than the one we are using, contradictions can result. You are probably familiar with some of the famous logical paradoxes. (*Paradox* is derived from the Greek words *para*, meaning beside, and *doxa*, meaning opinion. Thus a paradox is something that is not in agreement with prevailing opinion.) For example, the council of a certain village is said to have given orders that the village barber was to shave all the men in the village who did not shave themselves, and only those men. Who shaved the barber? A certain library undertook to compile a bibliographic catalogue listing all bibliographic catalogues that did not list themselves. Did the catalogue list itself? We now give a mathematical paradox of a similar nature that leads to a contradiction.

Notice first that it is possible for a set to have itself as one of its elements. For example, $S = \{1, 2, 3, S\}$ is such a set. Let R be the set of all sets that do not have themselves as an element; that is, $R = \{S: S \notin S\}$. Now if $R \in R$, then R is a set that does not have itself as an element, a contradiction. On the other hand, if $R \notin R$, then R is a set that has itself as an element, a contradiction! When this paradox was first discovered by Bertrand Russell, it created a minor revolution is mathematics that resulted in changes in the definition of a set. Since we will not deal with such wild sets in this book, our naive definition will be good enough.

Union

Let A and B be sets. The set consisting of the elements that belong to A or B (or both) is called the **union** of A and B, denoted $A \cup B$. Thus,

$$A \cup B = \{x: \ x \in A \text{ or } x \in B\}$$

For example,
$$\{1, 2, 3\} \cup \{3, 4, 5\} = \{1, 2, 3, 4, 5\}$$
$$\{1, 5, 7\} \cup \{a, b\} = \{1, 5, 7, a, b\}$$

Intersection

The set consisting of the elements that belong to both A and B is called the **intersection** of A and B, denoted $A \cap B$. Thus,

$$A \cap B = \{x: \ x \in A \text{ and } x \in B\}$$

For example,
$$\{1, 2, 3, 4\} \cap \{1, 2, 3, \ldots\} = \{1, 2, 3, 4\}$$
$$\{a, b, c\} \cap \{b, d, f\} = \{b\}$$
$$\{a, b, c\} \cap \{e, f\} = \emptyset$$

Venn Diagram

A useful way of picturing the relationships between sets, their unions, intersections, etc., is by depicting sets as circles in **Venn diagrams**. For example, Figure 1.12 gives Venn diagrams for $A \cap B$, $A \cup B$, and $A \subseteq B$.

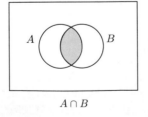
$A \cap B$

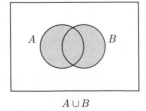
$A \cup B$

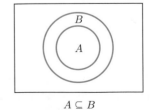
$A \subseteq B$

Figure 1.12

Equality of Sets

How can we tell if two sets are equal? Well, two sets are **equal** if they contain the same elements. Thus $A = B$ if $a \in A$ implies that $a \in B$ and if $b \in B$ implies that $b \in A$.

EXAMPLE 1   If $A = \{1, 2, 3\}$ and $B = \{3, 1, 2\}$, show that $A = B$.

  Solution.   Since A and B contain the same elements, they are equal. This shows that it does not make any difference what order we list the elements of a set in.   □

EXAMPLE 2   If $A \subseteq B$ and $B \subseteq A$, show that $A = B$.

  Solution.   Since $A \subseteq B$, if $a \in A$ then $a \in B$. Since $B \subseteq A$, if $b \in B$ then $b \in A$. Hence, $A = B$.   □

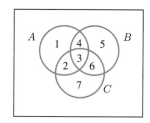

$$(A \cup B) \cap C = (A \cap C) \cup (B \cap C) = 2 \cup 3 \cup 6$$

Figure 1.13

There are important laws that hold for combinations of sets. One of these is the **distributive law**:

$$(A \cup B) \cap C = (A \cap C) \cup (B \cap C)$$

This law is depicted in the Venn diagram of Figure 1.13. In this diagram, each of the important regions is numbered. When $A = 1 \cup 2 \cup 3 \cup 4$ is written, it means that set A is the union of the sets numbered by 1, 2, 3, and 4. Now

$$A \cup B = 1 \cup 2 \cup 3 \cup 4 \cup 5 \cup 6$$

and
$$C = 2 \cup 3 \cup 6 \cup 7$$

Hence
$$(A \cup B) \cap C = 2 \cup 3 \cup 6$$

On the other hand,
$$A \cap C = 2 \cup 3 \quad \text{and} \quad B \cap C = 3 \cup 6$$

Hence
$$(A \cap C) \cup (B \cap C) = 2 \cup 3 \cup 6$$

which is the same as $(A \cup B) \cap C$.

  Another form of the distributive law is $(A \cap B) \cup C = (A \cup C) \cap (B \cup C)$. Can you illustrate this law using a Venn diagram?

Complements

In any particular context, the sets are usually subsets of one **universal set** U. This is the set represented by the rectangle in the Venn diagram.

> If $A \subseteq U$, then A' is the set consisting of those elements of U that are **not** elements of A. We call A' the **complement** of A.

Thus $A' = \{a \in U: a \notin A\}$. Notice that $U' = \emptyset$, $\emptyset' = U$, and $(A')' = A$ for all $A \subseteq U$. If $A, B \subseteq U$, the notation $A - B = A \cap B'$ is used. Notice that $A - B$ is the set consisting of those elements that are contained in A but are not contained in B. Figure 1.14a illustrates A and A'. Figure 1.14b shows that $A - B = 1$, $A \cap B = 2$, and $B - A = 3$.

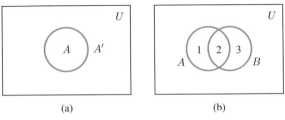

(a)   (b)

Figure 1.14

EXAMPLE 3   If $U = \{1, 2, 3, 4, 5, 6\}$ is the universal set, find the complements of
(a) $\{2, 4, 6\}$,   (b) $\{1, 2, 3, 5\}$.
  Solution.   (a) $\{2, 4, 6\}' = \{1, 3, 5\}$ (b) $\{1, 2, 3, 5\}' = \{4, 6\}$ □

EXAMPLE 4   If $U = \{a, b, c, d, e, f, g\}$, $A = \{a, c, e, g\}$, $B = \{b, c, d, e\}$, $C = \{a, f, g\}$, find (a) $A - B$, (b) $B - A$, (c) $A - C$, (d) $C - A$, (e) $B - C$, (f) $A - A$.
  Solution.   (a) $A - B = \{a, g\}$ (b) $B - A = \{b, d\}$ (c) $A - C = \{c, e\}$
(d) $C - A = \{f\}$ (e) $B - C = B$ (f) $A - A = \emptyset$ □

  There are some important laws that hold for complements. If $A \subseteq B$, then $B' \subseteq A'$. This is depicted in the Venn diagram in Figure 1.15a, where $A = 1$, $B = 1 \cup 2$, $B' = 3$, and $A' = 2 \cup 3$. The other laws are

> $(A \cup B)' = A' \cap B'$ and $(A \cap B)' = A' \cup B'$

These are called **De Morgan's laws**. In Figure 1.15b observe that $(A \cup B)' = 4$ while $A' = 3 \cup 4$, $B' = 1 \cup 4$, and so $A' \cap B' = 4$. Also $(A \cap B)' = 1 \cup 3 \cup 4$ and $A' \cup B' = 1 \cup 3 \cup 4$.

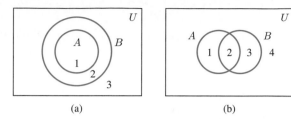

Figure 1.15

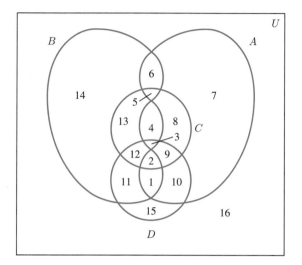

Figure 1.16

There is an interesting problem associated with Venn diagrams. In a Venn diagram with n sets, what is the largest number of regions into which U is divided? Referring to Figure 1.14a, observe that if there is one set A, then U is divided into two regions, namely A and A'. If there are two sets A and B, then U is divided into four regions, namely $A \cap B'$, $A \cap B$, $A' \cap B$, and $A' \cap B'$ (see Figure 1.14b). If there are three sets, then U is divided into eight regions (see Figure 1.13). Continuing this pattern, we conjecture that n sets can divide U into 2^n regions. This conjecture turns out to be true. Can you prove it? When one has four sets, the regions get fairly complicated and it is impossible to obtain the maximum number of regions using four circles. More complex figures must be used. The situation is depicted in Figure 1.16.

EXERCISES

1. Let $A = \{1, 2, 3\}$ and $B = \{2, 3, 4\}$. Find $A \cup B$ and $A \cap B$.
2. If $U = \{a, b, c\}$, $A = \emptyset$, $B = \{a\}$, $C = \{b, c\}$, and $D = \{a, b\}$, find A', B', C', D'.
3. If $U = \{1, 2, 3, 4, 5, 6, 7, 8\}$, find the complement of each of the following sets.
   (a) $\{1, 3, 5, 7\}$ (b) $\{2, 4, 6, 8\}$ (c) $\{3, 4, 5, 6\}$ (d) $\{1\}$ (e) U (f) $\emptyset$

4. If $U = \{1, 2, 3, 4, 5, 6, 7, 8\}$, $A = \{1, 3, 5, 7\}$, $B = \{3, 4, 5, 6\}$, and $C = \{2, 6, 7, 8\}$, find the following sets.
   (a) $A - B$ (b) $B - A$ (c) $A - C$ (d) $C - A$ (e) $B - C$ (f) $C - B$ (g) $U - A$
   (h) $B - \emptyset$ (i) $\emptyset - C$

5. Let $U = \{1, 2, 3, 4, 5, 6, 7\}$, $A = \{2, 4, 6, 7\}$, $B = \{1, 2, 4, 5, 7\}$. Draw a Venn diagram and place the elements of these sets in their proper location.

6. Draw Venn diagrams and shade each of the following sets.
   (a) $A - B$ (b) $A' \cup B$ (c) $A' \cap B'$ (d) $A' \cup B'$

In Exercises 7–12, describe each shaded region using the symbols A, B, C, $\cap$, $\cup$, $-$, and $'$. (There is more than one way to do these.)

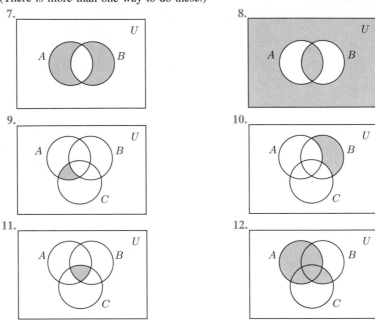

13. Show that $A \cup A = A$ and $A \cap A = A$ for any set A.

14. If $A \cap B = \emptyset$, what can you say about A and B?

15. Show that $A \cup B = B \cup A$ and $A \cap B = B \cap A$.

16. If $A \cup B = A \cap B$, what can you say about A and B?

17. Show that $A \cap B = \emptyset$ if $A \subseteq B'$. Show that $A \subseteq B'$ if $A \cap B = \emptyset$.

18. Let $A = \{a, b, c\}$, $B = \{b, c, e\}$, and $C = \{a, e, f\}$. Find:
    (a) $B \cup C$ (b) $A \cup (B \cap C)$ (c) $A \cap (B \cap C)$ (d) $(A \cap B) \cap C$ (e) $(A \cup B) \cap C$
    (f) $(A \cap C) \cup (B \cap C)$

19. Use Venn diagrams to illustrate the law $(A \cap B) \cup C = (A \cup C) \cap (B \cup C)$.

20. What is (a) $\emptyset \cup A$,   (b) $\emptyset \cap A$?

21. Use Venn diagrams to illustrate the **associative laws**

$$(A \cup B) \cup C = A \cup (B \cup C)$$

and
$$(A \cap B) \cap C = A \cap (B \cap C)$$

22. Show that $A \subseteq A$ for any set A. If $A \subseteq B$ and $B \subseteq C$, show that $A \subseteq C$.

23. If A and B are any sets, show that (a) $A \cap B \subseteq A$, (b) $A \cap B \subseteq B$, (c) $A \subseteq A \cup B$, and (d) $B \subseteq A \cup B$.

24. If $A \cap B = A$, show that $A \subseteq B$. If $A \cup B = A$, show that $B \subseteq A$.

25. Show that $A \cap B$ is the largest set that is contained in both A and B. Show that $A \cup B$ is the smallest set containing both A and B.

26. Show that $Q \cap I = \mathbb{N}$.

27. Use Venn diagrams to illustrate the following set equalities.
    (a) $A - B = A - A \cap B$
    (b) $A \cup B = (A - B) \cup (B - A) \cup (A \cap B)$
    (c) $(A - B) \cap (A \cap B) = \emptyset$

28. How would you resolve the barber and library paradoxes?

29. How many subsets does a four-element set have? List them.

30. What 3-tuples of 0s and 1s does each subset of $\{a, b, c\}$ correspond to?

31. In Figure 1.13, U is divided into eight regions. Describe these regions in terms of A, B, C, and their complements.

(Optional Exercises)

**32. In Figure 1.16, U is divided into 16 regions. Describe these regions in terms of A, B, C, D, and their complements.

**33. Let $S = \{1, 2, 3, 4, 5, 6\}$. How many zero-element subsets does S have? How many one-element subsets? How many two-element subsets? How many three-, four-, five-, and six-element subsets? Do these numbers add up to 2^6 ?

**34. Prove that n sets can divide U into 2^n regions.

**35. Let $S = \{1, 2, \ldots, n\}$. Give the number of subsets with $0, 1, 2, \ldots, n$ elements. Do these numbers add up to 2^n ?

1.4 | Surveys

There is no study that is not capable of delighting us after a little application of it.   — ALEXANDER POPE (1688–1744), English poet

In this section we use sets to solve problems in surveying groups of people. We shall consider only simple situations involving small amounts of data. However, similar methods are used for large-scale relational database systems employed in modern computers.

A Survey

A survey was taken of 100 people to determine their preferences for two breakfast cereals, Crisp and Crunch. The people were asked if they had tried Crisp or Crunch or neither during the past month. In response, 50 had tried Crisp, 40 had tried Crunch, and 20 had tried neither. The reason the numbers add up to more than 100 is that there were people who ate both. The survey team now wants to obtain additional information from these data. They would like to know how many had tried both brands, how many had tried Crisp but not Crunch, and how many had

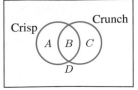

Figure 1.17

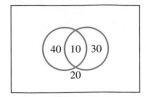

Figure 1.18

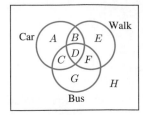

Figure 1.19

tried Crunch but not Crisp. How can we find this information? Well, we can draw a Venn diagram as in Figure 1.17. The various regions of the diagram are labeled A, B, C, and D. In this case, A represents the people who tried only Crisp, B the people who tried both, C the people who tried only Crunch, and D the people who tried neither.

We next label each region with the number of people represented by that region (Figure 1.18). Since 20 people tried neither brand, we place a 20 in region D. Since there were a total of 100 people, $A \cup B \cup C$ must contain 80 people. Since $A \cup B$ represents the people who tried Crisp, $A \cup B$ must contain 50 people. Hence C must contain $80 - 50 = 30$ people, so we place a 30 in region C. Since $B \cup C$ contains 40 people, B must contain $40 - 30 = 10$ people, so we place a 10 in region B. Finally, in region A we have $50 - 10 = 40$. We conclude that 10 tried both brands, 40 tried Crisp but not Crunch, and 30 tried Crunch but not Crisp.

Another Survey

In another survey, 65 employees were asked whether they go to work by car, bus, or on foot. It turned out that some people use other means such as cab, bike, or motorcycle. It was also found that some employees use more than one means of transportation at different times depending on the weather or their mood. The results of the survey showed:

36 take a car	16 take a bus and walk
32 take a bus	16 take a car and walk
32 walk	6 take a car and bus and walk
14 take a car and bus	

The reason these numbers sum to more than 65 is that there is overlap. For example, the 14 people who take a car and bus are also counted among the 36 who take a car and the 32 who take a bus. We can organize these data by constructing a Venn diagram as in Figure 1.19.

In Figure 1.19 we have labeled the various regions of the diagram $A, B, \ldots, H$. In this case, A represents the people who only take a car, B the people who take a car and walk but do not take a bus, and so on. We next label each region with the number of employees represented by that region. After this is done, we can extract any information we want from the diagram.

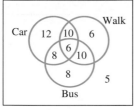

Figure 1.20

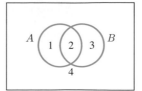

Figure 1.21

Number of Elements

These numbers are shown in Figure 1.20 and are computed as follows.

> For a given set, say A, we denote by $|A|$ the number of elements in A.

Since D represents the set of people who use all three transportation means, we have $|D| = 6$. Now 16 employees take a bus and walk, so $|D \cup F| = 16$. Since $|D| = 6$, we must have $|F| = 10$. Similarly, $|D \cup B| = 16$, so $|B| = 10$. Since $|D \cup C| = 14$, we have $|C| = 8$. Now $|A \cup B \cup C \cup D| = 36$ since these are the employees who take a car. Also $|B \cup C \cup D| = 10 + 8 + 6 = 24$, so we have $|A| = 36 - 24 = 12$. In a similar way, we conclude that $|G| = 32 - 24 = 8$ and $|E| = 32 - 26 = 6$. Since the total number represented by these seven regions is 60, we have $|H| = 65 - 60 = 5$.

EXAMPLE 1 Using the data from the transportation survey, answer the following questions. How many employees (a) only take a car? (b) do not walk? (c) take a bus but do not walk? (d) take a car or a bus?

Solution.   From Figure 1.20 we have (a) $|A| = 12$,   (b) $12 + 8 + 8 + 5 = 33$,
(c) $8 + 8 = 16$,   (d) $65 - 5 - 6 = 54$.   □

Formulas for the Number of Elements

We have seen that Venn diagrams can be used to find the number of elements $|A|$ in a set A from available data. An alternative method is to find a formula for $|A|$ and use this formula directly. Suppose we have two sets A and B and we know $|A \cup B|$, $|A|$, and $|B|$. How do we find $|A \cap B|$? In Figure 1.21 we have labeled the various regions of the Venn diagram for A and B. Since $|A| = |1| + |2|$, $|B| = |2| + |3|$, and $|A \cup B| = |1| + |2| + |3|$, we have

$$|A| + |B| = |1| + |2| + |2| + |3| = |A \cup B| + |2|$$

Now $2 = A \cap B$, so we have

$$|A| + |B| = |A \cup B| + |A \cap B| \qquad (1.1)$$

We can then solve (1.1) to obtain

$$|A \cap B| = |A| + |B| - |A \cup B| \qquad (1.2)$$

We say that two sets A and B are **disjoint** if they have no elements in common, that is, if $A \cap B = \emptyset$. If A and B are disjoint, we have from (1.1) that

$$|A| + |B| = |A \cup B| \qquad (1.3)$$

Of course, it is already clear that (1.3) should hold.

EXAMPLE 2   A club has two committees, A and B, and every club member belongs to at least one of the committees. If A has 10 members, B has 15 members, and there are three people who belong to both committees, how many people are in the club?

Solution.   Applying (1.1) gives

$$|A \cup B| = |A| + |B| - |A \cap B| = 10 + 15 - 3 = 22 \qquad \square$$

We can also obtain formulas for the number of elements in other sets. If U is the universal set, since $4 = A' \cap B'$ in Figure 1.21, we have

$$|U| = |A' \cap B'| + |A \cup B| \qquad (1.4)$$

Since $1 = A - B$ and $2 = A \cap B$ in Figure 1.21 and $|1| + |2| = |A|$, we have

$$|A - B| + |A \cap B| = |A|$$

or equivalently

$$|A - B| = |A| - |A \cap B| \qquad (1.5)$$

EXAMPLE 3   If 22 members of a club belong to at least one of two committees and five belong to neither, how many members are in the club?

Solution.   Applying (1.4) gives

$$|U| = 5 + 22 = 27 \qquad \square$$

EXAMPLE 4   In Example 2, how many members belong to committee A but not B?

Solution.   Applying (1.5) gives

$$|A - B| = 10 - 3 = 7 \qquad \square$$

A repeated application of (1.1) gives the following formula for three sets.

$$|A| + |B| + |C|$$
$$= |A \cup B \cup C| + |A \cap B| + |A \cap C| + |B \cap C| - |A \cap B \cap C| \qquad (1.6)$$

EXAMPLE 5   For the transportation survey, use (1.6) to find the number of employees who go to work by car, bus, or on foot.

Solution.   Let A = car, B = bus, C = walk. Solving (1.6) for $|A \cup B \cup C|$ gives

$$|A \cup B \cup C| = |A| + |B| + |C| - |A \cap B| - |A \cap C| - |B \cap C| + |A \cap B \cap C|$$
$$= 36 + 32 + 32 - 14 - 16 - 16 + 6 = 60 \qquad \square$$

EXERCISES

1. In a survey of 100 people, 40 had tried Crisp, 30 had tried Crunch, and 30 had tried neither. How many had tried both?

2. In a survey of 100 people, 50 had tried Crisp, 40 had tried Crunch, and 25 had tried neither.
   (a) How many had tried both?
   (b) How many tried Crisp but not Crunch?
   (c) How many tried Crunch but not Crisp?

3. For the survey of Example 1, how many employees (a) only took a bus? (b) did not take a car? (c) took a car but not a bus? (d) walked but did not take a car?

4. Suppose in a transportation survey of 100 employees, the following answers are given.

70 take a car	12 take a bus and walk
29 take a bus	23 take a car and walk
33 walk	8 take a car and bus and walk
15 take a car and bus	

   How many employees (a) only walk? (b) do not take a car? (c) take a bus but not a car? (d) only take a bus?

5. If A and B are disjoint and $|A \cup B| = 10$, $|A| = 5$, find $|B|$.

6. If $|A| + |B| = |A \cup B|$, prove that A and B are disjoint.

7. If $|A| = 0$, what is A?

8. Prove that $|A| + |B| \geq |A \cup B|$.

9. If $|A| = 7$, $|B| = 5$, $|A \cup B| = 10$, find $|A \cap B|$.

10. If $|A \cup B| = 15$, $|A \cap B| = 6$, $|B| = 8$, find $|A|$.

11. If $|A| = 8$, $|B| = 9$, $|A \cap B| = 5$, find $|A \cup B|$.

12. If $A \subseteq B$ and $|A| = |B|$, what would you conclude?

13. If $|A \cap B| = |A \cup B|$, how are A and B related?

14. If $|A| = 12$, $|A \cap B| = 5$, find $|A - B|$.

15. If $|A \cap B| = 3$ and $|A - B| = 7$, find $|A|$.

16. Prove that $|A'| = |U| - |A|$.

17. If $|U| = 38$, $|A| = 16$, $|A \cap B| = 12$, $|B'| = 20$, find (a) $|B|$,   (b) $|A \cup B|$,   (c) $|A - B|$, (d) $|B - A|$.

18. Draw a Venn diagram and use the following information to fill in the number of elements in each region: $|A'| = 28$, $|B| = 25$, $|A' \cup B'| = 45$, $|A \cap B| = 12$.

19. Repeat Exercise 18 for $|A'| = 17$, $|B| = 10$, $|A' \cup B'| = 22$, $|A \cap B| = 3$.

20. Repeat Exercise 18 for $|A \cup B| = 20$, $|A \cap B| = 2$, $|A| = 10$, $|A' \cup B'| = 33$.

21. Draw a Venn diagram and use the following information to fill in the number of elements in each region: $|A| = 27$, $|B| = 32$, $|C| = 22$, $|A \cap B| = 9$, $|B \cap C| = 7$, $|A \cap C| = 7$, $|A \cap B \cap C| = 4$, $|U| = 80$.

22. Repeat Exercise 21 with $|A| = 10$, $|A \cap B| = 7$, $|A \cup B| = 21$, $|A \cap B \cap C| = 3$, $|A \cap C| = 4$, $|B \cap C| = 9$, $|C| = 15$, $|B'| = 8$.

23. Repeat Exercise 21 with $|A| = 13$, $|A \cap B \cap C| = 4$, $|A \cap C| = 6$, $|A \cap B'| = 6$, $|B \cap C| = 6$, $|A' \cap B' \cap C'| = 5$, $|B \cap C'| = 11$, $|B \cup C| = 22$.

*24. Use a Venn diagram to illustrate (1.6).

*25. Prove equation (1.6).

1.5 | Functions

The theory that has had the greatest development in recent times is without any doubt the theory of functions.

— Vito Volterra, Ninteenth-Century Italian Scientist

Sets and functions are the two most basic concepts in mathematics. In every area of mathematics these two concepts are of fundamental importance. A function is a way of showing a correspondence between elements in a set or between elements in one set and elements in another. There are many examples of functions in everyday life. A dictionary is a function because it makes words correspond to their definitions. A telephone book matches people to their corresponding telephone numbers. Other examples of functions are social security numbers, zip codes, page numbers, names of people, and species of animals.

We now give a more precise definition. Let S and T be sets.

A **function** from S to T is a rule that assigns to each element of S a unique element of T.

A function from S to T is frequently written $f: S \to T$. If the rule assigns to the element $s \in S$ the element $t \in T$, we write $f(s) = t$ and call t the **value** of f at s.

Functions may be described by words, by listing all the elements of S and the assigned elements of T, or by a mathematical formula. For example, let S be the set of all people in the United States, let T be the set of all first names, and let $f: S \to T$ be the rule that assigns to each element of S that person's first name. As another example, let $S = \{1, 2, 3\}$, let $T = \{a, b\}$, and let $f: S \to T$ be the function given by $f(1) = f(2) = a$ and $f(3) = b$. In the first example the function was described

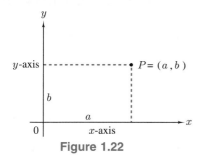

Figure 1.22

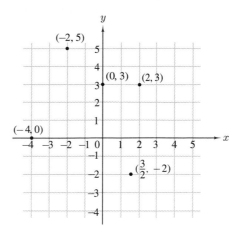

Figure 1.23

by words; in the second the function was described by listing the elements of S and the assigned elements of T. The following functions are described by mathematical formulas. Let $S = T = \mathbb{N}$, and let $f(n) = n + 1$ for all $n \in \mathbb{N}$, $g(n) = n^2$ for all $n \in \mathbb{N}$, and $h(n) = 2n + 3$ for all $n \in \mathbb{N}$. Thus $f(1) = 2$, $f(2) = 3, \ldots$; $g(1) = 1$, $g(2) = 4$, $g(3) = 9, \ldots$; $h(1) = 5$, $h(2) = 7$, $h(3) = 9, \ldots$.

Another example of a function is the following. Let $V = \{a, b, c, d\}$, let S be the set of all subsets of V, and let $f\colon S \to \mathbb{N} \cup \{0\}$ assign each subset in S to the number of elements of that subset. Thus $f(\emptyset) = 0$, $f(V) = 4$, $f\big(\{a, b\}\big) = 2$, $f\big(\{a, b, d\}\big) = 3$.

Cartesian Plane

If f is a function from S to T, and if S and T are both sets of numbers, we can graph f as follows. We begin by drawing a horizontal line called the **x-axis** and a vertical line called the **y-axis**. These lines cross at the **origin** 0 and determine a plane called the **Cartesian plane** (Figure 1.22). A point P in the Cartesian plane is determined by two numbers a and b, and we write $P = (a, b)$. We call a the **x-coordinate** of P and b the **y-coordinate** of P. To locate $P = (a, b)$, we move a units to the right of 0 along the x-axis if a is positive (to the left if a is negative) and b units up if b is positive (down if b is negative).

EXAMPLE 1   What are the x- and y-coordinates of (a) $(2, 3)$,   (b) $(-2, 5)$?

Solution.   (a) The x-coordinate is 2 and the y-coordinate is 3. (b) In this case the x-coordinate is -2 and the y-coordinate is 5.   □

EXAMPLE 2   Locate the following points in the Cartesian plane.
(a) $(2, 3)$ (b) $(-2, 5)$ (c) $(0, 3)$ (d) $(3/2, -2)$ (e) $(-4, 0)$

Solution.   These points are displayed in Figure 1.23.   □

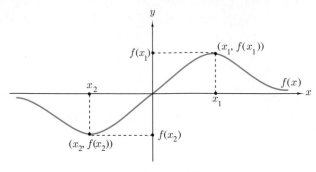

Figure 1.24

Graph of a Function

Notice that the origin corresponds to $(0,0)$, the x-axis to the set $\{(a,0)\colon a \in \mathbb{R}\}$, and the y-axis to the set $\{(0,b)\colon b \in \mathbb{R}\}$. To graph a function f, we plot the points $(x, f(x))$ on the Cartesian plane. Figure 1.24 illustrates the graph of a typical function. In this figure we have plotted two points, $(x_1, f(x_1))$ and $(x_2, f(x_2))$. The graph of a function provides a means for visualizing how a function acts. The graph of f clearly shows the correspondence between points x of the x-axis and points $f(x)$ of the y-axis determined by f.

EXAMPLE 3   Graph the function $f\colon \mathbb{N} \to \mathbb{N}$ given $f(n) = n + 1$.

    Solution.   We first make a table of values (Table 1.2). We then plot the points $(n, f(n))$ on the Cartesian plane (Figure 1.25). $\square$

Table 1.2

n	$f(n)$
1	2
2	3
3	4
4	5
5	6
6	7

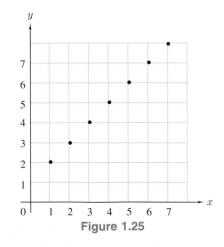

Figure 1.25

EXAMPLE 4   Graph the function $f: \mathbb{R} \to \mathbb{R}$ given by $f(x) = (1/2)x^2 - 2$.

Solution.   A table of values is given in Table 1.3. We then plot the points $(x, f(x))$ and connect these points with a curve in Figure 1.26.   □

Table 1.3

x	$f(x)$
0	-2
1	$-3/2$
2	0
3	$5/2$
-1	$-3/2$
-2	0
-3	$5/2$

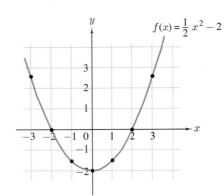

Figure 1.26

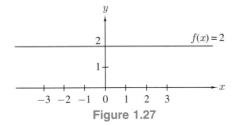

Figure 1.27

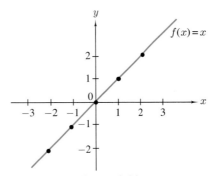

Figure 1.28

EXAMPLE 5   Graph the function $f(x) = 2$.

Solution.   This is a constant function that only has the value 2. Its graph is illustrated in Figure 1.27.   □

EXAMPLE 6   Graph the function $f(x) = x$.

Solution.   The graph consists of the set of points $\{(x, x): \ x \in \mathbb{R}\}$. This is a straight line as illustrated in Figure 1.28.   □

Linear and Quadratic Functions

There are many special functions that are important in mathematics and its applications. One of the simplest is a function of the form $f(x) = ax + b$ where a and b are constants in $\mathbb{R}$ and $x \in \mathbb{R}$. Such functions are called **linear** because their graphs are straight lines. Particular cases of linear functions are the functions in Examples 5 and 6. In Example 5, $a = 0$ and $b = 2$, while in Example 6, $a = 1$ and $b = 0$. Another type of special functions are the **quadratic** functions. Quadratic functions have the form $f(x) = ax^2 + bx + c$ where $a \neq 0$, b and c are constants in $\mathbb{R}$, and $x \in \mathbb{R}$. The function in Example 4 is a quadratic function with $a = 1/2$, $b = 0$, and $c = -2$.

Functions are frequently used to describe the dependence of one variable on another variable. For example, the area A of a square depends on the length x of one of its sides. We then say that A is a function of x and denote the area by $A(x)$. As you know, the exact formula is given by $A(x) = x^2$. Note that A is a quadratic function of x.

EXAMPLE 7   A car rental company offers to rent a car for $90 per week plus 15¢ per mile. If you use the car for a week, the cost C depends on the mileage x.
(a) Give a formula for C as a function of x.
(b) Find the cost if you drive 800 miles.
   Solution.   (a) The cost in dollars is 90 plus 0.15 times the number of miles driven. Hence,

$$C(x) = 90 + 0.15x$$

(b) If you drive 800 miles, then the cost is

$$C(800) = 90 + (0.15)(800) = \$210 \qquad \square$$

A function need not be given by a single formula and sometimes two or more formulas are used. For example, consider the following function.

$$f(x) = \begin{cases} x & \text{if } x < 1 \\ 2 & \text{if } x \geq 1 \end{cases}$$

This function is given by two formulas. The first formula holds when $x < 1$ and the second when $x \geq 1$. The graph of f is illustrated in Figure 1.29. Notice the filled and open dots on the graph at $x = 1$. These indicate that the function has the value 2 at $x = 1$ and does not have the value 1 at $x = 1$. That is, $f(1) = 2$ and $f(1) \neq 1$.

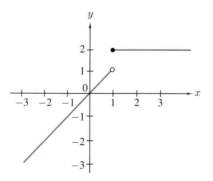

Figure 1.29

EXAMPLE 8   Postage for a first class letter depends on its weight x. Suppose the cost C is 29¢ for the first ounce (or less) and 24¢ for each additional ounce (or fraction thereof). Find the cost of a 4.5-ounce letter and graph the function $C(x)$.

   Solution.   We first construct a table of values as in Table 1.4. The graph of $C(x)$ is illustrated in Figure 1.30. Notice that $C(4.5) = \$1.25$.   □

Table 1.4

x	$C(x)$
$0 < x \leq 1$	29
$1 < x \leq 2$	53
$2 < x \leq 3$	77
$3 < x \leq 4$	101
$4 < x \leq 5$	125

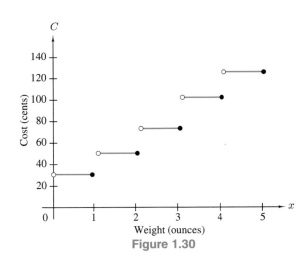

Figure 1.30

EXERCISES

1. Let $S = \{1, 2, 3\}$. Define three different functions from S to S.

2. Let $S = \{a, b, c, d\}$. Define three different functions from S to S.

3. For the set $S = \{1, 2, 3\}$ define the function $f: S \to S$ as $f(1) = 1$, $f(2) = 2$, and $f(3) = 3$. What is special about f? Why is f called the **identity function** for S? If A is an arbitrary nonempty set, how is the identity function for A defined?

4. If $S = \{1, 2, 3, 4\}$ and $T = \{2\}$, give an example of a function $f: S \to T$. How many functions are there from S to T? How many functions are there from T to S?

5. If S is an arbitrary nonempty set and $T = \{b\}$, give an example of a function $f: S \to T$. How many functions are there from S to T?

6. Give five examples of functions from everyday life that are not mentioned in the text.

7. Give an example of each of the following types of functions.
   (a) $f: \mathbb{N} \to \mathbb{R}$ (b) $f: \mathbb{N} \to \mathbb{N}$ (c) $f: \mathbb{R} \to \mathbb{N}$ (d) $f: Q \to \mathbb{N}$ (e) $f: I \to \mathbb{R}$

8. Find the x- and y-coordinates of the following points.
   (a) $(3, 5)$ (b) $(-2, 8)$ (c) $(0, -6)$ (d) $(3/2, -5/2)$

9. Plot the points given in Exercise 8.

10. Graph the following functions.
    (a) $f: \mathbb{R} \to \mathbb{R}$, $f(x) = 2x$
    (b) $f: \mathbb{R} \to \mathbb{R}$, $f(x) = 5x + 3$
    (c) $f: \mathbb{R} \to \mathbb{R}$, $f(x) = 5$

11. Let $f: \mathbb{R} \to \mathbb{R}$ be the function $f(x) = 2x^3 + x^2 - x + 1$. Find (a) $f(0)$,   (b) $f(1)$,   (c) $f(-1)$,   (d) $f(2)$,   (e) $f(-2)$.

12. Let f be the function defined by $f(x) = \sqrt{x}$ for $x \geq 0$. Find (a) $f(0)$,   (b) $f(1)$,   (c) $f(4)$, (d) $f(9)$,   (e) $f(2)$.

13. Let f be the function defined by $f(x) = 1/x$ for $x > 0$. Find (a) $f(1/3)$,   (b) $f(1/2)$, (c) $f(1)$,   (d) $f(2)$,   (e) $f(3)$.

14. Graph the following functions.
(a) $f: \mathbb{N} \to \mathbb{N}$, $f(n) = 2n - 1$ (b) $f: \mathbb{N} \to \mathbb{N}$, $f(n) = n^2 - 1$
(c) $f: \mathbb{R} \to \mathbb{R}$, $f(x) = 2x^2 - 3$

15. Graph the function $f(x) = \sqrt{x}$ for $x \geq 0$.

16. Graph the function $f(x) = 1/x$ for $x \neq 0$.

17. Graph the function $f(x) = 1/x^2$ for $x \neq 0$.

18. If $f(x) = 2x + 1$, find x_0 such that $f(x_0) = 0$.

19. If $f(x) = ax + b$, $a \neq 0$, find x_0 such that $f(x_0) = 0$.

20. If $f(x) = x^2 - 1$, find x_1, x_2 such that $f(x_1) = f(x_2) = 0$.

21. If $f(x) = x^2 + 2x + 1$, find x_0 such that $f(x_0) = 0$.

22. If $f(x) = x^2 - 3x + 2$, find x_1, x_2 such that $f(x_1) = f(x_2) = 0$.

23. If $f(x) = x^2 - x$, find x_1, x_2 such that $f(x_1) = f(x_2) = 0$.

24. Repeat Example 7 for a cost of $100 per week plus 20¢ per mile.

25. A car rental company offers to rent a car for $15 per day plus 10¢ per mile. (a) Give a formula for the cost C as a function of the number of days n and the number of miles x. (b) Find the cost of renting the car for 4 days and driving 500 miles.

26. Repeat Example 8 with a cost of 30¢ for the first ounce and 25¢ for each additional ounce.

27. Suppose the cost C of an airmail letter is 45¢ for the first one-half ounce and 40¢ for each additional one-half ounce. Find the cost of a 2-ounce letter and graph $C(x)$ as a function of weight.

28. Graph the following function. $f(x) = \begin{cases} 1 & \text{if } x \leq 0 \\ 2 & \text{if } x > 0 \end{cases}$

29. Graph the following function. $f(x) = \begin{cases} 1 & \text{if } x \neq 0 \\ 2 & \text{if } x = 0 \end{cases}$

30. Graph the following function. $f(x) = \begin{cases} 1 & \text{if } x \leq 1 \\ x & \text{if } x > 1 \end{cases}$

31. If $f: \mathbb{R} \to \mathbb{R}$ and $g: \mathbb{R} \to \mathbb{R}$, we define their **sum** $f + g$ by $(f + g)(x) = f(x) + g(x)$. For example, if $f(x) = 2x$ and $g(x) = x^2$, then $(f + g)(x) = 2x + x^2$. Find the sums of the following pairs of functions.
(a) $f(x) = x + 1$, $g(x) = 3x$ (b) $f(x) = 2x + 3$, $g(x) = 4x^2$
(c) $f(x) = x^2 + x + 1$, $g(x) = x^3$

32. Show that the sum of any two linear functions is a linear function. (See Exercise 31.)

33. Show that the sum of any two quadratic functions is a quadratic or linear function. (See Exercise 31.)

34. If $f: \mathbb{R} \to \mathbb{R}$ and $g: \mathbb{R} \to \mathbb{R}$, we define their **product** fg by $fg(x) = f(x)g(x)$. For example, if $f(x) = 2x + 1$ and $g(x) = x$, then $fg(x) = 2x^2 + x^2$. Find the products of the following pairs of functions.
(a) $f(x) = 2$, $g(x) = x$ (b) $f(x) = 3x + 2$, $g(x) = x$
(c) $f(x) = 5x - 1$, $g(x) = x + 1$

35. Show that the product of any two linear functions is a quadratic or linear function. (See Exercise 34.)

36. Can the product of a linear function and a quadratic function be a quadratic function? Can it be a linear function? (See Exercise 34.)

37. Can the product of two quadratic functions be a quadratic function? (See Exercise 34.)

*38. Why is the graph of a linear function a straight line?

39. If $S = \{a, b\}$ and $T = \{1, 2\}$, how many different functions are there from S to T?

40. If $S = \{a, b, c\}$ and $T = \{1, 2\}$, how many different functions are there from S to T?

**41. If S has m elements and T has n elements, how many functions are there from S to T? Can you prove this?

*42. Let V be a finite set, let S be the set of all subsets of V, and let $f: S \to \mathbb{N} \cup \{0\}$ assign each subset in S to the number of elements of that subset. Prove the following.
  (a) If $A \subseteq B \subseteq V$, then $f(A) \leq f(B)$.
  (b) If $f(A) = 0$, then $A = \emptyset$.
  (c) If $A \cap B = \emptyset$, then $f(A \cup B) = f(A) + f(B)$.
  (d) $f(A \cup B) + f(A \cap B) = f(A) + f(B)$.

1.6 | More About Functions

It isn't that they can't see the solution. It is that they can't see the problem. —G. K. CHESTERTON, *The Scandal of Father Brown*

Approach your problems from the right end and begin with the answers. Then one day, perhaps you will find the final question. —R. VAN GULIK, *The Chinese Maze Murders*

In Section 1.5 you had a tour of part of the function zoo. It soon became clear that this zoo contained a huge variety of different species. We now consider another function called the **logarithmic function**, $f(x) = \log x$. This function is very useful in applications. We shall use $\log x$ in the next section and later when we study geometry and consumer mathematics. In particular, $\log x$ is used for computing the dimension of a fractal, for finding the half-life of a radioactive element, for finding effective interest rates, and for measuring the strengths of earthquakes and sounds.

EXAMPLE 1   (a) If $10^n = 100$, what is n? (b) If $10^n = 1000$, what is n? (c) If $10^n = 500$, what is n?

Solution.   (a) This problem asks the following question. To what power must we raise 10 to get 100? Since $10^2 = (10)(10) = 100$, the answer is 2.

(b) To what power must we raise 10 to get 1000? Since $10^3 = (10)(10)(10) = 1000$, the answer is 3.

(c) To what power must we raise 10 to get 500? The answer to this question is much harder than those posed in (a) and (b). Since $10^2 = 100$ and $10^3 = 1000$, the answer is somewhere between 2 and 3, but where? Let's try $10^{2.5}$. What does this mean? There are formulas for computing fractional powers of 10 (or fractional powers of any number), but these are very complicated. The easiest way to find $10^{2.5}$ is to use

Table 1.5

x	10^x
-2.5	0.00316228
-1.5	0.0316228
-0.4	0.3981072
0.2	1.584893
0.4	2.511886
0.6	3.981072
0.8	6.309573

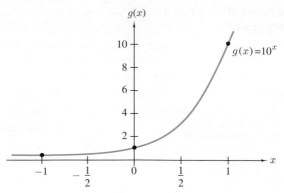

Figure 1.31

the 10^x key of a calculator. This gives $10^{2.5} = 316.23$, which is too small. If we try $10^{2.6}$ using a calculator we obtain $10^{2.6} = 398.11$, which is again too small. Trying another value, we obtain $10^{2.7} = 501.19$, so 2.7 is close to our desired result. This trial-and-error method is not very satisfactory. Is there a better way? The best method is to use the log key of a calculator. We then obtain $\log 500 = 2.69897$ and the answer to (c) is $n = 2.69897$. Thus, $\log 500$ is the power of 10 that equals 500. $\qquad \square$

In Example 1 we have introduced two functions, $f(x) = \log x$ and $g(x) = 10^x$. Let's look at these functions more closely. You already know that $g(0) = 1$, $g(1) = 10$, $g(2) = 100$, $g(3) = 1000$. You can even obtain values of $g(x)$ when x is a negative integer. For example,

$$g(-1) = 10^{-1} = \frac{1}{10} = 0.1, \qquad g(-2) = 10^{-2} = \frac{1}{10^2} = 0.01$$

and $\qquad g(-3) = 10^{-3} = \dfrac{1}{10^3} = 0.001$

Using the 10^x key of a calculator, you can obtain values of $g(x)$ when x is not an integer. Such values are given in Table 1.5. The graph of $g(x) = 10^x$ is illustrated in Figure 1.31.

Logarithmic Functions

> The function $f(x) = \log x$ is defined for all positive real numbers x and is the power of 10 that is needed to give x.

More precisely, $\log x = y$ if $10^y = x$. We can obtain $\log x$ for certain simple values of x. For example,

$\log 1 \quad = 0 \quad$ since $\quad 10^0 = 1 \qquad \log 10 = 1 \quad$ since $\quad 10^1 = 10$

$\log 100 = 2 \quad$ since $\quad 10^2 = 100 \qquad \log 0.1 = -1 \quad$ since $\quad 10^{-1} = 0.1$

The values of $\log x$ given in Table 1.6 were obtained using the $\log x$ key of a calculator. The graph of $f(x) = \log x$ is illustrated in Figure 1.32.

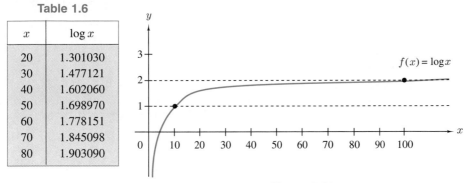

Table 1.6

x	$\log x$
20	1.301030
30	1.477121
40	1.602060
50	1.698970
60	1.778151
70	1.845098
80	1.903090

Figure 1.32

From the definitions of the functions $f(x) = \log x$ and $g(x) = 10^x$ we see that if $f(x) = y$ then $g(y) = x$. Roughly speaking, anything that f does, g undoes. Because of this, we say that f and g are **inverses** of each other. This inverse property can be written

$$\log 10^x = x$$
$$10^{\log x} = x$$

These two equations are very useful for computing logarithms.

EXAMPLE 2   Find (a) $\log 10^{0.4}$, (b) $\log 100^{0.4}$, (c) $\log 1000^{0.4}$, and (d) x if $10^x = 50$.
   Solution.   (a) $\log 10^{0.4} = 0.4$ (b) $\log 100^{0.4} = \log \left(10^2\right)^{0.4} = \log 10^{0.8} = 0.8$
(c) $\log 1000^{0.4} = \log \left(10^3\right)^{0.4} = \log 10^{1.2} = 1.2$
(d) Taking logarithms of both sides of $10^x = 50$ gives

$$x = \log 10^x = \log 50 = 1.69897 \qquad \square$$

Richter Scale

You frequently read in the newspaper that an earthquake has registered a certain amount on the Richter scale. This scale was introduced by the American seismologist Charles R. Richter in the 1930s and is a widely accepted measure of the strength of an earthquake. The Richter scale is a logarithmic scale for expressing the magnitude of an earthquake in terms of the energy dissipated in it. A reading of 1.5 indicates the smallest tremor that can be felt, 4.5 results from an earthquake causing slight damage, and 8.5 indicates a devastating earthquake. The strongest earthquake of this century

Photo 1.2
The 1989
San Francisco
earthquake
registered 7.2 on
the Richter scale.

occurred in 1960 in southern Chile with a Richter magnitude of 9.7. The strongest earthquake of this century in the United States was the famous San Francisco earthquake of 1906 with an 8.3 Richter magnitude. Table 1.7 gives the Richter magnitude of the major California earthquakes in this century.

> If x is the energy dissipated by an earthquake, then the corresponding Richter scale reading is $y = \log x$.

The Richter scale is used because the energy x can get very large and the logarithm, as can be seen from Figure 1.32, brings large numbers down to a manageable size.

EXAMPLE 3   The 1989 San Francisco earthquake measured 7.2 on the Richter scale, while an earthquake near Los Angeles in 1991 measured 6.3. How much stronger was the San Francisco earthquake than the Los Angeles earthquake?

Solution.   Let the energies of the San Francisco and Los Angeles earthquakes be x_1 and x_2, respectively. We then have $\log x_1 = 7.2$ and $\log x_2 = 6.3$. Hence,

$$x_1 = 10^{\log x_1} = 10^{7.2}$$
$$x_2 = 10^{\log x_2} = 10^{6.3}$$

This gives
$$\frac{x_1}{x_2} = \frac{10^{7.2}}{10^{6.3}} = 10^{(7.2-6.3)} = 10^{0.9} = 7.94$$

We conclude that $x_1 = 7.94\,x_2$, so the San Francisco earthquake was 7.94 times stronger than the Los Angeles earthquake.   □

Table 1.7

Year	Location	Richter scale
1906	San Francisco	8.3
1911	Coyote	6.6
1923	North Coast	7.2
1927	San Luis Obispo	7.7
1940	Imperial Valley	6.7
1952	Bakersfield	7.8
1968	Anza-Borrego Mountains	6.4
1979	Imperial Valley	6.4
1980	Eureka	7.0
1980	Mammoth Lakes (4 quakes)	6.0–6.6
1983	Coalinga	6.5
1989	Bay Area	7.2
1991	North Coast	7.1
1992	Eureka	6.9
1992	Northern California	6.9
1992	Yucca Valley	7.4
1992	Big Bear Lake	6.5

Decibels

Another important application of $\log x$ is in computing decibels. A decibel is a unit for expressing the intensity of sound on a scale from zero for the least perceptible sound to about 150 for severe ear damage. A decibel equals $1/10$ of a bel, a unit named after the famous inventor, scientist Alexander Graham Bell. A quiet room registers about 40 decibels, while a noisy street registers about 80. Table 1.8 illustrates the noise level of various decibel readings.

Photo 1.3
A crowded street corner can register a noise level of over 80 decibels.

Table 1.8

Decibels	Noise level
150	Severe ear damage
140	Jet take-off at close range
130	Threshhold of pain
120	Amplified rock band
110	Locomotive at close range
100	Circular saw
90	Dog barking
80	Vacuum cleaner
70	Telephone bell
60	Conversation
50	Hummingbird
40	Soft wind
30	Quiet stream
20	Whispering
10	Pin hitting floor
0	Inaudible

If x is the intensity of a sound, then the corresponding decibel reading is
$$y = 10 \log x$$

EXAMPLE 4 If a jet airplane gives a decibel reading of 120 and a truck gives a reading of 90, how much louder is the plane than the truck?

Solution. Let the sound intensities of the plane and truck be x_1 and x_2, respectively. Since $10 \log x_1 = 120$ and $10 \log x_2 = 90$, we have

$$x_1 = 10^{\log x_1} = 10^{12}$$
$$x_2 = 10^{\log x_2} = 10^9$$

Hence,

$$\frac{x_1}{x_2} = \frac{10^{12}}{10^9} = 10^3$$

We conclude that $x_1 = 10^3 x_2$, so the plane is 1000 times louder than the truck. □

Properties of Logarithms

The function $\log x$ has two important properties. These properties involve products, $\log ab$, and powers, $\log a^r$. Since $\log 10 = 1$, $\log 100 = 2$, and $\log 1000 = 3$, we see that

$$\log(10)(100) = 3 = \log 10 + \log 100$$

Let's try this for other values. Using a calculator we have

$$\log 4 = 0.602060$$
$$\log 5 = 0.698970$$
$$\log 20 = 1.301030$$

Hence,

$$\log(4)(5) = \log 20 = \log 4 + \log 5$$

Is it possible that $\log ab = \log a + \log b$ for any positive numbers a and b? The answer is yes, and all we need to verify this is the definition of $\log x$ and the law of exponents, $10^x 10^y = 10^{x+y}$. Indeed,

$$\log ab = \log \left(10^{\log a} 10^{\log b}\right) = \log \left(10^{\log a + \log b}\right)$$
$$= \log a + \log b$$

Let us next consider $\log a^r$. If $a = 10$, then

$$\log 10^r = r = r \log 10$$

If $a = 100$, then since $100^r = (10^2)^r = 10^{2r}$, we have

$$\log 100^r = \log 10^{2r} = 2r = r \log 100$$

Now suppose $r = 2$. Then by our previous formula,

$$\log a^2 = \log(a)(a) = \log a + \log a = 2 \log a$$

These examples seem to indicate that $\log a^r = r \log a$ for any positive number a. We can verify this using the other law of exponents, $(10^x)^y = 10^{xy}$. Indeed, we then have

$$\log a^r = \log \left(10^{\log a}\right)^r = \log \left(10^{r \log a}\right)$$
$$= r \log a$$

In summary, we have

$$\log ab = \log a + \log b \qquad (1.7)$$

and

$$\log a^r = r \log a \qquad (1.8)$$

B.C.   by Johnny Hart

EXAMPLE 5   Prove that $\log(a/b) = \log a - \log b$.

Solution.   By (1.7) and (1.8) we have

$$\log \frac{a}{b} = \log ab^{-1} = \log a + \log b^{-1} = \log a - \log b \qquad \square$$

EXAMPLE 6   If earthquakes A_1 and A_2 register y_1 and y_2 on the Richter scale, respectively, what is $y_1 - y_2$ so that A_1 is twice as strong as A_2?

Solution.   Let x_1 and x_2 be the energies for A_1 and A_2, respectively. If A_1 has twice the strength of A_2, then $x_1 = 2x_2$. Hence, by Example 5, we have

$$y_1 - y_2 = \log x_1 - \log x_2 = \log \frac{x_1}{x_2} = \log 2 = 0.301$$

Thus, a 6.301 earthquake is twice as devastating as a 6.000 earthquake. $\square$

EXAMPLE 7   The half-life in days of a certain radioactive material is the number x that satisfies $(0.99)^x = 1/2$. Find the half-life x.

Solution.   Taking the logarithm of both sides of the equation and using (1.8) gives

$$\log(0.5) = \log(0.99)^x = x \log(0.99)$$

Hence,

$$x = \frac{\log(0.5)}{\log(0.99)} = 68.97 \text{ days} \qquad \square$$

EXAMPLE 8   If you deposit funds in a savings account earning 4.5% interest compounded annually, the number of years n to double your money satisfies $(1.045)^n = 2$. Find this number of years n.

Solution. Applying (1.8) gives

$$\log 2 = \log(1.045)^n = n \, \log(1.045)$$

Hence,

$$n = \frac{\log 2}{\log(1.045)} = 15.75 \text{ yr} \qquad \square$$

EXERCISES

1. Write the following numbers in decimal form.
   (a) 2×10^2 (b) 3.5×10^2 (c) 0.5×10^2 (d) 2.5×10^3

2. Compute the following numbers.
   (a) $\log 1000$ (b) $\log 10,000$ (c) $\log(0.01)$ (d) $\log(0.001)$

3. Use a calculator to find the following numbers.
   (a) $10^{1.5}$ (b) $10^{2.3}$ (c) $10^{5.2}$ (d) $10^{-1.2}$ (e) $10^{-0.5}$

4. Use a calculator to find the following numbers.
   (a) $\log(0.5)$ (b) $\log(0.05)$ (c) $\log(90)$ (d) $\log(542)$

5. Find the following numbers. (a) $\log 10^{3.5}$ (b) $\log 100^{3.5}$ (c) $\log 10^{-0.4}$

6. Find the following numbers. (a) $10^{\log 7.2}$ (b) $10^{\log 0.5}$ (c) $10^{\log a}$

7. Graph the function $f(x) = 2^x$.

8. Graph the function $f(x) = 3^x$.

9. An earthquake that registers 5.9 on the Richter scale will leave virtually every building in an American city intact. However, such an earthquake has recently caused great devastation in Cairo, Egypt. Why is that?

10. An earthquake that registers 8.5 on the Richter scale would destroy practically every building in a city. How much stronger is such an earthquake than one that registers 6.5?

11. The dimension of a certain fractal is given by $\log 5 / \log 4$. Compute this number.

12. The dimension of a certain fractal is given by $\log 3 / \log 2$. Compute this number.

13. If earthquakes A_1 and A_2 register 7.0 and 6.5, respectively, on the Richter scale, how much stronger is A_1 than A_2?

14. If sounds A_1 and A_2 register 92 and 85 decibels, respectively, how much louder is A_1 than A_2?

15. Show that $10^{r \, \log a} = a^r$.

16. Use a calculator to find $\log 8.7$, $\log 12.5$, and $\log(8.7)(12.5)$. Then check to see if (1.7) holds.

17. Use a calculator to find $\log(15.4)^{2.5}$ and $2.5 \log 15.4$. Do these numbers agree?

18. Suppose you know that $\log 50 = 1.70$. Without using a calculator find
    (a) $\log 500$, (b) $\log 5000$, and (c) $\log 5$.

19. If earthquakes A_1 and A_2 register y_1 and y_2, respectively, on the Richter scale, what is $y_1 - y_2$ so that A_1 is four times as strong as A_2?

20. The half-life in days of a certain radioactive material is the number x that satisfies $(0.999)^x = 1/2$. Find the half-life x.

21. If you deposit funds in a savings account earning 5.5% interest compounded annually, how many years does it take to double your money?

22. The logarithmic function to the **base** 2 is defined by $\log_2 x = y$ if $2^y = x$. Show that $\log_2 x = \log x / \log 2$ for every positive number x.

23. Solve the equation $\log(2x + 1) = 1$ for x.

24. Solve the equation $2^x = 5$ for x.

25. Solve the equation $x^2 = 8$ for x.

26. Solve the equation $\log(x + 1) - \log(2x) = \log 3$ for x.

(Optional Exercises)

For the remaining exercises, you will need the following definitions. A function $f: S \to T$ is said to be **one-to-one** if $s_1 \neq s_2$ implies $f(s_1) \neq f(s_2)$. Thus a one-to-one function assigns different elements of S to different elements of T. A function $f: S \to T$ is **onto** if every element of T is a value of f. In other words $f: S \to T$ is onto if for any $t \in T$ there is an $s \in S$ such that $f(s) = t$.

27. Let $S = \{a, b, c\}$, $T = \{w, x, y, z\}$ and define $f: S \to T$, $g: S \to T$, and $h: T \to S$ as illustrated in Figure 1.33. Show that f is one-to-one but not onto, g is not one-to-one and not onto, h is not one-to-one but is onto.

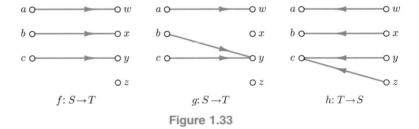

Figure 1.33

28. Define the function $f: \mathbb{R} \to \mathbb{R}$ by $f(x) = x^2$. Is f one-to-one? Is f onto?

29. Define the function $f: \mathbb{R} \to \{x \in \mathbb{R}: x \geq 0\}$ by $f(x) = x^2$. Show that f is onto.

30. Show that the function $f: \mathbb{R} \to \mathbb{R}$ defined by $f(x) = 2x + 1$ is one-to-one and onto.

31. Show that the function $g: \mathbb{R} \to \{x \in \mathbb{R}: x > 0\}$ defined by $g(x) = 10^x$ is one-to-one and onto.

32. For the set $S = \{1, 2, 3\}$ give an example of a function $g: S \to S$ that is not one-to-one.

33. If $f: \mathbb{N} \to \mathbb{N}$ is defined by $f(n) = 2n + 3$, what are $f(5)$ and $f(7)$? Is this function one-to-one? Is it onto?

34. If $f: \mathbb{N} \to \mathbb{N}$ is defined by $f(n) = n + 1$, is f one-to-one? Is f onto? If $g: \mathbb{N} \to \{2, 3, 4, \ldots\}$ is defined by $g(n) = n + 1$, is g one-to-one? Is g onto?

35. Let $S = \{1, 2, 3, 4\}$ and $T = \{a, b, c, d\}$ and define the function $f: S \to T$ by $f(1) = a$, $f(2) = b$, $f(3) = d$, $f(4) = c$ and the function $g: S \to T$ by $g(1) = g(2) = a$, $g(3) = b$, $g(4) = c$. Are f and g one-to-one? Are they onto?

36. If $S = \{1, 2, 3\}$ and $T = \{a, b\}$, then there are eight functions from S to T. Describe them. Which of these functions are one-to-one and which are onto?

37. If $S = \{a, b\}$ and $T = \{1, 2, 3\}$, then there are nine functions from S to T. Describe them. Which of these functions are one-to-one and which are onto?

38. By looking at the graph of $\log x$, can you tell if it is one-to-one?

39. By looking at the graph of $\log x$, can you tell if it is onto?

40. Show that the function $f: \{x \in \mathbb{R}: x \geq 0\} \to \mathbb{R}$ defined by $f(x) = \log x$ is one-to-one and onto.

**41. If S and T are finite sets with the same number of elements and $f: S \to T$, prove that f is one-to-one if and only if f is onto.

**42. If S and T are finite sets, prove the following statements.

   (a) There is a one-to-one function from S to T if and only if T has at least as many elements as S.

   (b) There is an onto function from S to T if and only if S has at least as many elements as T.

   (c) There is a one-to-one and onto function from S to T if and only if S and T have the same number of elements.

*43. If $f: A \to B$ and $g: B \to C$, then the **composition** of f and g is denoted $g \circ f: A \to C$ and is defined by $g \circ f(a) = g(f(a))$. If $f: \mathbb{N} \to \mathbb{N}$ is $f(n) = n + 1$ and $g: \mathbb{N} \to \mathbb{N}$ is $g(n) = n^2$, what are $g \circ f$ and $f \circ g$?

44. The **identity function i_A on the set A is defined by $i_A(a) = a$ for all $a \in A$. A function $f: A \to B$ is said to have an **inverse** if there is a function $g: B \to A$ such that $g \circ f = i_A$ and $f \circ g = i_B$ (see Exercise 43). Prove that f has an inverse if and only if f is one-to-one and onto.

1.7 Tower of Brahma; Growth and Decay (Optional)

When you can measure what you are talking about and express it in numbers, you know something about it.
— LORD KELVIN (1824–1907), British Physicist

There is no branch of mathematics however abstract which may not some day be applied to phenomena of the real world.
— NIKOLAS IVANOVITCH LOBACHEVSKI (1793–1856),
Russian Mathematician

Sequences

In this section some applications of functions to practical (and not so practical) situations are considered. In particular, functions of the type $f: \mathbb{N} \to \mathbb{R}$ are used. Such functions are called **sequences**. Instead of denoting the values of a sequence $f: \mathbb{N} \to \mathbb{R}$ by $f(n)$, the notation f_n is frequently used. An example of a sequence is $f_n = n^2$. This sequence has values $f_1 = 1$, $f_2 = 4$, $f_3 = 9$, $f_4 = 16, \ldots$. Sequences will first be used to solve the mystery of the *Tower of Brahma*.

Photo 1.4
Tower of Brahma
puzzle

Tower of Brahma

According to legend, in the great temple at Benares there is a large brass slab on which there are three diamond needles each a cubit high and as thick as the body of a bee. At the beginning of the world God placed 64 gold disks with holes in their centers on one of these needles, the largest resting on the brass slab and the others decreasing in size to the top. Day and night the priests transfer the disks from needle to needle without deviating from the immutable laws of Brahma. The priests may move only one disk at a time, and they must place this disk only on a free needle or on top of a larger disk. When the 64 disks have been transferred from the original needle to one of the other needles, the temple will crumble and the world will end. If the priests transfer one disk each second, what is the lifetime of the world?

Let f_n be the minimal number of moves needed to transfer n disks from one needle to another. Thus f_n is a sequence. We will solve the mystery if we can find the value f_{64}. Can we find a formula for f_n? For $n = 1$, the minimal number of moves needed to transfer one disk is 1; so $f_1 = 1$. For $n = 2$, the solution is illustrated in Figure 1.34, which shows that $f_2 = 3$.

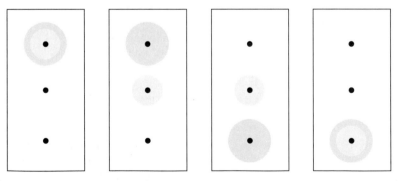

Figure 1.34

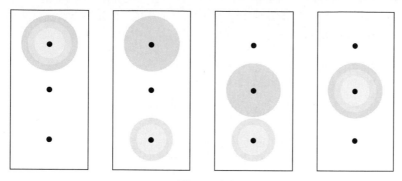

Figure 1.35

For $n = 3$, it takes 3 moves to transfer the top two disks to the third needle (see Figure 1.35), then 1 move to transfer the bottom disk to the second needle, and finally 3 moves to transfer the two disks on the third needle to the second needle. Thus $f_3 = 7$.

Is there a pattern emerging? Notice that $f_3 = 2f_2 + 1$, $f_2 = 2f_1 + 1$. Continuing this reasoning you might conjecture that $f_n = 2f_{n-1} + 1$ for $n = 2, 3, 4, \ldots$. In fact, this is true. Indeed, suppose there are n disks on the first needle. Transfer the $n - 1$ top disks to the third needle in f_{n-1} moves. Then transfer the remaining bottom disk to the middle needle in 1 move and finally transfer the disks from the third needle to the middle needle in f_{n-1} moves. The n disks have been transferred from the first to the second needle in $2f_{n-1} + 1$ moves, and this is the smallest number of moves possible. Why? We can reason backward to show this. If this were not the smallest number of moves possible for transferring n disks, then the n disks could be transferred in a smaller number of moves. But if this were possible, then the $n - 1$ top disks could be transferred in a smaller number of moves than we used. Hence these $n - 1$ disks can be transferred in less than f_{n-1} moves. But this contradicts the fact that f_{n-1} is the minimal number of moves needed to transfer $n - 1$ disks.

Difference Equations

The equation

$$f_n = 2f_{n-1} + 1, \qquad n = 2, 3, 4 \ldots$$

is called a **difference equation**. Can this difference equation help to find f_n? Yes it can. Since $f_{n-1} = 2f_{n-2} + 1$, we have

$$f_n = (2)(2f_{n-2} + 1) + 1 = 2^2 f_{n-2} + 2 + 1$$

Since $f_{n-2} = 2f_{n-3} + 1$, we have

$$f_n = (2^2)(2f_{n-3} + 1) + 2 + 1 = 2^3 f_{n-3} + 2^2 + 2 + 1$$

Continuing this process, we have

$$f_n = 2^p f_{n-p} + 2^{p-1} + 2^{p-2} + \cdots + 2^2 + 2 + 1$$

If we carry this out $p = n - 1$ times, we obtain

$$f_n = 2^{n-1} + 2^{n-2} + 2^{n-3} + \cdots + 2^2 + 2 + 1$$

But in Section 1.2 it was shown that

$$1 + 2 + 2^2 + \cdots + 2^{n-1} = 2^n - 1$$

and so $f_n = 2^n - 1$. This is the formula for f_n. In the *Tower of Brahma* mystery, $n = 64$, and so the lifetime of the universe is $f_{64} = 2^{64} - 1$ seconds.

Radioactive Decay

A more practical application of sequences is found in the study of radioactive decay. Radioactive materials lose their radioactivity by a fixed percentage, which depends only on the particular type of radioactive substance (such as radium or strontium-90) being considered.

Suppose a material has 100 units of radioactivity and it loses radioactivity at a rate of 10% per day. After the first day there will be $(0.9)(100)$ units of radioactivity left. After the second day there will be $(0.9)(0.9)(100) = (0.9)^2(100)$ units of radioactivity left. Observe that the amount of radioactivity will never be zero, although it will eventually become negligible. The amount of radioactivity that is negligible as far as human safety is concerned is an important problem that scientists have not completely resolved. One thing that can be determined is how long it takes for the substance to lose half its radioactivity. This time is called the **half-life** of the substance. To compute the half-life, the following equation must be solved for n:

$$(0.9)^n(100) = \left(\frac{1}{2}\right)(100) \qquad \text{or} \qquad (0.9)^n = \frac{1}{2}$$

Taking logarithms of both sides gives

$$\log\ 0.5 = \log\ (0.9)^n = n \log\ 0.9$$

so

$$n = \frac{\log\ 0.5}{\log\ 0.9} = 6.58$$

Thus the half-life of this particular substance is about $6\frac{1}{2}$ days.

Suppose that an accidental explosion at a nuclear power plant releases the radioactive substance discussed above at a radiation level 8 times that tolerable for human existence. How long should workers wait before entering the area to repair the damage? The answer can be found by using the fact that the radioactivity decreases by one-half every $6\frac{1}{2}$ days. After $6\frac{1}{2}$ days the radiation level is cut in half, after 13 days it will be one-fourth of what it was, and after $19\frac{1}{2}$ days it will be down by one-eighth.

These considerations can be used to derive another difference equation. Let f_n be the amount of radioactive units after n days. Then $f_{n+1} = 0.9f_n$ since the amount of radioactivity after $n + 1$ days is 0.9 times the amount after n days. The difference equation $f_{n+1} = 0.9f_n$ holds for $n = 0, 1, 2, \ldots$. The initial amount of radioactivity in this case is $f_0 = 100$ units. This difference equation can be solved to get a formula for f_n. First

$$f_1 = (0.9)(100)$$

Next

$$f_2 = 0.9f_1 = (0.9)^2(100)$$
$$f_3 = 0.9f_2 = (0.9)^3(100)$$

Continuing, we get $f_n = (0.9)^n(100)$ as the formula for f_n. In general, for a substance with A units of radioactivity that decreases at a rate of r each day, the difference equation is $f_{n+1} = (1-r)f_n$, $n = 0, 1, 2, \ldots$, with initial condition $f_0 = A$. Solving this equation as before, we see that the amount of radioactive units after n days becomes $f_n = A(1 - r)^n$.

Since a radioactive material loses radioactivity, this is called a **decay problem**. The opposite type of phenomenon gives rise to a **growth problem**.

Growth Problems

Suppose a certain germ culture grows at the rate of 10% per hour. What is the size of the culture after 24 hours if initially there were 1000 germs? After 1 hour there would be

$$1000 + (0.1)(1000) = (1.1)(1000) \text{ germs}$$

After 2 hours there would be

$$(1.1)(1000) + (0.1)(1.1)(1000) = (1.1)^2(1000) \text{ germs}$$

Continuing in this way, we see that after 24 hours there would be

$$(1.1)^{24}(1000) \approx 9850 \text{ germs}$$

Again this gives a difference equation. If f_n is the number of germs after n hours, then

$$f_{n+1} = f_n + 0.1f_n = (1 + 0.1)f_n \quad \text{for } n = 0, 1, 2, \ldots$$

with initial condition $f_0 = 1000$. As we have seen, the solution of this difference equation is $f_n = (1 + 0.1)^n(1000)$. In general, the difference equation $f_{n+1} = (1 + r)f_n$, $n = 0, 1, 2, \ldots$, with initial condition $f_0 = A$ has the solution $f_n = (1 + r)^n A$.

Interest

The advantage of knowing the solution of a general difference equation is that the solution can then be applied to diverse situations without repeatedly solving the equation. For example, suppose you were interested in solving the following problem. If you deposit $1000 in a savings account at 5% interest (compounded annually), how long will it take to double your money? Let f_n be the amount in your savings after n years. Since the amount after $n + 1$ years is $f_n + 0.05 f_n$, you have the difference equation

$$f_{n+1} = (1 + 0.05) f_n, \qquad n = 0, 1, 2, \ldots$$

with initial condition $f_0 = 1000$. The solution is

$$f_n = (1 + 0.05)^n (1000)$$

You would like the number n such that

$$(1 + 0.05)^n (1000) = 2000 \qquad \text{or} \qquad (1.05)^n = 2$$

Taking the logarithm of both sides gives

$$\log 2 = \log (1.05)^n = n \log 1.05$$

so

$$n = \frac{\log 2}{\log 1.05} = 14.21$$

Hence, it will take 14.21 years to double your money. You might now ask: What percent interest would double my money in 10 years? Answer: The solution for x of the equation $(1 + x)^{10} = 2$. This can be solved as follows:

$$1 + x = 2^{1/10}$$
$$x = 2^{1/10} - 1 = 1.0718 - 1 = 0.0718$$

or about 7.2% interest.

Population

There is much concern today about explosions. For example, there is the information explosion. We are producing information at such a rate that it more than doubles every decade. Our libraries and files are becoming buried under the bulk of books, papers, tapes, records, and microfilms. Of more importance, however, is the population explosion, which threatens our very existence. Concern about the rapid growth of the world population dates back to the eighteenth century. In 1798, Thomas Robert Malthus published an essay on population in which he claimed: "Population, when unchecked, increases in geometric ratio. Subsistence increases only in an arithmetic ratio."

When Malthus stated that the population increases in geometric ratio, he meant that after a certain time the population is a constant times what it was previously. For example, Table 1.9, which gives the U.S. census figures for the nineteenth century, shows that the population increased by roughly a factor of 1/3 every 10 years.

Table 1.9

Date	Population (millions)	Increase (%)
1810	7.2	
1820	9.6	33
1830	12.9	34
1840	17.1	33
1850	23.2	35
1860	31.4	35
1870	38.6	23
1880	50.2	30
1890	62.9	25
1900	76.0	21

Malthus based his conclusions on similar tables and also upon the fact that during a particular period, say 10 years, the population increases by a certain average number of children and a certain average (but smaller) number of people die, thus increasing the population by a roughly fixed proportion.

When Malthus stated that the subsistence increases in an arithmetic ratio, he meant that the difference in food production over fixed time periods is a constant. For example, suppose that in this country in 1800 we produced enough food for 10 million people and were able to increase our food production every 10 years to feed an additional 10 million people. Then in 1810 we could feed 20 million; in 1820, 30 million;...; in 1900, 110 million. In this case our food production would increase in an arithmetic ratio. Although the reasons for Malthus's statement about subsistence will not be gone into here, it should be mentioned that his conclusions seem to be roughly correct.

Malthusian Theory

What are the consequences of the Malthusian theory for the future? Let us examine the census and food-production figures for country X every 40 years starting with 1800. Let f_n denote the population after the nth 40-year period, so that f_0 is the population in 1800, f_1 the population in 1840, f_2 the population in 1880, and so forth. According to Malthus, the population f_{n+1} after the $(n + 1)$th period is some constant a (where $a > 1$) times the population f_n after the nth period. Thus f_n satisfies the difference equation $f_{n+1} = af_n$, $n = 0, 1, 2, \ldots$, with initial condition $f_0 = A$, where A is the population in 1800. We have seen that the solution of this difference equation is $f_n = a^n A$, $n = 0, 1, 2, \ldots$.

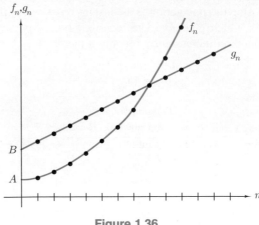

Figure 1.36

Now consider the food production. Let g_n denote the number of people country X can feed at the end of the nth period. According to Malthusian theory, the number of people g_{n+1} country X can feed at the end of the $(n+1)$th period minus the number of people g_n the country can feed at the end of the nth period is some constant C (where $C > 0$). Thus g_n satisfies the difference equation

$$g_{n+1} = g_n + C, \qquad n = 0, 1, 2, \ldots$$

Can we solve this difference equation? Let $g_0 = B$ be the initial number of people that country X can feed in 1800. Then

$$g_1 = B + C$$
$$g_2 = g_1 + C = B + 2C$$
$$g_3 = g_2 + C = B + 3C$$
$$\vdots$$
$$g_n = B + nC$$

This is the solution of the difference equation for g_n.

The important thing is the comparison of g_n with f_n. That is, how does the number of people country X can feed compare to country X's population? The ratio g_n/f_n gives the fraction of its population that country X can feed at the end of period n. Now $g_n/f_n = (B + nC)/a^n A$. The important point is that no matter how large B and C are, and no matter how close a is to 1, the denominator will eventually dominate the numerator and a large portion of the population will starve!

The situation is illustrated in Figure 1.36, which shows that although g_n increases linearly, f_n increases exponentially and is bound to overtake g_n.

Let us substitute some numbers to see how this goes. Suppose that in 1800 country X has a population of $A = 5$ million. Let us be optimistic and suppose that country X is an agricultural country that could feed $B = 50$ million people in 1800 and was therefore able to export a considerable amount of food (compare to the situation in the United States). Suppose that every 40 years country X increased production so that it could feed an additional 50 million people. Then $C = 50$ million. Finally, suppose the population doubled every 40 years so that $a = 2$. This last assumption is less than the population growth of the United States in the nineteenth century, and although our population growth is considerably smaller today, many developing countries have such population growths. Using these numbers, we find that

$$\frac{g_n}{f_n} = \frac{50 + n50}{(2^n)(5)} = \frac{10(n+1)}{2^n}$$

Table 1.10 gives the values of g_n / f_n from 1800 to 2080.

Table 1.10

Year	Population (millions)	g_n / f_n
1800	$f_0 = 5$	$g_0 / f_0 = 10$
1840	$f_1 = 10$	$g_1 / f_1 = 10$
1880	$f_2 = 20$	$g_2 / f_2 = \frac{30}{4} \approx 7.7$
1920	$f_3 = 40$	$g_3 / f_3 = \frac{40}{8} = 5$
1960	$f_4 = 80$	$g_4 / f_4 = \frac{50}{16} \approx 3.1$
2000	$f_5 = 160$	$g_5 / f_5 = \frac{60}{32} \approx 1.9$
2040	$f_6 = 320$	$g_6 / f_6 = \frac{70}{64} \approx 1.1$
2080	$f_7 = 640$	$g_7 / f_7 = \frac{80}{128} \approx 0.62$

Notice that although country X could feed 10 times its population in 1800 and export most of its food, in 2080 country X will be able to feed only 62% of its population. In 2080 country X must either import huge quantities of food or allow 38% of its population to starve!

EXERCISES

One of the arguments for limiting world population is that there simply is not room enough and we will soon be stacked on top of each other. Is this really true? Exercises 1 to 4 give an answer to this question.

1. There are now about 5 billion people on earth. Suppose each person is assigned a plot of land on which he or she could reasonably live, say 50 square feet (about the size of many prison cells). How many square feet would be needed to hold the earth's population?

2. How does the answer to Exercise 1 compare with the area of Massachusetts, which is 7867 square miles?

3. How many people would the State of Washington (96,000 square miles) hold if each person were given 50 square feet?

4. The land area of earth is approximately 52 million square miles. If this area is divided equally among its 5 billion people, how many acres will each person have? (An acre contains about 40,000 square feet.)

5. If you deposit $100 at 6% compounded annually, how much will you have in 2 years?

* 6. If you deposit $100 at 6% compounded semiannually, your total capital at the end of n years is given by $f_n = (100)(1.03)^{2n}$. How was this formula derived? What is your capital at the end of 2 years?

7. Suppose a *Tower of Brahma* puzzle originally contained five disks. How many additional moves will be required to complete the transfers if one more disk is added?

8. What is the answer to Exercise 7 if five is replaced by n?

9. According to the *Tower of Brahma* mystery, what is the lifetime of the universe in years?

10. List the moves for a *Tower of Brahma* with four disks.

11. If you placed a grain of sand on the first square of a checkerboard, two grains on the second, four on the third, eight on the fourth, etc., how many grains of sand would you have altogether?

12. If a radioactive substance decays at a rate of 20% a day, what is its half-life?

13. If a radioactive substance decays at a rate of 2% a day, what is its half-life?

14. If $(1.02)^n = 2$, find n.

15. If $(0.99)^n = 1/2$, find n.

16. If $f_n = (1.1)^n 100$, find f_0, f_1, f_2, f_3. Graph f_n for $n \in \mathbb{N}$.

17. If $g_n = 100 + 5n$, find g_0, g_1, g_2, g_3. Graph g_n for $n \in \mathbb{N}$.

*18. If there are initially 50 units of radioactivity in a substance and the substance decays at a rate of 1% a day, how many units of radioactivity are there after 10 days?

19. Solve the difference equation $f_n = 3f_{n-1} + 1$, $n = 2, 3, 4, \ldots$.

20. What is the solution of the difference equation $f_{n+1} = 2f_n$, $n = 0, 1, 2, \ldots$, with initial condition $f_0 = 3$?

21. If you get 10% interest compounded annually, how long does it take to double your money?

22. What percentage interest compounded annually will double your money in 3 years?

23. What is the solution of the difference equation $f_{n+1} = f_n + 1$, $n = 0, 1, 2, \ldots$, with initial condition $f_0 = 1$?

*24. (You will need tables or a calculator for this problem.) If you deposit $100 at 6% compounded continuously, then at the end of n years you will have $100\, e^{0.06n}$ dollars where $e \approx 2.72$. After 10 years, how would this amount compare with $100 at 6% compounded annually?

*25. Continue Table 1.10 until the year 2200.

*26. Peter Minuit bought Manhattan Island in 1626 for $24 worth of beads and trinkets. If instead he had put the $24 in a savings account that gave 5% interest a year, how much would his money be worth in 1994?

*27. What is the solution of the difference equation $f_{n+1} = f_n + ar^n$, $n = 0, 1, 2, \ldots$, with initial condition $f_0 = 0$?

28. Solve the difference equation $f_n = (f_{n-1})^2$, $n = 1, 2, \ldots$, with initial condition $f_0 = 2$.

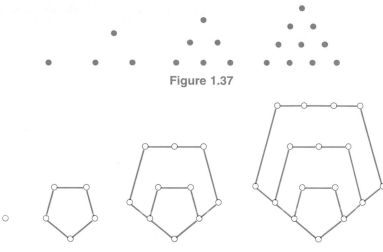

Figure 1.37

Figure 1.38

*29. The **triangular numbers** are shown in Figure 1.37. If f_n is the nth triangular number, then f_n satisfies the difference equation $f_{n+1} = f_n + n + 1$, $n = 1, 2, \ldots$, with initial condition $f_1 = 1$. Solve this difference equation.

30. The **pentagonal numbers are shown in Figure 1.38. If f_n is the nth pentagonal number, then f_n satisfies the difference equation $f_{n+1} = f_n + 3n$, $n = 1, 2, \ldots$, with initial condition $f_1 = 1$. Solve this difference equation. **Hint**: This equation has a solution of the form $f_n = an^2 + bn + c$. Substitute this expression into the difference equation and equate coefficients of like powers of n to find a and b. Use the initial condition to find c.

1.8 Sets and Politics (Optional)

> *A marvelous neutrality have these things mathematical, and also a strange participation between things supernatural, immortal, intellectual, simple and indivisible, and things natural, mortal, sensible, compounded, and divisible.* — JOHN DEE (1527–1608)

In Section 1.4 you saw an application of sets to survey problems. We now give another application of sets to obtain a mathematical model for voting procedures. Let A be a set of people who have to reach a decision through voting. We call A a **voting set**, and we assume that each member of A must vote either for or against the particular issue or candidate being considered. The members of A have various numbers of votes.

A subset $K \subseteq A$ is a **winning coalition** if the combined votes of the members of K are enough to win.

Photo 1.5
A joint session of
Congress. Sets
can be used to
model voting
procedures and
power politics.

For example, in some voting situations a simple majority is enough for a winning coalition, in others a two-thirds majority is necessary, and in others every vote but one (only one blackball) is needed.

EXAMPLE 1   Let $A = \{a, b, c, d\}$. Each member has one vote, and a simple majority decides. Hence a winning coalition must have at least three votes. The winning coalitions are $\{a, b, c\}$, $\{a, b, d\}$, $\{b, c, d\}$, and $\{a, b, c, d\}$.

EXAMPLE 2   Let A be the same set as in Example 1, with the same voting procedure except that a is a chairperson who has the deciding vote in case of a tie. Then in addition to the winning coalitions of Example 1, there are $\{a, b\}$, $\{a, c\}$, $\{a, d\}$. Instances of this type of voting procedure can be found in European barrister examinations and in certain courts that consist of three jurors and one judge.

EXAMPLE 3   Let $A = \{a, b, c, d\}$, and let the members a, b, c, and d have 5, 4, 3, and 1 votes, respectively. A majority of the possible 13 votes (that is, 7 votes) is necessary to win. The winning coalitions are $\{a, b\}$, $\{a, c\}$, $\{b, c\}$, $\{a, b, c\}$, $\{a, b, d\}$, $\{a, c, d\}$, $\{b, c, d\}$, and $\{a, b, c, d\}$. This procedure is used at stockholders' meetings in which each stockholder has a number of votes proportional to the number of stocks each one holds. This procedure is also used in coalition governments in which there are a number of minority parliamentary parties no one of which has a majority of the votes. For instance, suppose the socialist party has five seats, the labor party four seats, the conservative party three seats, and the republican party one seat. In order to reach a majority decision it is necessary for two or more of these parties to band together to form a winning coalition.

EXAMPLE 4 Let $A = \{a, b, c, d\}$, and let the members a, b, c, and d have 6, 5, 5, and 1 votes, respectively. A three-fourths majority is needed to win. The winning coalitions are $\{a, b, c\}$, and $\{a, b, c, d\}$.

EXAMPLE 5 Let $A = \{a, b, c, d\}$, and let the members a, b, c, and d have 5, 2, 1, and 1 votes, respectively. A vote of at least 5 is needed to win a decision. The winning coalitions are $\{a\}$ and all other sets containing a. An illustration would be a stockholders' meeting in which a owns a majority of the stock.

EXAMPLE 6 Let $A = \{a_1, a_2, \ldots, a_n\}$. Each of the n members has one vote. For success, n votes are needed. In this case A itself is the only winning coalition. This situation occurs when unanimity is required, as in certain juries.

Coalition Structures

Mathematics is frequently used to create models for describing real-life situations. These models are constructed by postulating certain axioms that correspond to properties of the real-life system. From these axioms, using logical deduction, further properties can be derived and predictions of how the system will act can be made. We now construct a model that describes voting procedures. This model is motivated by the properties of the six examples just considered. A **coalition structure** is a nonempty set A (called the **voting set**) together with a nonempty collection of subsets of A called **winning coalitions** that satisfy the following two axioms.

Axiom 1   If $K \subseteq A$ is a winning coalition, then any superset of K is a winning coalition.

Axiom 2   If K is a winning coalition, then K' is not a winning coalition.

Examples 1 to 6 are all coalition structures. As an illustration, consider Example 4. In this case $A = \{a, b, c, d\}$ is the voting set and the winning coalitions are the two sets $\{a, b, c\}$ and A. The supersets of $\{a, b, c\}$ are $\{a, b, c\}$ and A, while the only superset of A is A itself. Hence, supersets of winning coalitions are winning coalitions, so Axiom 1.1 holds. The complement of $\{a, b, c\}$ is $\{d\}$, and the complement of A is $\emptyset$. Since neither of these complements is a winning coalition, Axiom 1.2 holds.

If K is a winning coalition, we call K' a **losing coalition**. Notice that there can be subsets of A that are neither winning nor losing coalitions. We call such subsets **blocking coalitions**. When there is a blocking coalition, it is possible that no decision can be reached. For instance, in Example 1, $\{a, b\}$ is a blocking coalition. If the members of $\{a, b\}$ vote no and the members of $\{c, d\}$ vote yes, no decision can be reached. In Example 6, any nonempty subset of A except A itself is a blocking coalition.

A subset $K \subseteq A$ is a **minimal winning coalition** if K is a winning coalition and no subset of K is a winning coalition except K itself.

Thus minimal winning coalitions are winning coalitions that cannot lose a member and still remain winning. If K is a winning coalition, then members can be discarded from K if necessary until a minimal winning coalition results. Thus every winning coalition is a superset of a minimal winning coalition. We conclude that the minimal winning coalitions determine the collection of all winning coalitions. In Example 3, the minimal winning coalitions are $\{a, b\}$, $\{a, c\}$, and $\{b, c\}$. In Example 5, $\{a\}$ is the only minimal winning coalition. See if you can find the minimal winning coalitions in the other examples.

Notice that there may be members of a voting set that exert no influence whatsoever upon the decision. For instance, in Example 5, b, c, and d are such members. In Example 4, d is such a member. These members belong to no minimal winning coalition. An element $x \in A$ is called a **powerless figure** if x belongs to no minimal winning coalition.

We can now define power-political concepts. For example, a dictator would be the only person who has political power. We thus say that an element $x \in A$ is a **dictator** if all other members of A are powerless figures. Thus, in Example 5, a is a dictator. A person whose positive vote is needed to reach any decision would have the right of veto. Anything that such a person is against cannot be voted through. We thus say that $x \in A$ has the **right of veto** if x belongs to every minimal winning coalition. What is the difference between a dictator and a member with the right of veto? In Example 4, the only minimal winning coalition is $\{a, b, c\}$. Thus, a, b, and c all have the right of veto, yet none of these is a dictator. For example, a is not a dictator since b and c are not powerless figures.

We now prove some results that hold for any coalition structure A. First note that since A is a superset of any winning coalition, by Axiom 1.1, A is a winning coalition. Since $\emptyset = A'$, it follows from Axiom 1.2 that $\emptyset$ is a losing coalition. Suppose $x \in A$ and $\{x\}$ is the only minimal winning coalition. Then no other element of A is in a minimal winning coalition, so all other elements of A are powerless figures. Hence x is a dictator. We now prove the converse.

Theorem 1.1   If $x \in A$ is a dictator, then $\{x\}$ is the only minimal winning coalition.

Proof: Let $K \subseteq A$ be a minimal winning coalition. If $y \neq x$, then y is a powerless figure, so $y \notin K$. Since $K \neq \emptyset$, K must contain at least one element, so $K = \{x\}$. Hence $\{x\}$ is the only minimal winning coalition. $\square$

Notice that in all of our examples, two winning coalitions always have at least one member in common. For instance, in Example 3 two winning coalitions both contain either a, b, or c. It seems reasonable that for any coalition structure, two winning coalitions always have a nonempty intersection. Indeed, suppose K_1 and K_2 are winning coalitions and $K_1 \cap K_2 = \emptyset$. Now suppose everyone in K_1 votes for some decision and everyone in K_2 votes against that decision. Then since K_1 and K_2 are both winning coalitions, the decision will be simultaneously carried and defeated. Such a circumstance cannot reasonably occur. The next theorem shows that, in general, this is actually the case. The method of proof for this theorem is a technique that is frequently used in mathematics called *reductio ad absurdum* or **method of contradiction**. (We shall study this method in more detail in Chapter 3.) As in our previous discussion, we first assume that $K_1 \cap K_2 = \emptyset$. We then show that this implies that one of our axioms is contradicted. Since the axioms must hold in a coalition structure, we then conclude that our assumption is wrong, so $K_1 \cap K_2 \neq \emptyset$.

Theorem 1.2   If K_1 and K_2 are winning coalitions, then $K_1 \cap K_2 \neq \emptyset$.

Proof: Suppose $K_1 \cap K_2 = \emptyset$. Then $K_1 \subseteq K_2'$ (look at the Venn diagram or note that if $x \in K_1$, then $x \notin K_2$). Since K_2' is a superset of a winning coalition, by Axiom 1.1, K_2' is a winning coalition. Since K_2 is a winning coalition, this contradicts Axiom 1.2. Hence $K_1 \cap K_2 \neq \emptyset$.   □

We can prove many other theorems about coalition structures. Some of these appear in the exercises.

EXERCISES

1. Show that Axioms 1.1 and 1.2 hold in Example 1 and hence that this example gives a coalition structure.

2. Do Exercise 1 for Example 2.

3. Do Exercise 1 for Example 3.

4. Do Exercise 1 for Example 5.

5. Do Exercise 1 for Example 6.

6. Let $A = \{a, b, c\}$, where a, b, c have 3, 2, and 1 votes, respectively. What are the winning coalitions if a majority vote is needed to win? Show that Theorem 1.2 holds for this example.

7. Show that Theorem 1.2 holds for Example 1.

8. Show that Theorem 1.2 holds for Examples 2 to 6.

9. (a) Show that x is a dictator if $\{x\}$ is a minimal winning coalition.
   (b) Show that $\{x\}$ is a minimal winning coalition if x is a dictator.

10. Find the minimal winning coalitions in Examples 1, 2, 4, and 6.

11. Let $A = \{a, b, c, d\}$, where a, b, c, and d have 4, 3, 2, and 1 votes, respectively. If 6 votes are necessary to win a decision, what are the winning coalitions? The minimal winning coalitions?

12. A subset $L \subseteq A$ is a **maximal losing coalition** if L is a losing coalition and no superset of L is a losing coalition except L itself. Show that L is a maximal losing coalition if L' is a minimal winning coalition. Show that L' is a minimal winning coalition if L is a maximal losing coalition. Show that every losing coalition is a subset of a maximal losing coalition.

13. Find the blocking coalitions if any in Examples 2 to 5.

14. Find the powerless figures if any in Examples 1, 2, 3, and 6.

15. Find the dictators if any in Examples 1 to 6.

16. Find the members with the right of veto if any in Examples 1 to 6.

17. Suppose A has n members, each having one vote, and a simple majority wins. If n is even, what is the size of a minimal winning coalition? What if n is odd?

18. Do Exercise 17 when a two-thirds majority is needed to win.

19. Prove that if K is a blocking coalition, then so is K'.

20. Use Theorem 1.1 to prove that a dictator has the right of veto.

21. Prove that if $\{x\}$ is a minimal winning coalition, then $\{x\}$ is the only minimal winning coalition.

22. Prove that if x belongs to no losing coalition, then x has the right of veto. Prove that if x has the right of veto, then x belongs to no losing coalition.

23. Prove that if $\{x\}$ is a blocking coalition, then x has the right of veto.

*24. If there is no dictator and x has the right of veto, prove that $\{x\}$ is a blocking coalition.

25. Show that if L_1 and L_2 are losing coalitions, then $L_1 \cup L_2 \neq A$.

26. Prove that in a coalition structure there can be at most one dictator.

27. How can votes be distributed in Example 2 so that a simple majority vote to win determines the same winning coalitions?

28. Show that a blocking coalition in a coalition structure cannot be composed only of powerless figures.

29. Let $A = \{a, b, c, d\}$, and let a yes vote by any member win an issue. Does this give a coalition structure?

*30. For a coalition structure, prove or disprove the statement "Every blocking coalition is contained in a minimal winning coalition."

*31. Show that if x has the right of veto implies that $\{x\}$ is a blocking coalition, then there is no dictator.

32. In Example 2, a could be called a deciding leader. In general, if $x \in A$, we call x a **deciding leader if A has $2n$ elements ($n > 1$) and of the n-element subsets exactly those which contain x and of the $n + 1$ element subsets exactly those which do not contain x form minimal winning coalitions. Prove the following.
    (a) A deciding leader is not a dictator.
    (b) A deciding leader does not have the right of veto.
    (c) If x is a deciding leader, then $\{x\}$ is not a blocking coalition.
    (d) There is at most one deciding leader.

Chapter 1 Summary of Terms

n factorial $n! = n(n-1)(n-2)\cdots 3\cdot 2\cdot 1$

sum of natural numbers $1 + 2 + \cdots + n = \dfrac{n(n+1)}{2}$

geometric series $1 + r + r^2 + \cdots + r^n = \dfrac{r^{n+1}-1}{r-1}$, $r \neq 1$

fractions $\dfrac{a}{b} + \dfrac{c}{d} = \dfrac{ad+bc}{bd}$

membership symbol $a \in A$

subset symbol $A \subset B$

number of subsets of an n-element set 2^n

natural numbers $\mathbb{N}$

positive rational numbers Q

integers I

real numbers $\mathbb{R}$

union of sets $A \cup B$

intersection of sets $A \cap B$

complement of a set A'

set difference $A - B$

distributive laws $(A \cup B) \cap C = (A \cap C) \cup (B \cap C)$
$(A \cap B) \cup C = (A \cup C) \cap (B \cup C)$

associative laws $(A \cup B) \cup C = A \cup (B \cup C)$
$(A \cap B) \cap C = A \cap (B \cap C)$

De Morgan's laws $(A \cup B)' = A' \cap B'$
$(A \cap B)' = A' \cup B'$

Venn diagram

number of elements in a set $|A|$

number of elements in a union $|A \cup B| = |A| + |B| - |A \cap B|$

number of elements in a difference $|A - B| = |A| - |A \cap B|$

function $f\colon S \to T$

Cartesian plane $\{(x,y)\colon x,y \in \mathbb{R}\}$

graph of f $\{(x, f(x))\}$

logarithmic function $\log x$

properties of logarithms $\log 10^x = x$
$10^{\log x} = x$
$\log ab = \log a + \log b$
$\log a^r = r \, \log a$

sequence $f: \mathbb{N} \to \mathbb{R}$

difference equation $f_n = 2f_{n-1} + 1$

voting set A

winning coalition $K \subseteq A$

minimal winning coalition

powerless figure

dictator

right of veto

reductio ad absurdum

Chapter 1 Test

1. Evaluate $5!$, $10!/8!$, $100!/98!$.
2. If $n \in \mathbb{N}$, $n \geq 3$, show that $n!/(n-2)! = n(n-1)$.
3. Evaluate the sum $1 + 2 + \cdots + 200$.
4. Evaluate the sum $1 + \dfrac{1}{2} + \dfrac{1}{2^2} + \cdots + \dfrac{1}{2^{10}}$.
5. Find the value of $\dfrac{1}{2} + \dfrac{1}{3}$.
6. If $64 = 2^n$, find n.
7. Let $U = \{1, 2, 3, 4\}$, $A = \{1, 3, 4\}$, $B = \{2, 4\}$. Find $A \cup B$, $A \cap B$, B', and $A - B$.
8. If $A \subseteq B$, find $A \cup B$ and $A \cap B$.
9. If U is the universal set, find $A \cup U$ and $A \cap U$.
10. Prove that $A \cup A' = U$ and $A \cap A' = \emptyset$.
11. If $A \subseteq C$, use Venn diagrams to illustrate that $(A \cup B) \cap C = A \cup (B \cap C)$
12. If $A = \{1, 2, 3, 4, 5, 6\}$, how many subsets does A have? List the one-element subsets of A. List the five-element subsets of A.
13. If $|A \cup B| = 15$, $|A \cap B| = 5$, $|A| = 7$, find $|B|$.
14. In a survey of 50 students, the following data were collected. There are 19 taking English, 20 taking history, 19 taking biology, 7 taking history and biology, 8 taking English and history, 9 taking English and biology, and 5 taking all three subjects.
    (a) How many of the students are not taking any of the three subjects?
    (b) How many are taking only English?
    (c) How many are taking English and history but not biology?
15. A survey was taken of 30 students enrolled in three different subjects A, B, and C. Show that the following data are inconsistent: 18 in A, 10 in B, 9 in C, 3 in B and C, 6 in A and B, 9 in A and C, 2 in A, B, and C.

16. Graph the function $f \colon \mathbb{R} \to \mathbb{R}$ given by $f(x) = 6x - 3$.

17. Prove that $\log(1/a) = -\log a$.

18. Graph the function $f(x) = (x - 1)^2$.

19. Solve the equation $3^{2x+1} = 10$ for x.

20. If a whisper has decibel magnitude 24 and a conversation has decibel magnitude 62, how much louder is the conversation than the whisper?

21. What percentage interest compounded annually will double your money in 7 years?

22. Let $A = \{a, b, c, d, e, f\}$. Each member has one vote and a simple majority decides. List the minimal winning coalitions. How many members are in a blocking coalition?

23. Is each of the following statements true or false?
    (a) $4! = 24$
    (b) $1 + 2 + \cdots + 8 = 32$
    (c) If $r = 1$, then $1 + r + r^2 + \cdots + r^n = n + 1$
    (d) $(4^2)(4^3) = 4^6$
    (e) $(4^2)^3 = 4^5$
    (f) If $a \in A$, then $a \notin A'$
    (g) $(x + y)^2 = x^2 + 2xy + y^2$
    (h) $(-1)^2 = -1$
    (i) $x^2 + 3x + 2 = (x + 1)(x + 2)$
    (j) $-1 \in I$
    (k) $A \subseteq A \cap B$ for all sets A and B
    (l) $A \cap B \subseteq A \cup B$ for all sets A and B
    (m) $(A \cap B)' = A' \cup B'$ for all sets A and B
    (n) $(\log 5)(\log 6) = \log 30$
    (o) $1/(1/2) = 2$
    (p) $\log(1/2) = -\log 2$
    (q) $\log 20 = 1 + \log 2$

MATHEMATICS AND ART

2

A scientist worthy of the name, above all a mathematician, experiences in his work the same impression as an artist; his pleasure is as great and of the same nature.

— HENRI POINCARÉ (1854–1912), French Mathematician

Mathematizing may well be a creative activity of man, like language or music, of primary originality, whose historical decisions defy complete objective rationalization.

— HERMANN WEYL, Twentieth-Century Mathematician

The word "Verse" is used here as the term most convenient for expressing, and without pedantry, all that is involved in the consideration of rhythm, rhyme, meter, and versification... the subject is exceedingly simple; one tenth of it, possibly, may be called ethical; nine tenths, however, appertains to the mathematics....

— EDGAR ALLAN POE (1809–1849), American Poet

Although mathematics and art seem, at first sight, to be unrelated disciplines, there are amazing similarities between the two. Both fields are among the oldest intellectual pursuits of human beings, going back some 50,000 years. Artists, scientists, and mathematicians think in very similar ways. They experiment, alternate, and compromise between the rational and the irrational, the logical and the intuitive, the objective and subjective, the conscious and the unconscious, the systematic and the random. This is why Jean Cocteau's description of art as "science in the flesh" is very appropriate and why many believe that science and mathematics will have an increasing influence on art and also help to explain the appeal of some modern art.

Denmark's Piet Hein is a contemporary figure exemplifying the interaction between art, science, and mathematics. He is an inventor, poet, and physicist. Hein criticizes the artificial gulf that has been placed between the sciences and art. He says:

My field is across the borderlines. What I write and these other things I do all have the same stamp. They all flow from the same kind of imagination. Whether I am writing a poem or solving a technical problem, I think the same.

Speaking of art, Hein says:

After all, what is art? There is no way of defining art except from the inside. Art is the creative process and it goes through all fields. Einstein's theory of relativity—now that is a work of art! Einstein was more of an artist in physics than on his violin.

Hein calls art "the solution of a problem which cannot be expressed explicitly until it is solved." To Piet Hein, mathematics, science, and art are inseparable.

2.1 | Introduction

A mathematician, like a painter or a poet, is a maker of patterns. If his patterns are more permanent than theirs, it is because they are made of ideas. His patterns, like the painter's or the poet's, must be beautiful; the ideas, like the colors or the words, must fit together in a harmonious way. Beauty is the first test; there is no permanent place in the world for ugly mathematics.

— G. H. HARDY, Twentieth-Century Mathematician

Both mathematics and art are human creations, and they are greatly influenced by culture. The results of both of these fields are not just there to be discovered; they are products of human ingenuity and imagination. Cézanne once said, "I do not reproduce nature but represent it." This is also true of mathematicians. And there are frequently many ways to represent nature. For example, the idea of color may be subject to our judgment and our culture. The Choctaw Indians, for instance, do not distinguish between yellow and green. On the other hand, Eskimos have at least 30 words describing different types of white. In mathematics, there are certain African tribes that do not have numbers larger than 3. They count "1, 2, 3, many." As another example, Euclid postulated that parallel lines never meet. But how can we be sure? Have we ever followed parallel lines to infinity? There are now geometries in which parallel lines meet at infinity; they are just as good as Euclid's and lead to consistent results.

One of the important similarities between mathematics and art is that both employ self-consistent structures. For a work to be considered mathematics or art, it must be harmonious and free from contradiction. Seurat stressed the importance of "mental operations" in art. These are equally important in mathematics. He wrote: "Art is harmony. Harmony is the analogy of opposites, the analogy of similars, ..." This could equally well describe mathematics.

Both art and mathematics are a search for harmony, a search for structure, form, or gestalt, a search for pattern. As early as the Stone Age, patterns (in effect the first geometry) on pottery, weaving, and carpentry invoked a fusion of mathematics and art. The very word "line" probably derives from "linen," linking geometry and

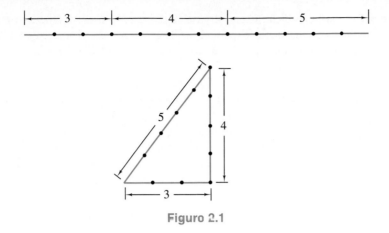

Figuro 2.1

weaving. In Egypt, lines were marked off with stretched rope or flax (from which linen is made). The rule of thumb for marking a right angle, say at the corner of a pyramid, was to tie knots in a rope at regular intervals and form a triangle with sides 3, 4, and 5 intervals long (Figure 2.1).

The ancient Greeks were the first to abstract number patterns. They considered triangular (Figure 2.2), square (Figure 2.3), rectangular, pentagonal (Figure 2.4), hexagonal (Figure 2.5), etc., numbers by arranging pebbles in different combinations. The word "calculate" comes from the Latin word for stone.

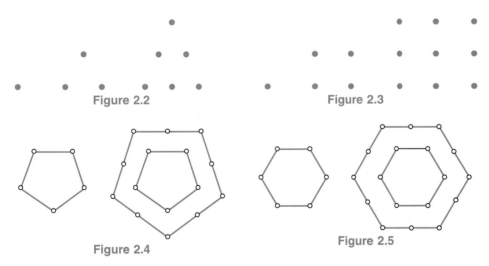

Figure 2.2

Figure 2.3

Figure 2.4

Figure 2.5

Mathematicians (like artists) usually develop their ideas in mental pictures although they write the final form in abstract mathematical shorthand. Recent research among mathematicians suggests that most think not in verbal terms but in images drawn inevitably from the culture of their time.

Mathematics and art have recently been brought even closer together with the advent of computer art. In the computer, people have created not just a useful inanimate tool but an active partner which, when fully exploited, can be used to produce entirely new art forms and new aesthetic experiences. The idea of creating artworks using machines may seem a little strange. If you have heard about experiments using digital computers in creative endeavors, you may have shrugged them off as being of no consequence. On the one hand, creativity has always been regarded as the personal domain of people; and on the other hand, the computer can do only what it has been programmed to do—which hardly anyone would call creative.

Nonetheless, artists have frequently used new scientific and technological developments in their work, and computers are no exception. Composers, film animators, and graphic artists use computers. Artistic experiments with computers have produced results that should make us reexamine our preconceptions about creativity and machines.

Computers are a new medium for artists. They lack the characteristics of the traditional media: oils, watercolors, tempera, etc. Nor are the messages that grow out of the artist's engagement with them likely to be similar to the messages of, for example, oil paintings. As with any new medium, new messages, new effects, new aesthetics are established.

Two of the new characteristics that computers introduce into art are extreme precision and extreme randomness, both of which are impossible for a person to obtain alone. The precision can be used to move, rotate, and transform figures many times by very small amounts, creating very pleasing designs. Random effects give surprising and often exciting results. For examples of computer art refer to the annotated bibliography.

EXERCISES

1. Examine the works of your favorite artist and see if you can find mathematical concepts in them.

2. What is your interpretation of "There is art in mathematics"?

3. What is your interpretation of the quote by G. H. Hardy at the beginning of this section?

4. The reason the knotted rope in Figure 2.1 forms a right triangle is that $3^2 + 4^2 = 5^2$. The converse of Pythagoras' theorem states that if $a^2 + b^2 = c^2$ then a, b, and c are the lengths of the sides of a right triangle. Give an example of three other natural numbers a, b, and c that satisfy $a^2 + b^2 = c^2$.

5. Figure 2.2 shows that the first three triangular numbers are 1, 3, and 6. What is the next triangular number? Draw a picture representing this number. Is this related to a popular sport?

6. What are the fifth and sixth triangular numbers?

7. The nth triangular number equals $n(n + 1)/2$. Why is this?

8. Figure 2.3 shows that the first three square numbers are 1, 4, and 9. What is the next square number? Draw a picture representing this number.

9. Why do all square numbers have the form n^2?

10. Draw pictures representing the rectangular numbers 6, 10, 14, and 15.

11. Draw pictures representing the cube numbers 8 and 27.

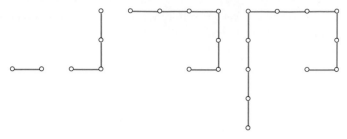

Figure 2.6

12. Figure 2.4 shows that the first two pentagonal numbers are 5 and 15. What is the next pentagonal number? Draw a picture representing this number.

13. The nth pentagonal number, of the type considered in this section, equals $5n(n + 1)/2$. Why is this?

14. Figure 1.38 of Chapter 1 shows another type of pentagonal number. Draw a picture representing the next number of this type. What is this number?

15. What is the hexagonal number after 6 and 18? Draw a picture representing this number.

16. Why does the nth hexagonal number equal $3n(n + 1)$?

17. Figure 2.6 illustrates the spiral numbers. What is the next spiral number? Draw a picture representing this number.

18. Find a formula for the nth spiral number.

19. What are the octagonal numbers? Draw some pictures that represent these numbers.

20. Draw a geometric figure and invent your own number.

21. Pythagoras' theorem states that if a and b are the lengths of the legs of a right triangle and c is the length of the hypotenuse (Figure 2.7), then $a^2 + b^2 = c^2$. If a right triangle has two legs of length 1, what is the length of the hypotenuse?

Figure 2.7

2.2 | Proportion and Order

Mathematics is music for the mind; music is mathematics for the soul. — ANONYMOUS

The basic principles underlying the greatest art so far produced in the world may be found in the proportions of the human figure and in the growing plant.
— CLAUDE BRAGDON, Twentieth-Century Author

He that holds fast the golden mean,
And lives contentedly between
The little and the great,
Feels not the wants that pinch the poor
Nor plagues that haunt the rich man's door,
Embittering all his state.
— WILLIAM COWPER (1731–1800), English Poet

In an age when specialization means isolation, you might be surprised that mathematics and art are intimately related. Yet it is a fact that these two disciplines have been closely identified since ancient times, have been fused harmoniously by such Renaissance geniuses as da Vinci and Dürer, and have exhibited startling interconnections in modern works.

Plato said in *The Statesman* that "all things which come within the province of art do partake in some sense of measure." Ancient records and monuments testify to effective calculation and precise measurement of lengths and angles. At least as early as 500 B.C., artists were searching for ideal proportions and mathematical principles of composition and order. An empirical and humanistic tradition is typified by the Greek sculptor Polycleitus, who made very careful measurements of human proportions the basis of design. The **golden ratio** was known and used by Euclid. This irrational number $(\sqrt{5} + 1)/2$ has properties that have intrigued mathematicians to this day. (Recall that $\sqrt{5}$ is a number whose square is 5.) A rectangle with sides approximately in this ratio, called the **golden rectangle** (for example, familiar 8 by 5 cards), was early recognized as pleasing, and this, coupled with the authority of Euclid, seems to have given it special status. Some 2000 years after Polycleitus, da Vinci and Dürer again made careful measurements of the human body and in typical Renaissance style fused the classical traditions with human knowledge. In this period, the golden ratio played a central part in philosophical discussions as well as in practical design.

Golden Ratio

The golden ratio $(\sqrt{5} + 1)/2 = 1.618033989\ldots$ has been a guide to proportion since ancient times. The Parthenon at Athens fits into a golden rectangle almost precisely. Leonardo da Vinci's famous painting *St. Jerome* fits so neatly into a golden rectangle that it is often said that Leonardo deliberately painted the figure to conform to those proportions. Given his well-known love of "geometrical recreations," as he described them, this is quite likely. Leonardo's notebooks devote many pages to the structure of proportion.

Dürer's work on the proportions of the human form combined a deep insight into anatomy, proportion, and geometry. His greatest contribution to geometry was a treatise published in 1523, which established him not only as an artist but also as a mathematician. According to Erwin Panofsky (Newman, 1956, p. 603):

> *It served as a revolving door between the temple of mathematics and the market square. While it familiarized the coopers and cabinet-makers with Euclid and Ptolemy, it also familiarized the mathematicians with what may be called "workshop geometry."*

The golden ratio is also used in modern abstract art. *La Parade* by Georges Seurat (Holt, 1971) contains many golden rectangles. Seurat is said to have approached every canvas in terms of the golden ratio. A similar analysis can be made of much of Piet Mondrian's work (Busignani, 1968). Juan Gris not only employed the golden ratio but also specifically praised its virtues (Canfield, 1965).

The same proportion can be seen in the modern apartment houses of Le Corbusier (Holt, 1971), who suggested that human life is "comforted" by mathematics. An exposition of the golden ratio appears in the modular theory of Le Corbusier in his book *Le Modular*. In his work the ratio of the height of a man to the height of his navel is the golden ratio.

Golden Rectangle

The golden rectangle was discovered by the ancient Greeks via geometry. The geometric construction of the golden rectangle begins with a square, which is then split in half as shown by the broken line in Figure 2.8. An arc of a circle is now drawn with center C, and the lower side of the square is extended to meet it. The resulting figure is the golden rectangle.

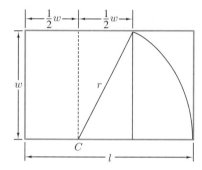

Figure 2.8

The ratio of length to width $\tau = l/w$ of the golden rectangle is the golden ratio. Can we find the numerical value of this ratio? We see that

$$l = \frac{w}{2} + r$$

and that by Pythagoras' theorem (this theorem is reviewed in Exercise 21 of the previous section)

$$r^2 = \frac{w^2}{4} + w^2 = \frac{5w^2}{4}$$

so that

$$l = \frac{w}{2} + \frac{\sqrt{5}\,w}{2} = \frac{(1 + \sqrt{5})w}{2}$$

Hence the golden ratio is

$$\tau = \frac{1 + \sqrt{5}}{2} = 1.618\ldots$$

The golden ratio satisfies the interesting equation $\tau = 1 + 1/\tau$. Indeed,

$$1 + \frac{1}{\tau} = 1 + \frac{2}{\sqrt{5}+1} = \frac{\sqrt{5}+3}{\sqrt{5}+1} = \frac{(\sqrt{5}+3)(\sqrt{5}-1)}{(\sqrt{5}+1)(\sqrt{5}-1)}$$

$$= \frac{5+3\sqrt{5}-\sqrt{5}-3}{5-1} = \frac{2+2\sqrt{5}}{4} = \frac{1+\sqrt{5}}{2} = \tau$$

It follows from this identity that

$$\frac{l}{w} = 1 + \frac{w}{l} = \frac{l+w}{l}$$

so that for a golden rectangle the ratio of the length to the width equals the ratio of half the perimeter to the length. Furthermore, it follows from the last equation that $l^2 = lw + w^2$. Solving for w^2, we obtain

$$w^2 = l^2 - lw = l(l-w) \qquad \text{and so} \qquad \frac{w}{l-w} = \frac{l}{w}$$

Thus we see that if the original square is removed, the rectangle left is still golden. It may be that the golden rectangle has aesthetic proportions because of these various relationships.

People have frequently marveled at the curious properties of the golden ratio τ. For instance, if you divide 1 by τ, you get the decimal part of τ, that is $0.618\ldots$, and if you square τ to get τ^2, the result is $2.618\ldots$; only the whole-number part is changed. But to anyone who knows the identity $\tau = 1 + 1/\tau$ these results are not at all surprising. Indeed, we see that $1/\tau = \tau - 1 = 0.618\ldots$, and if we multiply both sides of the equation by τ, we get $\tau^2 = \tau + 1 = 2.618\ldots$.

Fibonacci Sequence

There is a curious connection between the golden ratio, a certain sequence of numbers, and biology. The greatest medieval European mathematician, Leonardo of Pisa, nicknamed Fibonacci (son of Bonaccio), introduced the sequence of numbers

$$1, 1, 2, 3, 5, 8, 13, 21, 34, 55, \ldots$$

in 1202. This sequence was studied by the nineteenth-century number theorist Edouard Lucas, who called it the **Fibonacci sequence**. The sequence is constructed as follows. Start with 1 and 1; add them to get 2; add 1 to 2 to get 3; add 2 to 3 to get 5, etc. Thus each term is the sum of the previous two terms. Sunflower and daisy florets, pinecones, pineapples, the shells of some molluscs, and even patterns of paving stones and the mating habits of bees and rabbits exhibit properties given by the Fibonacci sequence. The seeds of sunflowers and daisies form a pattern consisting of two sets of spirals, one turning clockwise and the other counterclockwise. The numbers of spirals in the two sets are usually consecutive Fibonacci numbers. For example, in sunflowers there are usually 34 clockwise spirals and 55 counterclockwise spirals; in daisies, 21 clockwise and 34 counterclockwise.

Table 2.1

Ratio	Value
2/1	2.0
3/2	1.5
5/3	1.666
8/5	1.600
3/8	1.625
21/13	1.6153
34/21	1.6190
55/34	1.61764
89/55	1.6181818
144/89	1.617977
233/144	1.6180556
377/233	1.6180258
610/377	1.6180371

An interesting thing happens if we consider the ratios of consecutive Fibonacci numbers, given in Table 2.1. We see that this sequence

$$2, 3/2, 5/3, 8/5, \ldots$$

of successive ratios of Fibonacci numbers seems to be approaching the golden ratio $\tau = 1.618033989\ldots$. Why is this? The reason is as follows. Since $\tau = 1 + 1/\tau$, we can substitute $1 + 1/\tau$ for τ to get

$$\tau = 1 + \cfrac{1}{1 + \cfrac{1}{\tau}}$$

If we do it again, we get

$$\tau = 1 + \cfrac{1}{1 + \cfrac{1}{1 + \cfrac{1}{\tau}}}$$

Continuing in this way, it can be shown that the $1/\tau$ term becomes less and less significant. We then conclude that τ is approached by the sequence of numbers

$$1 + 1, \quad 1 + \cfrac{1}{1+1}, \quad 1 + \cfrac{1}{1 + \cfrac{1}{1+1}}, \quad 1 + \cfrac{1}{1 + \cfrac{1}{1 + \cfrac{1}{1+1}}}, \ldots$$

But this sequence is $2, 3/2, 5/3, 8/5, \ldots$, the successive ratios of Fibonacci numbers.

The golden rectangle can be used to construct certain curves that have a pleasing appearance and that occur in art and nature. One of these curves is the logarithmic spiral. Start with a golden rectangle. Attach a square based on the longer side to get another golden rectangle (see Figure 2.9).

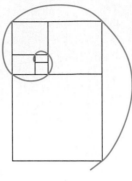

Figure 2.9

Continue this process, and draw a curve through alternate corners of these golden rectangles. This curve, the logarithmic spiral, represents the law of visible organic growth in shells, in arrangements of sunflower seeds, in spiral nebulae, and in many other cases.

Proportion

Proportion is an important concept that has connected art and mathematics from ancient to modern times. A closely related concept that has also tended to fuse the two disciplines is *order* or, as artists call it, the *style of formal order*. The style of formal order expresses an artist's preference for balance and stability in creative work. The Greeks sought order through measure and proportion, through the mathematical statements of relationships which they believed would produce harmony, balance, and beauty.

To the Greeks the immediate source of proportions for order and hence for beauty was the nude human figure. They concluded that what we call beautiful is the result of a harmonious relationship between the parts of any whole and that the mathematical ratios based on the human figure lie behind every instance of visual harmony. Always, in classical art or in the style of formal order, there is a rule of number or measure applied to the problems of artistic creation. Art is a search for the "right" proportions, for the "right" order.

Western culture has not been unswerving in its devotion to classical ideas of measure, order, and moderation. Periodically, it has been caught up in violent convulsions such as our own era has frequently known. Today's art has been characterized by chaotic and violent style features as well as by rationality and order. Often, in the career of a single artist (Picasso, for instance) both the style of formal order and that of spontaneous emotional expression may be present.

[1] Lyonel Feininger, *Church of the Minorites II,* 1926. Oil on canvas, $42\frac{3}{4} \times 36\frac{5}{8}$ in. (Collection Walker Art Center, Minneapolis, Gift of the T. B. Walker Foundation, Gilbert M. Walker Fund, 1943.)

The 1926 painting *Church of the Minorites II* [1] by Lyonel Feininger is an example of the style of formal order by an artist much influenced by cubism. There is a quality in Feininger's work which suggests an effort to impose the geometric order of crystalline structures upon the visible world. It was cubism, with its tendency to convert all forms into geometric shapes, which created the breakthrough Feininger exploited. "Feininger's work implies that our universe has an underlying structure which is fixed, ordered and perhaps, divinely decreed" (Feldman, 1972).

It is not difficult to see that the style of formal order, in the hands of certain artists, has religious implications. This is understandable, for the great protagonists of the style, the Greeks, had an almost mystical view of mathematics: it possessed a divine beauty for them; art created according to the rules of proportion and order would therefore share the divinity of mathematics.

Some of the best-known contemporary examples of order are found in the work of Piet Mondrian. His *Composition with Red, Yellow, and Blue* [Ch. 5, 1] has been created entirely with vertical and horizontal lines, with areas of unvarying color, and with no representational elements whatever. It is the interaction between the forms which suggests an intellectual order of mathematical complexity. By reducing and refining his pictorial vocabulary, Mondrian tried to employ the pure activity of the viewer's eye to create impressions of motion, opposition of weight, and asymmetrical balance.

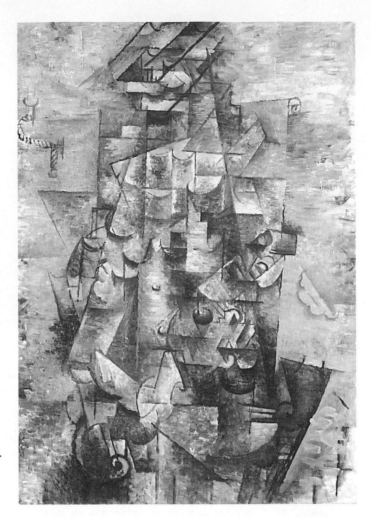

[**2**] Georges Braque,
Man with a Guitar, 1911.
Oil on canvas, $45\frac{3}{4} \times 31\frac{7}{8}$ in.
(The Museum of Modern
Art, New York, Lillie P. Bliss
Bequest.)

Cubism

One of the earliest twentieth-century forms of order appeared in cubism, especially in the stage called analytical cubism (1907–1912). Braque and Picasso worked together at this time. In Braque's *Man with a Guitar* [2], straight lines, a narrow range of color, and a kind of slicing of the figure into geometric shapes suggest a highly impersonal and unemotional approach to the figure. "The artist appears to be saying that all the variations we see in the visible world are superficial—shells, as it were, underneath which there is an order like the order of mathematics" (Feldman, 1972). Also, it is interesting to note that important theoretical developments in mathematics with implications for physics were being made at this time, notably by Einstein, Hilbert, and Lebesgue.

[**3**] Richard Lippold, *Variation Number 7: Full Moon*, 1949–1950. Brass rods, nickel-chromium and stainless steel wire, $10 \times 6 \times 6$ ft. (The Museum of Modern Art, New York, Mrs. Simon Guggenheim Fund.)

The work of Richard Lippold, for example, *Variation Number 7: Full Moon* (1949–1950) in [3], is sculpture of great formal order. The subject is light rays as represented by fine filaments of metal radiating from several centers, forming complex geometric shapes and seemingly hanging suspended in space.

EXERCISES

1. Can you think of any familiar objects that are the approximate shape of a golden rectangle?
2. How many golden rectangles can you find in *La Parade*?
3. What were some of Juan Gris's ideas concerning the golden ratio?
4. What do we mean when we say that the golden ratio is irrational?
5. Find $\sqrt{4}$, $\sqrt{9}$, and $\sqrt{16}$.
6. Use a calculator to find $\sqrt{5}$, $\sqrt{6}$, and $\sqrt{7}$.

7. If $a \geq 0$, what is $\sqrt{a^2}$?

8. What does $(\sqrt{2}+1)(\sqrt{2}-1)$ equal? What does $(\sqrt{5}+1)(\sqrt{5}-1)$ equal?

9. Show that $\dfrac{1}{\sqrt{2}+1} = \sqrt{2}-1$ and $\dfrac{1}{\sqrt{5}+1} = \dfrac{\sqrt{5}-1}{4}$.

10. If $a \geq 0$ and $a \neq 1$, show that $\dfrac{1}{\sqrt{a}+1} = \dfrac{\sqrt{a}-1}{a-1}$.
    What if $a = 1$?

11. Show that $(1 - \sqrt{5})/2$ also satisfies the equation $\tau = 1 + 1/\tau$.

12. How can you solve the equation $\tau = 1 + 1/\tau$ for τ using the quadratic formula? (If you haven't had the quadratic formula, skip this problem.)

13. If a rectangle has width w and length l, what is the rectangle's perimeter?

*14. If a_n is the nth term of the Fibonacci sequence, what is the difference equation satisfied by a_n?

15. A sequence of numbers starts with 1, 2, and each term is the sum of the previous two terms. What are the first 10 terms of the sequence? Answer this question if the sequence starts with 1, 3.

16. The sequence of ratios of consecutive Fibonacci numbers $1/2, 2/3, 3/5, 5/8, 8/13, \ldots$ approaches the number $(\sqrt{5}-1)/2$. Why?

17. Divide the circumference of a circle into five equal parts and draw a pentagon and a five-pointed star. You will find a pentagon inside the star. Draw another star inside the smaller star. This can be continued indefinitely. The result is the magic symbol of the ancient Pythagoreans, called the **pentagram** (Figure 2.10). It can be proved that $bc/ab = bc/dc = dc/ec = \tau$. Measure these distances and see if you agree.

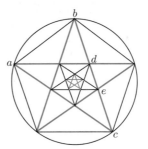

Figure 2.10

18. Draw a line segment with endpoints a and b. The ancient Greeks found that there is only one point p on the segment with the property that $bp/ap = ab/bp$. Show that $bp/ap = \tau$. If $ab = 1$, where is p?

19. In the branching of a tree, the following assumptions frequently hold. A branch grows indefinitely but produces no new branch during its first year. At the end of the second year and each succeeding year a new branch is produced. Starting with a single branch, how many branches are there at the beginning of year 1, 2, 3, 4, 5? Draw a picture.

*20. Starting with the fifth Fibonacci number, which ones do you find are prime? Make a conjecture about which Fibonacci numbers are prime.

[4] Albrecht Dürer, Woodcut, c. 1527. (The British Museum.)

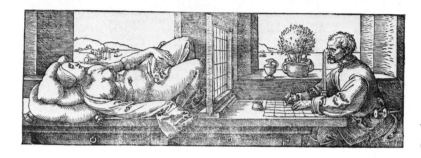

[5] Albrecht Dürer, Woodcut, c. 1527. (The British Museum.)

2.3 Perspective

Mighty is geometry; joined with art, resistless.
— EURIPIDES (480–406 B.C.), Greek Tragedian

No branch of mathematics competes with projective geometry in originality of ideas, co-ordination of intuition in discovery and rigor in proof, purity of thought, logical finish, elegance of proofs and comprehensiveness of concepts. The science born of art proved to be an art. — MORRIS KLINE, Twentieth-Century Mathematician

Before the Renaissance most paintings were flat and lacked depth. One of the significant advances in art was the full exploitation of perspective by sixteenth- and seventeenth-century Renaissance painters. *The Last Supper*, by da Vinci, *The School of Athens* and *The Marriage of the Virgin*, by Raphael, and *The Interior of the Pantheon*, by Panini are just a few examples of the use of perspective.

As mentioned earlier, there is not one geometry but several. The first move in the creation of a new geometry was made by Alberti in his 1435 book *Della Pittura*. His work was studied by Dürer, da Vinci, and others. The sixteenth-century woodcuts of Dürer, [4] and [5], showed experiments in perspective. One experiment is the following. Stand at arm's length before a window and view a distant scene. With one eye closed, trace the scene with a marker on the glass. The image on the pane marks the points at which light rays reflecting from the distant scene to your eye pierce the pane. (The word "perspective" comes from the Latin for "see through.") The cone of

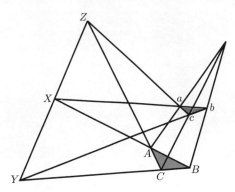

Figure 2.11

light rays is called a **projection** and the set of points marked on the pane is called a **section**. As Leonardo and others discovered, the section creates the same impression on the viewer's eye as the scene itself.

Projective Geometry

One has only to look at the beams in da Vinci's *Last Supper* to see that parallel lines meet at a point in a painting (a section). As a result, Euclidean geometry is not the tool to handle the properties of various sections in a projection. What geometrical properties, the Renaissance artists asked, in the real scene and in the section remain unchanged so that they produce the same effect on the eye? The answers to such questions were given by a new branch of mathematics inspired by art, called **projective geometry**. The study of projective geometry was begun by the mathematicians Desargues and Pascal.

An example of the incredible qualities of projective geometry can be seen from a simple theorem (Desargues' theorem), which you can test for yourself. Draw a triangle of any size or shape and a projection of it (Figure 2.11). The theorem says that any pair of corresponding sides of a triangle ABC and a section abc, for example, the sides AB and ab, will meet at a point. The theorem continues to say that all three points X, Y, and Z will lie on a line. The remarkable fact is that this works for any triangle and any section. What happens, you say, if the corresponding sides of the triangles happen to be parallel? Desargues replied that all parallel lines, by convention, meet at a common point, which may be taken at infinity. Thus we have a non-Euclidean geometry.

William Hogarth's 1754 engraving *False Perspective* [6] is an excellent spoof on how to mishandle perspective. The principal fisherman, who has his line crossed with that of the more remote angler, seems about to slip down the floor that is sloping toward the viewer. The old woman lights the pipe of the man on the far-off hill. The hunter in the boat aims at the swan through the solid bricks of a bridge that doesn't even span the river. We see both ends of the church. The trees dwindle improbably into the foreground; the nearby inn sign hangs between two of them. The wagon wheels are oval although seen face on.

William Hogarth, Engraving, *False Perspective*, 1754. (*The British Mu-*

[**6**] William Hogarth, *False Perspective*, 1754. Engraving. (The British Museum.)

Hogarth has created an impossible world by breaking the mathematical principles of perspective. Others, for example Maurits Escher, have done the same by employing (or abusing) perspective and other mathematical principles. His lithograph *Ascending and Descending* [7] both embodies a sense of eternity and presents the eye with an impossible situation that it cannot resolve. The laws of perspective are deadlocked. When the psychologists L. S. Penrose and R. Penrose published their impossible triangle (Figure 2.12), on which some of Escher's works are based, he wrote of it (M. C. Escher, 1967, p. 16):

If we follow the various parts of this construction one by one we are unable to discover any mistake in it. Yet it is an impossible whole because changes suddenly occur in the interpretation of distance between our eye and the object.

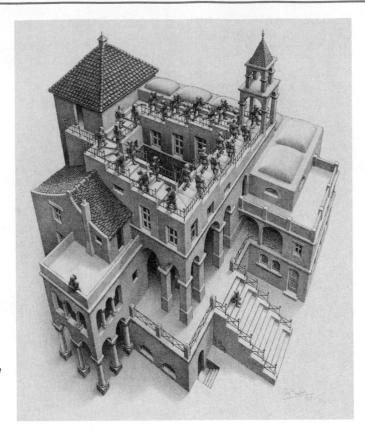

[7] M. C. Escher, *Ascending and Descending*, 1960. (M. C. Escher Foundation, Baarn, Holland.)

EXERCISES

1. Carry out Dürer's experiment in perspective.

2. Test Desargues' theorem for two triangles and two projections.

3. Can you find any other contradictions in Hogarth's engraving [6]?

4. In what way is Escher's print [7] based on Figure 2.12?

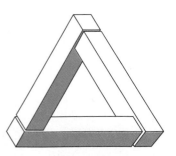

Figure 2.12

2.4 Symmetry

God, Thou great symmetry,
Who put a biting lust in me
From whence my sorrows spring,
For all the frittered days
That I have spent
in shapeless ways,
Give me one perfect thing.

— ANNA WICKHAM, Twentieth-Century Poet

The chief forms of beauty are order and symmetry and precision
which the mathematical sciences demonstrate in a special degree.
— ARISTOTLE (384–322 B.C.), Greek Philosopher

Symmetry in art is almost as old as art itself. Bilateral symmetry, the symmetry of left and right, appeared in early works of art. For example, the design which appears on the silver vase of King Entemena, who ruled in the city of Lagash, Sumeria, around 2700 B.C., shows bilateral symmetry (Weyl, 1952). The Sumerian eagle was given two heads facing in either direction (Figure 2.13), the principle of bilateral symmetry thus overwhelming the imitative principle of truth to nature. This heraldic design can be followed to Persia, Syria, Byzantium, and finally to the double-headed eagle of Czarist Russia and the Austro-Hungarian empire.

Figure 2.13

If a vertical line is drawn down the center of the Sumerian eagle, the images on either side of the line are identical. The image on a mirror placed on this centerline and viewed from the left will give the right side of the design. This is why bilateral symmetry is sometimes called mirror symmetry. If a point a distance d to the left of the centerline is transformed to a distance d to the right of the centerline, it appears at the same spot in the design. Another way of looking at bilateral symmetry is the following. If the design is flipped or reflected 180° clockwise about the centerline, the design on the left will fall exactly on its counterpart on the right.

There are other types of symmetry besides bilateral symmetry, for example, translational and rotational symmetry. Symmetry has continued to appear in art down through the ages in ornamental borders, tile patterns, star polygons, spirals, balanced design, Greek sculpture and architecture, Renaissance paintings, and finally modern

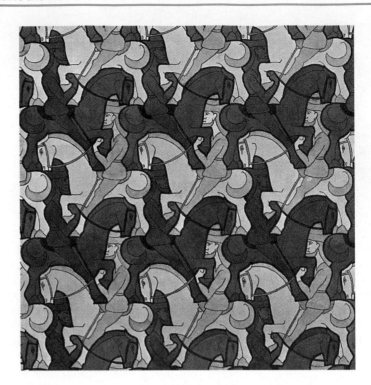

[**8**] M. C. Escher,
Symmetry Drawing E 67.
(M. C. Escher Foundation,
Baarn, Holland.)

art. The mathematical theory of symmetry, which is part of the theory of groups, did not develop until the late nineteenth and twentieth centuries. For example, it was not until 1891 that the Russian crystallographer E. S. Fedorov proved that there are precisely 17 essentially different two-dimensional patterns, yet the ancients included all 17 in their mosaics in the Alhambra. It is fascinating that the mathematical theories underlying some of Escher's symmetrical designs were not developed until after his works were completed.

Let us examine some of Escher's work involving symmetry. In *Symmetry Drawing E 67* [8] there is translational symmetry; in *Day and Night* [9] there is a type of bilateral symmetry. Use the annotated bibliography as a source for other examples of Escher's work. Can you find other types of symmetry? For example, do you see the rotational symmetry in *Circle Limit IV* [Ch. 5, 2]? It is fascinating that a book (Macgillavry, 1965) containing many reproductions of Escher's work and illustrating most of the important symmetry transformations has been used as a textbook for university courses in chemistry and crystallography.

Escher had the following to say about these works (M. C. Escher, 1967, p. 9):

In the course of the years I designed about a hundred and fifty of these tessellations. In the beginning I puzzled quite instinctively, driven by an irresistible pleasure in repeating the same forms, without gaps, on a piece of paper. These first drawings were tremendously time-devouring because I had never heard of crystallography; so I did not know that my game was based on rules which have been scientifically investigated. Nor had I visited the Alhambra at that time.

[**9**] M. C. Escher, *Day and Night*, 1938. (M. C. Escher Foundation, Baarn, Holland.)

EXERCISES

1. What is meant by translational symmetry? Draw a figure that has translational symmetry.
2. What is meant by rotational symmetry? Draw a figure that has rotational symmetry.

What types of symmetry occur in the following figures? Describe them.

3. Figure 2.14a   4. Figure 2.14b   5. Figure 2.14c
6. Figure 2.14d   7. Figure 2.14e   8. Figure 2.14f

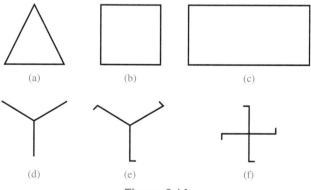

Figure 2.14

9. Why does Figure 2.14b have more symmetry than Figure 2.14c?
10. Why does Figure 2.14d have more symmetry than Figure 2.14e?
11. Of all the parts of Figure 2.14, which has the most symmetry?
12. Discuss the symmetry of a circle.

2.5 | Surrealism

Through and through the world is infested with quantity: To talk sense is to talk quantities. It is no use saying the nation is large— How large? It is no use saying radium is scarce—How scarce? You cannot evade quantity. You may fly to poetry and music, and quantity and number will face you in your rhythms and your octaves.
— ALFRED NORTH WHITEHEAD (1861–1947), English Mathematician

Creativity is characterized by the overlap or convergence of two otherwise unrelated ideas.
— KENNETH HIERBERT, Twentieth-Century Designer/Graphic Artist

Surrealism is a modern art form that frequently has curious mathematical qualities. Two techniques employed by surrealists are fantasy and space-time transformations. It is the space-time transformations that give the mathematical quality. These transformations distort space and time to give eerie and unusual locations. Figures are displaced and mutilated. Size, proportion, and volume are distorted. Time is warped. Figure 2.15 shows how a face can be distorted in various ways.

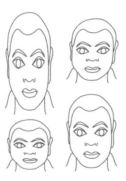

Figure 2.15

Coordinates

To show how such distortions can be created mathematically, we must first see how to locate points on a plane. (This was already discussed in Section 1.5, but we briefly review it here.) To start with, we locate an origin, say the bottom left-hand corner of a piece of paper. We denote the origin by 0. If a point is, say, 2 units to the right and 3 units up from the origin, we denote this point by $(2, 3)$. In general a point that is x units to the right and y units up from the origin is denoted (x, y) (Figure 2.16). In this case x and y are called **coordinates** of the point (x, y).

The origin need not be at the bottom left-hand corner but can be anywhere. In this case we may have negative coordinates. For example, $(-2, -3)$ is located 2 units to the left and 3 units down from the origin (see Figure 2.17).

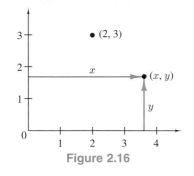

Figure 2.16

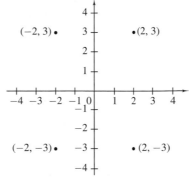

Figure 2.17

Transformations

Now that we know how to locate points according to coordinates, we can transform them using mathematical formulas. Such transformations give surrealistic effects. For example, in Figure 2.18 we have drawn a crude face. We now transform this face according to the transformation formula $(x, y) \rightarrow \left(2x, (x + y)/2\right)$. This means that if a point originally had the coordinates (x, y), then after transformation it has the coordinates $\left(2x, (x + y)/2\right)$. For example, the bottom left-hand corner of the chin is initially located at the point $(1, 4)$. After transformation, this point becomes $(2, 2.5)$. The bottom left-hand corner of the nose is initially at the point $(0, 9)$ and after transformation it is at the point $(0, 4.5)$. Figure 2.19 shows the transformed face.

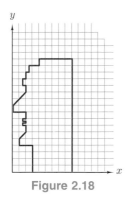

Figure 2.18

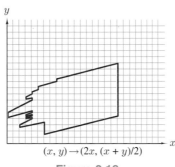

$(x, y) \rightarrow (2x, (x + y)/2)$

Figure 2.19

In Figure 2.20 we have an ordinary-looking house. After transforming according to the formula $(x, y) \rightarrow (x, y^2/20)$ we obtain an eerie, surrealistic house (Figure 2.21).

Many important transformations can be simply described using mathematics. For example, if $a > 1$, a dilation (or expansion) by the factor a is described by $(x, y) \rightarrow (ax, ay)$. If $0 < a < 1$, a contraction by the factor a is given $(x, y) \rightarrow (ax, ay)$. For numbers a and b, the transformation $(x, y) \rightarrow (x + a, y + b)$ gives a translation of a units in the x direction and b units in the y direction. A reflection about the x

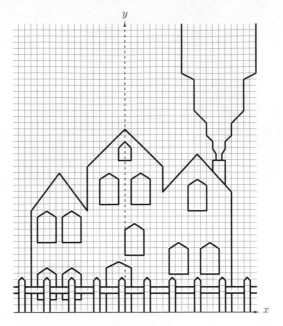

Figure 2.20

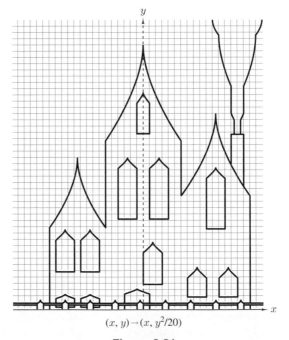

$(x, y) \rightarrow (x, y^2/20)$

Figure 2.21

[**10**] Salvador Dali, *Soft Construction with Boiled Beans; Premonition of Civil War*, 1936. Oil on canvas, $39\frac{3}{8}$ in. $\times$ 39 in. (Philadelphia Museum of Art, Louise and Walter Arensberg Collection.)

axis and a reflection about the y axis are given by $(x, y) \rightarrow (x, -y)$ and $(x, y) \rightarrow (-x, y)$, respectively. Rotations of 90° counterclockwise and 180° counterclockwise are given by $(x, y) \rightarrow (-y, x)$ and $(x, y) \rightarrow (-x, -y)$, respectively. You can convince yourself of the validity of these statements by transforming some points using the given transformations.

Matta, Dali, and Magritte are good representatives of surrealist artists. Matta uses many mathematical forms in his art. Dali and Magritte are concerned with creating images of striking abnormality.

In Dali's *Soft Construction with Boiled Beans; Premonition of Civil War* (1936) we see a highly naturalistic technique combined with truncation and distortion of figure [10]. A small figure in the lower left portion of the picture establishes the monstrous scale of the central subject. We recognize portions of female anatomy, stretched skeletal forms, and a horrible grimacing head attached to a breast that is being squeezed by a clawlike hand. A supporting element seems to be a tree trunk, which emerges into a foot. The melting of objects—the synthetic into the natural, or the organic into the inorganic—is characteristic of surrealistic art.

In Magritte's *Le Soir Qui Tombe* (Evening Fall) (Larkin, 1972) we might first be reminded of Dürer's experiments in perspective. Then we notice that time has been transformed; it has stopped and the image is permanently emblazoned on the broken windowpane. Magritte's *Le Chateau des Pyrénées* (The Castle of the Pyrenees)

(Larkin, 1972) shows a huge rock and castle which are transformed in space to hang motionless and timeless over the moving sea. In *La Reproduction Interdite* (reproduction forbidden) (Larkin, 1972) there is a violation of symmetry. This results not in a space-time transformation but a symmetry transformation.

It is interesting that space-time transformations in art have come at a time when they are important in mathematical physics of general and special relativity.

EXERCISES

1. Locate the following points on a piece of paper: $(1, 2)$, $(3, 4)$, $(-2, 5)$, $(-3, -3)$, $(0.5, 2.5)$.

2. Are the points $(2, 3)$ and $(3, 2)$ at the same location?

3. Transform the points in Exercise 1 using the transformation formula $(x, y) \rightarrow \left(2x, (x + y)/2\right)$.

4. Transform the points in Exercise 1 using the transformation formula $(x, y) \rightarrow (x, y^2/20)$.

5. What is the relationship between transformations, as we have discussed them, and functions?

6. Apply the transformation $(x, y) \rightarrow (x/2, 2y - x/2)$ to the face in Figure 2.18.

7. Apply the transformation $(x, y) \rightarrow (x/2, 2y - x/2)$ to the face in Figure 2.19.

8. What is the relationship between the transformation $(x, y) \rightarrow (2x, (x + y)/2)$ and the transformation $(x, y) \rightarrow (x/2, 2y - x/2)$?

9. What is the relationship between the transformation $(x, y) \rightarrow (x, y^2/20)$ and the transformation $(x, y) \rightarrow (x, \sqrt{20y})$?

10. Draw a square and apply the transformation $(x, y) \rightarrow (x, 2y)$ to it. What do you obtain?

11. Answer Exercise 10 using the transformation $(x, y) \rightarrow (2x, 2y)$ instead.

12. Answer Exercise 10 using the transformation $(x, y) \rightarrow (x^2, y^2)$ instead.

13. Apply the transformations in Exercises 10 and 11 to the face in Figure 2.18.

14. Draw your own version of a face on graph paper and apply the transformation $(x, y) \rightarrow (2x, (x + y)/2)$ to it.

15. Transform the points $(1, 0)$ and $(0, 1)$ using (a) a dilation, (b) a contraction, (c) a translation.

16. Show that $(x, y) \rightarrow (-x, y)$ gives a reflection about the y-axis by transforming the points $(1, 0)$ and $(0, 1)$.

17. Show that $(x, y) \rightarrow (x, -y)$ gives a reflection about the x-axis by transforming the points $(1, 0)$ and $(0, 1)$.

18. Show that $(x, y) \rightarrow (-y, x)$ gives a 90° counterclockwise rotation by transforming the points $(1, 0)$ and $(0, 1)$.

19. Show that $(x, y) \rightarrow (-x, -y)$ gives a 180° counterclockwise rotation by transforming the points $(1, 0)$ and $(0, 1)$.

20. What transformation gives a 90° clockwise rotation?

*21. Show that $(x, y) \rightarrow ((x - y)/\sqrt{2}, (x + y)/\sqrt{2})$ gives a 45° counterclockwise rotation by transforming the points $(1, 0)$ and $(0, 1)$.

*22. What transformation gives a 45° clockwise rotation?

*23. Show that $(x, y) \rightarrow ((x - \sqrt{3}\,y)/2, (\sqrt{3}x + y)/2)$ gives a 60° counterclockwise rotation by transforming the points $(1, 0)$ and $(0, 1)$.

*24. What transformation gives a 30° counterclockwise rotation?

What portion of the plane is given by the sets in Exercises 25–31?

25. $\{(x, y)\colon x \geq 0\}$ **26.** $\{(x, y)\colon x \geq 0, y \geq 0\}$

27. $\{(x, y)\colon x \geq 0\} \cap \{(x, y)\colon y \geq 0\}$ **28.** $\{(x, y)\colon x \geq 0, y \leq 0\}$

29. $\{(x, y)\colon x = y\}$ **30.** $\{(x, y)\colon x \geq y\}$

31. $\{(x, y)\colon x \geq y\} \cup \{(x, y)\colon x \geq 0\}$

32. How would you locate a point in three-dimensional space? Four-dimensional space?

*33. The location of a point in space-time is represented by (x, y, z, t), where x, y, z are the space coordinates and t is the time coordinate. The path of a point in space-time is called the **world line** of the point. Discuss your own world line. For example, what happens if you go up in an elevator? In an airplane? As you get older? What would happen in a time machine?

2.6 Art and Gestalt Psychology

> *But mathematics is the sister, as well as the servant, of the arts and is touched with the same madness and genius.*
>
> — MARSTON MORSE, Twentieth-Century Mathematician

Visual art is a perception. It is a response by our brain to what our eyes see. As such it is related to psychology, in particular to gestalt psychology. "Gestalt" is German for form and gestalt psychology is the organized study of perception. Gestalt psychology was introduced in 1910 by three German psychologists, Max Wertheimer, Wolfgang Köhler, and Kurt Koffka. According to Wertheimer, "A Gestalt is a whole whose characteristics are determined, not by the characteristics of its individual elements, but by the internal nature of the whole." Koffka said, "To apply the Gestalt category means to find out which parts of nature belong as parts to functional wholes."

Although the investigations of the gestalt school dealt mainly with visual perception, the gestalt viewpoint has now been blended into virtually all branches of psychology. To the artist it represents the most important and fruitful field of inquiry in this century. It gives us psychological explanations for the visual effectiveness of such concepts as proportion, order, perspective, and symmetry. It also points out relationships between mathematics and art.

Gestalt Psychology

Gestalt psychology has reappraised and discovered much about people and the process and motivations that give order to their world. As Köhler puts it, "Organization in the sensory field originates as a characteristic achievement of the nervous system." Beauty (or ugliness) is not out *there* in our environment but *here* within our brain. Where formerly we strove to find laws of symmetry, balance, harmony, and the like in nature, we now may be sure, thanks to gestalt psychology, that such are our endowments, not nature's. Beauty, harmony, rhythm, proportion, color, form, and space are not the properties of things but of human perception. What you see literally or what actually exists in the external world is not the true substance of perception but only the stimulus to it. There is no sound if a tree falls in a forest when no one is near. There may be

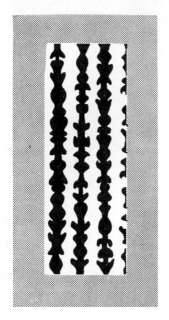 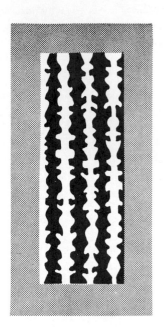

[11] From Faber Birren, *Color, Form and Space*, figures 14A and B, p. 18, Reinhold, New York, 1961. (Reprinted by permission of the publisher.)

vibrations in the air, but the word "sound" specifically relates to sensation. The same is true of color. A surface may reflect light in certain ways, but color itself is a human experience.

What kind of conclusions have the gestalt psychologists arrived at from their studies of human visual perception? For one thing, the most satisfactory forms are the simple ones—the square or cube, circle or sphere, triangle or pyramid, etc. The perceptually simple forms are the mathematically simple forms. The visual process seems to demand and prefer simplicity. It strives to make irregular shapes appear regular, to look for symmetry and balance. An experiment described by Koffka, showing the insistence of simple and symmetrical forms over asymmetrical forms, is illustrated in [11]. The eye forever seeks symmetrical forms and concentrates on them. In the left drawing the eye prefers the black objects and in the right drawing the eye prefers the white objects because they are more symmetrical.

Figure 2.22

Figure-Ground Phenomenon

Another important observation of gestalt psychologists is that of the figure-ground phenomenon. M. D. Vernon writes, "The first essential stage in perception is the emergence of one principal part of the field which is the 'figure' from the remainder which is the 'ground.' " In each of the three sketches in Figure 2.22 the geometrical floral pattern is seen as the figure, while the space around it appears as the ground, regardless of size. This involves a basic law of perception. The figure looks superimposed, solid, and highly structured, while the ground may be more filmy and vague. The figure, in short, stands out, and the ground retires.

In Figure 2.23, (a) looks more natural than (b). In (a), the figure has detail and seems properly formed and well defined. In (b), the detail is in the wrong place and the tree shape looks as if it were cut and empty. There may be some relationship between the figure-ground phenomenon and the fact that central vision is clearer and more distinct than peripheral vision.

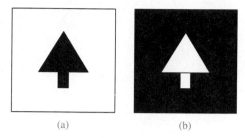

(a)   (b)

Figure 2.23

Convexity and Concavity

Related to the figure-ground phenomenon is the matter of convexity and concavity. It has been found in psychological experiments that convexity tends to win out over concavity. In Figure 2.24, convexity makes for a better figure than concavity. (a) looks better than (b); (a) looks superimposed, whereas (b) resembles a hole. Also (a) is softer and has a better tactile quality, (b) is sharp and pointed. Convexity tends to expand, to be soft and billowing; concavity tends to contract and have sharp edges.

(a)

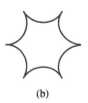

(b)

Figure 2.24

Traditionally most buildings of monumental design have been based on convex shapes. The first four outlines in Figure 2.25 are characteristic of ancient buildings, Roman, Gothic, Oriental. The last, or concave, form, not often found in large buildings, resembles a tent.

Figure 2.25

How can convexity be defined precisely? This is where mathematics comes in handy.

> A set S is said to be **convex** if for any two points x, y in S the straight-line segment from x to y is in S.

Thus (a) in Figure 2.26 is convex but (b) is not. Some of the pleasing figures in art are not strictly convex in this sense but are approximately convex. Convexity is a concept that has been studied by mathematicians for many years.

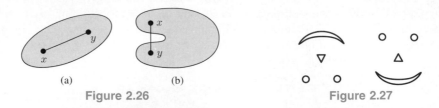

(a)   (b)

Figure 2.26   **Figure 2.27**

Gestalt psychologists have also found that perception not only tries to simplify what is seen but also strives to give meaning to it. For example, in Figure 2.27 the forms to the left give the mere impression of a set comprising a crescent, a triangle, and two circles. If the same grouping is turned around, one sees a face. According to F. C. Bartlett, "The experiments repeatedly demonstrate that temperament, interest and attitudes often direct the course and determine the content of perceiving."

Physiognomic Perception

An important concept developed by the gestalt psychologist Heinz Werner is physiognomic perception. We see the world in a dynamic rather than a static way. The "pictures" which the eye projects on the brain are not like photographs in an album. On the contrary, the brain works on these pictures and adds much to them. They take on meaning that is affected and influenced by the whole emotional and psychic makeup.

Figure 2.28

As an example, Figure 2.28 shows the dynamic qualities of fire and water expressed by Claude Bradgon. Köhler made up two nonsense words, "takete" and "maluma," and created the two abstract designs shown in Figure 2.29. He found that virtually all subjects tested agreed on which word applied to which design.

In physiognomic perception there is a dynamization of things. Perception adds to what we see. Viewers become artists in their own right, in that they take part in what they view. It is easy for the average person to translate thoughts, feelings, and moods into designs and forms. Pointed things may be sharp and cruel. Sagging things may be tired or lazy. Bulging things may be soft and jolly. In Figure 2.30, tensed will, longing, and fury are physiognomically expressed. Although the end product may be quite complex, physiognomic perception is based on the brain's interpretation of simple mathematical curves and figures.

Figure 2.29

Tensed will   Longing   Fury

Figure 2.30

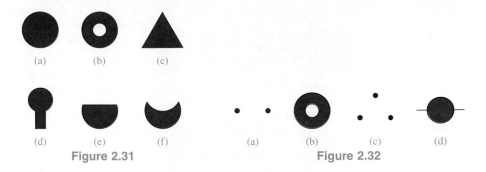

Figure 2.31   Figure 2.32

EXERCISES

1. In Figure 2.29, which is takete and which is maluma? Why?

2. Draw physiognomic figures representing man, woman, peace, hate, danger, joy, sadness.

3. Which of the sets in Figure 2.31 are convex?

4. Prove that the intersection of two convex sets is convex.

5. Prove that the intersection of any number of convex sets is convex.

6. Give an example of two convex sets whose union is not convex.

7. Suppose that $A_1, A_2, \ldots$ are convex sets and $A_1 \subseteq A_2 \subseteq \cdots$. Prove that $A_1 \cup A_2 \cup \cdots$ is convex.

The **convex hull** of a set A is the smallest convex set containing A.

* 8. Show that A is convex if and only if A equals its convex hull.

9. Find the convex hull of the sets in Figure 2.32.

10. Find the convex hull of the set $\{(0,0), (0,1), (1,0)\}$.

11. Find the convex hull of the set $\{(0,0), (0,1), (1,0), (1,1)\}$.

12. Can a convex set have a hole in it? An indentation? What does the convex hull do to holes and indentations?

2.7 | Optical Art

> *Today in the Twentieth century, the science of art cannot be separated from the art of science.*
> — RENE PAROLA, Twentieth-Century Artist

> *An artist does not skip steps; if he does, it is a waste of time because he has to climb them later.*
> — JEAN COCTEAU, Twentieth-Century Artist

The elements of time, motion, space, optics, and perception are the concerns of artist, scientist, and mathematician. Our scientific age is reflected in optical or perceptual abstraction.

Optical art is comprehensive in scope and places a binding emphasis on perception. Basically it distills the principles of art, using them singly and with force and commitment. It is the art of essentials, relying on total abstraction. To the layman, op

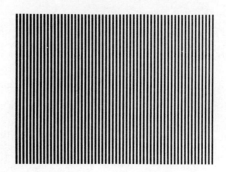

[**12**] From Rene Parola, *Optical Art*, figure 1-1, p. 14, Reinhold, New York, 1969. (Reprinted by permission of the publisher.)

art is sudden and immediate. The viewer does not have to depend on the critic, the connoisseur, the artist, or the scientist to "understand."

Op art is based on gestalt principles. The gestaltists believed that we have an innate tendency to perceive multiple objects as a group or totality and that we base this interpretation on certain perceptual clues, namely proximity, similarity, and good figure. Contemporary psychologists believe that two other organizing principles play a fundamental role in perception: assimilation and contrast. These, in conjunction with gestalt theory, help us see how images group, fluctuate, and move to become new entities.

Assimilation and Contrast

Assimilation is our tendency to minimize stimuli and create uniformity. It is a simplifying process. For example, in [12] we perceive the image as a middle-tone gray even though there is a 33 percent increase from dark to light as our eyes move from left to right. Our first response is one of assimilation.

If the dark and light are increased, attaining a full scale of graduation, there comes a critical point at which assimilation disappears and contrast takes over. **Contrast**, the other basic organizing principle, is the antithesis of assimilation and may be described as the accentuation of differences. Contrast is prominent in [13]. With the increase of dark and light, each end of the graded scale provides us with a perceptual clue. A definite progression of grays has been established, and assimilation is negligible.

If a bold line or any type of boundary is interjected into the field, we have a reversion to assimilation. The area divides into two separate fields, one dark and the other light, each section having its own average value [14].

What happens when we go beyond simple differentiation, when we are confronted by several stimuli? Proximity, similarity, and good figure, the three principles defined by the gestaltists to explain how we interpret multiple stimuli, are more sophisticated grouping faculties than assimilation and contrast only because they deal with larger numbers.

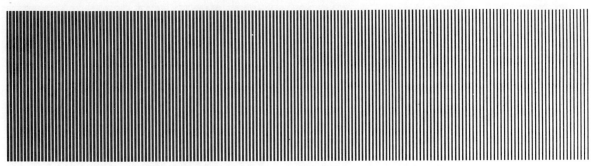

[**13**] From Rene Parola, *Optical Art*, figure 1-2, p. 15, Reinhold, New York, 1969.
(Reprinted by permission of the publisher.)

[**14**] From Rene Parola, *Optical Art*, figure 1-3, p. 15, Reinhold, New York, 1969.
(Reprinted by permission of the publisher.)

Proximity and Similarity

The definition of **proximity** is simple: Objects that are near one another tend to group. In [15] the squares regroup into two or four sections because of distance. Proximity is also an important mathematical concept. Distance functions, or metrics, are defined on mathematical spaces to tell how far apart different points are. Proximity is also important in mathematical topology and approximation theory.

The definition of **similarity** is also simple: All similar things have a tendency to group. In [16] the dots group or form lines because of their similar values of light-dark quality. Similarity is also an important mathematical concept. Similar types of mathematical structures are said to be isomorphic, and this concept is quite important in the classification of mathematical structures. You will see this concept again in our studies of graphs and groups. There is also the mathematical concept of similarity for geometrical objects.

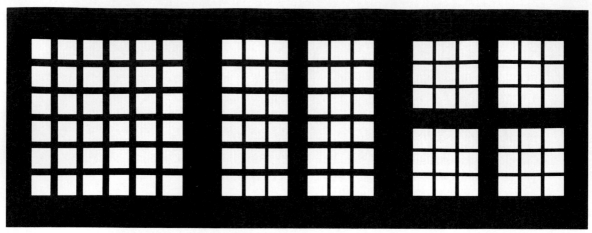

[**15**] From Rene Parola, *Optical Art*, figure 1-12, p. 23, Reinhold, New York, 1969. (Reprinted by permission of the publisher.)

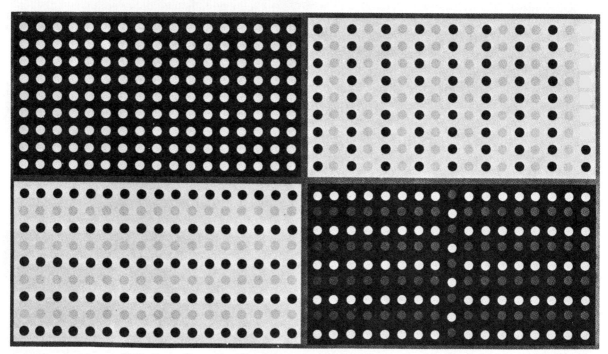

[**16**] From Rene Parola, *Optical Art*, figure 1-13, p. 24, Reinhold, New York, 1969. (Reprinted by permission of the publisher.)

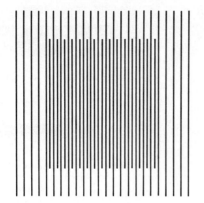 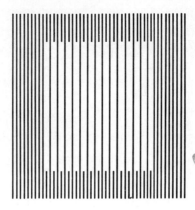

[17] From Rene Parola, *Optical Art*, figure 1-16, p. 27, Reinhold, New York, 1969. (Reprinted by permission of the publisher.)

Good Figure

The last principle of perceptual grouping is **good figure**. This is the perceptual ability to predict a total entity from a minimum amount of information or stimuli. Within the concept of good figure there are three indicators that help determine "goodness": symmetry, which has already been discussed, closure, and common fate.

Closure is the grouping of elements to make a more complete or closed form. Because of closure, images may exist even when important parts are missing. In [17] there are no squares, yet we perceive them. Proximity undoubtedly helps to form closure.

Again, closure is an important mathematical concept which has been given much attention. It is related to "openness" in mathematical topology.

A set S in the plane is said to be **open** if for any point $x \in S$ there is a disk D whose center is x such that $D \subseteq S$. A set is **closed** if it is the complement of an open set.

In Figure 2.33 the triangular region (a), which does not include its boundary lines, is open, while the triangular region (b), which includes its boundary lines, is closed. (A boundary that is not included in a region is denoted by a broken line; a boundary that is included is denoted by an unbroken line.)

Common fate is the tendency for elements that move or shift in the same direction to group. Those elements which move in another direction also will group. Common fate is closely related to similarity, the only difference being the implied movement or direction.

Perceptual gyrations keep an op painting constantly alive. This is because op paintings accommodate the basic principles of perception. When its primary assessments have been made for it, the mind is like a pinball machine run amok. Lights keep flashing, new groupings evolve, motions occur.

(a)

(b)

Figure 2.33

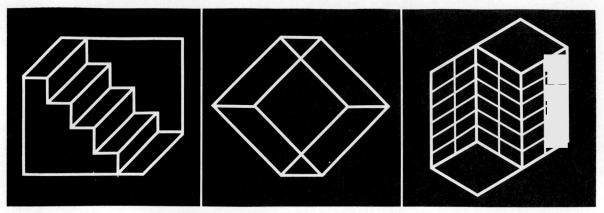

[**18**] From Rene Parola,*Optical Art*, figure 3-5, p. 40, Reinhold, New York, 1969.
(Reprinted by permission of the publisher.)

Illusions

Illusions are commonly made using reversible images of simple geometric figures. If you fixate on any of the figures in [18], you find that two separate visual "pictures" emerge. At times Schröder's staircase seems to be a regular set of stairs with the lowest level on the right. After visual concentration the stairs reverse and appear to be upside down, with the highest level on the right. Some figures, like Necker's cube, reverse more easily because they have symmetry and good figure. After prolonged viewing and a little practice, you can control the reversal with startling efficiency. Many explanations and opinions have been offered to explain why these figures reverse. Although there is no proof, scientists presently support the theory of satiation. In figures that have innate reversal qualities, we tend to choose first a dominant, or good, figure. After extended inspection we seek a weaker variant. Illusions also can make straight lines seem curved [19].

Another type of illusion is negative-positive reversal. Previously we considered the relationship between negative space (or ground) and positive space (or figure), stressing that the most consistent image is the positive space if both object and background are equal in size. What happens if we exclude the consistency factor and make negative and positive space equally important? When our perceptions are deprived of clues, we have ambiguous reactions ([20] and [21]). Whether we choose the dark or light image as the dominant figure, we have only to wait for visual fixation to bring forth the latent image. We even see subtle pastel colors in black-and-white drawings.

The influence of angles is another mathematical concept that creates illusions. Today, scientists reinforce the artist's knowledge of this subject. Extensive experiments by the psychologists Marshall Segal, Donald Campbell, and Melville Hershovits (see, for example, *Science*, February 1962) revealed that we tend "to interpret acute and obtuse angles on a two-dimensional surface as representations of rectangular objects in three-dimensional space." Thus, another mathematical concept is brought into play, namely the dimension of space. The vertical lines in [22] do not appear parallel because of the small intersecting angles.

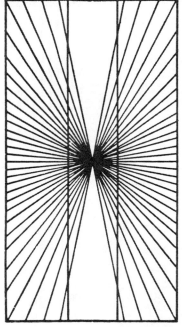

[**19**] From Faber Birren, *Color, Form and Space*, plate II, p. 32, Reinhold, New York, 1961. (Reprinted by permission of the publisher.)

[**20**] From Rene Parola, *Optical Art*, figure 4-11(A), p. 68, Reinhold, New York, 1969. (Reprinted by permission of the publisher.)

[**21**] From Rene Parola, *Optical Art*, figure 4-2. (first illustration), p. 62, Reinhold, New York, 1969. (Reprinted by permission of the publisher.)

[**22**] From Rene Parola, *Optical Art*, figure 3-15, p. 46, Reinhold, New York, 1969. (Reprinted by permission of the publisher.)

[**23**] Bridget Riley, *Current*, 1964. Synthetic polymer paint on composition board, $58\frac{3}{8}$ in. $\times$ $58\frac{3}{8}$ in.. (Collection, The Museum of Modern Art, New York, Philip Johnson Fund.)

Pattern

Another important concept in op art is pattern. **Pattern** is nothing more than the exact, orderly, symmetrical repetition of a figure. As we repeat a unit, we replace individual subject matter with organization. No single unit exists for itself. Each becomes a fragment that functions by dependency and association to form an indivisible arrangement [23].

Victor Vasarely is one of the great op artists. What mathematical and gestalt principles do you see in his work in [24]?

EXERCISES

1. State whether the sets in Figure 2.34 are open or closed or neither.
2. Prove that the intersection of two open sets is open.
* 3. Prove that the intersection of a finite number of open sets is open.
4. Give an example of an infinite number of open sets whose intersection is not open.
5. Prove that the union of two open sets is open.
6. Prove that the union of any number of open sets is open.
7. What can be said about the union and intersection of closed sets?

The **closure** of a set A is the smallest closed set containing A.

* 8. Prove that a set A is closed if and only if A equals its closure.

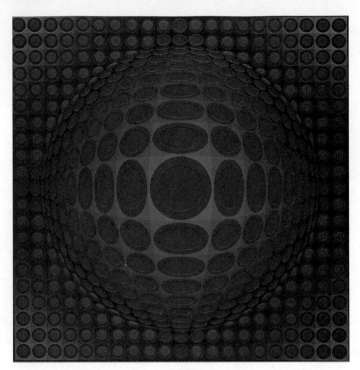

[**24**] Victor Vasarely,
Vega 222. (Courtesy Sidney
Janis Gallery, New York.)

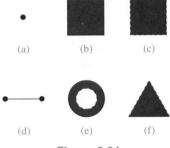

(a)   (b)   (c)

(d)   (e)   (f)

Figure 2.34

9. Find the closure of the sets (c), (e), and (f) in Figure 2.34.

*10. Find the closure of the set $A = \{1, 1/2, 1/3, 1/4, \ldots\}$.

*11. How would you define an open set in three-dimensional space? In one-dimensional space, that is, on a line?

*12. If (x_1, y_1) and (x_2, y_2) are two points in the plane, the distance between them is

$$\left[(x_2 - x_1)^2 + (y_2 - y_1)^2\right]^{1/2}$$

Use Pythagoras' theorem to explain why this is so. What is the distance between $(1, 2)$ and $(4, 6)$?

2.8 | M. C. Escher

Although I am absolutely innocent of training or knowledge in the exact sciences, I often seem to have more in common with mathematicians than with my fellow artists.

—M. C. ESCHER, Twentieth-Century Dutch Artist

By keenly confronting the enigmas that surround us, and by considering and analyzing the observations that I had made, I ended up in the realm of mathematics. —M. C. ESCHER

The modern mind has adventured far and fearlessly in the new realms of thought opened up by research and discovery, but it has left no trail of beauty. That it has not done so is the fault of the artist, who has failed to interpret and portray the movement of the modern mind. . . . The world order is most perfectly embodied in mathematics.

—CLAUDE BRAGDON, Twentieth-Century Author

M. C. Escher has exhibited more mathematical principles in his works than any other artist. He has also been influenced by the principles of gestalt psychology (*Scientific American*, July 1974). In particular, positive-negative reversal played an important part in some of his works (for example, [9]). A related concept is that of a tessellation.

Tessellation

Tessellation, filling space with repetitive geometric figures, is related to patterns, periodic drawings, and symmetry. Such patterns appeared in Islamic and ancient Egyptian art, in African Bakuba art, and in the work of Moorish artists. The Moors occupied Spain from 711 to 1492. Forbidden by their religion to depict animate objects, they devoted their artistic energies to developing geometric patterns. Escher, the modern master of tessellation, has said (M. C. Escher, 1967, p. 9):

This is the richest source of inspiration that I have ever struck; nor has it yet dried up. . . a surface can be regularly divided into, or filled up with, similar-shaped (congruent) figures which are contiguous to one another, without leaving any open spaces. The Moors were past masters of this. They decorated walls and floors, particularly the Alhambra in Spain, by placing congruent, multi-colored pieces of majolica [tiles] together without leaving any space between. What a pity it is that Islam did not permit them to make "graven images." They always restricted themselves, in their massed tiles, to designs of an abstract geometrical type. . . . This restriction is all the more unacceptable to me in that the "recognizability" of the components of my own designs is the reason for my unfailing interest in this sphere.

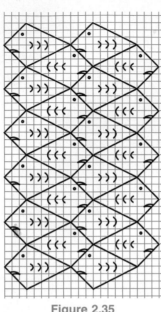

Figure 2.35

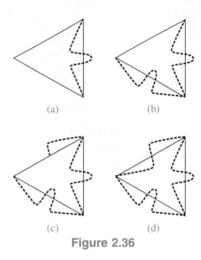

Figure 2.36

Escher's works in Section 2.4, namely [8] and [9], are examples of tessellations. How can you produce tessellations yourself? A simple way is to first fill the space with repetitions of a simple geometric figure. Only certain figures will work—for example, triangles, rectangles, and hexagons. Figure 2.35 is an example of such a tessellation. For more complicated tessellations the following method can be used. First draw an equilateral (all sides are equal) triangle. Draw a curve on one side (Figure 2.36a). Reproduce this curve on another side as in Figure 2.36b. Mark the midpoint of the remaining third side and draw a curve between the vertex and this midpoint (Figure 2.36c). Draw the reflection of this curve about the triangle's side from the other vertex to the midpoint (Figure 2.36d). The resulting figure can be used in a tessellation (Figure 2.37).

Infinity

There are other mathematical concepts in Escher's work besides tessellations, periodicity, and symmetry. In *Circle Limit IV* [Ch. 5, 2] we see the concept of infinity. The number of devils and angels is increasing to infinity. There is also the concept of limit and the concept of convergence. Although the figures never reach the boundary of the circle, they converge to limits on the circle. Notice that there are an infinite number of limits on the circle. This is in contrast to *Smaller and Smaller* [Ch. 5, 3], in which there is a single limit at the center of the drawing, or *Whirlpools* [Ch. 5, 5], in which there are two limits.

Figure 2.37

Dimension

Day and Night [9] exhibits the mathematical concept of dimension. Notice how the two-dimensional fields are gradually transformed into three-dimensional birds. This lithograph also illustrates the concept of positive-negative reversal, with the white and black birds interchanging roles as figure and ground. Perspective is also very strong in this work.

Positive-Negative Reversal

Positive-negative reversal is evident in Escher's *Sky and Water I* (Escher, 1970; Locher, 1971). Look at the boundary line between sky and water. What is fish and what is sky? What is bird and what is water? Another concept in this work is that of set. What is the set of birds? What is the set of fish? Are these well-defined sets?

Reptiles [Ch. 5, 8] illustrates many mathematical concepts. First there is the tessellation and symmetry of the reptiles drawn on the pad. Then there is the gradual change from two-dimensional drawn reptiles to three-dimensional "live" reptiles. There is also the notion of progression in time. Furthermore, Escher's love of geometry is revealed in the triangle and dodecahedron (see Chapter 12 for more details).

Möbius Strip

Another mathematical concept is illustrated in *Moebius Strip II* and *Moebius Strip I* [Ch. 5, 4]. The Möbius strip is an unusual surface studied in mathematical topology. It is unusual in that it is a two-dimensional surface with only one side (see [25]). *Moebius Strip II* shows that it has only one side. Think of this figure as portraying

[**25**] M. C. Escher, *Moebius Strip II.*
(M. C. Escher Foundation, Baarn,
Holland.)

one ant crawling over the surface. The ant crawls over the entire surface without crossing an edge. Hence the surface has only one side. If you cut a Möbius strip with a scissors down its center all the way around, you will not obtain two strips, as you might expect, but one single strip. You can convince yourself of this by studying *Moebius Strip I.*

The Möbius strip has an interesting history. It was once believed to have been discovered by the German mathematician A. F. Möbius in 1865, but it has since been found in a Roman mosaic dating back to the third century (*American Scientist*, September-October, 1973). This is quite remarkable, since a search of the mosaics of that time has failed to yield another example of a Möbius strip. The reason this is remarkable is that geometric motifs used in Roman mosaics rarely depart from a limited number of frequently used diagrams. In fact, it was common practice for the mosaicist to wander from place to place with a notebook of ready-made patterns to offer potential customers.

[**26**] M. C. Escher, *Belvedere*, 1958.
(M. C. Escher Foundation, Baarn,
Holland.)

Relativity

Relativity (Chapter 2 Opener) illustrates the mathematical-physical concept in its
name. This work represents people sharing different planes of existence simultaneously. They cannot possibly communicate with each other because they live in different
worlds. Notice the two figures on the stairs in the upper portion of the picture. One is
walking down and the other up, yet they are walking in the same direction! The floor
of the figure seated on the box is the wall of the figure holding the bag. The door of
the seated figure is an anomaly to the other figure. There are many such inconsisten-
cies in the lithograph. For example, what relationship do the people "outside" have to
each other?

Impossible Buildings

Belvedere [26] is one of Escher's impossible buildings. It is based on the impossible
three-dimensional geometric figure held by the man in the lower left-hand corner. Can
you see why this building is impossible?

    Cube with Magic Ribbons [27] is an example of an illusion that can be obtained
with geometric figures. Are those protrusions or holes in the ribbon? You can change
them at will.

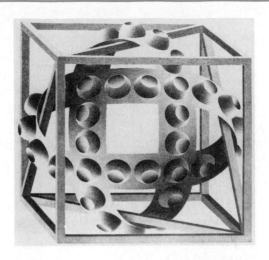

[**27**] M. C. Escher, *Cube with Magic Ribbons*. (M. C. Escher Foundation, Baarn, Holland.)

[**28**] M. C. Escher, *Order and Chaos*, 1950. (M. C. Escher Foundation, Baarn, Holland.)

In *Order and Chaos* [28] you see the order and symmetry of the geometric solid contrasted with the chaos of the randomly placed and chosen objects. On the one hand there is the ordered structure of great symmetry; on the other there is almost complete randomness as studied in probability theory.

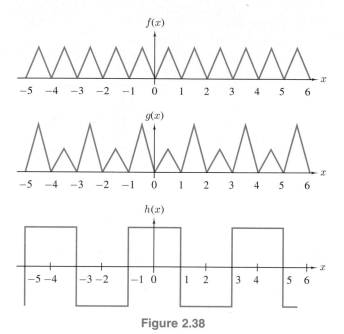

Figure 2.38

EXERCISES

1. Construct a tessellation using (a) triangles, (b) rectangles, (c) hexagons.

2. Can you construct a tessellation using pentagons?

3. Using the method of this section, construct a tessellation consisting of recognizable figures.

4. A function $f\colon \mathbb{R} \to \mathbb{R}$ is **periodic** with period a if $f(x + a) = f(x)$ for all $x \in \mathbb{R}$. Show that a constant function is periodic with any period. Which of the functions in Figure 2.38 are periodic? For those that are periodic, what is their period?

5. To what limits do the following sequences converge?
   (a) $1, 1/2, 1/3, 1/4, \ldots$
   (b) $1/2, 2/3, 3/4, 4/5, \ldots$
   (c) $1/2, -1/2, 1/3, -1/3, 1/4, -1/4, \ldots$
   (d) $1, 2, 3, 4, 5, \ldots$

6. Examine any one of Escher's works and see how many mathematical concepts you can find.

7. Construct a Möbius strip by giving a strip of paper a half-twist and taping the ends together. Cut the Möbius strip lengthwise down the center and see what you get. Next, cut it again. What do you get now? Cut a Möbius strip lengthwise one-third of the distance from its edge. What do you get?

8. Construct a Möbius strip by giving a strip of paper three half-twists and taping the ends together. Perform the steps in Exercise 7.

2.9 | Art in Mathematics

A mathematician who is not also something of a poet will never be a complete mathematician.
— KARL WEIERSTRASS (1815–1897), German Mathematician

The science of Pure Mathematics, in its modern developments, may claim to be the most original creation of the human spirit.
— ALFRED NORTH WHITEHEAD (1861–1947), English Mathematician

So far we have mainly considered mathematical concepts in art. But is there also art in mathematics? Many feel that there is. Mathematics studies a vast world of objects and their transformations, their static and dynamic characters, intricate and tangled yet systematic and ordered to a degree far greater than that of any other known world. This world has been created by the minds of people, partly to enable them to handle the world of natural phenomena, but also for the elegant beauty of its objects.

Looking over the history of mathematics in every age, we always come to the same conclusion. We first encounter the vague, the chaotic, the tangled, the unformed, but under the magic wand of mathematics we find the vague assuming form, the chaotic appearing in order, the tangled turning into beautiful lace, the unformed showing first a line, then a net, then a structure. Modern scientists have discovered that what they are studying is really not the world of phenomena but the webs spun by the mind to hold these phenomena together. And mathematicians do not stop with phenomena; they create structure for the pure joy of creating. They are carrying out Sandburg's definition of poetry as the "achievement of the synthesis of hyacinths and biscuits."

The difference between poetry and mathematics is mainly one of subject matter. Whitehead said that "mathematics is the greatest invention of the human mind unless it is perhaps music."

Said Poincaré (James B. Shaw, "Mathematics as a fine art," *The Mathematics Teacher*, November 1967, pp. 738–747):

Mathematics has a triple end. It should furnish an instrument for the study of nature. Furthermore it has a philosophic end. And I venture to say it, an aesthetic end. It incites the philosopher to search into the notions of number, space and time; and above all, adepts find in mathematics delights analogous to those that painting and music give. They admire the delicate harmony of number, of form; they are amazed at the unexpected vistas opened by a new discovery; and has not the joy they thus experience an aesthetic character even though the senses have no part therein? Only the privileged few have a chance to enjoy it fully, it is true, but is it not so with all the noblest arts? Hence I do not hesitate to say that mathematics deserves to be cultivated for its own sake, and theories not admitting any application to physics deserve to be studied as much as do the others.

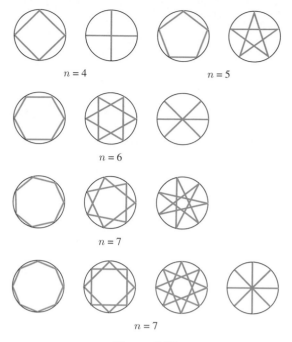

$n = 4$ $n = 5$

$n = 6$

$n = 7$

$n = 7$

Figure 2.39

Even though only the privileged few can enjoy mathematics to the fullest, anyone can enjoy it to a certain extent. There is art in mathematics. But to appreciate this fact one must know some mathematics. One must know enough to see its beauty, its form, its structure. And this is what this book is all about.

In this section and the next you will see some of the art in mathematics. At this stage, because your mathematical knowledge is limited, you cannot go too deep. However, as you read subsequent chapters, look for the art in mathematics. You will see the art and beauty in mathematical thought, in mathematical structure, in mathematical discovery, and in mathematical proofs.

Poinsot Stars

In 1809 while Napoleon was invading Austria and putting Europe in turmoil, the French mathematician Louis Poinsot (1777–1859) was quietly doodling with star-shaped objects. Not only were these stars artistically interesting but 150 years later they also proved useful to research involving the use of electronic computers in solving complex problems.

Poinsot began by drawing a number of equally spaced points around the circumference of a circle. He then joined the points with line segments. First he joined the points consecutively. Then he joined every other point, then every third point, and so forth (Figure 2.39). The resulting geometrical figures are called **Poinsot stars**.

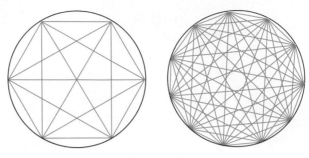

Figure 2.40

A Poinsot star with n vertices is said to have **order** n. Notice that one of the Poinsot stars of order n is just the regular polygon with n sides. Interesting designs can be obtained by superimposing all Poinsot stars of a given order. In Figure 2.40 we have such superpositions for orders 6 and 11. (These superpositions are complete graphs studied in Chapter 12.) Some other Poinsot stars are illustrated in Figure 2.41.

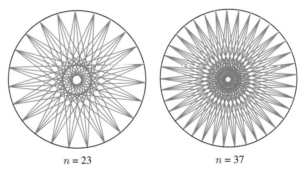

$n = 23$ $n = 37$

Figure 2.41

There are some interesting mathematical questions concerning Poinsot stars. The first that might come to mind is: How many Poinsot stars are there with a given order n? Table 2.2 gives this number for the orders 4 to 11. Looking at this table you might conjecture that for n even there are $n/2$ Poinsot stars of order n and for n odd there are $(n-1)/2$. This conjecture is true. How would you prove it? Let us consider, for example, Poinsot stars of order 7. If you join consecutive points, this is the same as joining every sixth point, since $1 + 6 = 7$. If you join every second point, this is the same as joining every fifth point, since $2 + 5 = 7$. Similarly, joining every third point is the same as joining every fourth point, since $3 + 4 = 7$. Thus there are three Poinsot stars of order 7 because 7 can be written as the sum of two positive integers in only three different ways: $7 = 1 + 6 = 2 + 5 = 3 + 4$. In a similar way, there are five Poinsot stars of order 11 because 11 can be written as a sum of two positive integers in only five different ways: $11 = 1 + 10 = 2 + 9 = 3 + 8 = 4 + 7 = 5 + 6$. In general, n can be written as the sum of two positive integers in $n/2$ different ways if n is even and

Table 2.2

Order	Number of Poinsot stars
4	2
5	2
6	3
7	3
8	4
9	4
10	5
11	5

$(n-1)/2$ different ways if n is odd. Indeed, if n is even, then

$$n = 1 + (n-1) = 2 + (n-2) = 3 + (n-3) = \cdots = \frac{n}{2} + \frac{n}{2}$$

and if n is odd, then

$$n = 1 + (n-1) = 2 + (n-2)$$

$$= 3 + (n-3) = \cdots = \frac{n-1}{2} + \frac{n+1}{2}$$

Notice that there can be two types of Poinsot stars of a given order. For example, consider order 4. If you start at some point and join consecutive points, you get a star with four vertices. However, if you start at some point and join every other point, you never reach some of the points and get a star with only two vertices. You must now start at another point to include all the points as vertices. The first type of Poinsot star is called **regular** and the second **compound**. Table 2.3 gives the number of regular and compound Poinsot stars of each order.

Table 2.3 Poinsot Stars

Order	Number of regular stars	Number of compound stars
4	1	1
5	2	0
6	1	2
7	3	0
8	2	2
9	3	1
10	2	3
11	5	0

We thus arrive at the question: How many regular Poinsot stars are there of order n? To answer this question we need to answer the following. What causes a star to be compound? In the second star of order 4 we joined every second point. Let $d = 2$. Now some points are never reached since 2 is a factor of both d and 4; so after joining every second point 2 times we come back to our starting point and can then reach no new points. In the second star of order 6 we joined every second point. Again let $d = 2$. Some points are never reached since 2 is a factor of both d and 6; so after joining every second point 3 times we come back to our starting point and then can reach no new points. In the third star of order 6 we joined every third point. Now let $d = 3$. But 3 is a factor of both d and 6; so again there are points that are never reached.

Reasoning in this way, Poinsot came to the following conclusion. A star of order n in which every dth point is joined is regular if and only if d and n have no factors in common except 1 (we then say that d and n are **relatively prime**). Poinsot then concluded that the number of regular stars of order n is the number of positive integers that are less than or equal to $n/2$ and relatively prime to n.

Let us test Poinsot's result on the stars we have considered (Table 2.4). We see that the table does indeed give the number of regular Poinsot stars of each order considered.

Table 2.4

Order n	Positive integers $\leq n/2$ relatively prime to n	Number of such integers
4	1	1
5	1, 2	2
6	1	1
7	1, 2, 3	3
8	1, 3	2
9	1, 2, 4	3
10	1, 3	2
11	1, 2, 3, 4, 5	5

Notice, in particular, that if n is prime (the only factors of n are 1 and n), then *every* integer $\leq n/2$ is relatively prime to n and so every Poinsot star of order n is regular. Hence there are $(n - 1)/2$ regular stars of order n if n is prime.

Poinsot even gave a formula for the number $S(n)$ of regular stars of order n. Let $p_1, p_2, \ldots, p_m$ be the prime factors of n. Then

$$S(n) = \left(\frac{n}{2}\right) \left(1 - p_1^{-1}\right) \left(1 - p_2^{-1}\right) \cdots \left(1 - p_m^{-1}\right)$$

Let us check this formula for some special cases. If n is prime, then the only prime factor of n is n itself (1 is, by convention, not prime); so

$$S(n) = \left(\frac{n}{2}\right) \left(1 - n^{-1}\right) = \frac{n - 1}{2}$$

which is what it should be. As another example, if $n = 9$, then

$$S(9) = \left(\frac{9}{2}\right)\left(1 - 3^{-1}\right) = \left(\frac{9}{2}\right)\left(\frac{2}{3}\right) = 3$$

As a final example, if $n = 10$, then

$$S(10) = \left(\frac{10}{2}\right)\left(1 - 2^{-1}\right)\left(1 - 5^{-1}\right) = \left(\frac{10}{2}\right)\left(\frac{1}{2}\right)\left(\frac{4}{5}\right) = \frac{40}{20} = 2$$

EXERCISES

1. How many Poinsot stars are there of order 23? Of order 24?
2. Draw the Poinsot stars of order 9, 10, and 11. Which are regular, and which are compound?
3. How many regular Poinsot stars are there of order 23? Of order 24?
4. Draw the Poinsot stars of order 12. Which are regular and which are compound?
5. Write 12 as the sum of positive integers in as many ways as you can. Do the same for 13. Do these agree with what we have obtained in this section?
* 6. Find $S(n)$ for $n = 12, 13, \ldots, 20$.
7. Are 6 and 14 relatively prime? What about 8 and 21?
8. Continue Table 2.4 to $n = 20$.
9. List the first 10 prime numbers.
10. Show that 2 is the only even prime number.

2.10 | Residue Designs

A single curve, drawn in the manner of the curve of prices of cotton, describes all that the ear can possibly hear as the result of the most complicated musical performance. . . . That to my mind is a wonderful proof of the potency of mathematics.
                    —LORD KELVIN (1827–1907), British Physicist

Mathematics has a light and wisdom of its own, above any possible applications to science, and it will richly reward any intelligent human being to catch a glimpse of what mathematics means to itself. This is not the old doctrine of art for art's sake; it is art for humanity's sake.   —E. T. BELL, Twentieth-Century Mathematician

In Poinsot stars we joined consecutive points or every second point or every third point, etc., in a regular fashion. What if we join the points differently? If we join points randomly, we will probably end up with a rather chaotic design that is not very pleasing. However, if we join points according to a definite mathematical rule, the results can be quite pleasing. We will join points according to residue tables. The resulting designs have a very pleasing appearance, possibly because of their intrinsic mathematical structure.

Residues

> Two integers a and b are said to be **congruent modulo an integer n** if n is a factor of $a - b$.

If a and b are congruent modulo n, we write $a = b$ (mod n). For example, $12 = 2$ (mod 5) because $12 - 2 = 10$ is divisible by 5 and $39 = 6$ (mod 11) because $39 - 6 = 33$ is divisible by 11. Because any given integer is congruent (mod n) to exactly one of the **residues** $0, 1, 2, \ldots, n - 1$, congruence theory is a **finite mathematics**.

EXAMPLE 1   What is the residue of (a) 7 (mod 5), (b) 24 (mod 7)?

    Solution.   (a) Since $7 = 2$ (mod 5), the residue is 2.

(b) Since $24 = 3$ (mod 7), the residue is 3.   □

EXAMPLE 2   What values do the residues (mod 6) have?

    Solution.   The residues have the values 0, 1, 2, 3, 4, 5.   □

EXAMPLE 3   Give three numbers that have residue 3 (mod 5).

    Solution.   The numbers 3, 8, 13 have residue 3 (mod 5).   □

Modular Arithmetic

Tables 2.5 to 2.8 are addition and multiplication tables modulo 5 and 6. In multiplication tables we leave out multiplication by zero because such multiplications are not very interesting.

Table 2.5 Addition mod 5

+	0	1	2	3	4
0	0	1	2	3	4
1	1	2	3	4	0
2	2	3	4	0	1
3	3	4	0	1	2
4	4	0	1	2	3

Table 2.6 Multiplication mod 5

·	1	2	3	4
1	1	2	3	4
2	2	4	1	3
3	3	1	4	2
4	4	3	2	1

Table 2.7 Addition mod 6

+	0	1	2	3	4	5
0	0	1	2	3	4	5
1	1	2	3	4	5	0
2	2	3	4	5	0	1
3	3	4	5	0	1	2
4	4	5	0	1	2	3
5	5	0	1	2	3	4

Table 2.8 Multiplication mod 6

·	1	2	3	4	5
1	1	2	3	4	5
2	2	4	0	2	4
3	3	0	3	0	3
4	4	2	0	4	2
5	5	4	3	2	1

Notice that there are essential differences between the multiplication table for a prime modulus (say 5) and that for a composite (not prime) modulus (say 6). For example, in the mod 6 multiplication table there are zeros present and there is a repetitive nature to some of the rows and columns. This difference is reflected in the fact that $1, 2, 3, 4$ (mod 5) forms what is called a multiplicative group whereas $1, 2, 3, 4, 5$ (mod 6) does not. In the addition tables, $0, 1, 2, 3, 4$ (mod 5) and $0, 1, 2, 3, 4, 5$ (mod 6) both form groups.

Since multiplication tables give more interesting residue designs, we shall use only multiplication tables for our constructions. By a $(5, 3)$ residue design is meant a residue design constructed with a mod 5 multiplication table and a constant multiplier 3 as follows. Draw a circle and divide the circumference into four equal parts, labeling the points of division 1, 2, 3, 4. The multiplier is 3, so from the third row of the mod 5 multiplication table we read $(3)(1) = 3$, $(3)(2) = 1$, $(3)(3) = 4$, $(3)(4) = 2$ (mod 5). Draw chords joining the points 1 to 3, 2 to 1, 3 to 4, and 4 to 2. After these chords have been drawn, complete the design by "coloring in" (Figure 2.42). Of course, more interesting designs are obtained from larger moduli.

In general, an (n, m)-residue design is constructed by dividing the circumference of a circle into $n - 1$ equal parts, labeling the points of division $1, 2, \ldots n - 1$, and drawing a chord from each point to its m-multiple (mod n). Other residue designs are shown in [29], [30], [31], and [32].

The residue designs reveal interesting mathematical properties. For example, the $(5, 3)$ and $(5, 2)$ designs are identical because $(3)(2) = 1$ (mod 5), so that 3 and 2 are **reciprocal multipliers**. For a composite modulus, zero divisors (numbers whose product is zero) must be excluded as multipliers. For example, 3 cannot be a multiplier for a mod 6 design. In general, the (n, m) design exists only if n and m are relatively prime. The types like $(21, 10)$ are called *star of David* designs. The $(19, 18)$ design is a barred type. All $(n, n - 1)$ designs are of this type since $n - 1 = -1$ (mod n) implies $(n - 1)i = -1$ (mod n) for all integers i; so the chords are drawn between i and $-i$ (which is the same as $n - i$). The $(65, 2)$ **cardioid** design is a **one-cusp epicycloid** and $(65, 3)$ is a **two-cusp epicycloid**.

Figure 2.42

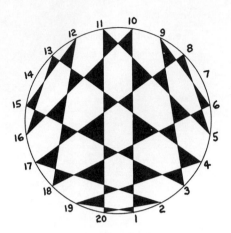

[**29**] Phil Locke, *Mathematics Teacher*, figure 4, p. 262. (Reprinted by permission of the artist.)

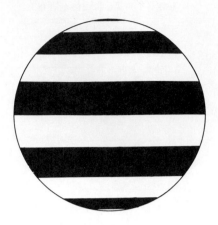

[**30**] Phil Locke, *Mathematics Teacher*, figure 5, p. 262. (Reprinted by permission of the artist.)

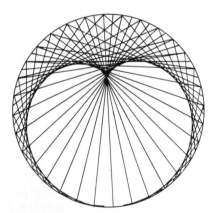

[**31**] Phil Locke, *Mathematics Teacher*, figure 7, p. 263. (Reprinted by permission of the artist.)

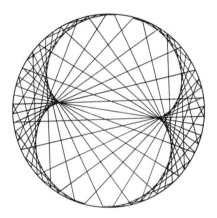

[**32**] Phil Locke, *Mathematics Teacher*, figure 8, p. 263. (Reprinted by permission of the artist.)

Roughly speaking, if n is large and m is small, then (n, m) gives an $(m - 1)$-cusp epicycloid.

Residue tables can be used to make other interesting designs. If in the multiplication mod 5 table we replace each number by a design, we obtain a multiple design (Figure 2.43).

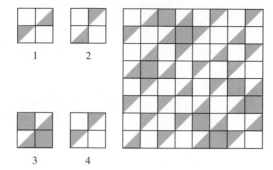

Figure 2.43

Figure 2.44

If we reflect the previous design about the right vertical line and then reflect the resulting design about the horizontal base line, we obtain the design of Figure 2.44. A variation of the previous design can be constructed by replacing the standard square grid by a converging grid to obtain Figure 2.45.

EXERCISES

1. What is the residue of
   (a) 12 (mod 6) (b) 25 (mod 6) (c) 35 (mod 7) (d) 64 (mod 10)?

2. What values do the residues (mod 7) have?

3. Give three numbers that have residue 4 (mod 7).

4. Construct the addition and multiplication tables modulo 2.

5. Construct the addition and multiplication tables modulo 3.

6. Construct the addition and multiplication tables modulo 7. Find the reciprocal multipliers.

7. Construct the addition and multiplication tables modulo 8. Are there any zero divisors?

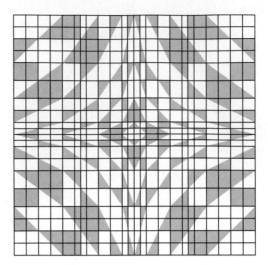

Figure 2.45

8. Construct the $(11, 3)$ residue design.

9. Construct a design by replacing each number in the multiplication modulo 11 table by a design of your own.

10. Why is any given integer congruent (mod n) to exactly one of the residues $0, 1, 2, \ldots,$ $n - 1$?

11. Check to see if the $(15, 14)$ residue design is a barred type.

12. A slight modification of addition modulo 12 is sometimes called **clock arithmetic**. How would you define this addition? If it is now 5:00, use your addition to find the time (a) 17 hours later, (b) 33 hours later.

Chapter 2 Summary of Terms

types of numbers   triangular
square
pentagonal
hexagonal

golden ratio $\tau = \dfrac{1 + \sqrt{5}}{2} = 1.618033989\ldots$

Fibonacci sequence $1, 1, 2, 3, 5, 8, 13, 21, 34, 55, \ldots$

perspective

symmetry

transformations

gestalt psychology

convexity and concavity

optical art

open and closed sets

tesselations

Poinsot stars

congruences modulo n

residues

Chapter 2 Test

1. State Pythagoras' theorem.
2. State the converse of Pythagoras' theorem.
3. What is the seventh triangular number?
4. Find $\sqrt{64}$, $\sqrt{100}$, $\sqrt{169}$.
5. Evaluate $(\sqrt{7}+1)(\sqrt{7}-1)$.
6. What is the next term of the Fibonacci sequence after 55?
7. If a and b are natural numbers, then $\sqrt{a^2+2ab+b^2}$ is a natural number. Verify this for three different pairs of natural numbers a, b. Why is this always true?
8. Does the human body possess bilateral symmetry? First discuss the external human body, then the internal human body.
9. Give an example of a convex set whose complement is also convex.
10. Give an example of a set that is neither open nor closed.
11. Using the transformation $(x, y) \rightarrow (x+y, x-y)$, transform the following points.
    (a) $(0,0)$ (b) $(1,2)$ (c) $(-1,-1)$ (d) $(-2,3)$
12. What portion of the plane is given by the set $\{(x, y): x < y\}$?
13. Show that the set $\{(x, y): x \neq 0, y \neq 0\}$ is not convex.
14. Give an example of a set that is both open and closed.
15. The nth term of a sequence is $(n+1)/n$. Find the first five terms. To what number does this sequence converge?
16. How many Poinsot stars are there of order 100? Of order 111?
17. Is 91 a prime number? Why?
18. Construct the multiplication table modulo 9.
19. What is the residue of 13 (mod 5)?

LOGIC

You can fool all of the people some of the time, and you can fool some of the people all of the time, but you can't fool all of the people all of the time.

— ABRAHAM LINCOLN (1809–1865), American President

Logic is indispensable not only for mathematical reasoning but for everyday real-life thinking as well. You do not have to be Sherlock Holmes to require logical thought. As an illustration of everyday logical thinking, consider the statements made by some politicians. Politicians sometimes use logic in a very shrewd way and we must be aware of this fact. When a politician states an implication, many people unwittingly believe that the inverse implication also holds. But this is far from the case. For example, consider the following implication: If the American people drive less, then pollution will decrease. Does this mean that the inverse implication (if the American people do not drive less, then pollution will not decrease) also holds? Not necessarily. Another example is the following. If I am elected, there will be no recession. Does this mean that if I am not elected, then there will be a recession? Not at all. In this chapter we shall study methods for analyzing logical arguments and for deciding whether a conclusion derived from an argument is true.

3.1 Statements

logic 1. the science which investigates the principles governing correct or reliable inference. **2.** reasoning or argumentation. **3.** the system or principles of reasoning applicable to any branch of knowledge. **4.** the study of valid arguments involving strings of statements.
— *The Hammond International Dictionary* (Random House)

A **statement** is a declarative sentence that is either true or false.

Two examples of declarative sentences are

> The window is open.
> $3 < 5$.

Each of these sentences is either true or false. The truth or falsity of the first sentence depends on the particular window to which we are referring. The second sentence is mathematical shorthand for the true statement "three is less than five." Unlike mathematics, the English language is very imprecise. Consider the sentence "The sun is shining." Although we consider this a statement, more clarification might be needed to determine whether it is true or false. This statement is usually interpreted as meaning that the sun is visible at this instant at this location. If another interpretation is intended, then the truth or falsity of the statement may change. The following sentences are not statements.

> Eat your vegetables.
> Is the window open?
> She's not a nice person.
> This sentence is false.

We cannot identify these sentences as being true or false. The first is a command, the second a question, and the third an opinion. The fourth sentence is a paradox. If it is true, then it is false; if it is false, then it is true.

Negation

> Every statement has a **negation**. The negation of a true statement is false and the negation of a false statement is true.

For example, the negation of the statement "My name is John" is "My name is not John." The negation of the statement "All pencils are orange" is "Not every pencil is orange." This negation can also be phrased "Some pencils are not orange."

Connectives

> We can combine two or more statements to form a **compound statement**. Such combinations are obtained using **connectives** such as "and," "or," and "if...then."

Examples of compound statements are

> Mares eat oats and does eat oats and little lambs eat ivy.
> It's you or me, baby.
> If it rains, then it pours.
> You Jane and me Tarzan.

When we combine two statements with an "and" connective, the result is a **conjunction** of the two statements; and when we combine them with an "or" connective, the result is a **disjunction**. When we combine statements with an "if...then" connective, the resulting statement is a **conditional statement** or an **implication**.

In symbolic logic, we represent statements by letters such as p, q, or r. We also represent negation ("not") by the symbol $\sim$ and we represent the connectives "and," "or," and "if...then" by $\wedge$, $\vee$, $\rightarrow$, respectively. Table 3.1 summarizes these symbols.

Table 3.1

Logical operation	Symbol	Type of statement
not	$\sim$	Negation
and	$\wedge$	Conjunction
or	$\vee$	Disjunction
if...then	$\rightarrow$	Conditional

Components

The statements that appear within a compound statement are called the **components** of the compound statement.

For example, the components of $p \vee q$ are p and q; the components of $p \wedge q \rightarrow r$ are p, q, and r.

EXAMPLE 1   We can write the statement "You Jane and me Tarzan" in a more grammatical form as "You are Jane and I am Tarzan." This compound statement has the two components

p = You are Jane.
q = I am Tarzan.

Since the connective in this compound sentence is "and," we represent it symbolically as $p \wedge q$.

EXAMPLE 2   The conditional statement "If it rains, then it pours" has the two components

p = It rains.
q = It pours.

Since the connective is "if...then," we represent this compound statement as $p \rightarrow q$.

EXAMPLE 3   "It's you or me, baby" has the two components

p = It's you, baby.
q = It's me, baby.

This compound statement is represented symbolically as $p \lor q$ because the connective is "or."

EXAMPLE 4   "Mares eat oats and does eat oats and little lambs eat ivy" has the three components

p = Mares eat oats.
q = Does eat oats.
r = Little lambs eat ivy.

We represent this compound statement symbolically as $p \land q \land r$.

EXAMPLE 5   There may be different ways to represent a compound statement symbolically. Consider the statement "It's neither here nor there." If we let

p = It's here.
q = It's there.

Then we can represent the statement by $\sim p \land \sim q$. That is, this compound statement is the same as the statement "It's not here and it's not there." Another equivalent way of writing this statement is "It is not the case that it's here or it's there." Thus, we can also represent this statement by $\sim(p \lor q)$. For still another approach, we could let

r = It's not here.
s = It's not there.

We can then represent the statement as $r \land s$.

EXAMPLE 6   Let p represent "There are clouds" and let q represent "It is raining." Write each of the following symbolic statements in words: (a) $p \land q$, (b) $p \rightarrow q$, (c) $q \rightarrow p$, (d) $\sim p \land q$, (e) $\sim(p \lor q)$.
  Solution.   (a) There are clouds and it is raining.
  (b) If there are clouds, then it is raining.
  (c) If it is raining, then there are clouds.
  (d) There are no clouds and it is raining.
  (e) It is not the case that there are clouds or it is raining.   □

Tax Forms

So far we have considered very simple compound statements. However, in practice, compound statements can be quite complicated combinations of many components with many connectives. This is evident to anyone who has tried to read a legal document or fill out a federal income tax form. Table 3.2 gives the criteria for determining whether you must file a federal income tax form.

Table 3.2

Filing status	Age	Gross income
Single and not head of a household	Under 65 65 or older	at least $5,500 at least $6,400
Single and head of a household	Under 65 65 or older	at least $7,150 at least $8,000
Married, joint return	Under 65 (both spouses) 65 or older (one spouse) 65 or older (both spouses)	at least $10,000 at least $10,650 at least $11,300
Married, separate returns	Any	at least $2,150

To keep the situation manageable, let us suppose that you are single. We then need consider only the top half of Table 3.2. (You will be asked to treat the married status in an exercise.) We would now like to construct a conditional statement that determines whether you must file a federal income tax form. We shall need the following components.

p = My gross income is at least $5,500.
q = My gross income is at least $6,400.
r = My gross income is at least $7,150.
s = My gross income is at least $8,000.
t = I am the head of a household.
u = My age is less than 65.
v = I must file a federal income tax form.

You should check that the desired conditional statement is the following:

$$\{\sim t \wedge [(u \wedge p) \vee (\sim u \wedge q)]\} \vee \{t \wedge [(u \wedge r) \vee (\sim u \wedge s)]\} \rightarrow v$$

Statements and Sets

We now consider the relationship between statements and sets. As you will see, these two concepts have close analogies. Let U be a fixed universal set. In the following discussion, we shall assume that x is an element of U and that the sets A, B, and C are subsets of U. Corresponding to a set A we have the statement p = "x is a member of A" or in mathematical notation p = "$x \in A$." The statement p is either true or false depending on whether x is or is not a member of A. The complement A' of A corresponds to the statement p' = "$x \in A'$." We can rewrite p' in the form p' = "$x \notin A$." Since p' is true precisely when p is false, we conclude that $p' = \sim p$. Thus, A' corresponds to the statement $\sim p$.

Table 3.3

Set operations	Logical operations
A'	$\sim p$
$A \cap B$	$p \wedge q$
$A \cup B$	$p \vee q$
$A \subseteq B$	$p \rightarrow q$

Continuing the analogy between sets and statements, let the set A correspond to statement p and let set B correspond to statement q. Now $A \cap B$ corresponds to the statement $r =$ "$x \in A \cap B$." But $x \in A \cap B$ precisely when $x \in A$ and $x \in B$. Hence, we can rewrite r as $r =$ "$x \in A$ and $x \in B$" or $r = p \wedge q$. We conclude that $A \cap B$ corresponds to the statement $p \wedge q$. In a similar way, $A \cup B$ corresponds to the statement $p \vee q$. Suppose $A \subseteq B$. This corresponds to the statement $s =$ "If $x \in A$ then $x \in B$," which can be written $s = p \rightarrow q$. In this way, negation and the logical connectives correspond to the basic operations on sets. These observations are summarized in Table 3.3.

In Section 1.3 we studied various laws of set theory. For example, the distributive law stated that $A \cap (B \cup C) = (A \cap B) \cup (A \cap C)$. Using the correspondence between set operations and logical operations, the distributive law corresponds to the logical equation $p \wedge (q \vee r) = (p \wedge q) \vee (p \wedge r)$. Does this latter equation hold in symbolic logic? We shall see in Section 3.4 that in a certain sense it does hold.

EXAMPLE 7   Write the symbolic logic equation corresponding to De Morgan's law $(A \cup B)' = A' \cap B'$.

Solution.   Applying the correspondences in Table 3.3 we have $\sim(p \vee q) = \sim p \wedge \sim q$. In Section 3.4, we show that this equation holds in a certain sense.   □

EXERCISES

Decide whether or not each of the following is a statement.
1. Get off the phone.   2. This sentence has five words.
3. This sentence does not have seven words.
4. How much is that doggie in the window?
5. $3 + 5 = 7$.   6. $2 \leq 1$.
7. If you don't take a bath for long enough, even the fleas will leave you alone.
8. Today is Friday.

What type of compound statements are the following?
9. Roses are red and violets are blue.   10. His name is George or his name is Jack.
11. If you want to see a rainbow, then you got to stand a little rain.
12. If $x + 2 = 3$, then $x = 1$.   13. $x < 1$ or $x > 3$.

Write the negation of the following statements.
14. Today is Monday.   15. $2 < 3$ 16. $x = 5$
17. $a \in A$ 18. All dogs have fleas.

Let p represent the statement "She is tall" and let q represent the statement "She has blue eyes."
Convert each of the following statements into words. *tall and she has blue eyes*

19. $p \vee q$ 20. $p \wedge q$ *She is tall*   21. $\sim p$

22. $\sim q$ 23. $p \vee \sim q$ 24. $\sim p \wedge q$ *She is not tall and she has blue eyes*

25. $\sim p \wedge \sim q$ 26. $\sim(p \vee q)$ 27. $\sim(\sim p \wedge q)$

28. Define the two components in Exercise 9 and represent the statements in symbolic form.

29. Repeat Exercise 28, but for Exercise 10 components.

30. Repeat Exercise 28, but for Exercise 11 components.

31. Repeat Exercise 28, but for Exercise 12 components.

32. Repeat Exercise 28, but for Exercise 13 components.

For the income tax discussion write the following symbolic statements in words.

33. $\sim u \wedge q$ 34. $(u \wedge r) \vee (\sim u \wedge s)$

35. $t \wedge [(u \wedge r) \vee (\sim u \wedge s)]$ 36. $\sim t \wedge [(u \wedge p) \vee (\sim u \wedge q)]$

*37. Construct a conditional statement that determines whether a married person must file a federal income tax form. Be sure that you have defined the components.

Write the symbolic logic equation corresponding to each of the following laws of set theory.

38. $(A')' = A$ 39. $A \cup (B \cap C) = (A \cup B) \cap (A \cup C)$

40. $(A \cap B)' = A' \cup B'$ 41. $A \cup (B \cup C) = (A \cup B) \cup C$

42. $A \cap (B \cap C) = (A \cap B) \cap C$ 43. If $A \subseteq B$, then $B' \subseteq A'$.

*44. Write the negation of the following statement in words without using *not*: Some dogs have fleas.

*45. Write the negation of the following statement in words without using *not*: No dogs have fleas.

*46. Discuss whether the following statement is compound: Jane and Jack are married.

*47. Discuss whether the following statement is compound: Paula and Jim are sister and brother.

48. Discuss the following headlines excerpted from *Squad Helps Dog Bite Victim and Other Flubs from the Nation's Press* (edited by the *Columbia Journalism Review*; Dolphin, New York, 1980).

   (a) SQUAD HELPS DOG BITE VICTIM (*Herald Independent* 4/29/76)

   (b) BAN ON SOLICITING DEAD IN TROTWOOD (*Dayton Daily News* 4/7/76)

   (c) STUD TIRES OUT (*Ridgewood (N.J.) News* 3/30/78)

   (d) DR. TACKETT GIVES TALK ON MOON (*Indiana Evening Gazette* 3/13/76)

   (e) MRS. GANDHI STONED AT RALLY IN INDIA (*Toronto Star* 1/18/71)

   (f) JUMPING BEAN PRICES AFFECT POOR
                  (*Eugene (Oregon) Register-Guard* 2/27/74)

   (g) MAN ROBS, THEN KILLS HIMSELF (*The Washington Post* 12/19/75)

   (h) CHOU REMAINS CREMATED (*Peoria Journal Star* 1/12/76)

   (i) MAULING BY BEAR LEAVES WOMAN GRATEFUL FOR LIFE
                  (*Herald-Dispatch (Hunting, W. Va.)* 9/8/77)

   (j) GOV. MOORE MEETS MINERS' DEMANDS, TWO PICKETS SHOT
                  (*Cleveland Plain Dealer* 3/14/74)

   (k) MBA STUDIES MUSHROOM (*SBA News (Youngstown, Ohio)* Fall 1975)

   (l) SCSC GRADUATES BLIND SENIOR CITIZEN
                  (*Journal Inquirer (Manchester, Conn.)* 5/24/76)

   (m) CARTER PLANS SWELL DEFICIT (*Houston Tribune* 3/17/77)

   (n) TAX CUT DUEL IN STORE (*Palm Beach Times* 7/27/78)

(o)   NEW HOUSING FOR ELDERLY NOT YET DEAD
                            (The Times-Argus (Barre-Montpelier, Vt.) 5/31/74)

(p)   POLICE CAN'T STOP GAMBLING *(Detroit Free Press 7/1/75)*

(q)   TOWN OKs ANIMAL RULE *(Ashville (S.C.) Citizen 3/2/77)*

(r)   GREEKS FINE HOOKERS *(Contra Costa (Calif.) Times 5/31/77)*

(s)   GENETIC ENGINEERING SPLITS SCIENTISTS
                            (The Washington Post 11/29/75)

(t)   POLICE KILL MAN WITH AX *(Charlotte Observer 11/27/76)*

3.2 True or False

"I know what you're thinking about," said Tweedledum, "but it isn't so, nohow." "Contrawise," continued Tweedledee, "if it was so, it might be; and if it were so, it would be, but as it isn't, it ain't. That's logic."

—CHARLES DODGSON (LEWIS CARROLL) (1832–1898)
English Logician and Author

How do you decide whether a fairly complicated compound statement p is true or false? First you should decide whether each of its components is true or false. This is called finding the truth values of the components. The present section shows how to construct a table called a truth table. The truth values of the components can be entered into the truth table for p and this determines the truth value of p.

Truth Values

Table 3.4

p
T
F

> A statement has two possible **truth values**, T (true) and F (false).

For a statement p, we symbolize this as in Table 3.4. From the truth values of p, we can find the corresponding truth values of $\sim p$, as in Table 3.5.

Table 3.5

p	$\sim p$
T	F
F	T

It is sometimes not so easy to decide the truth value of a compound statement when we know the truth values of its components. For instance, let p = "There are clouds," let q = "It is raining," and consider the compound statement $\sim(p \vee q)$. Suppose there are clouds so p is true but it is not raining so q is false. Is $\sim(p \vee q)$ true or false? Well, $\sim(p \vee q)$ is false. This is because the statement $(p \vee q)$ "There are clouds or it is raining" is true since p is true. Since $(p \vee q)$ is true, $\sim(p \vee q)$ must be false. What if both p and q are false? Using similar reasoning, we would conclude that $\sim(p \vee q)$ is true. To avoid reasoning out each special case or unraveling long tedious compound statements using English, a symbolic method can be used which is much easier. This method is to use a **truth table**. A truth table shows all the possible truth values of a compound statement in terms of the corresponding truth values of its components.

Truth Tables

Let's find the truth table for the conjunction $p \wedge q$. Now for $p \wedge q$ to be true, both p and q must be true. The truth table for conjunction is given in Table 3.6.

Table 3.6

p	q	$p \wedge q$
T	T	T
T	F	F
F	T	F
F	F	F

Table 3.7

p	q	$p \vee q$
T	T	T
T	F	T
F	T	T
F	F	F

To find the truth table for the disjunction $p \vee q$ we must be clear about what $p \vee q$ means. In this respect, the English language is ambiguous. Let us agree that $p \vee q$ is true if p is true or q is true or both are true. For example, the statement "I go by car or train" is true if I go by car or if I go by train or if I go by both car and train. In the English language, this statement is sometimes considered false if I first take the car and then board a train, so we must be precise about what we mean. The truth table for disjunction is shown in Table 3.7.

We can now use Tables 3.5, 3.6, and 3.7 to construct truth tables for more complicated compound statements.

EXAMPLE 1   Construct a truth table for $\sim(p \vee q)$.

Solution.   We first list the truth values of $p \vee q$ and use these to find the truth values of $\sim(p \vee q)$. This is illustrated in Table 3.8. Notice that we obtained the truth values in the $\sim(p \vee q)$ column by taking the values opposite those in the $p \vee q$ column.   □

Table 3.8

p	q	$p \vee q$	$\sim(p \vee q)$
T	T	T	F
T	F	T	F
F	T	T	F
F	F	F	T

EXAMPLE 2   Construct a truth table for $(\sim p \vee q) \wedge \sim q$.

Solution.   We first list the truth values of $\sim p$, then those of $\sim p \vee q$ and $\sim q$, and finally combine the last two. This is done in Table 3.9. The $\sim p$ column is just the opposite of the p column. The $\sim p \vee q$ column is found by combining the $\sim p$ column and the q column using the disjunction table, Table 3.7. The $\sim q$ column is the opposite of the q column. The $(\sim p \vee q) \wedge \sim q$ column is found by combining the $\sim p \vee q$ column and the $\sim q$ column using the conjunction table, Table 3.6.   □

Table 3.9

p	q	$\sim p$	$\sim p \lor q$	$\sim q$	$(\sim p \lor q) \land \sim q$
T	T	F	T	F	F
T	F	F	F	T	F
F	T	T	T	F	F
F	F	T	T	T	T

Strictly speaking, the truth table for $(\sim p \lor q) \land \sim q$ is given by Table 3.10, but we frequently include the intermediate steps to show how it is constructed.

Table 3.10

p	q	$(\sim p \lor q) \land \sim q$
T	T	F
T	F	F
F	T	F
F	F	T

EXAMPLE 3   (a) If p and q are both false, find the truth value of $(\sim p \lor q) \land \sim q$.
(b) If p represents the statement "$2 < 3$" and q represents the statement "$1 + 4 = 6$," is the statement $(\sim p \lor q) \land \sim q$ true or false?

Solution.   (a) Look at the last row of Table 3.10 and find that the compound statement is true.
(b) Since p is true and q is false, the second row of Table 3.10 shows that the compound statement is false.   □

EXAMPLE 4   Construct a truth table for $\sim [\sim q \land (p \lor q)]$.

Solution.   We use methods similar to those used for Example 2. The truth table is shown in Table 3.11.   □

Table 3.11

p	q	$\sim q$	$p \lor q$	$\sim q \land (p \lor q)$	$\sim [\sim q \land (p \lor q)]$
T	T	F	T	F	T
T	F	T	T	T	F
F	T	F	T	F	T
F	F	T	F	F	T

Truth tables can also be constructed for statements containing three or more components. If a statement has three components p, q, and r, the truth table requires eight rows to list all possible combinations of truth values for p, q, and r. If a statement has n different letters, a truth table requires 2^n rows.

Table 3.12

p	q	r	$\sim p$	$\sim p \wedge q$	$\sim q$	$\sim q \wedge r$	$(\sim p \wedge q) \vee (\sim q \wedge r)$
T	T	T	F	F	F	F	F
T	T	F	F	F	F	F	F
T	F	T	F	F	T	T	T
T	F	F	F	F	T	F	F
F	T	T	T	T	F	F	T
F	T	F	T	T	F	F	T
F	F	T	T	F	T	T	T
F	F	F	T	F	T	F	F

EXAMPLE 5 Construct a truth table for $(\sim p \wedge q) \vee (\sim q \wedge r)$.

   Solution. The construction (Table 3.12) is similar to that of Example 2 except that more rows are required. □

EXAMPLE 6 If p and r are true and q is false, find the truth value of $(\sim p \wedge q) \vee (\sim q \wedge r)$.

   Solution. The third row of Table 3.12 gives the truth value T. □

EXAMPLE 7 If p, q are true and r, s are false, find the truth value of $(p \vee \sim r) \wedge (q \vee s) \wedge (\sim q \vee \sim r)$.

   Solution. Since there are four components, the complete truth table requires 16 rows. But since we know the truth values of the components, all we need is one row. This is shown in Table 3.13. □

Table 3.13

p	q	r	s	$\sim q$	$\sim r$	$p \vee \sim r$	$q \vee s$	$\sim q \vee \sim r$	$(p \vee \sim r) \wedge (q \vee s) \wedge (\sim q \vee \sim r)$
T	T	F	F	F	T	T	T	T	T

EXERCISES

Give the truth values of the following statements.

1. $1 = 2$ 2. $1 \neq 2$ 3. $1 = 2$ or $1 \neq 2$
4. $1 = 2$ and $1 \neq 2$ 5. $2 \leq 3$ or $3 \leq 1$ 6. $\frac{2}{3} \leq \frac{7}{10}$
7. $x \in A$ or $x \notin A$ 8. $x \in A$ and $x \notin A$ 9. $\frac{1}{2} + \frac{1}{3} = \frac{5}{6}$
10. $A \cap B \subseteq A$ 11. If $x \leq 2$ then $x \leq 3$ 12. $A \cap A' \neq \emptyset$

If p is true and q is false, find the truth value of the following statements.

13. $\sim q$ 14. $p \vee q$ 15. $p \wedge q$
16. $\sim p \vee q$ False   17. $\sim (p \wedge q)$ 18. $\sim [\sim q \wedge (p \vee q)]$ False
19. $\sim [\sim p \vee (p \wedge \sim q)]$

Construct a truth table for the following compound statements.

20. $p \wedge \sim q$

21. $\sim p \vee \sim q$

22. $(p \vee q) \wedge \sim q$

23. $(\sim p \vee \sim q) \wedge q$

24. $(p \wedge \sim q) \vee \sim p$

25. $(p \vee q) \vee (q \wedge r)$

26. $(p \wedge q) \wedge r$

27. $(\sim p \wedge q) \vee \sim r$

28. $(p \vee q) \wedge (\sim q \vee \sim r)$

29. $(\sim p \vee q) \wedge (\sim q \vee \sim r)$

30. If p, q are true and r, s are false, find the truth value of
$[(p \wedge r) \vee q] \wedge [(\sim p \wedge s) \vee \sim q]$.

31. If p, q, r are true and s is false, find the truth value of
$[\sim(p \vee q) \wedge r] \vee [(p \vee \sim s) \wedge \sim q]$.

32. A boy and a girl are sitting on the front steps of their house. "I'm a boy," said the one with black hair. "I'm a girl," said the one with red hair. If at least one of them is lying, what color is the girl's hair?

*33. "Frank owns at least a thousand books," said John. "He owns fewer than that," said George. "Surely he owns at least one book," said Janet. If only one statement is true, how many books does Frank own?

34. If a statement contains four different letters p, q, r, and s, why are 16 rows required in its truth table?

*35. If a statement contains n different letters, show that 2^n rows are required in its truth table.

3.3 | Conditionals

God exists since mathematics is consistent, and the Devil exists since we cannot prove it. — A. WEIL, Twentieth-Century Mathematician

In Section 3.2 we constructed truth tables for conjunction and disjunction but we did not study the other important connective, the conditional. The conditional or implication "if... then" is a common connective in everyday usage. For example,

If it goes up, then it comes down.
If it's cold outside, then I'll take my coat.

"If p, then q" is symbolized as $p \rightarrow q$. This is interpreted as meaning that if p is true then q is true. In other words, q is true under the condition that p is true (this is why it's called a **conditional statement**). Another way of expressing this is to say that p implies q (this is why it's also called an **implication**).

> In the conditional $p \rightarrow q$, the statement p is the **antecedent** (or **hypothesis**) and q is the **consequent** (or **conclusion**).

Truth Table for $p \rightarrow q$

In order to decide on the proper truth values of $p \rightarrow q$, suppose p is the statement "It's cold outside" and q is the statement "I'll take my coat." We then have the four possibilities listed in Table 3.14.

Table 3.14

Is it cold outside?		Do I take my coat?		Is $p \rightarrow q$ true?	
yes	p is T	yes	q is T	yes	T
yes	p is T	no	q is F	no	F
no	p is F	yes	q is T	yes	T
no	p is F	no	q is F	yes	T

In the first possibility it is cold outside and I do take my coat, so $p \rightarrow q$ is true. In the second possibility it is cold outside and I do not take my coat, so $p \rightarrow q$ is false. For the third and fourth possibilities, it is not cold outside, in which case I did not say what I would do. In these cases the statement is not incorrect so it is considered to be true. We conclude that $p \rightarrow q$ is false only when the antecedent p is true and the consequent q is false. The truth table for $p \rightarrow q$ is given in Table 3.15.

Table 3.15

p	q	$p \rightarrow q$
T	T	T
T	F	F
F	T	T
F	F	T

It should be noted that a conditional connective does not necessarily imply a cause-and-effect relationship. Any two statements can be combined using the conditional connective to form a compound statement. For example,

If my cat is black, then Grant is buried in Grant's tomb.

is true because the consequent is true even though it has no relationship to the color of my cat.

EXAMPLE 1 Let p be the statement "$1/3 < 1/2$" and q the statement "$3^2 = 8$." Are the following statements true or false?
(a) $p \rightarrow q$ (b) $\sim p \rightarrow q$ (c) $q \rightarrow p$ (d) $\sim q \rightarrow p$

   Solution.   (a) Since p is true and q is false, $p \rightarrow q$ is false.
(b) In this case, $\sim p$ is "$1/3 \nless 1/2$," which is false. Hence, $\sim p \rightarrow q$ is true.
(c) Since q is false, $q \rightarrow p$ is true.
(d) In this case, $\sim q$ is "$3^2 \neq 8$," which is true. Since p is true, $\sim q \rightarrow p$ is true.   □

We can construct truth tables for statements involving conditionals using the methods of the last section. This is illustrated in the following examples.

EXAMPLE 2   Construct a truth table for the statement $p \to (\sim p \land q)$.
   Solution.   This truth table is constructed in Table 3.16. $\square$

Table 3.16

p	q	$\sim p$	$\sim p \land q$	$p \to (\sim p \land q)$
T	T	F	F	F
T	F	F	F	F
F	T	T	T	T
F	F	T	F	T

EXAMPLE 3   Construct a truth table for the statement $(p \to q) \land (q \to p)$.
   Solution.   This truth table is constructed in Table 3.17. $\square$

Table 3.17

p	q	$p \to q$	$q \to p$	$(p \to q) \land (q \to p)$
T	T	T	T	T
T	F	F	T	F
F	T	T	F	F
F	F	T	T	T

EXAMPLE 4   Construct a truth table for the statement $(p \to \sim q) \to (q \lor \sim p)$.
   Solution.   This truth table is constructed in Table 3.18. $\square$

Table 3.18

p	q	$\sim p$	$\sim q$	$p \to \sim q$	$q \lor \sim p$	$(p \to \sim q) \to (q \lor \sim p)$
T	T	F	F	F	T	T
T	F	F	T	T	F	F
F	T	T	F	T	T	T
F	F	T	T	T	T	T

Open Statements

In mathematics we frequently encounter statements that contain a variable. Such statements are called **open statements**.

For example, the statement "$x + 2 < 5$" is either true or false but it depends on the value of x. For instance, if $x = 2$ then the statement is true, but if $x = 3$ then the statement is false. Open statements of the form $p \to q$ are usually treated as follows. We first assume that the antecedent p is true. Then if the consequent q is necessarily true, the statement $p \to q$ is true. Otherwise, $p \to q$ is false. For example,

If $x < 5$, then $x < 7$

is true because $5 < 7$ so whenever $x < 5$ we have $x < 7$ is true. However, the statement

If $x > 0$, then $x > 3$

is false because the antecedent can be true without the consequent being true (for example, $x = 2$).

EXAMPLE 5 Is the following statement true or false? If $x < 3$, then $x + 2 < 5$.
  Solution. The statement is true because whenever $x < 3$ is true, $x + 2 < 5$ is also true. □

Biconditional

The statement $(p \to q) \land (q \to p)$ in Example 3 is called a **biconditional** and is denoted $p \leftrightarrow q$. From the truth table of Example 3, we see that $p \leftrightarrow q$ is true when p and q have the same truth value and is false otherwise. The conditional $p \to q$ is sometimes written "p implies q" and the biconditional $p \leftrightarrow q$ is written "p implies q and q implies p." We also write "p if and only if q" for $p \leftrightarrow q$.

EXAMPLE 6 Are the following biconditionals true or false?
(a) $2 + 3 = 6$ if and only if $3 < 2$.
(b) $x + 2 = 3$ if and only if $x = 1$.
(c) $x^2 = 5$ if and only if $x = 2$.
  Solution. (a) Both $2 + 3 = 6$ and $3 < 2$ are false, so the biconditional is true.
(b) If the first statement is true, so is the second. If the second statement is true, so is the first. Hence, the biconditional is true.
(c) If the second statement is true, then the first is false. Hence, the biconditional is false. □

Corresponding to a conditional statement $p \to q$ there are three other statements called the converse, inverse, and contrapositive. These are defined as follows:

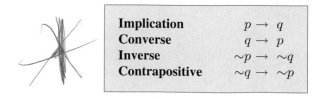

Implication	$p \to q$
Converse	$q \to p$
Inverse	$\sim p \to \sim q$
Contrapositive	$\sim q \to \sim p$

Notice that a conditional statement is the converse of the converse statement. Another way of viewing this is that the converse changes the direction of the arrow in the implication and if we change the direction of the arrow twice we are back to the original implication. We thus say that $p \to q$ and $q \to p$ are converses of each other. Notice also that the inverse and contrapositive are converses of each other.

EXAMPLE 7   Write the converse, inverse, and contrapositive of $p \to (p \land q)$.
     Solution.   Converse: $(p \land q) \to p$ Inverse: $\sim p \to \sim(p \land q)$
Contrapositive: $\sim(p \land q) \to \sim p$.   □

EXAMPLE 8   Write the converse, inverse, and contrapositive of the statement

     If it is snowing, then it is winter.

     Solution.   Converse: If it is winter, then it is snowing.
The converse can also be written "It is snowing only if it is winter."
Inverse: If it is not snowing, then it is not winter.
Contrapositive: If it is not winter, then it is not snowing.   □

     The truth tables for the four statements $p \to q$, $q \to p$, $\sim p \to \sim q$, and $\sim q \to \sim p$ are given in Table 3.19.

Table 3.19

				Implication	Converse	Inverse	Contrapositive
p	q	$\sim p$	$\sim q$	$p \to q$	$q \to p$	$\sim p \to \sim q$	$\sim q \to \sim p$
T	T	F	F	T	T	T	T
T	F	F	T	F	T	T	F
F	T	T	F	T	F	F	T
F	F	T	T	T	T	T	T

EXERCISES

Let p represent the statement "It rains" and let q represent the statement "I wash my car." Express each of the following compound statements in words.

1. $p \to q$ 2. $q \to p$ 3. $\sim p \to \sim q$
4. $\sim q \to \sim p$ 5. $p \leftrightarrow q$ 6. $\sim p \leftrightarrow q$
7. $p \leftrightarrow \sim q$ 8. $\sim p \leftrightarrow \sim q$ 9. $p \to (q \lor \sim p)$
10. $(q \land p) \to \sim p$ 11. $(p \lor q) \to (p \land q)$ 12. $p \lor (\sim p \to q)$

If p and q are true and r is false, find the truth value of the following statements.

13. $\sim p \to q$ 14. $\sim q \to \sim r$ 15. $r \to \sim q$
16. $p \to (q \land \sim r)$ 17. $(p \to q) \lor (r \to \sim q)$ 18. $(p \to q) \to (q \to r)$
19. $(p \lor q) \to (\sim q \land r)$ 20. $(p \to \sim q) \to \sim r$ 21. $(p \to q) \to r$

Write the converse, inverse, and contrapositive of each of the following statements.

22. If is smells, then it is spoiled.

23. If I win the lottery, then I'll quit my job.

24. If I get an A, then I'll celebrate.

25. If I get a flu shot, then I will not get the flu.

Construct a truth table for each of the following statements.

26. $p \rightarrow \sim q$

27. $\sim p \rightarrow \sim q$

28. $\sim p \leftrightarrow q$

29. $(p \wedge \sim q) \rightarrow q$

30. $(p \rightarrow q) \rightarrow (p \wedge \sim q)$

31. $(p \vee q) \rightarrow (q \wedge p)$

32. $(p \rightarrow q) \rightarrow r$

33. $(p \rightarrow \sim q) \rightarrow (q \vee r)$

34. $(p \leftrightarrow q) \rightarrow r$

Identify each of the following statements as true or false.

35. If $x + 5 \leq 10$, then $x \leq 5$.

36. If $x < 7$, then $x < 9$.

37. If $2x + 5 = 9$, then $x = 2$.

38. If $x = 4$, then $x + 3 = 8$.

39. $ab = 0$ if and only if $a = 0$.

40. $ab = 0$ if and only if $a = 0$ or $b = 0$.

41. If $x = 2$, then $x^2 = 4$.

42. If $x = -2$, then $x^2 = 4$.

43. $x^2 = 4$ if and only if $x = 2$.

44. If $x = -1$, then $-x = 1$.

45. If $x < -3$, then $-x > 3$.

3.4 Tautologies and Equivalence

Professor: "Give me an example of a paradox."
Student: "A person walks a mile and only moves two feet."

> A statement is called a **tautology** if it is always true no matter what the truth values of its components.

For example, $p \rightarrow p$ and $p \vee \sim p$ are tautologies. Their truth tables are given in Tables 3.20 and 3.21.

Table 3.20

p	$p \rightarrow p$
T	T
F	T

Table 3.21

p	$\sim p$	$p \vee \sim p$
T	F	T
F	T	T

An example of $p \rightarrow p$ is: If it rains, then it rains.
An example of $p \vee \sim p$ is: It rains or it does not rain.

There are many other tautologies; for example, $p \rightarrow p \vee q$ and $p \wedge q \rightarrow p$. Their truth tables are given in Tables 3.22 and 3.23.

Table 3.22

p	q	$p \vee q$	$p \rightarrow p \vee q$
T	T	T	T
T	F	T	T
F	T	T	T
F	F	F	T

Table 3.23

p	q	$p \wedge q$	$p \wedge q \rightarrow p$
T	T	T	T
T	F	F	T
F	T	F	T
F	F	F	T

A statement that is false no matter what the truth values of its components is called an **absurdity**.

The negation of any tautology is an absurdity. An example of an absurdity is $p \wedge \sim p$. The truth table of $p \wedge \sim p$ is given in Table 3.24. An absurd English statement is

It is raining and it is not raining.

Table 3.24

p	$\sim p$	$p \wedge \sim p$
T	F	F
F	T	F

Two statements are **equivalent** if they have the same truth values in every possible situation.

Thus, two statements s and t are equivalent if the last column is the same in both their truth tables. If s and t are equivalent, we write $s \equiv t$. It is clear that the equivalence relationship $\equiv$ has the following three properties:

(1) $s \equiv s$.
(2) If $s \equiv t$, then $t \equiv s$.
(3) If $s \equiv t$ and $t \equiv u$, then $s \equiv u$.

Two examples of obviously equivalent statements are

$$p \vee q \equiv q \vee p, \qquad p \wedge q \equiv q \wedge p$$

Other simple examples of equivalent statements are

$$p \equiv p \vee p, \qquad p \equiv p \wedge p, \qquad \sim(\sim p) \equiv p$$

The truth tables for these statements are given in Tables 3.25(a), (b), and (c).

Table 3.25

p	$p \vee p$
T	T
F	F

(a)

p	$p \wedge p$
T	T
F	F

(b)

p	$\sim p$	$\sim(\sim p)$
T	F	T
F	T	F

(c)

If $s \equiv t$, then s and t are logically the same, so one may be substituted for the other in a statement. We can sometimes use this to simplify statements. For example,

$$\sim(\sim p) \vee q \equiv p \vee q$$

and
$$(p \vee p) \wedge q \equiv p \wedge q$$

The following two equivalences are called **De Morgan's laws**:

$$\sim(p \vee q) \equiv \sim p \wedge \sim q$$
$$\sim(p \wedge q) \equiv \sim p \vee \sim q$$

De Morgan's first law is verified in Tables 3.26 and 3.27 and the verification of the second is similar.

Table 3.26

p	q	$p \vee q$	$\sim(p \vee q)$
T	T	T	F
T	F	T	F
F	T	T	F
F	F	F	T

Table 3.27

p	q	$\sim p$	$\sim q$	$\sim p \wedge \sim q$
T	T	F	F	F
T	F	F	T	F
F	T	T	F	F
F	F	T	T	T

EXAMPLE 1   Write the negation of

I will go to a movie or to a restaurant.

Solution.   The clearest way to state the negation is to use De Morgan's law $\sim(p \vee q) \equiv \sim p \wedge \sim q$. We thus have for the negation

I will not go to a movie and will not go to a restaurant.   □

EXAMPLE 2   Simplify the statement $\sim(\sim p \wedge \sim q)$.
Solution.   By De Morgan's laws,

$$\sim(\sim p \wedge \sim q) \equiv \sim(\sim p) \vee \sim(\sim q) \equiv p \vee q$$ □

Two other important equivalences are the **distributive laws**. These are the following:

$$p \wedge (q \vee r) \equiv (p \wedge q) \vee (p \wedge r)$$
$$p \vee (q \wedge r) \equiv (p \vee q) \wedge (p \vee r)$$

We shall verify the first distributive law in Tables 3.28 and 3.29. You will verify the second in an exercise.

Table 3.28

p	q	r	$q \vee r$	$p \wedge (q \vee r)$
T	T	T	T	T
T	T	F	T	T
T	F	T	T	T
T	F	F	F	F
F	T	T	T	F
F	T	F	T	F
F	F	T	T	F
F	F	F	F	F

Table 3.29

p	q	r	$p \wedge q$	$p \wedge r$	$(p \wedge q) \vee (p \wedge r)$
T	T	T	T	T	T
T	T	F	T	F	T
T	F	T	F	T	T
T	F	F	F	F	F
F	T	T	F	F	F
F	T	F	F	F	F
F	F	T	F	F	F
F	F	F	F	F	F

We can also obtain an equivalent form for the conditional $p \rightarrow q$. This is given by

$$p \rightarrow q \equiv \sim p \vee q \tag{3.1}$$

To verify (3.1), compare the last columns of Tables 3.30 and 3.31.

EXAMPLE 3   Write the following conditional without using "if...then."

If given the opportunity, then he will succeed.

Table 3.30

p	q	$p \to q$
T	T	T
T	F	F
F	T	T
F	F	T

Table 3.31

p	q	$\sim p$	$\sim p \lor q$
T	T	F	T
T	F	F	F
F	T	T	T
F	F	T	T

Solution.   Since $p \to q \equiv \sim p \lor q$, we can restate the conditional as

He will not be given the opportunity or he will succeed.   □

In Section 3.3 we defined the converse, inverse, and contrapositive of the conditional $p \to q$. If you examine Table 3.19 you can see that

$$p \to q \equiv \sim q \to \sim p$$
$$q \to p \equiv \sim p \to \sim q$$

In other words, an implication and its contrapositive are equivalent. Moreover, the converse and the inverse are equivalent.

EXAMPLE 4   The converse of the statement

(a) If you brush your teeth, then you will not get cavities,

is the statement

(b) If you do not get cavities, then you brush your teeth.

Write an equivalent "if…then" statement for statements (a) and (b).
  Solution.   The contrapositive of (a) is equivalent to (a) and is written

If you get cavities, then you do not brush your teeth.

The inverse of (a) is equivalent to (b) and is written

If you do not brush your teeth, then you get cavities.   □

It is important to note that while an implication and its contrapositive are equivalent, an implication and its converse (or inverse) are not equivalent. Indeed, if you examine Table 3.19, you will see that $p \to q$ and $q \to p$ have different truth values.

EXAMPLE 5   Show that the following statement is true but its converse is false. If $x = 1$, then $x^2 = 1$.
  Solution.   If $x = 1$, then $x^2 = 1^2 = 1$. However, if $x^2 = 1$, then $x = 1$ or $x = -1$, so x need not be 1.   □

Let T represent a tautology and F an absurdity. We use the letter T because a tautology is always true and the letter F because an absurdity is always false. We then have, for example, $p \vee \sim p \equiv$ T and $p \wedge \sim p \equiv$ F. Note that

$$p \vee \text{T} \equiv \text{T}, \qquad p \wedge \text{T} \equiv p, \qquad p \vee \text{F} \equiv p, \qquad p \wedge \text{F} \equiv \text{F}$$

These can be seen from the truth tables in Table 3.32.

Table 3.32

p	T	F	$p \vee \text{T}$	$p \wedge \text{T}$	$p \vee \text{F}$	$p \wedge \text{F}$
T	T	F	T	T	T	F
F	T	F	T	F	F	F

We can sometimes use T and F to simplify a statement by finding a simpler equivalent statement.

EXAMPLE 6   Show that $(p \vee q) \wedge (p \vee \sim q) \equiv p$.

Solution.   Applying the distributive law gives

$$(p \vee q) \wedge (p \vee \sim q) \equiv p \vee (q \wedge \sim q) = p \vee \text{F} = p$$

Of course, this can also be shown using a truth table.   □

EXAMPLE 7   Show that $p \leftrightarrow q \equiv (\sim p \wedge \sim q) \vee (q \wedge p)$.

Solution.   Applying the distributive law and (3.1) gives

$$
\begin{aligned}
p \leftrightarrow q = (p \rightarrow q) \wedge (q \rightarrow p) &\equiv (\sim p \vee q) \wedge (\sim q \vee p) \\
&\equiv [(\sim p \vee q) \wedge \sim q] \vee [(\sim p \vee q) \wedge p] \\
&\equiv [(\sim p \wedge \sim q) \vee (q \wedge \sim q)] \vee [(\sim p \wedge p) \vee (q \wedge p)] \\
&\equiv [(\sim p \wedge \sim q) \vee \text{F}] \vee [\text{F} \vee (q \wedge p)] \\
&\equiv (\sim p \wedge \sim q) \vee (q \wedge p)
\end{aligned}
$$

Again, this can also be shown using a truth table.   □

You may have noticed the similarity between De Morgan's laws and the distributive laws for statements and for sets. The reason for this is that there is a close relationship between statements and sets. This relationship has already been studied in Section 3.1.

EXERCISES

Write the negation of each statement.

1. This bed is too hard and this bed is too soft.
2. Five foot two and eyes of blue.
3. If you had told me, then I would have known.
4. It's raining cats and dogs.
5. I can take it or leave it.
6. It's neither here nor there.
7. If we are here, then we are not all there.

Use truth tables to show that the following are tautologies.

8. $\sim p \rightarrow (\sim p \vee q)$
9. $p \rightarrow (p \vee \sim q)$
10. $(p \wedge \sim q) \rightarrow p$
11. $(\sim p \wedge q) \rightarrow \sim p$
12. $(p \wedge q) \rightarrow (p \vee q)$
13. $(\sim p \wedge \sim q) \rightarrow (\sim p \vee \sim q)$

Use truth tables to verify the following equivalences.

14. $\sim(p \wedge q) \equiv \sim p \vee \sim q$
15. $\sim(\sim p \wedge q) \equiv p \vee \sim q$
16. $\sim(\sim p \vee \sim q) \equiv p \wedge q$
17. $\sim(p \vee \sim q) \equiv \sim p \wedge q$
18. $p \vee (q \wedge r) \equiv (p \vee q) \wedge (p \vee r)$
19. $p \wedge (\sim q \vee r) \equiv (p \wedge \sim q) \vee (p \wedge r)$
20. $p \rightarrow \sim q \equiv \sim p \vee \sim q$
21. $\sim p \rightarrow q \equiv p \vee q$
22. $p \equiv p \wedge (p \vee q)$
23. $p \equiv p \vee (p \wedge q)$
24. $p \wedge (p \vee q) \equiv p \vee (p \wedge q)$
25. $p \leftrightarrow q \equiv (\sim p \wedge \sim q) \vee (q \wedge p)$
26. $p \equiv (p \wedge q) \vee (p \wedge \sim q)$
27. $p \equiv (p \vee q) \wedge (p \vee \sim q)$

Verify the following without using truth tables.

28. $p \rightarrow q \equiv \sim q \rightarrow \sim p$
29. $q \rightarrow p \equiv \sim p \rightarrow \sim q$
30. $\sim(\sim p \wedge q) \equiv p \vee \sim q$
31. $\sim(\sim p \vee \sim q) \equiv p \wedge q$
32. $\sim p \rightarrow q \equiv p \vee q$
*33. $p \equiv (p \wedge q) \vee (p \wedge \sim q)$
*34. $p \equiv (p \vee q) \wedge (p \vee \sim q)$
35. $p \vee \sim p \equiv q \vee \sim q$

Show that the statements in Exercises 36–40 are true but their converses are false.

36. If $a = 0$, then $ab = 0$.
37. If $x = 2$, then $x^2 = 4$.
38. If $x < 2$, then $x < 3$.
39. If $x = 1$, then $x < 2$.
40. If $x = 2$, then $x + 3 \leq 5$.

3.5 Logic and Circuits (Optional)

> *God is a child; and when he began to play, he cultivated mathematics.*
> *It is the most godly of man's games.*
> — V. ERATH, Twentieth-Century German Novelist

Symbolic logic can be applied as an aid to the design of electric switching circuits. Since such circuits are an integral component of high-speed electronic computers, the simplification and efficient design of their circuitry can lead to great economical savings as well as decreased size. This application of symbolic logic was originated by Claude Shannon in 1937 and he also applied it to work on telephone relays at the Bell Laboratories.

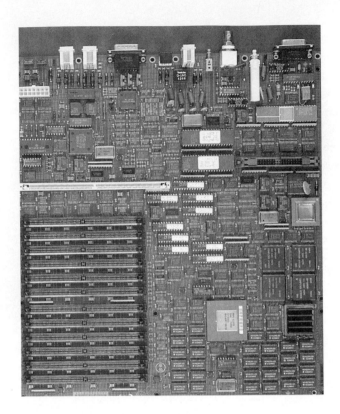

Photo 3.1
Symbolic logic is used to
design electronic circuits.

Switches

A rough illustration of an electric switch is shown in Figure 3.1, where a portion of an electric circuit is exhibited. An electric current enters at point S and leaves at point T. If the switch p is on (closed), current will flow from S to T and if the switch p is off (open), current will not flow.

In modern electronic circuits, instead of a mechanical switch we have much faster and more efficient electronic switches constructed from a solid-state device such as a transistor. An electronic switch has its own circuitry which signals it to be on or off (Figure 3.2a). In the sequel, we denote a switch as illustrated in Figure 3.2b.

Coupled Switches

We can also have **coupled** switches as illustrated in Figure 3.3a. Such switches are either on or off in unison. In the mechanical case, they can be thought of as being controlled by the same handle; in the electronic case, as being signaled to be simultaneously on or off. Coupled switches are labeled by the same letter. Another type of coupled switches are **complementary** switches as shown in Figure 3.3b. The **complement** $\sim p$ of a switch p is on when p is off and vice versa.

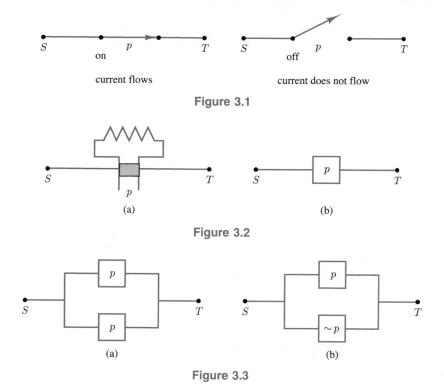

current flows   current does not flow

Figure 3.1

Figure 3.2

Figure 3.3

Series

Figure 3.4 shows two switches connected in **series**. In such a circuit, current will flow from S to T only when both switches are on. Note the analogy between a series circuit and the logical conjunction $p \wedge q$. We know that $p \wedge q$ is true only when both p and q are true. For this reason we use the notation $p \wedge q$ for two switches in series.

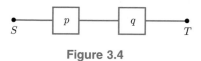

Figure 3.4

Parallel

A circuit corresponding to the disjunction $p \vee q$ is a **parallel** circuit illustrated in Figure 3.5. In this case, current flows from S to T if either p or q is on or if both p and q are on. We again use the notation $p \vee q$ for two switches in parallel.

Just as we had a truth table for a logical statement, we now construct a **current table** for a switching circuit. Instead of truth values T and F, we assign **current values** 1 (on) and 0 (off). The four possible 1–0 (on–off) arrangements of the switches p and q and the resulting current flow from S to T for the series circuit $p \wedge q$ and the parallel circuit $p \vee q$ are given in Table 3.33. The relationship between a switch p and its complement $\sim p$ is shown in Table 3.34.

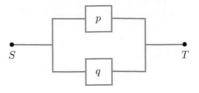

Figure 3.5

Table 3.33

p	q	$p \wedge q$	$p \vee q$
1 (on)	1 (on)	1 (current)	1 (current)
1 (on)	0 (off)	0 (no current)	1 (current)
0 (off)	1 (on)	0 (no current)	1 (current)
0 (off)	0 (off)	0 (no current)	0 (no current)

Table 3.34

p	$\sim p$
1 (on)	0 (off)
0 (off)	1 (on)

If the 1 and 0 were replaced by T and F, respectively, then Tables 3.33 and 3.34 would be the basic truth tables. Therefore, we can apply the laws of symbolic logic to solve switching circuit problems. In particular, we can use symbolic logic to design and simplify switching circuits. In analogy with equivalent statements, if two current tables give the same current values, then the corresponding circuits result in the same current flow from S to T. Moreover, a **tautology**, which we now denote by I, is a circuit in which current always flows from S to T and an **absurdity**, denoted by 0, is a circuit in which current never flows from S to T.

As a simple example, consider the circuits in Figure 3.3. The circuit in Figure 3.3a has the form $p \vee p$. Since $p \vee p \equiv p$, we can replace Figure 3.3a by the circuit in Figure 3.6a. The circuit in Figure 3.3b has the form $p \vee \sim p$. Since $p \vee \sim p \equiv I$, we can replace Figure 3.3b by the circuit in Figure 3.6b.

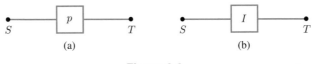

Figure 3.6

EXAMPLE 1   Simplify the circuit in Figure 3.7.

Solution.   This circuit can be written as $(p \wedge q) \vee (p \wedge r)$. By the distributive law we have

$$(p \wedge q) \vee (p \wedge r) \equiv p \wedge (q \vee r)$$

which has the circuit of Figure 3.8.   □

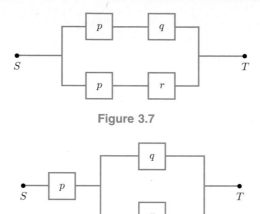

Figure 3.7

Figure 3.8

EXAMPLE 2   Construct a diagram for the switching circuit

$$[p \wedge (q \vee r)] \vee (\sim p \wedge q)$$

Solution.   We construct circuits for $p \wedge (q \vee r)$ and $\sim p \wedge q$ and then put them in a parallel arrangement. The circuit for $p \wedge (q \vee r)$ is shown in Figure 3.9a and the circuit for $\sim p \wedge q$ is shown in Figure 3.9b. Placing these two circuits in parallel, we obtain Figure 3.10.   □

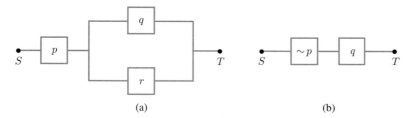

(a)   (b)

Figure 3.9

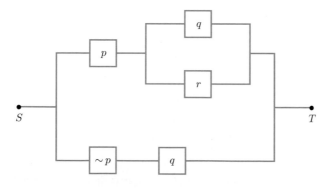

Figure 3.10

EXAMPLE 3   Construct a diagram for the switching circuit $p \leftrightarrow q$.

    Solution.   Since $p \leftrightarrow q \equiv (\sim p \vee q) \wedge (\sim q \vee p)$, we obtain the circuit in Figure 3.11.   □

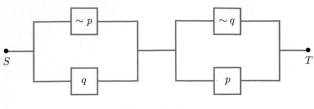

Figure 3.11

EXAMPLE 4   Write a symbolic expression for the circuit in Figure 3.12.

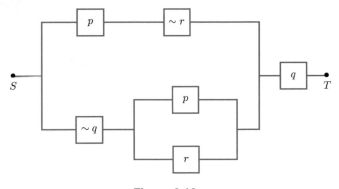

Figure 3.12

    Solution.   We have $p \wedge \sim r$ in parallel with $\sim q \wedge (p \vee r)$, giving

$$(p \wedge \sim r) \vee [\sim q \wedge (p \vee r)]$$

This circuit is in series with q, so we obtain

$$q \wedge \{(p \wedge \sim r) \vee [\sim q \wedge (p \vee r)]\}$$ □

EXAMPLE 5   What switches must be on for current to flow through the circuit in Figure 3.13?

    Solution.   This circuit has the symbolic expression $(p \vee \sim q) \wedge (q \vee r)$. The current table is given in Table 3.35. We see that current will flow from S to T when switches p and q are on, or when p and r are on, or when p and q are off and r is on.   □

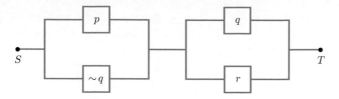

Figure 3.13

Table 3.35

p	q	r	$\sim q$	$p \vee \sim q$	$q \vee r$	$(p \vee \sim q) \wedge (q \vee r)$
1	1	1	0	1	1	1
1	1	0	0	1	1	1
1	0	1	1	1	1	1
1	0	0	1	1	0	0
0	1	1	0	0	1	0
0	1	0	0	0	1	0
0	0	1	1	1	1	1
0	0	0	1	1	0	0

EXAMPLE 6   Find a simpler circuit to replace the circuit in Figure 3.14.

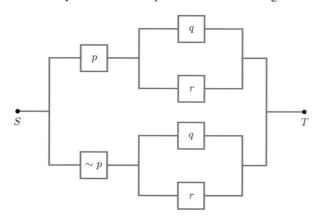

Figure 3.14

Solution.   This circuit has the logical expression

$$[p \wedge (q \vee r)] \vee [\sim p \wedge (q \vee r)]$$

By the distributive law we have

$$[p \wedge (q \vee r)] \vee [\sim p \wedge (q \vee r)] \equiv (p \vee \sim p) \wedge (q \vee r)$$
$$\equiv I \wedge (q \vee r) = q \vee r$$

Hence, we can replace the given circuit by Figure 3.15. $\square$

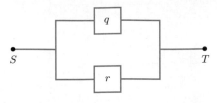

Figure 3.15

EXERCISES

1. We can apply the methods of this section to water-pipe systems and railroad switching systems. Why is this?

2. If a series circuit contains coupled switches p and p, why can one of these switches be removed? *Because p and p operate on same switch*

3. If a series circuit contains complementary switches p and $\sim p$, why will current never flow?

4. Construct the current table for the circuit in (a) Figure 3.8, (b) Figure 3.10, (c) Figure 3.11.

5. What switches must be on for current to flow through the circuit $p \wedge q \wedge r \wedge s$?

6. What switches must be on for current to flow through the following circuit?

$$(p \wedge q \wedge r \wedge s) \vee (\sim p \wedge q \wedge r \wedge s)?$$

7. Simplify the circuit $(p \wedge \sim p \wedge q \wedge r \wedge s) \vee t$.

8. Simplify the circuit $(p \wedge p \wedge p \wedge q \wedge r \wedge s) \vee t$.

Draw circuits representing each of the following statements.

9. $p \wedge (\sim p \vee q)$ 10. $(p \wedge q) \vee (\sim p \wedge \sim q)$

11. $(p \vee q) \wedge (\sim p \vee \sim q)$ 12. $[(p \wedge q) \vee r] \wedge \sim p$

13. $p \rightarrow (q \wedge r)$ 14. $p \rightarrow (q \rightarrow r)$

Write logical statements representing each of the following circuits. Simplify each circuit when possible.

15.

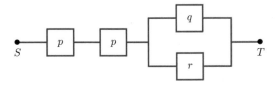

16.

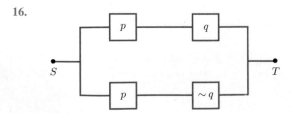

17.

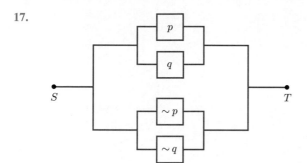

18.

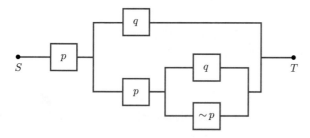

19.

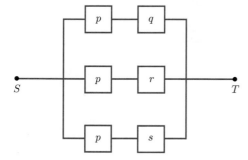

3.6 Validity

Marriage is an institution.
Marriage is love.
Love is blind.
Therefore, marriage is an institution for the blind.
— AUTHOR UNKNOWN

Most of human knowledge has come in the course of trying to prove
false hypotheses. — JOHN FOWLER, Twentieth-Century Author

Until now we were concerned with discovering the truth values of individual statements and with determining whether two statements are equivalent. We now study
sequences of statements called arguments and we shall be concerned with their
validity.

Photo 3.2
Is that really a valid argument?

An **argument** is a sequence of statements called **premises** and a final statement called the **conclusion**. The conclusion is said to follow from the premises. An argument is **valid** if whenever the premises are true, the conclusion is also true. In some arguments, true premises will lead to a conclusion that is false or is sometimes true and sometimes false. In this case, the argument is **invalid**. An invalid argument is also called a **fallacy**.

In speaking about the validity of an argument we are not concerned with the actual truth or falsity of the premises. In many cases, we are not able to determine these. We are only concerned with the structure of the argument itself.

EXAMPLE 1   In the following argument the first two statements are the premises and the last statement is the conclusion.

If you work hard, then you succeed.
You do work hard.
Therefore, you succeed.

We do not know for sure if the first premise is true. Sometimes it is and sometimes it is not. Nevertheless, whenever the premises are true the conclusion is also true. Therefore, this is a valid argument. This argument method is called *modus ponens* (or **direct reasoning**).

EXAMPLE 2   We shall later show that the following argument is invalid. This is called the **fallacy of the inverse**.

If you work hard, then you succeed.
You do not work hard.
Therefore, you do not succeed.

EXAMPLE 3 The next argument method is called *modus tollens* (or the **method of contraposition** or **indirect reasoning**). We shall show later that this is a valid argument.

If you work hard, then you succeed.
You do not succeed.
Therefore, you do not work hard.

EXAMPLE 4 The following invalid argument is called the **fallacy of the converse**.

If you cannot stand the heat, then you get out of the kitchen.
You get out of the kitchen.
Therefore, you cannot stand the heat.

EXAMPLE 5 The next argument is valid and is called the **disjunctive syllogism**.

You can stand the heat or you get out of the kitchen.
You cannot stand the heat.
Therefore, you get out of the kitchen.

To determine whether an argument is valid or invalid, it is best that we do not examine it in its original form because we can be deceived by the wording of the statements. To avoid this danger, we represent an argument symbolically. In symbolic form, we first list the premises, then draw a line and write the conclusion below. Examples 1 through 5 have the following forms:

(1) *Modus ponens*

$$p \rightarrow q$$

$$\underline{p}$$

$$q$$

(2) *Fallacy of the inverse*

$$p \rightarrow q$$

$$\underline{\sim p}$$

$$\sim q$$

(3) *Modus tollens*

$$p \rightarrow q$$

$$\underline{\sim q}$$

$$\sim p$$

(4) *Fallacy of the converse*

$$p \rightarrow q$$

$$\underline{q}$$

$$p$$

(5) *Disjunctive syllogism*

$$p \vee q$$

$$\underline{\sim p}$$

$$q$$

Table 3.36 Modus Ponens *Direct reasoning*

		Premises		Conclusion
p	q	$p \rightarrow q$	p	q
T	T	T	T	T
T	F	F	T	F
F	T	T	F	T
F	F	T	F	F

To prove the validity or invalidity of an argument, we first construct a truth table that contains the truth values of the premises and conclusion. If the conclusion is true whenever the premises are all true, then the argument is valid. Otherwise the argument is invalid. Table 3.36 shows the truth table for *modus ponens*. The first row of Table 3.36 is the only row in which both premises are true. Since the conclusion is also true, the argument is valid.

Table 3.37 shows the truth table for the fallacy of the inverse. In the last two rows of Table 3.37, both premises are true. However, in the third row the conclusion is false. Hence, the argument is invalid.

Table 3.37 Fallacy of the Inverse

movie popcorn problem

		Premises		Conclusion
p	q	$p \rightarrow q$	$\sim p$	$\sim q$
T	T	T	F	F
T	F	F	F	T
F	T	T	T	F
F	F	T	T	T

That modus tollens is valid, the fallacy of the converse is invalid, and disjunctive syllogism is valid are shown in Tables 3.38, 3.39, and 3.40.

indirect reasoning

Table 3.38 Modus Tollens

		Premises		Conclusion
p	q	$p \rightarrow q$	$\sim q$	$\sim p$
T	T	T	F	F
T	F	F	T	F
F	T	T	F	T
F	F	T	T	T

Table 3.39 Fallacy of the Converse

		Premises		Conclusion
p	q	$p \rightarrow q$	q	p
T	T	T	T	T
T	F	F	F	T
F	T	T	T	F
F	F	T	F	F

Table 3.40 Disjunctive Syllogism

		Premises		Conclusion
p	q	$p \vee q$	$\sim p$	q
T	T	T	F	T
T	F	T	F	F
F	T	T	T	T
F	F	F	T	F

EXAMPLE 6 Is the following argument valid? I will watch television or I will read a book. If I read a book, then I will fall asleep. I will not fall asleep. Therefore, I will watch television.

Solution. We first replace the component statements by letters:

p = I will watch television.
q = I will read a book.
r = I will fall asleep.

We next write the argument symbolically:

$$p \vee q$$
$$q \rightarrow r$$
$$\underline{\sim r}$$
$$p$$

We finally form the truth table as in Table 3.41. The fourth row of Table 3.41 is the only row in which all three premises are true. Since the conclusion is also true, the argument is valid. □

Reasoning by Transitivity

Another common argument method is called **reasoning by transitivity** or the **hypothetical syllogism**. Such an argument has the form

$$p \rightarrow q$$
$$\underline{q \rightarrow r}$$
$$p \rightarrow r$$

Table 3.41

p	q	r	Premises			Conclusion
			$p \lor q$	$q \rightarrow r$	$\sim r$	p
T	T	T	T	T	F	T
T	T	F	T	F	T	T
T	F	T	T	T	F	T
T	F	F	T	T	T	T
F	T	T	T	T	F	F
F	T	F	T	F	T	F
F	F	T	F	T	F	F
F	F	F	F	T	T	F

Table 3.42

p	q	r	Premises		Conclusion
			$p \rightarrow q$	$q \rightarrow r$	$p \rightarrow r$
T	T	T	T	T	T
T	T	F	T	F	F
T	F	T	F	T	T
T	F	F	F	T	F
F	T	T	T	T	T
F	T	F	T	F	T
F	F	T	T	T	T
F	F	F	T	T	T

To show that this argument is valid, consider Table 3.42. We see from Table 3.42 that the conclusion is true whenever the premises are true.

Reasoning by transitivity applies for any number of conditionals. For example, the following argument is valid:

$$p \rightarrow q$$
$$q \rightarrow r$$
$$\underline{r \rightarrow s}$$
$$p \rightarrow s$$

Why is that? Well, all we have to show is that the conclusion is true whenever all the premises are true. Hence, assume that $p \rightarrow q$, $q \rightarrow r$, and $r \rightarrow s$ are all true. Then, reasoning by transitivity, we conclude that $p \rightarrow r$ is true. Reasoning by transitivity again for $p \rightarrow r$ and $r \rightarrow s$, we finally conclude that $p \rightarrow s$ is true.

EXAMPLE 7   Is the following argument, due to Lewis Carroll, valid?

Babies are illogical.
Nobody is despised who can manage a crocodile.
Illogical persons are despised.
Therefore babies cannot manage a crocodile.

Solution.   We first rewrite the statements as conditionals.

If you are a baby, then you are illogical.
If you can manage a crocodile, then you are not despised.
If you are illogical, then you are despised.
Therefore, if you are a baby, then you cannot manage a crocodile.

We next represent the component statements by letters:

p = You are a baby.
q = You are illogical.
r = You can manage a crocodile.
s = You are despised.

The symbolic form of the argument becomes

$$p \rightarrow q$$
$$r \rightarrow \sim s$$
$$\underline{q \rightarrow s}$$
$$p \rightarrow \sim r$$

Now we can rearrange the premises in a different order and replace the conditional $r \rightarrow \sim s$ by its equivalent contrapositive $s \rightarrow \sim r$. We now rewrite the argument as

$$p \rightarrow q$$
$$q \rightarrow s$$
$$\underline{s \rightarrow \sim r}$$
$$p \rightarrow \sim r$$

Reasoning by transitivity, we conclude that the argument is valid. We could also prove this by constructing a truth table, but this would be a lot of work because we would need 16 rows.   □

EXERCISES

Give an example, other than that in the text, of each of the following arguments.
1. Modus ponens.
2. Fallacy of the inverse.
3. Modus tollens.
4. Fallacy of the converse.
5. Disjunctive syllogism.
6. Reasoning by transitivity.

Use the truth tables to determine the validity of each of the following arguments.

7. $p \vee q$
   $\underline{\sim q}$
   p

8. $p \wedge q$
   $\underline{p}$
   q

9. $p \to \sim q$
   $\underline{q}$
   $\sim p$

10. $p \to \sim q$
    $\underline{\sim q}$
    p

11. $p \to q$
    $\underline{q \to p}$
    $\sim p \wedge \sim q$

12. $p \vee \sim q$
    $\underline{q}$
    p

13. $p \to q$
    $q \to r$
    $\underline{}$
    $\sim p \vee r$

14. $(p \vee q) \wedge (p \to q)$
    $\underline{\sim p}$
    $\sim q$

*15. $(p \wedge r) \to (\sim p \vee q)$
    $\underline{(\sim r \to \sim p)}$
    $q \to r$

*16. $p \to \sim q$
    $q \to r$
    $\underline{\sim r}$
    $\sim p$

Determine the validity of each of the arguments in Exercises 17–25. If the argument is valid, indicate, if possible, which of our valid argument forms it follows. If not, name the fallacy it exhibits, if possible.

17. If you think, then you exist. You think. Therefore, you exist.

18. All men are mortal. Socrates is a man. Therefore, Socrates is mortal.

19. Stoplights are red or green. This stoplight is not red. Therefore, this stoplight is green.

20. Stoplights are red, yellow, or green. This stoplight is not red. Therefore, this stoplight is green.

21. If I go to a movie, then I buy popcorn. I do not buy popcorn. Therefore, I do not go to a movie.

22. If I go to a movie, then I buy popcorn. I do not go to a movie. Therefore, I do not buy popcorn.

23. If I go to a movie, then I buy popcorn. If I buy popcorn, then I get thirsty. Therefore, if I go to a movie, then I get thirsty.

24. If Joan is promoted, then John is not promoted. If John is promoted, then George will quit. George will not quit. Therefore, Joan is not promoted.

25. If I work this problem, then I pass this course. If I pass this course, then I graduate. If I graduate, then I get a job. Therefore, if I work this problem, then I get a job.

**26. Suppose an argument has premises $p_1, p_2, \ldots, p_n$ and conclusion q. Show that the argument

   p_1
   p_2
   $\vdots$
   $\underline{p_n}$
   q

is valid if and only if $(p_1 \wedge p_2 \wedge \cdots \wedge p_n) \to q$ is a tautology.

3.7 | Mathematical Proofs

Mathematical proofs, like diamonds, are hard as well as clear, and will be touched with nothing but strict reasoning.
— John Locke (1632–1704), English Philosopher

Reductio ad absurdum is one of a mathematician's finest weapons. It is a far finer gambit than any chess gambit: a chess player may offer the sacrifice of a pawn or even a piece, but the mathematician offers the game.
— G. H. Hardy, Twentieth-Century English Mathematician

What is now proved was once only imagined.
— William Blake (1757–1827), British Poet and Artist

> There are two main types of reasoning,
> **inductive reasoning** and **deductive reasoning**.

In inductive reasoning we examine many specific examples, notice that they share some common property, and then conjecture that this property always holds (or always holds under certain restrictions). Inductive reasoning is the type of reasoning that is used in the physical sciences. We can never prove a law of nature but can only test it in many cases. The more cases for which it holds, the more confident we become that the law is "true." Of course, if one case is found for which it does not hold, the law must be either discarded or modified.

Inductive Reasoning

Inductive reasoning is also used in mathematics. For example, we have noticed that

$$1 + 2 + 3 + \cdots + 100 = (50)(101)$$

and also that

$$1 + 2 + 3 + \cdots + 1000 = (500)(1001)$$

and so forth. Using inductive reasoning, we conjecture that

$$1 + 2 + 3 + \cdots + n = \frac{n(n+1)}{2} \tag{3.2}$$

for any natural number n. We have not proved our conjecture in this way; inductive reasoning cannot be used for proofs. However, inductive reasoning was very valuable in arriving at the conjecture. To actually prove that the conjecture is true we must use deductive reasoning. In deductive reasoning we begin with certain axioms or postulates and then proceed, using logical steps until the result is proved. For example, in Section 1.2 we proved that (3.2) holds by using deductive reasoning.

HERMAN®

"I've just proved I don't exist."

Deductive Reasoning

Equation (3.2) gives a formula for the sum of the first n natural numbers. What about the sum of the first n even natural numbers? Recall that a natural number is **even** if it is a multiple of 2. Thus the first few even natural numbers are $2 = 2(1)$, $4 = 2(2)$, $6 = 2(3)$, $8 = 2(4), \ldots$. Let's look at the sums of the first few even natural numbers:

$$2 + 4 = 6 = 2(3)$$
$$2 + 4 + 6 = 12 = 3(4)$$
$$2 + 4 + 6 + 8 = 20 = 4(5)$$
$$2 + 4 + 6 + 8 + 10 = 30 = 5(6)$$

Do you notice a pattern? Of course, you do! Inductive reasoning leads to the conjecture that the sum of the first n even natural numbers is $n(n + 1)$. That is,

$$2 + 4 + 6 + \cdots + 2n = n(n + 1)$$

Can you prove this? The proof goes as follows. Using (3.2) we have

$$2 + 4 + 6 + \cdots + 2n = 2(1 + 2 + 3 + \cdots + n) = \frac{2n(n + 1)}{2}$$
$$= n(n + 1) \tag{3.3}$$

What about the sum of the first n odd natural numbers? A natural number is **odd** if it has the form $2n - 1$ for some natural number n. Thus, $1 = 2(1) - 1$, $3 = 2(2) - 1$,

$5 = 2(3) - 1$, $7 = 2(4) - 1$, are all odd. The sums of the first few odd natural numbers look as follows:

$$1 + 3 = 4 = 2^2$$
$$1 + 3 + 5 = 9 = 3^2$$
$$1 + 3 + 5 + 7 = 16 = 4^2$$
$$1 + 3 + 5 + 7 + 9 = 25 = 5^2$$

This pattern leads us to the conjecture that the sum of the first n odd natural numbers is n^2. How do we prove this? Well, let T be the sum of the first $2n$ natural numbers, let E be the sum of the first n even natural numbers, and let S be the sum of the first n odd natural numbers. We then have

$$T = 1 + 2 + 3 + 4 + 5 + 6 + \cdots + 2n$$
$$= 1 + 3 + 5 + \cdots + (2n - 1) + 2 + 4 + 6 + \cdots + 2n = S + E$$

Hence, $S = T - E$. But we can use (3.2) to find T and (3.3) to find E. We then have

$$S = T - E = \frac{2n(2n + 1)}{2} - n(n + 1) = n(2n + 1) - n(n + 1)$$
$$= 2n^2 + n - n^2 - n = n^2$$

As another example, notice that $2^2 + 3^2 \geq (2)(2)(3)$, $3^2 + 4^2 \geq (2)(3)(4)$, and $2^2 + 5^2 \geq (2)(2)(5)$. Try some of your own sample cases. Do you see some common property?

After trying many cases you may be willing to use inductive reasoning and conjecture that $a^2 + b^2 \geq 2ab$ for any two numbers a and b. Inductive reasoning has enabled you to make an interesting conjecture. You cannot prove this conjecture using inductive reasoning because you cannot verify every case individually. To prove the conjecture you must use deductive reasoning. The deductive proof goes as follows.

$$0 \leq (a - b)^2 = a^2 - 2ab + b^2$$

and so
$$a^2 + b^2 \geq 2ab$$

Method of Contradiction

In mathematics an important method of deductive reasoning is the **method of contradiction** or *reductio ad absurdum* (reduction to absurdity). This method is closely related to the modus tollens argument of the previous section and is based upon the assumption that mathematics is self-consistent. That is, mathematical systems are so constructed that there are no contradictions.

Thus if we start with a true statement and reason in a logical fashion, we cannot come to a contradiction. The method of contradiction proceeds as follows. Suppose we want to prove that statement p is true. Assume p is false. Reason to a contradiction. Since there are no contradictions in mathematics, the original assumption (that p is false) is false. Hence p is true.

Let us illustrate this method by a nonmathematical example. We will prove that you exist. Suppose you do not exist. Then you cannot be reading this book. But you are reading this book. Contradiction! Therefore, you exist.

EXAMPLE 1   Prove that there is no largest natural number.

Solution.   Assume there is a largest natural number N. Now $N + 1$ is a natural number. But $N + 1 > N$, which contradicts the fact that N is the largest. Since this is a contradiction, the original assumption must be false. Hence there is no largest natural number.   □

EXAMPLE 2   If the product of two natural numbers is greater than 64, prove that at least one of the numbers is greater than 8.

Solution.   Assume that both of the numbers are less than or equal to 8. Then their product is less than or equal to 64. This contradicts the fact that their product is greater than 64. Hence the assumption was false, and so at least one of the numbers is greater than 8.   □

EXAMPLE 3   The following is an ancient riddle. An aging rich man has three sons, Tom, Dick, and Harry. He wants to leave his fortune to the smartest of the three and decides to give them a test. He draws a colored circle, either white or blue, on the forehead of each son. Each son cannot see the color of his own circle but can see the circles of his brothers. He then gives his sons the following instructions. "If you see two blue circles you must raise your hand and the first to shout out his color wins." No one raises his hand and there is silence for a long time. Finally, Harry shouts out, "I'm white." How did he know?

Solution.   Applying the method of contradiction, Harry reasoned as follows. Suppose I'm blue. Then Tom would know he is white, for if he were blue then Dick would see two blues and raise his hand. But then Tom would have proclaimed he was white. Since he did not, I cannot be blue. Hence I must be white.   □

Symbolic logic can be used as an aid to understanding mathematical proofs. In a mathematical proof we usually try to prove a statement or an implication (conditional). For instance, in Example 1, we proved the statement "there is no largest natural number." In Example 2, we proved the implication "if the product of two natural numbers is greater than 64, then at least one of the numbers is greater than 8." If the implication $p \rightarrow q$ is true, that is, if q is true whenever p is true, we say that p **implies** q or that $p \rightarrow q$ **holds**. For example, let p be the statement "n is an even natural number." This is an open statement which is true or false depending on the value of n. Let q be the statement "n has the form $n = 2k$, where k is a natural number." Again, q can be true or false, depending on n. In this case, q is true whenever p is true so $p \rightarrow q$ holds. It also happens that $q \rightarrow p$ holds. We then write $p \leftrightarrow q$ and call p and q **equivalent** statements. Moreover, we say that p is true if and only if q is true.

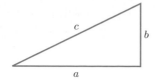

Figure 3.16

EXERCISES

1. What is the difference between inductive and deductive reasoning?

2. Compute the sum $2 + 4 + 6 + \cdots + 20$.

3. Compute the sum $1 + 3 + 5 + \cdots + 21$.

4. Compute the sum $3 + 6 + 9 + \cdots + 3n$.

5. Prove that there is no largest even number.

6. A natural number is a **square number** if it has the form n^2 for some $n \in \mathbb{N}$. Prove that there is no largest square number.

7. If the product of two numbers is zero, prove that one of the numbers must be zero.

8. Does the converse of the implication in Example 2 hold?

9. If the sum of two numbers is less than 100, prove that one of the numbers is less than 50.

10. What is the contrapositive of the implication in Exercise 9? Does the contrapositive hold?

11. What are the inverse and converse of the implication in Exercise 9? Do they hold?

12. Prove that there is no smallest positive rational number. (These are the numbers of the form m/n, $m, n \in \mathbb{N}$.)

13. Pythagoras' theorem states that if a and b are the lengths of the legs of a right triangle and c is the length of the hypotenuse, then $a^2 + b^2 = c^2$ (Figure 3.16). What is the converse of this theorem? Do you think the converse is true?

14. What is the contrapositive of Pythagoras' theorem (Exercise 13)? Is the contrapositive true?

15. Let p, q, and r be statements. In order to prove that these statements are mutually equivalent, show that it is enough to prove that $p \rightarrow q$, $q \rightarrow r$, and $r \rightarrow p$ hold.

16. Prove the following by the method of contradiction. Let n be a two-digit natural number. If $5n$ has two digits, then the first digit on the left of n is 1.

17. Prove the converse of the implication in Exercise 16.

18. Observe that
$$\frac{1}{2} - \frac{1}{3} = \frac{1}{6}, \qquad \frac{1}{3} - \frac{1}{4} = \frac{1}{12}, \qquad \frac{1}{4} - \frac{1}{5} = \frac{1}{20}$$

Try some other cases. Use inductive reasoning to make a conjecture about $1/(n-1) - 1/n$. Prove your conjecture.

19. Notice that
$$1^2 + 2^2 = \frac{(2)(3)(5)}{6}$$
$$1^2 + 2^2 + 3^2 = \frac{(3)(4)(7)}{6}$$
$$1^2 + 2^2 + 3^2 + 4^2 = \frac{(4)(5)(9)}{6}$$

Try some other cases. Use inductive reasoning to make a conjecture about
$$1^2 + 2^2 + 3^2 + \cdots + n^2$$

20. Notice that

$$(1)(2) + (2)(3) = \frac{(2)(3)(4)}{3}$$

$$(1)(2) + (2)(3) + (3)(4) = \frac{(3)(4)(5)}{3}$$

$$(1)(2) + (2)(3) + (3)(4) + (4)(5) = \frac{(4)(5)(6)}{3}$$

Try some other cases. Use inductive reasoning to make a conjecture about

$$(1)(2) + (2)(3) + (3)(4) + \cdots + n(n + 1)$$

21. Compute the values of the following sums.

$$1 + \frac{1}{2}, \qquad 1 + \frac{1}{2} + \frac{1}{4}, \qquad 1 + \frac{1}{2} + \frac{1}{4} + \frac{1}{8}$$

Use inductive reasoning to make a conjecture.

For Exercises 22 and 23, observe each set of equations. What is the next step in these patterns? Use inductive reasoning to make a conjecture. Prove your conjecture.

22. $2(2) + 2^2 = 3^2 - 1$
$2(3) + 3^2 = 4^2 - 1$
$2(4) + 4^2 = 5^2 - 1$

23. $2^2 + 2(3) = 3^2 + 1$
$3^2 + 2(4) = 4^2 + 1$
$4^2 + 2(5) = 5^2 + 1$

24. Prove the following implication. If m and n are even, then $m + n$ is even.

25. Does the converse of Exercise 24 hold?

26. Prove that if m and n are odd, then $m + n$ is even.

27. If $m, n \in \mathbb{N}$, prove by the method of contradiction that $mn \geq n$.

*28. If $m, n \in \mathbb{N}$ with $m, n \geq 2$, prove by the method of contradiction that $mn \geq m + n$.

*29. Prove that if n^2 is even, then n is even.

Chapter 3 Summary of Terms

Logical operation	Symbol	Type of statement
not	$\sim$	Negation
and	$\wedge$	Conjunction
or	$\vee$	Disjunction
if... then	$\rightarrow$	Conditional
if and only if	$\leftrightarrow$	Biconditional

Basic Truth Table

p	q	$p \wedge q$	$p \vee q$	$p \rightarrow q$	$p \leftrightarrow q$
T	T	T	T	T	T
T	F	F	T	F	F
F	T	F	T	T	F
F	F	F	F	T	T

implication $p \rightarrow q$

converse $q \rightarrow p$

inverse $\sim p \rightarrow \sim q$

contrapositive $\sim q \rightarrow \sim p$

equivalent statements $p \vee q \equiv q \vee p, \quad p \wedge q \equiv q \wedge p, \quad p \equiv p \vee p, \quad p \equiv p \wedge p, \quad \sim(\sim p) \equiv p$

De Morgan's laws $\sim(p \vee q) \equiv \sim p \wedge \sim q$
$\sim(p \wedge q) \equiv \sim p \vee \sim q$

distributive laws $p \wedge (q \vee r) \equiv (p \wedge q) \vee (p \wedge r)$
$p \vee (q \wedge r) \equiv (p \vee q) \wedge (p \vee r)$

conditional $p \rightarrow q \equiv \sim p \vee q$

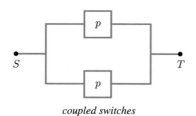

coupled switches

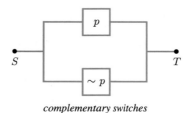

complementary switches

series circuit

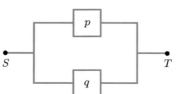

parallel circuit

arguments: valid

modus ponens	*modus tollens*	*disjunctive syllogism*	*reasoning by transitivity*
$p \rightarrow q$	$p \rightarrow q$	$p \vee q$	$p \rightarrow q$
$\underline{p}$	$\underline{\sim q}$	$\underline{\sim p}$	$\underline{q \rightarrow r}$
q	$\sim p$	q	$p \rightarrow r$

arguments: invalid

fallacy of the inverse

$$p \rightarrow q$$

$$\sim p$$

$$\sim q$$

fallacy of the converse

$$p \rightarrow q$$

$$q$$

$$p$$

inductive reasoning

deductive reasoning

method of contradiction

Chapter 3 Test

1. Construct a truth table for the statement $(p \vee \sim q) \wedge \sim r$.

2. If p and q are true and r and s are false, what is the truth value of the following statement? $(p \vee \sim q) \wedge (\sim r \vee s)$.

3. Consider the conditional "If it snows, then I ski."
   (a) What is the antecedent? What is the consequent?
   (b) What is the converse? Inverse? Contrapositive?
   (c) If the original conditional is true, which of the conditionals in part (b) are necessarily true?

4. Construct a truth table for the statement $(p \rightarrow q) \wedge (\sim q \vee r)$.

5. Is the following statement true or false? If $x \leq y$, then $-x \leq -y$.

6. Use a truth table to verify the equivalence $p \equiv p \wedge (p \vee \sim q)$.

7. Verify the equivalence $p \equiv p \wedge (q \vee \sim q)$ without using a truth table.

8. Give the pairs of equivalent statements among the following.
   (a) $p \rightarrow q$ (b) $q \rightarrow p$ (c) $\sim p \rightarrow \sim q$ (d) $\sim q \rightarrow \sim p$

9. Give a statement r whose truth table is Table 3.43.

10. Give a statement r whose truth table is Table 3.44.

Table 3.43		
p	q	r
T	T	F
T	F	F
F	T	F
F	F	T

Table 3.44		
p	q	r
T	T	T
T	F	F
F	T	T
F	F	T

11. We have seen that sets correspond to statements. What statement corresponds to the empty set $\emptyset$? What statement corresponds to the universal set U?

12. Draw a circuit representing the statement $[(p \wedge q \wedge r) \vee (\sim p \wedge \sim r)] \wedge q$.

13. Write a logical statement representing the circuit in Figure 3.17.

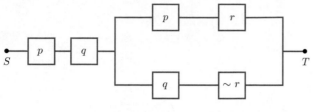

Figure 3.17

14. Use a truth table to determine the validity of the following argument.

$$p \to q$$
$$\underline{\sim r \to \sim q}$$
$$p \to r$$

15. Is the following argument valid?

$$p \to q$$
$$r \to \sim s$$
$$\underline{\sim s \to \sim q}$$
$$p \to \sim r$$

16. Prove that the sum of an even number and an odd number is odd. Does the converse hold?

17. Prove that a natural number is odd if and only if it has the form $2n+1$ for some $n \in \mathbb{N} \cup \{0\}$.

18. Show that $(p \to q) \vee (q \to p)$ is a tautology.

19. Are the following statements true or false?
    (a) The symbol for the logical operation "and" is $\vee$.
    (b) $p \to q$ and $\sim q \vee p$ are logically equivalent.
    (c) The statement "p only if q" can be symbolized as $q \to p$.
    (d) The contrapositive of $p \to q$ is $\sim p \to \sim q$.
    (e) "If you squeeze here, then it comes out there" is a conditional statement.
    (f) "It's raining cats and dogs" is a disjunction.
    (g) *Modus tollens* is an invalid argument form.
    (h) The *fallacy of the inverse* is a valid argument form.
    (i) An example of a tautology is $p \wedge (\sim p)$.
    (j) The argument $p \to q$, $q \to r$, therefore $p \to r$ is called reasoning by transitivity.
    (k) The set $A - B$ corresponds to the symbolic logic statement $p \wedge (\sim q)$.

20. What well-known quotation is represented by the statement "$2B \vee \sim 2B = ?$"?

21. Discuss the following headlines.
    (a) FARMER BILL LOSES IN CONGRESS
    (b) ANTIBUS RIDER KILLED IN SENATE

22. Discuss the following statement. It was not a good day for the people who oppose the foes of anti-abortion.

23. What is the truth value of q in the following cases?
    (a) p and $p \wedge q$ are true.
    (b) p is true and $p \to q$ is false.
    (c) p is true and $p \leftrightarrow q$ is true.

NUMBER THEORY

4

God made the integers; all the rest is the work of man.
— Leopold Kronecker (1823–1891), German Mathematician

The natural numbers $1, 2, 3, \ldots$ are objects everyone is acquainted with. In fact, you probably knew how to count before you entered grade school and you learned how to add and multiply natural numbers early in your education. Since the elementary properties of natural numbers are well known to you, the material of this chapter is based upon familiar ground.

The study and use of the natural numbers is one of the oldest intellectual pursuits of humankind. Undoubtedly even the early cave people used the first few natural numbers to keep track of their possessions. Fairly large natural numbers were used by traders, calendar makers, and surveyors over 4000 years ago. The natural numbers were analyzed by ancient Greek philosophers and mythologists, who originated the art of numerology, the study of the mystic properties of numbers. Although the natural numbers were originally invented for practical purposes, later studies developed into a purely intellectual pursuit. The ancients discovered fascinating patterns in numbers and investigated these patterns for the pure joy of satisfying their curiosity and their search for truth. Recently the cycle has been closed through applications of number theory to computer science and cryptography.

4.1 | Number Sequences

Arithmetic is one of the oldest branches, perhaps the very oldest branch, of human knowledge; and yet some of its most abstruse secrets lie close to its tritest truths.
— H. J. S. Smith, Twentieth-Century Mathematician

Number patterns were discovered early in the history of mathematics. Many of these patterns were in the form of number sequences. For example, the even numbers

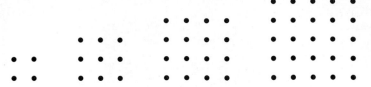

Figure 4.1

$2, 4, 6, 8, 10, \ldots$ and the odd numbers $1, 3, 5, 7, 9, \ldots$ are number sequences. Another interesting sequence consists of the square numbers $1, 4, 9, 16, 25, \ldots$. The ancients experimented with square numbers by arranging pebbles such as in Figure 4.1.

Using arrangements of pebbles, such as in Figure 4.2, the ancient Greeks discovered the triangular numbers. From Figure 4.2 we see that the first few triangular numbers are $1, 3, 6, 10, 15$. Can you find the next triangular number? Can you find a formula for the nth triangular number? (Try your hand at this before you read on.)

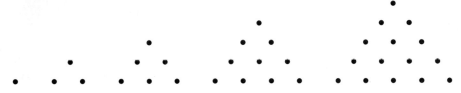

Figure 4.2

In order to find a formula for the nth triangular number, examine Figure 4.2 closely. If you look at the fourth arrangement, for example, you see that it consists of one pebble on the top, two at the second level, three at the third, and four at the fourth. We thus see that the nth triangular number is $1 + 2 + 3 + \cdots + n$. But we have seen in Chapter 1 that this sum is $n(n+1)/2$.

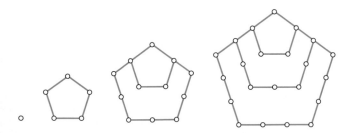

Figure 4.3

It was a natural step for the ancient Greeks to examine other number sequences. For example, they considered the pentagonal numbers illustrated in Figure 4.3. The first few pentagonal numbers are $1, 5, 12, 22$. Can you find the next pentagonal number? Can you find a formula for the nth pentagonal number?

An important number sequence results from the solution to the following problem. A branch of a tree grows indefinitely but produces no new branch during its first year. At the end of the second year and each succeeding year each existing branch produces a new branch. Starting with a single branch, how many branches are there at the beginning of year $1, 2, 3, \ldots$? Figure 4.4 illustrates nine years of growth and the number of branches.

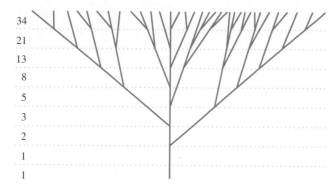

Figure 4.4

Fibonacci Sequence

The sequence whose first few terms are $1, 1, 2, 3, 5, 8, 13, 21, 34$ is called the **Fibonacci sequence**. It was discovered by the Italian mathematician Leonardo of Pisa, nicknamed Fibonacci (son of Binaccio) in 1202. Do you see a pattern in these numbers? What is the next number in this sequence? (Try to answer these questions before continuing.)

The key to the pattern in this sequence is found by adding successive numbers

$$1 + 1 = 2$$
$$1 + 2 = 3$$
$$2 + 3 = 5$$
$$3 + 5 = 8$$
$$5 + 8 = 13$$
$$8 + 13 = 21$$
$$13 + 21 = 34$$

Thus starting with $1, 1$, each term of the Fibonacci sequence is the sum of the previous two terms. We have already discussed the Fibonacci sequence in Section 2.2.

EXAMPLE 1   What is the next number in the Fibonacci sequence?
   Solution.   The next number is $21 + 34 = 55$.   □

Another interesting number sequence was discovered by Eugene Charles Catalan in the nineteenth century. In Figure 4.5 we see that a square can be divided into two triangles in two ways.

Before reading on, can you tell how many ways a pentagon can be divided into triangles?

Figure 4.5

Photo 4.1
The number of tree branches
frequently gives a Fibonacci
sequence.

Figure 4.6 shows that the answer is five.

Figure 4.6

What is the corresponding answer for a hexagon? Figure 4.7 shows that a hexagon can be divided into triangles in 14 different ways.

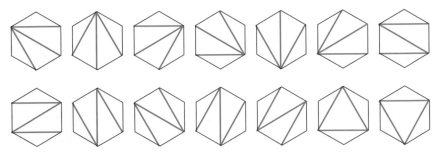

Figure 4.7

Catalan Sequence

The number of ways a 7-gon can be divided into triangles is 42, but drawing all these would obviously be quite tedious. The sequence $1, 2, 5, 14, 42, \ldots$ is called the **Catalan sequence** (the first term comes from the fact that a triangle can be divided into triangles in only one way).

The Catalan sequence comes up in many situations. For another example, consider the following problem. Candidates A and B both get n votes in an election. How many ways can the votes be tallied so that A never trails? (Can you answer this for $n = 1, 2, 3$?)

For $n = 1$, A's vote must be tallied first, so we represent this tally by AB. In this case we have only one possible tally. For $n = 2$, there are two possible tallies

$$AABB \qquad ABAB$$

For $n = 3$ there are five possible tallies

$$AAABBB \qquad AABABB \qquad AABBAB \qquad ABAABB \qquad ABABAB$$

Can you show that there are 14 possible tallies for $n = 4$?

Is there a formula for the nth Catalan number C_n ? There is such a formula and it turns out to be

$$C_n = \frac{(2n)!}{(n + 1)!n!}$$

The derivation of this formula is a little complicated and we shall omit it.

EXAMPLE 2 What is the sixth Catalan number?
Solution.

$$C_6 = \frac{(12)!}{7!6!} = \frac{(12)(11)(10)(9)(8)}{(6)(5)(4)(3)(2)} = 132 \qquad \square$$

EXERCISES

1. What is the sixth even number?
2. Give a formula for the even numbers.
3. What is the sixth odd number?
4. Give a formula for the odd numbers.
5. How would you define the cube numbers? List the first four cube numbers. Give a formula for the cube numbers.
6. How can you arrange pebbles to represent the cube numbers?
7. The nth term of a number sequence is given by $3n$. List the first five numbers of this sequence.
8. The nth term of a number sequence is given by 2^n. List the first five numbers of this sequence.
9. What are the sixth and seventh triangular numbers?
10. What is the fifth pentagonal number of the type discussed in this section?
11. Give a formula for the square numbers.

12. Show that the formula for C_n gives the first five Catalan numbers.

13. List the first 15 Fibonacci numbers.

14. Draw 10 ways in which a 7-gon can be divided into triangles.

15. How many ways can an octagon be divided into triangles?

16. What is the ninth term of the Catalan sequence?

17. In a simplified model, a pair of rabbits in their first month of life do not produce young, but in the second and each ensuing month they produce a new pair. Assume that deaths do not occur during the period considered and that each pair consists of a male and female. Starting with a single pair, how many pairs will there be at the beginning of months $1, 2, 3, \ldots, 12$?

*18. If A and B both get four votes, list the 14 possible tallies in which A never trails.

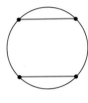

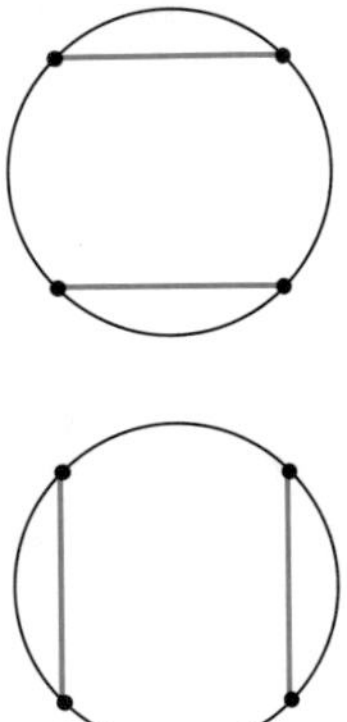

19. If you put four points on a circle, then there are two ways of drawing nonintersecting chords connecting each pair of points (see Figure 4.8). How many ways are there for six points?

20. Answer Exercise 19 for eight points. How is this related to the Catalan sequence?

21. Let t_n be the nth triangular number. Show that $t_{n+1} = t_n + n + 1$.

22. Let p_n be the nth pentagonal number. Show that $p_{n+1} = p_n + 3n + 1$.

23. Add the first three Fibonacci numbers. The first four. The first five. The first six. Do you see a pattern? Make a conjecture about the sum of the first n Fibonacci numbers.

24. A number sequence starts with $1, 2$, and each term is the sum of the previous two terms. What are the first 10 terms of the sequence?

25. Answer Exercise 24 for a sequence beginning with $1, 3$.

Figure 4.8

26. A **Lucas sequence** is like a Fibonacci sequence except that it need not begin with $1, 1$. Thus a Lucas sequence begins with any two natural numbers and each later term is the sum of its two predecessors. Show that the following is the beginning of a Lucas sequence: $7, 11, 18, 29, 47, 76, \ldots$. What are the next three terms of this sequence?

27. (Continuation of 26) Give the next three equations for the following sequence.

$$18 = (7)(1) + (11)(1)$$
$$29 = (7)(1) + (11)(2)$$
$$47 = (7)(2) + (11)(3)$$
$$76 = (7)(3) + (11)(5)$$

28. Using Exercise 27, make a conjecture that gives any Lucas sequence in terms of its first two numbers and the Fibonacci sequence.

29. Find two natural numbers whose sum equals their product.

30. Repeat Exercise 29 for three natural numbers.

31. Repeat Exercise 29 for four natural numbers.

32. Repeat Exercise 29 for five natural numbers.

33. Repeat Exercise 29 for six natural numbers.

34. If $n \geq 2$ is a natural number, show that there are n natural numbers whose sum equals their product.

*35. Astronomers have found that solar and lunar eclipses show certain repeating patterns every $6, 41, 47, 88, 135, 223$, and 358 years. What kind of a sequence is this? Write this sequence in terms of $6, 41$, and the Fibonacci sequence.

*36. Notice the following pattern, involving cubic and square numbers:

$$1^3 = 1^2$$
$$1^3 + 2^3 = 9 = 3^2$$
$$1^3 + 2^3 + 3^3 = 36 = 6^2$$
$$1^3 + 2^3 + 3^3 + 4^3 = 100 = 10^2$$

(a) What is the next term in this sequence?

(b) What is the nth term in this sequence?

*37. Using the pattern in Exercise 36, we see that

$$1^3 = 1^2 - 0^2$$
$$2^3 = 3^2 - 1^2$$
$$3^3 = 6^2 - 3^2$$
$$4^3 = 10^2 - 6^2$$

(a) What about 5^3?

(b) Show that any cube is the difference of two squares.

**38. Find a formula for the pentagonal numbers.

**39. Let $a_1, a_2, \ldots$ be the Fibonacci sequence. Compare $a_4^2 - a_3^2$ and $a_4 a_3$. Do the same for $a_5^2 - a_5^2$ and $a_5 a_4$. For $a_6^2 - a_5^2$ and $a_6 a_5$. Try some others. Make a conjecture concerning $a_n^2 - a_{n-1}^2$ and $a_n a_{n-1}$.

4.2 Partitions

Mathematics is the queen of sciences and number theory is the queen of mathematics.
— KARL FRIEDRICH GAUSS (1777–1855), German Mathematician

One of the beauties of number theory is that it has many unsolved problems which are extremely easy to state. It is this quality of number theory that makes it accessible to amateur mathematicians and puzzle enthusiasts around the world. Some of these unsolved problems have been investigated for hundreds of years without success. Yet number theory involves the most elementary of mathematical objects, the natural numbers $1, 2, 3, \ldots$, and many of its problems are stated in terms of the simplest mathematical operations, namely addition and multiplication.

A large class of problems in number theory ask whether certain natural numbers are the sum of certain other natural numbers.

When we represent a natural number n as a sum of other natural numbers, we call this a **partition** of n.

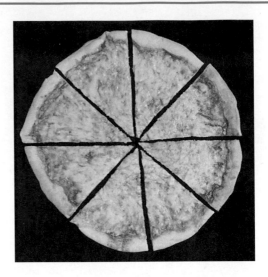

Photo 4.2
A pizza is partitioned into eight slices.

For example, the number 2 has the partition $2 = 1 + 1$ and this is the only partition of 2. We call the numbers in the partition **summands**. Thus, the summands in $2 = 1 + 1$ are 1 and 1.

The number 3 has two partitions, $3 = 1 + 1 + 1$ and $3 = 1 + 2$. Notice that we do not call $3 = 2 + 1$ a different partition than $3 = 1 + 2$ as they involve only a difference in order. For convenience we usually write the summands of a partition in increasing order; for example, we write $4 = 1 + 1 + 2$ instead of $4 = 1 + 2 + 1$ or $4 = 2 + 1 + 1$. If there are m summands in a partition, we call it an m-**partition**. Thus, $5 = 1 + 2 + 2$ is a 3-partition of 5 and $6 = 2 + 4$ is a 2-partition of 6. Table 4.1 exhibits the partitions of the numbers 2 through 6.

Table 4.1

Number	2-Partitions	3-Partitions	4-Partitions	5-Partitions	6-Partitions
2	$1 + 1$				
3	$1 + 2$	$1 + 1 + 1$			
4	$1 + 3$ $2 + 2$	$1 + 1 + 2$	$1 + 1 + 1 + 1$		
5	$1 + 4$ $2 + 3$	$1 + 1 + 3$ $1 + 2 + 2$	$1 + 1 + 1 + 2$	$1 + 1 + 1 + 1 + 1$	
6	$1 + 5$ $2 + 4$ $3 + 3$	$1 + 1 + 4$ $1 + 2 + 3$ $2 + 2 + 2$	$1 + 1 + 1 + 3$ $1 + 1 + 2 + 2$	$1 + 1 + 1 + 1 + 2$	$1 + 1 + 1 + 1 + 1 + 1$

Do you see any patterns in Table 4.1? For example, look at the upper left to lower right diagonals. In these diagonals an entry is frequently formed by adding 1 to the previous entry as we move down to the right. Notice the pattern in the 2-partition

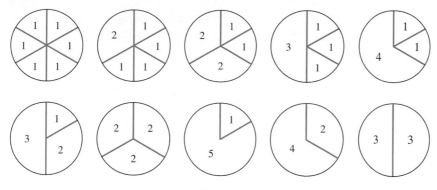

Figure 4.9

column. The numbers of 2-partitions of 2, 3, 4, 5, and 6 are 1, 1, 2, 2, 3, respectively. How many 2-partitions of 7 would you expect? How many 2-partitions of n are there?

Figure 4.9 illustrates the partitions of 6 by partitioning a pie into slices.

Four Squares Problem

A problem stated by the ancient Greeks some 2500 years ago is the four squares problem. Can every natural number greater than 1 be partitioned into four or fewer squares? Table 4.2 shows the plausibility of a positive answer to this question.

Table 4.2

Number	Partition into squares
2	$1^2 + 1^2$
3	$1^2 + 1^2 + 1^2$
4	2^2
5	$1^2 + 2^2$
6	$1^2 + 1^2 + 2^2$
7	$1^2 + 1^2 + 1^2 + 2^2$
8	$2^2 + 2^2$
9	3^2
99	$1^2 + 1^2 + 4^2 + 9^2$
140	$2^2 + 6^2 + 10^2$
141	$2^2 + 4^2 + 11^2$

This problem remained unsolved for over 2000 years. The answer is yes, but it was only proved about 200 years ago by the French mathematician Joseph Louis Lagrange.

Fermat's Last Conjecture

A very famous partitioning problem is Fermat's (1601–1665) last conjecture. First of all, there are many squares that can be 2-partitioned into squares. For example,

$5^2 = 3^2 + 4^2$, $10^2 = 6^2 + 8^2$, $13^2 = 5^2 + 12^2$. One might now ask, can a cube be 2-partitioned into cubes? That is, are there natural numbers x, y, and z such that $z^3 = x^3 + y^3$? More generally, for any natural number $n > 2$, are there natural numbers x, y, and z such that $z^n = x^n + y^n$? Fermat asserted that the answer was no. In fact, he wrote in the margin of a number theory book he was reading that he had found a truly remarkable proof of this fact, but there was not room enough in the margin to write it. This was very unfortunate, for many mathematicians have tried in vain to reproduce such a proof. Using very sophisticated methods, the assertion has been shown to be true for many special values of n, but not in general.

An important class of natural numbers is the set of prime numbers.

A **prime** number is a natural number other than 1 which is not wholly divisible by any natural number except 1 and itself.

For example, the first few prime numbers are $2, 3, 5, 7, 11, 13, 17, 19, 23, \ldots$.

Goldbach's Conjecture

We shall consider prime numbers in detail later, but for now they are needed to state a partitioning problem proposed by Christian Goldbach in 1742. Goldbach conjectured that every even number greater than 2 has a prime 2-partition. That is, every even number greater than 2 can be represented as the sum of two primes. Table 4.3 shows the plausibility of this conjecture.

Table 4.3

Even number	Prime 2-partition
4	$2 + 2$
6	$3 + 3$
8	$3 + 5$
10	$5 + 5$
12	$5 + 7$
14	$7 + 7$
16	$5 + 11$
18	$5 + 13$
20	$3 + 17$
22	$11 + 11$
68	$31 + 37$
84	$41 + 43$

Goldbach's conjecture has not been proved or disproved to this date. Notice that this conjecture is not true for odd numbers. In fact, 3 has no prime 2-partition. Also, the 2-partitions of 11 are

$$11 = 1 + 10 = 2 + 9 = 3 + 8 = 4 + 7 = 5 + 6$$

so 11 has no prime 2-partition.

EXERCISES

1. List the partitions of 7.

2. List the 2-partitions of 7, 8, and 9.

3. How many 2-partitions are there for n?

4. How many n-partitions are there for n? What are they?

5. How many $(n-1)$-partitions are there for n? What are they?

6. List the 3-partitions of 10.

7. Prove that if $m < n$, then the number of partitions of m is less than the number of partitions of n.

8. Draw the sliced pies corresponding to 2, 3, 4, and 5 as in Figure 4.9.

9. Figure 4.9 shows that the number of ways of slicing a pie into pieces whose angles are a multiple of $60°$ equals the number of partitions of 6. What is the corresponding statement for the partitions of n?

10. Expand Table 4.2 for the numbers 10 through 20.

11. Partition 143 into four or fewer squares.

12. Partition 1000 into four or fewer squares.

13. Give another example of a 2-partition of a square into squares.

14. Does 2^2 have a 2-partition into squares? What about 3^2 and 4^2?

15. Find four natural numbers w, x, y, z which satisfy $w^3 + x^3 + y^3 = z^3$. Does this disprove Fermat's last conjecture?

16. Try to express each natural number between 2 and 30 as the sum of two consecutive natural numbers.

17. Make a conjecture using Exercise 16.

18. Prove your conjecture in Exercise 17.

19. Expand Table 4.3 for the numbers 24, 26, 28, 30.

20. Give five examples of odd numbers that have prime 2-partitions.

21. Give five examples of odd numbers that do not have prime 2-partitions.

22. Exactly which odd numbers have prime 2-partitions?

23. Prove that there are an infinite number of squares.

24. Samuel Isaac Krieger claimed to have found a counterexample to Fermat's last conjecture. Krieger announced that it was $1324^n + 731^n = 1961^n$ where n is a certain positive integer greater than 2, which Krieger refused to disclose. Why can't this be right?

*25. Find three natural numbers n such that $1 + n(n+1)(n+2)(n+3)$ is square.

*26. Using Exercise 24, make a conjecture.

**27. Prove your conjecture in Exercise 25.

Photo 4.3
Numbers game:
Mathematics students—or
just plain numbers fans—
have it easy when approaching Snowmass at Aspen.
Some enterprising person
added up all numbers on this
sign, giving a total that has
absolutely no significance.

snowmass	
Population	934
Elevation	8624
Established	1967
Total	11,525

4.3 Prime Factorization

*All results of the profoundest mathematical investigation must ulti-
mately be expressed in the simple form of properties of integers.*
— LEOPOLD KRONECKER (1823–1891), German Mathematician

The last section discussed the representations of natural numbers as sums of other
natural numbers. We now discuss representation of natural numbers as products of
simpler natural numbers. See if you can solve the following problem before looking
at the solution.

EXAMPLE 1   Suppose you want to tile a floor measuring 12 by 15 feet and you want
to use only whole tiles which are all the same size. If tiles are available in squares
which are 4, 5, 6, 8, 9, and 12 inches to a side, which size tiles can you use?
    Solution.   Since a 12 by 15 foot floor is 144 by 180 inches, if a tile is to work,
its length must divide both 144 and 180 without leaving a remainder. Thus 4, 6, 9,
and 12 inch tiles will work.   □

Divisors

> A natural number k **divides** (or is a **factor** of, or is a **divisor** of) a natural number
> n if there is no remainder when n is divided by k.

If k divides n, then a line segment of length n can be divided into k equal parts
(Figure 4.10).

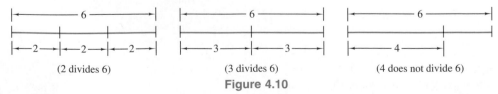

| (2 divides 6) | (3 divides 6) | (4 does not divide 6) |

Figure 4.10

If k divides n, we write $k|n$. Thus $k|n$ if there is a natural number t such that $n = kt$. For example, in Figure 4.10 since the line segment of length 6 is divided into three line segments of length 2, we have $2|6$ and $6 = 2 \times 3$. The notation $k|n$ denotes a relationship between the numbers k and n and should not be confused with the fraction k/n. Some examples are $3|6$, $4|16$, $2|8$.

Prime and Composite Numbers

A natural number greater than 1 whose only divisors are 1 and itself is **prime**. A natural number greater than 1 which is not prime is **composite**. Thus every natural number greater than 1 is either prime or composite; it is convenient not to classify 1 itself as either.

It is quite easy to decide whether a small natural number is prime or composite, but for very large numbers this decision can be difficult. Table 4.4 classifies the first few natural numbers as prime or composite.

Table 4.4

Number	Prime (p) Composite (c)	Number	Prime (p) Composite (c)
2	p	16	c
3	p	17	p
4	c	18	c
5	p	19	p
6	c	20	c
7	p	21	c
8	c	22	c
9	c	23	p
10	c	24	c
11	p	25	c
12	c	26	c
13	p	27	c
14	c	28	c
15	c		

Suppose n is a composite number. Since n is not prime, n must have a divisor k different from n and 1. Thus $n = kt$ for some natural number $t \neq 1$. We see that n has been broken down into a product of simpler numbers. In mathematics there is an old saying: If you find a good thing, do it as much as you can. So try to break k and t down into simpler numbers. Continue this process until the remaining numbers cannot be broken down any further, in which case n is a product of primes. For example, 36 is composite. Take one of the divisors of 36, say 4, so $36 = (4)(9)$. Now 4 and 9 are composite: $4 = (2)(2)$ and $9 = (3)(3)$; so $36 = (2)(2)(3)(3)$. These numbers cannot be broken down any further, and so 36 becomes a product of primes.

Prime Factorization

When a natural number n is written as a prime or a product of primes, this is called the **prime factorization** of n (abbreviated pf of n).

For example, the pf of 36 is $36 = (2)(2)(3)(3)$, the pf of 3 is $3 = 3$, the pf of 12 is $12 = (2)(2)(3)$. The pf of 12 could also be written as $(2)(3)(2)$ or $(3)(2)(2)$. These different ways all involve the same primes; only the order has changed. Is there any other pf of 12? If there were, we would have to use primes other than 2 and 3 or a different number of 2s and 3s. But there are no primes other than 2 and 3 that divide 12. Also we could not have two 3s in the pf of 12 because $(3)(3) = 9$ does not divide 12. The following theorem summarizes these observations.

Theorem 4.1   Every natural number greater than 1 is either a prime or can be written as a product of primes in only one way (except for the order of the factors).

Theorem 4.1 is called the **prime factorization theorem**. This result shows that the primes are the building blocks from which any natural number (other than 1) can be constructed (Figure 4.11). If the order in which the blocks are placed is ignored, then there is only one such construction for a given natural number.

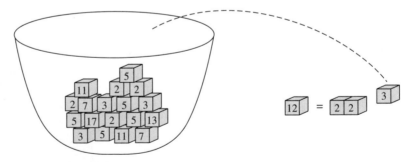

Figure 4.11

It is frequently of interest to know the number of divisors of a given natural number. For example, 9 has the three divisors 1, 3, and 9. In this case we write $d(9) = 3$. Similarly, 8 has four divisors, 1, 2, 4, and 8, and so we write $d(8) = 4$. In general, we let $d(n)$ denote the number of divisors of n. Thus d is a function from the set of natural numbers to itself. Notice that $d(n) = 2$ if and only if n is prime. Table 4.5 gives the first few values of d.

Table 4.5

n	Divisors of n	$d(n)$
1	1	1
2	1, 2	2
3	1, 3	2
4	1, 2, 4	3
5	1, 5	2
6	1, 2, 3, 6	4
7	1, 7	2
8	1, 2, 4, 8	4
9	1, 3, 9	3
10	1, 2, 5, 10	4
11	1, 11	2
12	1, 2, 3, 4, 6, 12	6
$\vdots$	$\vdots$	$\vdots$
24	1, 2, 3, 4, 6, 8, 12, 24	8

Table 4.6

n	Prime factorization	$d(n)$
3	3	2
4	$(2)(2) = 2^2$	3
6	$(2)(3)$	4
8	$(2)(2)(2) = 2^3$	4
9	$(3)(3) = 3^2$	3
10	$(2)(5)$	4
12	$(2)(2)(3) = (2^2)(3)$	6
24	$(2)(2)(2)(3) = (2^3)(3)$	8

Formula for $d(n)$

Can you find a formula for $d(n)$? To get a clue for such a formula, let us list the pf of some numbers and compare these with $d(n)$; see Table 4.6. Do you see any pattern in Table 4.6? Notice that the exponent of the prime in the pf of 4 is 2, and $d(4) = 3 = 2 + 1$; the exponents of the primes in the pf of 6 are 1 and 1, and $d(6) = 4 = (1 + 1)(1 + 1)$; the exponents of the primes in the pf of 12 are 2 and 1, and $d(12) = 6 = (2 + 1)(1 + 1)$; and the exponents of the primes in the pf of 24 are 3 and 1, and $d(24) = 8 = (3 + 1)(1 + 1)$. Can you now make a conjecture for the value of $d(n)$?

Did you conjecture that $d(n)$ is found by taking the exponents in the pf of n, adding 1 to them, and multiplying the resulting numbers? Let us check this conjecture on some larger numbers. For example, $36 = (2^2)(3^2)$, so if the conjecture is correct, $d(36) = (2 + 1)(2 + 1) = 9$. Since the divisors of 36 are $1, 2, 3, 4, 6, 9, 12, 18, 36$, the conjecture is correct for this case. As another example, $48 = (2^4)(3)$ and the divisors of 48 are $1, 2, 3, 4, 6, 8, 12, 16, 24, 48$, which agrees with $d(48) = (4 + 1)(1 + 1) = 10$. The following theorem summarizes this result.

Theorem 4.2   If the pf of n is $n = p^a q^b r^c \ldots$, then

$$d(n) = (a + 1)(b + 1)(c + 1) \cdots$$

Let us work an example to illustrate the proof of Theorem 4.2. Suppose we want $d(720)$. The pf of 720 is $720 = (2^4)(3^2)(5)$. By the prime factorization theorem (Theorem 4.1), every divisor of 720 must have the form $(2^i)(3^j)(5^k)$, where

$i = 0, 1, 2, 3, 4$; $j = 0, 1, 2$; $k = 0, 1$; and every number of this form is a divisor of 720. For example, $144 = (2^4)(3^2)(5^0)$ is a divisor of 720 in which $i = 4$, $j = 2$, and $k = 0$. Thus to find the number of divisors of 720 we can count the number of possible combinations of exponents. For example, $i = 0$, $j = 0$, $k = 0$; $i = 0$, $j = 1$, $k = 0$; and $i = 4$, $j = 1$, $k = 1$ are possible combinations of exponents giving different divisors. Now for each of the five possible values of i there are three possible values of j and two possible values of k, giving a total of $(4 + 1)(2 + 1)(1 + 1) = 30$ divisors of 720.

Divisor Diagram

A useful way of visualizing the divisors of a natural number is by a "divisor diagram." Figures 4.12 through 4.16 show the divisor diagrams for 4, 10, 12, 24, and 36.

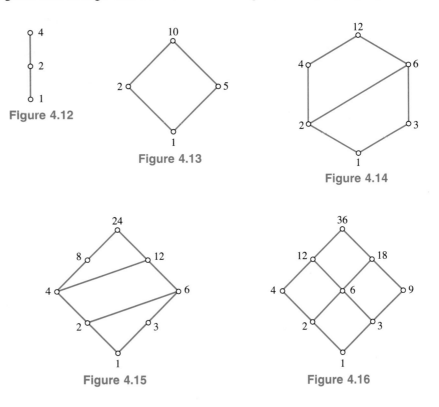

Figure 4.12

Figure 4.13

Figure 4.14

Figure 4.15

Figure 4.16

Do you see how these diagrams are constructed? If k and n are divisors of a particular natural number and $k|n$ but there is no other divisor m such that $k|m$ and $m|n$, then k and n are connected by a line segment and n is placed higher than k. Thus, k and n are connected by a rising line segment if there is no other divisor that divides n and is a multiple of k, that is, if there is no other divisor "between" k and n.

EXERCISES

1. What would the answer to Example 1 be if the floor measured 6 by 10 feet?

2. Find the divisors of 12. For each divisor, draw a figure like Figure 4.10.

3. Which of the following are true? (a) $4|6$ (b) $5|10$ (c) $5|11$ (d) $7|21$

4. It was stated in the text that if $k|n$, then $n = kt$ for some natural number t. Find the corresponding t for $3|9$, $4|12$, and $6|48$.

5. Show that $1|n$ and $n|n$ for every natural n.

6. Show that $n|2n$ for every natural number n.

7. Continue Table 4.4 to the number 40.

8. Find the pf of 360.

9. Fill in Table 4.5 between 12 and 24.

10. Fill in Table 4.6 between 12 and 24.

11. Find the divisors of 36. How many did you find? Compare this number with that given in Theorem 4.2.

12. What is the pf of 60? Find the divisors of 60, compute the pf of these divisors, and compare these to the pf of 60.

13. Find the pf's of 840, 4800, 960. Compute $d(840)$, $d(4800)$, $d(960)$.

14. Find two different natural numbers that have 36 divisors.

15. What is the smallest natural number that has 2, 3, 5, and 7 among its divisors? How many divisors does this number have?

16. Find $d(468)$, $d(210)$, $d(360)$.

17. Prove that the product of any two primes greater than 2 is odd.

18. Show that n is the square of a prime if and only if $d(n) = 3$.

19. Draw the divisor diagrams for 3 and 8.

20. Draw the divisor diagrams for 144.

21. Draw the divisor diagrams for 30.

22. Draw the divisor diagrams for 48.

23. If $k|m$ and $m|n$, prove that $k|n$.

24. If $k|n$ and $n|k$, prove that $k = n$.

25. Prove that $k|n$ if and only if n/k is a natural number.

26. Let p be a prime and suppose $p|mn$. Prove that $p|m$ or $p|n$.

27. If $k|mn$, does $k|m$ or $k|n$? Why?

28. Let p be a prime. Prove that n has the form $n = q^{p-1}$, where q is prime, if and only if $d(n) = p$.

*29. Find the three smallest natural numbers with exactly 10, 18, and 24 divisors, respectively.

*30. Show that if $d(n)$ is odd, then n is a square.

*31. Show that if $d(n)$ is even, then n is not a square.

*32. Prove the following statement. If the pf of n is $n = p^a q^b r^c \cdots$ and d is a divisor of n, then $d = p^i q^j r^k \cdots$, where $0 \le i \le a$, $0 \le j \le b$, $0 \le k \le c, \ldots$.

*33. Prove the converse of Exercise 32.

**34. Prove Theorem 4.2.

Photo 4.4
The size of the tiles is a common divisor of the length and the width of the rectangular area.

4.4 Greatest Common Divisor and Least Common Multiple

People who don't count won't count.
— ANATOLE FRANCE (1844–1924), French Author

You probably learned the concepts of greatest common divisor and least common multiple in elementary school. At that time you most likely computed these quantities using trial and error. In this section a systematic method will be given for finding these quantities using prime factorizations.

Since 2 divides both 4 and 8, we call 2 a **common divisor** for 4 and 8.

Common Divisor

> In general, if $k|m$ and $k|n$, then k is a **common divisor** of m and n.

Table 4.7 displays some number pairs and their common divisors.

A number k is a common divisor for m and n precisely when squares of length k can be used to tile a rectangle of length m and width n. This is illustrated in Figures 4.17 and 4.18 for the number pairs $(4, 8)$ and $(6, 15)$.

Table 4.7

Number pairs	Common divisors
(4, 8)	1, 2, 4
(6, 15)	1, 3
(12, 90)	1, 2, 3, 6
(36, 90)	1, 2, 3, 9, 18
(32, 48)	1, 2, 4, 8, 16

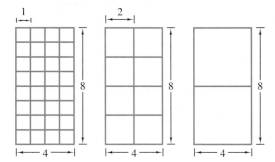

Figure 4.17

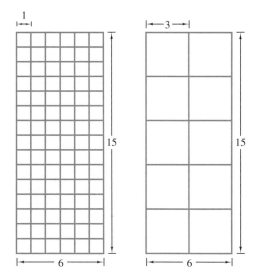

Figure 4.18

Greatest Common Divisor

The greatest common divisor of two natural numbers m and n is denoted $\gcd(m, n)$. From Table 4.7 we see that

$$\gcd(4, 8) = 4, \ \gcd(6, 15) = 3, \ \gcd(12, 90) = 6, \ \gcd(36, 90) = 18, \ \gcd(32, 48) = 16$$

According to our observation in the previous paragraph, $\gcd(m, n)$ is the length of the largest square that can tile an $m \times n$ rectangle.

Common Multiple

We also have the concept of a common multiple of two natural numbers.

> We say that k is a **common multiple** for m and n if $m|k$ and $n|k$.

For example, 6 is a common multiple for 2 and 3. Unlike common divisors, there are infinitely many common multiples for two natural numbers. For example, any multiple of 6 is a common multiple for 2 and 3. Table 4.8 displays some number pairs and their common multiples.

Table 4.8

Number pairs	Common multiples
$(2, 3)$	$6, 12, 18, 24, \ldots$
$(4, 6)$	$12, 24, 36, 48, \ldots$
$(4, 8)$	$8, 16, 24, 32, \ldots$
$(6, 15)$	$30, 60, 90, 120, \ldots$
$(6, 10)$	$30, 60, 90, 120, \ldots$

A number k is a common multiple for m and n precisely when a square of length k can be tiled by squares of length m and squares of length n. This is illustrated in Figure 4.19 for the number pair $(2, 3)$.

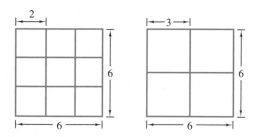

Figure 4.19

Least Common Multiple

The least common multiple of two natural numbers m and n is denoted $\operatorname{lcm}(m, n)$. Table 4.8 shows that $\operatorname{lcm}(2, 3) = 6$, $\operatorname{lcm}(4, 6) = 12$, $\operatorname{lcm}(4, 8) = 8$, $\operatorname{lcm}(6, 15) = 30$, $\operatorname{lcm}(6, 10) = 30$. You can always find a common multiple for two natural numbers merely by multiplying them. However, this may not give the *least* common multiple.

For example, $4 \times 6 = 24$ and yet $\text{lcm}(4,6) = 12$. Notice that $\text{lcm}(m,n)$ is the length of the smallest square that can be tiled by squares of length m and squares of length n.

You can easily compute the gcd and lcm for small numbers in your head. But what do you do for large numbers? For example, what is $\gcd(2400, 25{,}200)$? One method would be to find all the common divisors for 2400 and 25,200 and pick out the largest of these. That would be very tedious, so a better approach is needed. Will the pf's of these numbers help? The pf of 2400 is $(2^5)(3)(5^2)$ and the pf of 25,200 is $(2^4)(3^2)(5^2)(7)$. By the prime factorization theorem, if k is a common divisor for 2400 and 25,200, then the pf of k cannot contain primes other than 2, 3, and 5. If k is the greatest common divisor, then the powers of 2, 3, and 5 must be as large as possible yet not larger than the corresponding powers for 2400 and 25,200. Thus

$$\gcd(2400, 25{,}200) = (2^4)(3)(5^2) = 1200$$

The least common multiple of 2400 and 25,200 is treated in a similar way. If k is a common multiple for 2400 and 25,200, then the pf of k must contain the primes 2, 3, 5, and 7. If k is the least common multiple, then the powers of 2, 3, 5, and 7 must be as small as possible yet not smaller than the corresponding powers for 2400 and 25,200. Thus

$$\text{lcm}(2400, 25{,}200) = (2^5)(3^2)(5^2)(7) = 50{,}400$$

To see if you understand this method, work the following problems before checking the solution.

EXAMPLE 1   Find $\gcd(126, 360)$ and $\text{lcm}(126, 360)$.
   Solution.   Since $126 = (2)(3^2)(7)$ and $360 = (2^3)(3^2)(5)$, it follows that $\gcd(126, 360) = (2)(3^2) = 18$ and $\text{lcm}(126, 360) = (2^3)(3^2)(5)(7) = 2520$.   □

EXAMPLE 2   Maria shopped at the Italian market today. She bought merchandise from the wine merchant, the vegetable merchant, and the bakery. At this market the wine merchant is open only every fourth day, the vegetable merchant every twelfth day, and the bakery every ninth day. How long will she have to wait until all three shops are again open on the same day?
   Solution.   She will have to wait $\text{lcm}(4, 12, 9)$ days. Since $4 = 2^2$, $12 = (2^2)(3)$, $9 = 3^2$,
$$\text{lcm}(4, 12, 9) = (2^2)(3^2) = 36 \text{ days}$$ □

Greatest common divisors and least common multiples can be read off from divisor diagrams. For example, consider the divisor diagram for the number 48 (Figure 4.20). Then for instance, $\gcd(8, 12) = 4$ and $\text{lcm}(8, 12) = 24$. In the diagram, 4 is the highest number below both 8 and 12 which is connected to 8 and 12 by a sequence of line segments. Also, 24 is the lowest number above both 8 and 12 which is connected to 8 and 12 by a sequence of line segments. In a similar way, we see from the diagram that $\gcd(3, 8) = 1$ and $\text{lcm}(3, 8) = 24$.

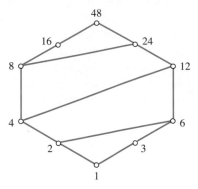

Figure 4.20

EXERCISES

1. Find the common divisors for 60 and 100.

2. Find four common multiples for 6 and 20.

3. Diagram the divisors of 36. By examining the diagram, find $\gcd(12, 18)$, $\text{lcm}(12, 18)$, $\gcd(9, 12)$, $\text{lcm}(9, 12)$.

4. Diagram the divisors of 60. By examining the diagram, find $\gcd(4, 30)$, $\text{lcm}(4, 30)$, $\gcd(6, 20)$, $\text{lcm}(6, 20)$.

5. If $m|n$, what is $\gcd(m, n)$? Why?

6. If $m|n$, what is $\text{lcm}(m, n)$? Why?

7. Find $\gcd(332, 526)$, $\text{lcm}(332, 526)$.

8. Find $\gcd(86, 124)$, $\text{lcm}(86, 124)$.

9. If $\gcd(m, n) = m$, what can you say about m and n?

10. If $\text{lcm}(m, n) = m$, what can you say about m and n?

11. If $\gcd(m, n) = m$, prove that $\text{lcm}(m, n) = n$.

12. What is $\gcd(n, n)$? What is $\text{lcm}(n, n)$?

13. If p is prime, what is $\gcd(p, n)$?

14. Scholars discovered 2520 in hieroglyphs engraved on the stone lid of a tomb in an Egyptian pyramid. Why was such an honor paid this number? Perhaps because it is the smallest natural number divisible by every natural number from 1 through 10. Prove this.

15. Show that $\gcd(\gcd(6, 9), 12) = \gcd(6, \gcd(9, 12))$ and $\text{lcm}(\text{lcm}(6, 9), 12) = \text{lcm}(6, \text{lcm}(9, 12))$.

*16. Which of the following statements are always true? For those statements that are false give an example to show that they are false.
    (a) If $k|(m + n)$, then $k|m$ or $k|n$.
    (b) If $k|mn$, then $k|m$ or $k|n$.
    (c) If $k|n$ and $m|n$, then $k|m$ or $m|k$.

*17. If k, m, n are primes, answer Exercise 16.

4.5 | Prime Numbers

The reasoning of mathematics is founded on certain infallible princi-
ples. Every word they use conveys a determinate idea, and by accu-
rate definitions they excite the same ideas in the mind of the reader
that were in the mind of the writer.
— JOHN ADAMS (1735–1826), Second President of the United States

It has been shown that prime numbers are building blocks from which all natural numbers can be constructed. For this reason, if for no other, prime numbers are certainly important. But even besides this importance, people have been fascinated by prime numbers ever since they began thinking about the natural numbers. Of the following questions posed concerning prime numbers, some have been answered and others have not. After reading these questions, try your hand at them. Then think up your own questions about the primes.

Questions about Primes

1. Is there a formula that will give all the prime numbers?

2. Is it possible to find a million consecutive composite numbers?

3. How many primes are between 1 and 1000, between 1000 and 2000, between 2000 and 3000, etc.?

4. Is the set of primes finite or infinite?

5. How many pairs of primes are there whose difference is exactly 2? (These are called **twin primes**.)

These questions have been around for hundreds and even thousands of years. Let us look at the state of these questions today.

1. Formula for Primes

Many people have tried to find such a formula but without success. There are simple formulas for other special numbers. For example, $F(n) = n^2$ is a formula that gives all the square numbers, $F(n) = 2n$ gives all the even numbers, $F(n) = 2n - 1$ gives all the odd numbers, $F(n) = n(n+1)/2$ gives the triangular numbers, $F(n) = (2n)!/(n+1)!n!$ gives the Catalan numbers, and so forth.

Finding a formula that gives all the primes seems to be so difficult that researchers have relaxed their expectations and attempted to find a formula that gives only primes but not necessarily all of them. A simple example of such an attempt is the formula $F(n) = n^2 - n + 41$. If you evaluate $F(n)$ for the first few natural numbers you find $F(1) = 41$, $F(2) = 43$, $F(3) = 47$, $F(4) = 53, \ldots$, which are indeed prime. In fact, $F(1)$ to $F(40)$ are all prime. Unfortunately, since $F(41) = 41^2$ is not prime, this attempt goes down in failure. What is wanted is a useful formula that generates many primes. Useless formulas that generate one or a couple of primes are easily constructed. Examples of such formulas are $2 + 1^n$, $5 + 0^n$, $2 + n/n$. Fermat thought he had discovered a formula that gives only primes: $G(n) = 2^{2^n} + 1$. It was very hard for Fermat's contemporaries to check the accuracy of this formula because the numbers got large very quickly. But since $G(1) = 5$, $G(2) = 17$, $G(3) = 257$, and $G(4) = 65,537$ are all prime, it was assumed for many years that Fermat's formula generated primes. About a century later, though, Leonhard Euler showed that $G(5) = 2^{32} + 1 = 4,294,967,297 = (641)(6,700,417)$, thus shooting down another attempt.

In 1644 Father Marin Mersenne (1588–1648), an amateur mathematician who corresponded with the great mathematicians of his time, asserted that $2^p - 1$ is prime for $p = 2, 3, 5, 7, 13, 17, 31, 67, 127, 257$. This assertion stood unchallenged for 250 years. In October 1903 an American number theorist, Frank Cole, gave a lecture before the American Mathematical Society. He walked up to the chalkboard, raised 2 to the 67th power, and subtracted 1 from this 21-digit number. Afterwards he proceeded, without saying a word the entire lecture, to multiply $(193,707,721)(761,838,257,287)$ and got the same number, thus destroying Mersenne's assertion. This was one of the few times an audience of the society ever gave the author of a paper a standing ovation, all without a word being uttered. This result, of course, was interesting; but more important, many people wondered what ingenious method Cole had used to factor $2^{67} - 1$. The answer, disappointingly enough, was that it took him many years of Sunday afternoons to prove Mersenne wrong—essentially by trial and error. In 1952 it was proved that Mersenne was correct for most of his numbers. That is, $2^p - 1$ is prime for $p = 2, 3, 5, 7, 13, 17, 19, 31, 61, 89, 107$, and 127. It can also be shown that $2^p - 1$ is not prime for any other p, where $p < 127$. For example,

$$2^4 - 1 = 15$$
$$2^6 - 1 = 63$$
$$2^8 - 1 = 255$$
$$2^9 - 1 = 511 = (7)(73)$$
$$2^{10} - 1 = (2^5)^2 - 1 = (2^5 - 1)(2^5 + 1)$$
$$2^{11} - 1 = 2047 = (23)(89)$$
$$2^{12} - 1 = (2^6 - 1)(2^6 + 1)$$

Primes of the form $2^p - 1$ have become quite important in number theory (see Section 4.6 for one reason) and are called **Mersenne primes**.

2. Consecutive Composite Numbers

Consider the number $N = (6)(5)(4)(3)(2) = 6! = 720$. Now N is not prime. Also

$$N + 2 = 2\,[(6)(5)(4)(3) + 1] = 722$$
$$N + 3 = 3\,[(6)(5)(4)(2) + 1] = 723$$
$$N + 4 = 4\,[(6)(5)(3)(2) + 1] = 724$$
$$N + 5 = 5\,[(6)(4)(3)(2) + 1] = 725$$
$$N + 6 = 6\,[(5)(4)(3)(2) + 1] = 726$$

are composite. This gives five consecutive composite numbers. Similarly, if $M = 10!$, then $M+2, M+3, \ldots, M+10$ give nine consecutive composite numbers. In general, if n is a natural number, then $n!+2, n!+3, \ldots, n!+n$ give $n-1$ consecutive composite numbers. Since n could be 1,000,001, there exist a million consecutive composite numbers. In the same way there exist a billion, trillion, etc., consecutive composite numbers. Thus if all the natural numbers are listed, there will eventually be huge gaps between primes. We might be tempted to conjecture that since these gaps become so immense, the primes will eventually cease to occur; thus there are only finitely many primes.

Table 4.9

Primes	*Total*
2 **3 5 7 11 13 17 19** 23 **29 31** 37 **41 43** 47 53 **59 61** 67 **71 73** 79 83 89 97	25
173 **101 103 107 109** 113 127 131 **137 139 149 151** 157 163 167 **179 181 191 193 197 199**	21
211 223 **227 229** 233 **239 241** 251 257 263 **269 271** 277 **281 283** 293	16
307 **311 313** 317 331 337 **347 349** 353 359 367 373 379 383 389 397	16
401 409 **419 421 431 433** 439 443 449 457 **461 463** 467 479 478 491 499	17
503 509 **521 523** 541 547 557 563 **569 571** 577 587 593 **599**	14
601 607 613 **617 619** 631 **641 643** 647 653 **659 661** 673 677 683 691	16
701 709 719 717 733 739 743 751 757 761 769 773 787 797	14
809 811 821 823 827 829 839 853 **857 859** 863 877 881 883 887	15
907 911 919 929 937 941 947 953 967 971 977 983 991 997	<u>14</u>
	168

3. Distribution of Primes

To get an idea of the distribution of primes among the other natural numbers, the primes in the first 1000 natural numbers are listed in Table 4.9. The twin primes are boldfaced. The number of primes drops at a fairly steady pace. There are 168 primes less than 1000, 135 between 1000 and 2000, and 127 between 2000 and 3000. These numbers steadily decrease, again showing that the number of primes is going down as one reaches very large numbers. Since the primes get sparser with larger numbers, we might be further tempted to conclude that there are only finitely many primes.

One of the great unsolved problems in number theory is to find the number of primes $P(n)$ less than n. Although a formula for $P(n)$ has not yet been found, Legendre and Gauss discovered an approximate formula in 1800. They studied tables of values of $P(n)$ for large n and noticed that $P(n)$ was approximately equal to $n/\ln n$

BLOOM COUNTY   by Berke Breathed

where $\ln n$ is the natural logarithm of n ($\ln n$ is the power of $2.718\ldots$ that equals n). Look at Table 4.10 and see if you agree. Legendre and Gauss conjectured that $P(n)$ approaches $n/\ln n$ as n goes to infinity. In 1896, Hadamard and de la Vallee Poussin proved this conjecture.

Table 4.10

n	$P(n)$	$n/\ln n$
100	25	21.7
1,000	168	144.8
1,000,000	78,498	72,382.0
1,000,000,000	50,847,478	48,254,942.0

4. Infinitely Many Primes

The evidence in Questions 2 and 3 notwithstanding, Euclid proved over 2000 years ago that there are an infinite number of primes. His proof stands as a classic in mathematical thought which has not been improved in the ensuing 2000 years. (The proof is worked out in Exercises 24 through 26.)

Theorem 4.3   **(Euclid)** There are infinitely many primes.

---

5. Twin Primes

You can see from our list of the primes among the first 1000 natural numbers (Table 4.9) that twin primes are getting fewer and farther between. Are there an infinite number of them? No one knows. Some very large twin primes are known, but whether they stop or not is an open question. The largest twin primes known are $1,706,595 \times 2^{11235} \pm 1$.

A weaker result has recently been proved by the Chinese mathematician Jingrun Chen. A number n is **almost prime** if it is a product of two or fewer primes. For example, 4, 6, 9, and 15 are almost prime. Chen proved that there are an infinite number of pairs p, $p + 2$ where p is prime and $p + 2$ is almost prime. The proof is extremely difficult. To write it down in a form comprehensible to a mathematician who is not a specialist in the problem would require 600 pages of condensed text.

EXERCISES

1. Is every odd number prime?

2. Is every even number composite?

3. Does the formula $H(n) = 3^n - 1$ give primes for $n = 1, 2, 3, 4$?

* 4. Let $F(n) = n^2 - n + 41$. Show that $F(n)$ is prime for $n = 1$ through $n = 40$.

5. Let $P(n)$ be the number of primes less than n. What is $P(n)$ for $n = 200, 300, 400, 500, 600, 700, 800,$ and 900?

6. Give an example of five consecutive composite numbers less than 60.

7. Show that there exist 1 billion consecutive composite numbers.

8. Triplet primes are three primes of the form n, $n + 2$, $n + 4$. Give an example of triplet primes. If it were known whether or not there exist an infinite number of triplet primes, would this settle the twin-primes problem?

9. Show that there are infinitely many odd numbers.

10. Show that if n is a prime greater than 2, then $n + 1$ is not prime.

11. Show by examples that if n is prime, then $n + 2$ may or may not be prime.

12. Show that $n^2 - 4$ is not prime for $n > 3$.

13. Find five natural numbers n such that $n^2 + 1$ is prime.

14. It is unknown whether the set of primes of the form $n^2 + 1$ is an infinite set. How many primes are there of the form $n^2 - 1$?

15. Show that there are infinitely many composite numbers.

16. Find some other useless formulas that generate primes.

17. Find a formula that gives composite numbers.

18. Is a prime number almost prime?

19. Is an almost prime number prime?

20. List the almost prime numbers between 2 and 50.

21. List the number pairs between 2 and 50 of the form p, $p + 2$ where p is prime and $p + 2$ is almost prime.

22. How many divisors can an almost prime number have?

23. If a natural number has 2, 3, or 4 divisors, is it almost prime?

In Exercises 24–26 we shall give the proof of Euclid's theorem (Theorem 4.3) that there are infinitely many primes.

24. Suppose there are finitely many primes. Then there is a largest prime t. Let x be the product of all the primes plus 1, $x = 1 + (2)(3)(5)(7) \cdots (t)$. Since $x > t$, x is composite. Use Theorem 4.1 to show that x has the form $x = ps$ where p is prime and $s > 1$.

25. Use Exercise 24 to show that $1 = p[s - (2)(3)(5) \cdots (t)]$, where no p appears in the expression $(2)(3)(5) \cdots (t)$.

26. Use Exercise 25 to conclude that p divides 1, which is impossible.

*27. The Sieve of Eratosthenes can be used to find primes. List all the natural numbers from 2 to 200. Circle 2 to indicate that it is prime. Then cross out every second number from then on, as they are composite (they are divisible by 2). The next uncrossed, uncircled number is 3; circle it to designate it as prime. Then cross out every third number from then on, as they are composite (they are divisible by 3). The next uncrossed, uncircled number is 5; circle it to designate it as prime. Continue this process to find all primes between 2 and 200.

*28. To test whether a natural number n is prime or not you only have to show that it is not divisible by any natural number less than or equal to $\sqrt{n}$ (except 1). Why?

29. Give three examples of primes p for which $1 + p!$ is not prime.

30. Give three examples of primes p for which $1 + p!$ is prime.

*31. Prove that 3, 5, 7 are the only triplet primes (see Exercise 8). [*Hint*: If $3 \nmid n$, $3 \nmid (n + 2)$, then $3 | (n + 1)$.] (The author is indebted to Elisabeth Gambler for this problem.)

4.6 | Perfect Numbers

Six is a number perfect in itself, and not because God created all things in six days; rather the converse is true; God created all things in six days because this number is perfect, and it would have been perfect even if the work of the six days did not exist.
— Aurelius Augustinus (St. Augustine) (A.D. 354–430)

> A natural number is said to be **perfect** if it is the sum of its divisors other than itself.

Thus 6 and 28 are perfect numbers because $6 = 3+2+1$, $28 = 14+7+4+2+1$. The first four perfect numbers are 6, 28, 496, and 8128. The ancient Greek and early Judaic philosophers knew about these four perfect numbers and endowed them with mystical properties. Why did God create the universe in 6 days? Why does a cycle of the moon take 28 days? Because these numbers are perfect. In classical Greek numerology 6 was considered the most beautiful of all numbers; 6 represented marriage, health, and beauty, being the sum of its own parts. It also represented the goddess of love, Venus, for it is the product of 2, which represents female, and 3, which represents male.

Conjectures

After examining the first four perfect numbers and after much philosophizing, the ancients were led to make the following conjectures:

> 1. There are infinitely many perfect numbers.
> 2. The perfect numbers end alternately in 6 and 8.
> 3. There is exactly one perfect number with a given number of digits.
> 4. Every perfect number is even.

According to Nicomachus' *Arithmetica* (A.D. 100),

> But it happens that, just as the beautiful and the excellent are rare and easily
> counted but the ugly and the bad are prolific, so also excessive and defective
> numbers are found to be many and in disorder, their discovery being unsystem-
> atic. But the perfect are both easily counted and drawn up in a fitting order; for
> only one is found in the units, 6; and only one in the tens, 28; and a third in the
> depth of the hundreds, 496; as a fourth one, on the border of the thousands, that
> is short of the ten thousand, 8128. It is their uniform attribute to end in 6 or 8
> and they are invariably even.

It is not known whether these conjectures were based on proofs or were purely empirical and philosophical. However, the latter was probably the case since some of
these conjectures have turned out to be false. It is known that the ancients searched
for years to find the fifth perfect number but were unsuccessful. This could have been
because they expected the fifth perfect number to be between 10,000 and 100,000,
but, as we will see, it is much larger. Let us see how these four conjectures stand
today.

1. It is not known whether there are infinitely many perfect numbers or not. Even
   today only 32 perfect numbers have been found, and perfect numbers are now
   difficult to discover.
2. This conjecture is false. Although they do not end alternately in 6 and 8, it will
   be shown later that all even perfect numbers end in 6 or 8.
3. This conjecture is false.
4. It is not known whether every perfect number is even. All perfect numbers found
   so far have been even, and it has been proved that if there are odd perfect numbers,
   they must be greater than 10^{300}.

Although, as already noted, there is no known formula that gives all the primes,
it turns out that there is a formula that gives all the even perfect numbers. Let us see
if we can discover this formula. The prime factors of the first four perfect numbers
are

$$6 = (2)(3) = (2)(2^2 - 1)$$
$$28 = (2^2)(7) = (2^2)(2^3 - 1)$$
$$496 = (2^4)(31) = (2^4)(2^5 - 1)$$
$$8128 = (2^6)(127) = (2^6)(2^7 - 1)$$

After examining these equations you might conjecture that every even perfect number has the form $2^{p-1}(2^p - 1)$, where $2^p - 1$ is prime. It turns out that not only is
this conjecture true but its converse is also true, that is, every number of the form
$2^{p-1}(2^p - 1)$, where $2^p - 1$ is prime, is an even perfect number.

The following theorem was obtained by Euler in 1757. Part of the proof is outlined
in Exercises 30 through 32 at the end of this section.

Theorem 4.4   **(Euler)** An even number is perfect if and only if it has the form $2^{p-1}(2^p - 1)$, where $2^p - 1$ is prime.

Hence to find all the even perfect numbers it suffices to find all the primes of the form $2^p - 1$, that is, all the Mersenne primes. Although Mersenne asserted that $2^p - 1$ is prime for certain values of p, these numbers were not rigorously proved to be prime until much later.

Known Perfect Numbers

For $p = 2, 3, 5, 7$ we have the first four perfect numbers:

$$2(2^2 - 1) = 6$$
$$2^2(2^3 - 1) = 28$$
$$2^4(2^5 - 1) = 496$$
$$2^6(2^7 - 1) = 8128$$

In 1603 Cataldi proved that $2^{13} - 1$, $2^{17} - 1$, and $2^{19} - 1$ are prime; so the next three perfect numbers are

$$2^{12}(2^{13} - 1) = 33,550,336$$
$$2^{16}(2^{17} - 1) = 8,589,869,056$$
$$2^{18}(2^{19} - 1) = 137,438,691,328$$

It is no wonder that the ancient Greeks could not find the fifth perfect number! The next perfect number was found by Euler in 1772. He proved that $2^{31} - 1$ is prime; so

$$2^{30}(2^{31} - 1) = 2,305,483,008,139,952,128$$

is perfect. The size of this number led Barlow to claim in his number theory book in 1811 that this perfect number "is the greatest that will ever be discovered for as they are merely curious without being useful it is not likely that any person will attempt to find one beyond it." However, this claim did not do justice to the tremendous power of human curiosity.

Over a century after Euler's discovery, Seelhoff and Pervusin in 1886 proved that $2^{61} - 1$ is prime, and so the next even perfect number is

$$(2^{60})(2^{61} - 1) = (1,152,921,504,609,846,976)(2,305,843,009,213,693,951)$$

This is truly a large number, larger than practically any known physical constant, rivaling in size the number of electrons in the universe. But this isn't anything! In 1912, Powers and Cunningham proved that $(2^{88})(2^{89} - 1)$ is perfect.

No progress was made for the next 40 years until the advent of the high-speed computer. In 1952, Uhler showed that the next even perfect numbers are $(2^{106})(2^{107}-1)$ and $(2^{126})(2^{127} - 1)$, and in the same year Robinson proved that $(2^{520})(2^{521} - 1)$,

$(2^{606})(2^{607} - 1)$, $(2^{1278})(2^{1279} - 1)$, $(2^{2202})(2^{2203} - 1)$, and $(2^{2280})(2^{2281} - 1)$ are perfect. A decade passed during which no new perfect numbers were found. Then in 1963, utilizing an advanced second-generation transistorized computer, the ILLIAC at the University of Illinois, with an investment of hundreds of thousands of dollars in computer time, the next even perfect numbers were discovered: $(2^{3216})(2^{3217} - 1)$, $(2^{4252})(2^{4253} - 1)$, $(2^{4422})(2^{4423} - 1)$, $(2^{9688})(2^{9689} - 1)$, $(2^{9940})(2^{9941} - 1)$, and $(2^{11212})(2^{11213} - 1)$. This last number has 6957 digits and would take 10 pages of this book to write out in full.

It was thought for a while that the end had come. But in 1971, another even perfect number was found with the help of a third-generation microcircuit computer. Its value, $(2^{19936})(2^{19937} - 1)$, is so large it would take a small volume to write out in full. Three more perfect numbers were then found using the latest computers of the time. Teenagers Laura Nickel and Curt Noll showed that $(2^{21700})(2^{21701} - 1)$ is perfect in 1978. In 1979, Curt Noll showed that $(2^{23208})(2^{23209} - 1)$ is perfect. The next largest perfect number was found by David Slowinski and Harry Nelson in 1979. Its value is $(2^{44496})(2^{44497} - 1)$. New perfect numbers were discovered in 1982, 1983, 1985, and 1988. Finally, the largest known perfect number was found in 1992 by a group of British mathematicians. This feat required 19 hours of searching by a Cray-2 supercomputer. This number has 455,663 digits and it would take almost this entire book to write out in full. Its value is $(2^{756838})(2^{756839} - 1)$.

Finding perfect numbers is a mathematical pursuit much like mountain climbing. It is a sport for cerebral thrill-seekers, combining equal parts ingenuity, luck, and mathematical brute force. Table 4.11 lists the values of p for the known even perfect numbers, their last digit, and the year in which they were shown to be perfect.

EXERCISES

1. Show that 12 is not perfect by adding its divisors less than itself.

2. Show that there are no perfect numbers between 6 and 28.

3. Verify that 496 is perfect by adding its divisors less than itself.

4. Verify that 8128 is perfect by adding its divisors less than itself.

5. Find $d(6)$, $d(28)$, and $d(496)$.

6. Find $d(8128)$ and $d(33,550,556)$.

7. If $n = (2^{p-1})(2^p - 1)$, where $2^p - 1$ is prime, what is $d(n)$?

For Exercises 8–22 you will need the following definitions. For a natural number n, $s(n)$ is the sum of the divisors of n. Two natural numbers are **relatively prime** if they have no common divisors except 1.

8. Are 2 and 4 relatively prime? Are 4 and 9 relatively prime?

9. Find $s(n)$ for $n = 1, 2, \ldots, 10$.

10. Find $s(6)$, $s(28)$, and $s(496)$.

11. (a) Are two different primes relatively prime?
    (b) Do two natural numbers have to be prime to be relatively prime?

12. Show that n is prime if $s(n) = n + 1$.

13. If n is prime, show that $s(n) = n + 1$.

Table 4.11

p	Last digit of $(2^{p-1})(2^p - 1)$	Date
2	6	2500 B.C.
3	8	2500 B.C.
5	6	2500 B.C.
7	8	2500 B.C.
13	6	1603
17	6	1603
19	8	1603
31	8	1772
61	6	1886
89	6	1912
107	8	1952
127	8	1952
521	6	1952
607	8	1952
1,279	8	1952
2,203	8	1952
2,281	6	1952
3,217	6	1963
4,253	6	1963
4,423	8	1963
9,689	6	1963
9,941	6	1963
11,213	6	1963
19,937	6	1971
21,701	6	1978
23,209	6	1979
44,497	6	1979
86,243	8	1982
110,503	8	1988
132,049	6	1983
216,091	8	1985
756,839	8	1992

14. Show that m and n are relatively prime if and only if $\gcd(m, n) = 1$.

15. Show that n is perfect if and only if $s(n) = 2n$.

16. Give three examples of natural numbers n that satisfy $s(n) < 2n$.

17. Give three examples of natural numbers n that satisfy $s(n) > 2n$.

18. Find $s(8128)$, $s(33,550,336)$.

19. Two natural numbers n and m are called **friendly** if $s(m) - m = n$ and $s(n) - n = m$. Show that 220 and 284 are friendly.

20. Compare $s(4)s(9)$ with $s(4 \times 9)$.

21. For the following pairs of numbers, check to see if $s(mn) = s(m)s(n)$.
    (a) $(6, 35)$ (b) $(3, 6)$ (c) $(3, 5)$ (d) $(4, 6)$

*22. Prove that if m and n are relatively prime, then $s(mn) = s(m)s(n)$.

23. Show that $(2^{63})(2^{64} - 1)$ is not perfect.

24. A natural number is called **magic** if when we add its digits and then add the digits of the resulting sum, etc., we obtain 1. For example, 298 is magic. Show that the first six perfect numbers except 6 are magic.

25. Find 10 magic numbers that are not perfect (see Exercise 24 for the definition of magic).

26. A natural number is **multiplicatively perfect** if it equals the product of its divisors other than itself. Show that 6, 8, 27, 35, 125, and 134 are multiplicatively perfect.

27. If n is the product of two distinct primes, show that n is multiplicatively perfect (see Exercise 26).

28. If $n = p^3$, where p is prime, show that n is multiplicatively perfect (see Exercise 26).

29. Are there infinitely many multiplicatively perfect numbers (see Exercise 26)?

*30. Let $n = 2^{p-1}(2^p - 1)$ where $2^p - 1$ is prime. List the divisors of n other than n itself.

*31. Add the divisors found in Exercise 30 and use a result in Section 1.2 to simplify.

*32. Show that n of Exercise 30 is perfect.

*33. Prove that the last digit of 2^{p-1} must be 2, 4, 6, or 8.

*34. If $2^p - 1$ is prime, prove that $2^{p-1}(2^p - 1)$ must have last digit 6 or 8.

*35. It can be shown that a perfect number $2^{p-1}(2^p - 1)$ ends in 6 if and only if $4 \mid (p - 1)$ (except for the case of $p = 2$). Verify the last digit entries in Table 4.11.

4.7 Base 2

It is India that gave us the ingenious method of expressing all numbers by means of ten symbols, each receiving a value of position as well as an absolute value; a profound and important idea which appears so simple to us now that we ignore its true merit.
— PIERRE SIMON DE LAPLACE (1749–1827), French Mathematician

After reading the previous sections of this chapter you might be tempted to comment: "I can see that number theory is an important intellectual endeavor for cultural historical reasons, but what use is it? Does it have any practical applications?" The answer is yes. Many practical applications are derived from base 2, or binary, numeration. Applications of binary numeration to computers will be discussed in this section and also in Chapter 7, on computer science. In Section 4.8, an application of base 2 to game theory will be discussed.

Centuries before the invention of the computer, number theorists studied numeration systems other than our usual decimal system. In the decimal system we use 10 symbols, 0, 1, 2, 3, 4, 5, 6, 7, 8, and 9, and the places represent powers of 10. For example, 876 means 6 ones (or 10^0s), 7 tens (or 10^1s), and 8 hundreds (or 10^2s); thus 876 is a shorthand notation for $8 \times 10^2 + 7 \times 10^1 + 6 \times 10^0$. A numeration system need not be based on 10; it can be based on any natural number greater than 1.

Base 2 System

In a numeration system based on 2 we use two symbols, 0 and 1, and the places represent powers of 2.

For example, 1110 means 0 ones (or 2^0s), 1 two (or 2^1s), 1 four (or 2^2s), and 1 eight (or 2^3s); thus 1110 is shorthand notation for $1 \times 2^3 + 1 \times 2^2 + 1 \times 2^1 + 0 \times 2^0$, which in base 10 notation is 14.

The first mathematician to consider the binary system was Gottfried Leibniz in about 1670. Although Leibniz saw the great simplicity and mathematical beauty of such a system, he did not find any important mathematical uses for it. Even though he did not recognize its importance in computing machines, Leibniz had a great interest in such machines. In fact, he drew plans for a mechanical calculator, which was built in 1695 but proved unreliable. Leibniz, who was also a statesman, philosopher, and theologian, saw the binary system mainly in terms of its religious significance. He saw God represented by 1 and nothing represented by 0. Leibniz used binary numeration as proof that God created the world out of nothing.

When computers were invented, this binary (two-symbol) numeration system fitted them perfectly. A computer is composed of many electronic components in which electric current is either flowing or not flowing or of magnetic fields having one direction or its opposite. Thus each component is in one of two possible configurations. These two configurations can be used to represent two symbols, say 0 and 1. Thus sequences of 0s and 1s form the language that a computer most easily understands.

It is unfortunate that people have 10 fingers. If we had only one finger on each hand, we would probably count in the binary system and it would be much easier to communicate with computers. For practice, Table 4.12 gives the first 16 natural numbers and their binary representations. For example, 1001 in the table is the binary representation of $1 \times 2^3 + 0 \times 2^2 + 0 \times 2^1 + 1 \times 2^0$, which equals 9 in decimal notation. See if you can verify that the other entries of Table 4.12 are correct.

Figure 4.21 illustrates part of the "binary tree." Any binary number can be represented if the tree is made large enough.

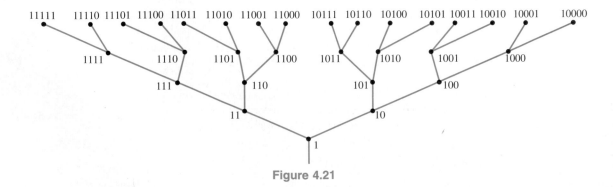

Figure 4.21

Table 4.12

Number	Binary representation
1	1
2	10
3	11
4	100
5	101
6	110
7	111
8	1000
9	1001
10	1010
11	1011
12	1100
13	1101
14	1110
15	1111
16	10000

Addition

Your first reaction might be: Why take a simple number like 9 and write it in the ridiculous form, 1001? This somewhat cumbersome way of writing numbers is greatly compensated for by the resulting ease of manipulation. For example, to perform addition you do not have to remember complicated addition table entries like $3 + 6 = 9$ and $8 + 7 = 15$; all you have to remember is that $0 + 0 = 0$, $1 + 0 = 1$, and $1 + 1 = 10$. Table 4.13 is the binary addition table. For instance, consider the addition problem

$$
\begin{array}{r}
1\,10 \\
+111 \\
\hline
1101
\end{array}
$$

In the first column (on the right) you have $0 + 1 = 1$; in the second column $1 + 1 = 10$, so put down 0 and "carry" the 1; in the third column $1 + 1 = 10$ and $10 + 1 = 11$, so the answer is 1101. Of course, in base 10 calculation this is $6 + 7 = 13$, but using binary numeration you had to memorize practically nothing to obtain the result.

Table 4.13

+	0	1
0	0	1
1	1	10

Table 4.14

×	0	1
0	0	0
1	0	1

Multiplication

Multiplication is also easy. All you have to remember, besides the simple addition table, is $(0)(0) = 0$, $(0)(1) = 0$, $(1)(1) = 1$. Table 4.14 is the binary multiplication table. For example, consider the multiplication

$$
\begin{array}{r}
1101 \\
\times 101 \\
\hline
1101 \\
11010 \\
\hline
1000001
\end{array}
$$

Of course, in base 10 calculation this is $(5)(13) = 65$.

The ease with which binary arithmetic is performed also makes binary numbers especially suitable for computers. Electronic circuits can be designed to perform this simple arithmetic very efficiently and economically.

Binary to Decimal Conversion

It is useful to have simple methods for converting binary numbers to decimal numbers and vice versa. For example, suppose you want to convert the binary number 10101 to its decimal form. One method would be to write this number as

$$
10101 = 1 \times 2^0 + 0 \times 2 + 1 \times 2^2 + 0 \times 2^3 + 1 \times 2^4
$$

This equals $1 + 4 + 16 = 21$.

There is another method called **double-dabble**. This method is easier, especially for large numbers. Double-dabble works as follows. Double the first digit, add the result to the second digit, double the sum, add the result to the third digit, double the sum, and so forth. This method is illustrated in Figure 4.22 for binary number 10101. In Figure 4.23 the method is illustrated for the binary number 11011010.

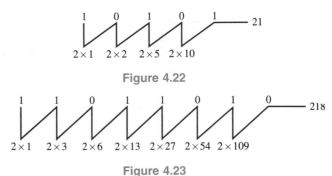

Figure 4.22

Figure 4.23

Decimal to Binary Conversion

Methods for converting decimal numbers to their binary form are now given. Suppose you want to convert decimal 21 to its binary form. One method is the following. Find the largest power of 2 that is less than or equal to 21, in this case 16. Write 21 as 16 plus what is left over, $21 = 16 + 5$. Find the largest power of 2 that is less than or equal to 5, in this case 4. Then

$$21 = 16 + 4 + 1 = 1 \times 2^4 + 0 \times 2^3 + 1 \times 2^2 + 0 \times 2 + 1 \times 2^0$$

Hence 21 is 10101 in binary notation.

A simpler method, especially for large numbers, is called **inverse double-dabble**. Using this method to convert decimal 21 to binary, you divide 21 by 2 and note the remainder, divide the result by 2 and note the remainder, and so forth. Write down the remainders in reverse order; this is the binary form of 21. Figure 4.24 illustrates this method; the remainders are circled.

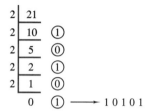

Figure 4.24

Information Retrieval

Binary numbers are useful in information retrieval systems. For example, the following experiment shows how binary numbers can be used to sort cards. Take 64 cards of the same size and punch six holes across the top as in Figure 4.25. Leave a larger space to the left so that you know which way is face up. Make these cards correspond to the numbers 0 to 63 according to base 2 notation by cutting through the edge where a 0 appears and leaving the holes as they are for a 1. Card 41 (binary 101001) is illustrated in Figure 4.26. Shuffle the cards, keeping them face up. Now with a pencil or stiff wire poke through the first holes on the right and lift out the cards with a 1 on the right. Place these on the bottom of the deck, still face up, and repeat with the second hole from the right. Place the cards you lift out on the bottom face up. Continue from right to left until you do the sixth hole. When you finish, all 64 cards will be in order after only six steps. Why does this work?

Figure 4.25

Figure 4.26

EXERCISES

1. We can write 39 as a sum of multiples of powers of 10 as $39 = 3 \times 10 + 9 \times 10^0$. Write 47 as a sum of multiples of powers of 10.

2. Write 239 as a sum of multiples of powers of 10.

3. Write 1239 as a sum of multiples of powers of 10.

4. Write 1 million as a power of 10.

5. Write 1 billion as a power of 10.

6. How many zeros does 10^{23} have in decimal notation?

7. How many zeros does 10^n have in decimal notation?

8. Continue the binary tree (Figure 4.21) to one more level.

9. Write the following numbers in binary notation. (a) 2   (b) 4   (c) 8   (d) 16

10. Write the following numbers in binary notation. (a) 32   (b) 64   (c) 128

11. Write 2^{15} in binary notation.

12. How many zeros does 2^n have in binary notation?

13. Find the decimal representation of each of the following binary numbers.
    (a) 10   (b) 100   (c) 1000   (d) 10000

14. Using two different methods, find the decimal representation of each of the following binary numbers. (a) 1000000   (b) 101010   (c) 11111   (d) 100111

15. Using two different methods, find the binary representation of each of the following decimal numbers. (a) 25   (b) 64   (c) 220   (d) 371

16. Find the binary representation of the first three perfect numbers.

17. Perform the following additions and check your answers using base 10.

    (a)   10111   (b)   11011   (c)   1011
         +11010   +11000   11001
                                                  +1001

18. Perform the following multiplications and check your answers using base 10.

    (a)   1111   (b)   110   (c)   1110
         ×101   ×1011   ×1001

19. What does base 3 numeration mean? Find the base 3 (**ternary**) representation of the numbers in Exercise 9. Find the base 10 representation of the ternary numbers 12, 1021, and 1211.

20. Find the binary representation of the fourth and fifth perfect numbers.

*21. Explain why the card-sorting experiment works.

*22. Discuss base n enumeration for $n > 1$ a natural number.

*23. In this exercise you will show why double-dabble works. Let $abcde$ be a binary number, where a, b, c, d, and e are 0 or 1. The corresponding decimal number is

$$2^4 a + 2^3 b + 2^2 c + 2d + e \tag{4.1}$$

where the sum is given in decimal notation. Carry out the double-dabble steps on $abcde$ and show that you obtain (4.1).

*24. Do Exercise 23 for the binary number $abcdefg$.

*25. In this exercise you will show why inverse double-dabble works. Let $2^4 a + 2^3 b + 2^2 c + 2d + e$ be a decimal number. Carry out the steps of inverse double-dabble to obtain the binary number $abcde$.

*26. Do Exercise 24 for the decimal number $2^6 a + 2^5 b + 2^4 c + 2^3 d + 2^2 e + 2f + g$.

4.8 | Nim (Optional)

In fact, everything that can be known has number, for it is not possible
to conceive of or to know anything that has not.
— PHILOLAUS OF CROTON (500 B.C.), Greek Philosopher

Binary numeration can be applied to finding a winning strategy for the game of nim.
Games like nim are part of a general theory in mathematics called **game theory**.
Although nim is merely an amusing game played for pleasure, general game theory
is a deep and serious subject. It can be used to find best strategies in trade, business,
defense, economic and political competition, operations research, and other important
fields.

The game of nim is played with two people using three piles of sticks (toothpicks
or matches are usually the most convenient). The game starts with an arbitrary number
of sticks in each pile. Each player in turn takes at least one stick from one and only
one of the piles. A player may take any number of sticks as long as they are from the
same pile. The player who removes the very last stick wins the game.

Position

To analyze this game, suppose there are a, b, and c sticks in the first, second, and third
piles, respectively. We represent this by (a, b, c) and call it a **position**. For example,
suppose we begin with 7, 12, and 15 sticks in the three piles. Then $(7, 12, 15)$ is the
initial position (Figure 4.27).

Figure 4.27

Let us play a sample game. Player A takes 7 sticks from the first pile, giving position
$(0, 12, 15)$ (Figure 4.28).

Figure 4.28

Player B takes 3 from the third pile, giving position $(0, 12, 12)$ (Figure 4.29).

Figure 4.29

Player B has now locked into a winning strategy. No matter how many sticks player A
takes from the second or third pile, B takes the same number from the opposite pile.

Figure 4.30   Figure 4.31   Figure 4.32

For example, if A takes 10 from the second pile, position $(0, 2, 12)$ in Figure 4.30, B takes 10 from the third pile, giving position $(0, 2, 2)$ (Figure 4.31).

What can A do? A cannot take 2 sticks from a pile, for then B will take the remaining 2 and win. So A takes 1 from the third pile, giving position $(0, 2, 1)$ (Figure 4.32). Now B takes 1 from the second pile, giving position $(0, 1, 1)$. Player A is forced to take a stick, and B takes the remaining stick to win.

Let us replay the game with a smarter A (Figure 4.33). Player A takes 4 sticks from pile 1, giving position $(3, 12, 15)$. Player B takes 13 from pile three to give $(3, 12, 2)$. Player A takes 11 from pile 2, giving $(3, 1, 2)$. Player B takes 1 from the first pile, giving $(2, 1, 2)$. Player A takes 1 from pile 2, giving $(2, 0, 2)$. As we saw before, B is now bound to lose. How did A find this winning strategy? There is a simple method, using binary numeration, for A to make the right moves so that A will always win starting with the initial position $(7, 12, 15)$.

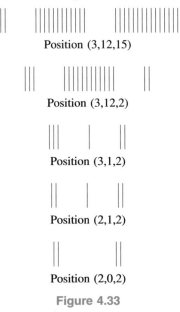

Position (3,12,15)

Position (3,12,2)

Position (3,1,2)

Position (2,1,2)

Position (2,0,2)

Figure 4.33

Strategic Position

A position (a, b, c) is said to be **held** by a player if this position results after that player makes his play. A position is called a **strategic position** if the player who holds it can force his opponent to lose no matter how the opponent plays from then on.

Thus if player A holds a strategic position, his opponent B cannot capture a strategic position on his next play; but on A's next play A can hold a new strategic position regardless of what B has done. One of the important studies of general game theory is to ascertain whether a particular game has strategic positions. It turns out that some games do and others do not. Nim is a game that does have strategic positions. This fact will be proved, and the strategic positions will be determined. If (a, b, c) is a position, write the numbers a, b, c in their binary representation in three rows. For example, if the position is $(6, 14, 9)$, represent this as

$$110$$
$$1110$$
$$1001$$

Even and Odd Positions

If there are an even number of 1s in each column, then (a, b, c) is called an **even position**; otherwise (a, b, c) is an **odd position**.

Thus $(6, 14, 9)$ is an odd position; while $(7, 14, 9)$, which is represented

$$111$$
$$1110$$
$$1001$$

is an even position. It will eventually be shown that the even positions are the strategic ones. Moral? Never place yourself in an odd position.

Lemma 4.5   If (a, b, c) is an even position, then any position that can be reached in one play from (a, b, c) is odd.

This lemma is now illustrated with an example. Suppose the even position is $(5, 8, 13)$, represented in binary as

$$101$$
$$1000$$
$$1101$$

No matter which row is changed, the resulting position is odd. Suppose it is decided to change the first row, that is, reduce the number 5. If 5 is changed to 4, 3, 2, 1, or 0, the first row becomes 100, 11, 10, 1, or 0, respectively. In each of these cases an odd position results.

Lemma 4.6   If (a, b, c) is an odd position, then an even position can be reached in the next play.

Again the lemma is illustrated with an example. Suppose the odd position is $(7, 11, 9)$, which is written in the standard form

$$111$$
$$1011$$
$$1001$$

Notice that the second column from the left is the first one with an odd number of 1s. Since the first row has a 1 in this column, change this 1 to 0. Also change the last 1 in the first row to 0, obtaining 10 (leave the second 1 alone in this row because there already are an even number of 1s in that column). The resulting position is even:

$$10$$
$$1011$$
$$1001$$

Now suppose player A holds an even position. By Lemma 4.5, after the next play B holds an odd position. By Lemma 4.6, A can hold an even position on the next play. This continues until A holds the smallest even position $(0, 0, 0)$ and wins. Conversely, if A holds an odd position, then after the next play B can hold an even position. Since B can retain even positions no matter what A does, B can eventually win. We have therefore proved the following theorem.

Theorem 4.7   A position is strategic if and only if it is even.

Armed with this knowledge, let us see how player A discovered the winning strategy in Figure 4.33. The initial position $(7, 12, 15)$ has the form

$$111$$
$$1100$$
$$1111$$

Since this is an odd position, A can hold a strategic position on the first play. In order to hold an even position, A must cause the second column to have an even number of 1s. A simple way of doing this is to replace 111 in the first row by 11. This would change 7 sticks in the first pile to 3 sticks, so A removes 4 sticks from pile 1. This gives the even position $(3, 12, 15)$, which has the form

$$11$$
$$1100$$
$$1111$$

No matter what B's first play is, B will hold an odd position. In this case B takes 13 sticks from pile 3 to obtain the position $(3, 12, 2)$

$$11$$
$$1100$$
$$10$$

To hold an even position, A must change the two 1s in the second row to 0s and the last 0 to 1. Thus A removes 11 sticks from pile 2 to obtain position $(3, 1, 2)$:

$$11$$
$$1$$
$$10$$

Player B takes 1 stick from the first pile, giving position $(2, 1, 2)$:

$$10$$
$$1$$
$$10$$

Player A must remove 1 stick from the second pile to hold the even position $(2, 0, 2)$:

$$10$$
$$0$$
$$10$$

As we have seen, A can now win easily.

Strategy in Nim

The game of nim has thus been completely analyzed. If the initial position is even, then the second player has the advantage since the first player must begin by holding an odd position, after which the second player can hold a strategic position. However, since there are more odd positions than even ones, usually the initial position is odd. In this case the first player has the advantage because this player can immediately hold a strategic position.

A list of the strategic positions other than those of the form $(0, n, n)$—which we know are strategic—for the case when each pile has 10 or fewer sticks is

$$
\begin{array}{lll}
(1, 2, 3) & (2, 4, 6) & (3, 4, 7) \\
(1, 4, 5) & (2, 5, 7) & (3, 5, 6) \\
(1, 6, 7) & (2, 8, 10) & \\
(1, 8, 9) & &
\end{array}
$$

Of course, the numbers may apply to any of the three piles; that is, $(1, 2, 3)$ not only represents $(1, 2, 3)$ but also $(2, 1, 3)$, $(2, 3, 1)$, $(1, 3, 2)$, $(3, 1, 2)$, and $(3, 2, 1)$.

If you memorize these positions, you can always win a nim game that has 10 or fewer sticks to a pile if the initial position is odd. Even if the initial position is even, you have a good chance of winning against an uninformed player.

EXERCISES

1. Why is a position of the form $(0, n, n)$ even?

2. Show that (n, n, n) is an odd position if $n \neq 0$.

3. Which of the following positions are strategic?
   (a) $(5, 8, 11)$ (b) $(7, 13, 6)$ (c) $(10, 11, 12)$ (d) $(9, 12, 17)$

4. How would our analysis of nim change if there were four piles of sticks? If there were n piles?

5. If the initial position is $(15, 13, 11)$ and you are the first player, what would you do?

6. Show that the following positions are strategic.
   (a) $(1, 2, 3)$ (b) $(2, 4, 6)$ (c) $(3, 4, 7)$ (d) $(1, 4, 5)$ (e) $(2, 5, 7)$
   (f) $(3, 5, 6)$ (g) $(1, 6, 7)$ (h) $(2, 8, 10)$ (i) $(1, 8, 9)$

* 7. Prove Lemma 4.5.

* 8. Prove Lemma 4.6.

Chapter 4 Summary of Terms

number sequences   triangular numbers
pentagonal numbers
Fibonacci sequence
Catalan sequence

partitions   four squares problem
Fermat's last conjecture
Goldbach's conjecture

prime numbers

prime factorization

composite numbers

divisors

greatest common divisor

least common multiple

Mersenne primes

Euclid's theorem

twin primes

perfect numbers

base 2

double-dabble

nim

strategic positions

Chapter 4 Test

1. Prove that the product of two even numbers is even.
2. Prove that the product of an even number and an odd number is even.
3. Prove that the product of two odd numbers is odd.
4. What are the seventh and eighth terms of the Catalan sequence?
5. List the partitions of 8.
6. Find the pf of 120.
7. Find $d(480)$.
8. Draw the divisor diagram for 120.
9. Find $\gcd(664, 526)$ and $\operatorname{lcm}(664, 526)$.
10. Show that 93 is not prime.
11. Write 4200 as a product of primes.
12. Prove that $\gcd(m, n) \mid \operatorname{lcm}(m, n)$.
13. If a number n is a product of three different primes, what is $d(n)$?
14. (a) Give an example of two primes whose sum is also prime.
    (b) Give an example of three primes whose sum is also prime.
15. Find the binary representation of 197.
16. Find the decimal representation of the binary number 1011011.
17. Are the following statements true or false?
    (a) 55 is a number in the Fibonacci sequence.
    (b) An odd number times an even number is always odd.
    (c) $3 = 1 + 2$ and $3 = 2 + 1$ are different partitions of 3.
    (d) $\gcd(32, 48) = 8$.
    (e) There are no numbers that satisfy $x^3 + y^3 = z^3$.
    (f) 91 is a prime number.
    (g) 881 and 883 are twin primes.
    (h) 8128 is a perfect number.
    (i) $2^{756839} - 1$ is the largest prime number.
    (j) The binary representation of 32 is 100000.
18. "I was n years old in the year n^2," said John in 1971. When was he born?
19. Give two different examples of five natural numbers whose sum equals their product.
20. Show that if an odd number has a prime 2-partition, then one of the summands must be 2.

GEOMETRY

5

Geometry will draw the soul toward truth and create the spirit of philosophy. — PLATO (429–347 B.C.), Greek Philosopher

What science can there be, more noble, more excellent, more useful for men, more admirably high and demonstrative, than this mathematics? —BENJAMIN FRANKLIN (1706–1790) American Statesman and Inventor

A comprehensive study of geometry began with the ancient Greeks. To the Greeks of antiquity, mathematics consisted of two main branches, geometry and measure (the study of numbers). Although the Greeks were preceded in their studies by the Egyptians and Babylonians, they were the first to insist that geometrical statements be given rigorous proofs.

The most basic concepts in geometry are point, line, and plane. The properties of these concepts were summarized by the Greek mathematician Euclid in his book *Elements*, written about 300 B.C. Although we understand intuitively what a point, line, and plane are, we cannot give precise definitions for these terms. Euclid described a point as "that which has no part," but this description is so vague that it is of no use. Euclid described a line as "that which has breadthless length" and a plane as "a surface which lies evenly with the straight lines of itself." We know what Euclid was intending and we can sympathize with his difficulties, but we cannot accept these as precise definitions. How would you define these concepts? We can visualize a point as a very small dot, a line as a set of points with no gaps extending straight in both directions indefinitely, and a plane as an infinite flat surface. To go beyond these imprecise descriptions we can only list the properties of these concepts and study their implications.

According to Euclid, geometry has five main properties. These are the postulates or axioms of plane geometry. Can you draw pictures to illustrate each of these postulates?

Euclid's postulates

1. Two points determine one and only one straight line.

2. A straight line extends indefinitely in either direction.

3. A circle may be drawn with any given center and any given radius.

4. All right angles are equal.

5. Given a straight line ℓ and a point P not on ℓ, there is one and only one straight line through P that is parallel to ℓ.

A geometrical structure satisfying Euclid's five postulates is called a **Euclidean geometry**. Using these postulates, Euclid was able to prove many theorems about geometric figures. In this chapter we shall discuss some of these theorems as well as some non-Euclidean geometries which fail to satisfy all the above postulates.

5.1 | Analytic Geometry

One would state with great confidence that he could convince any sane child that the simpler propositions of Euclid are true; but, nevertheless, he would fail, utterly, with one who should deny the definitions and axioms.
— ABRAHAM LINCOLN (1809–1865), American President

As long as algebra and geometry proceeded along separate paths, their advance was slow and their applications limited. But when these sciences joined company they drew from each other fresh vitality and henceforth marched on at a rapid pace toward perfection.
— JOSEPH LOUIS LAGRANGE (1736–1813), French Mathematician

To shorten the terminology, we shall refer to a straight line as simply a **line**. Theorems are proved in Euclidean geometry by starting with Euclid's postulates and proceeding using logical deduction. Our first example illustrates this method.

EXAMPLE 1   Prove that two distinct nonparallel lines intersect at precisely one point.
  Solution.   By definition, two nonparallel lines must intersect. If the lines have two points in common, then by Postulate 1, the lines must coincide. But this contradicts the fact that the lines are distinct. Hence, the lines have precisely one point in common.   □

Many other results can be derived abstractly from Euclid's five postulates. However, we shall not pursue this line of thought and shall leap 2000 years to more modern and efficient methods. These methods are due to the French mathematician René Descartes (1596–1650). In 1637, Descartes introduced a way of unifying algebra and geometry called **analytic geometry**. Using analytic geometry, we can represent

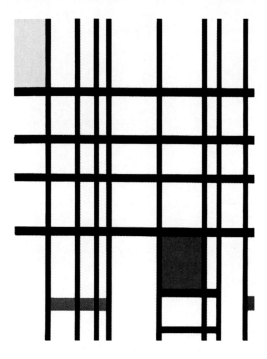

[1] Piet Mondrian (1872-1944), *Composition with Red, Yellow and Blue*, c. 1926. (Tate Gallery, London/Art Resource)

points, lines, and other geometric objects by numbers. We can then derive algebraic equations to prove results about these objects.

Cartesian Plane

Starting with the real numbers $\mathbb{R}$, Descartes defined a plane as the set $\mathbb{R}^2$ of ordered pairs of real numbers

$$\mathbb{R}^2 = \{(x, y): \; x, y \in \mathbb{R}\}$$

The pairs are called ordered because their order is important. For example, $(2, 3)$ is not the same as $(3, 2)$.

> The set $\mathbb{R}^2$ is called the **Cartesian plane** and a pair (x, y) represents a point in the plane.

In this way, Descartes was able to describe Euclid's points and plane by numbers. We have already considered the Cartesian plane in Section 1.5. In particular, we have discussed how to locate points in the Cartesian plane and how to graph functions.

Standard Form

One of the first things that Descartes accomplished with his analytic geometry was the description of lines using algebraic equations. In general, we have the following definition.

A **line** is the graph of an equation of the form
$$Ax + By = C \qquad (5.1)$$
where A and B are constants which are not both 0.

That is, a line is a set of points $(x, y) \in \mathbb{R}^2$ where x and y satisfy (5.1). Equation (5.1) is called the **standard form** for the equation of a line.

Intercepts

Various important properties of a line can be deduced from its equation.

The **x-intercept** of a line is the x-coordinate (if any) of the point where the line crosses the x-axis. The **y-intercept** is the y-coordinate (if any) of the point where the line crosses the y-axis.

EXAMPLE 2   Discuss the intercepts of the line given by the equation $y = 2$.

Solution.   This line consists of the graph of $y = 2$ as shown in Figure 5.1. The line has no x-intercept since it never crosses the x-axis. Its y-intercept is $y = 2$.   □

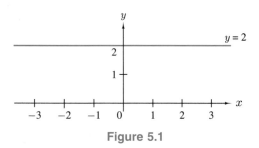

Figure 5.1

EXAMPLE 3   Discuss the intercepts of the line given by the equation $x = -1$.

Solution.   This line consists of the graph of $x = -1$ as shown in Figure 5.2. The line has no y-intercept since it never crosses the y-axis. Its x-intercept is $x = -1$.   □

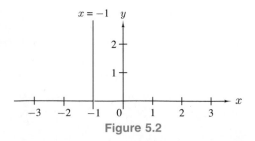

Figure 5.2

Horizontal and Vertical Lines

The equations in Examples 2 and 3 are in standard form. In Example 2, $A = 0$, and when this is the case the line is called **horizontal**. In Example 3, $B = 0$, and when this is the case the line is called **vertical**. The x-intercept of a line is found by setting $y = 0$ and the y-intercept is found by setting $x = 0$. Thus, in the standard form (5.1), if $A \neq 0$ the x-intercept is $x = C/A$ and if $B \neq 0$ the y-intercept is $y = C/B$.

EXAMPLE 4   Find the x-intercept and y-intercept of the line given by the equation $2x - 5y = 10$. Graph the line.

    Solution.   Setting $y = 0$ gives the x-intercept $x = 5$. Setting $x = 0$ gives the y-intercept $y = -2$. We conclude that the two points $(5, 0)$ and $(0, -2)$ lie on the line. The line can now be constructed by connecting these two points as in Figure 5.3. $\square$

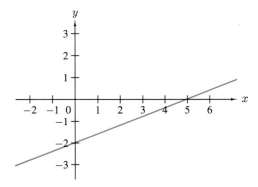

Figure 5.3

Slope

Another important property of a line is its steepness or **slope**. The slope of a line is denoted by m from the French word *monter*, which means "to climb." The number m is the ratio of the vertical change of the line to its horizontal change and is often referred to as "rise over run." We thus have

$$\text{slope } = m = \frac{\text{rise}}{\text{run}} = \frac{\text{difference of } y\text{-coordinates}}{\text{difference of } x\text{-coordinates}}$$

If (x_1, y_1) and (x_2, y_2) are two different points on a line and $x_1 \neq x_2$, then we can compute the slope as follows:

$$m = \frac{y_2 - y_1}{x_2 - x_1} \tag{5.2}$$

Equation (5.2) is illustrated in Figure 5.4.

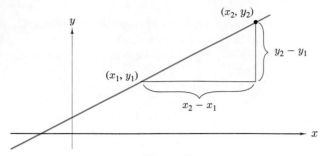

Figure 5.4

Suppose a line has an equation in standard form (5.1) where $A \neq 0$ and $B \neq 0$. We have seen that the x-intercept is C/A and the y-intercept is C/B. In this case the line rises an amount $-C/B$ as it runs an amount C/A. Hence, the slope is $m = (-C/B)/(C/A) = -A/B$. In short,

$$m = -\frac{A}{B} \qquad (5.3)$$

For example, in Figure 5.3, the x-intercept is 5 and the y-intercept is -2. In this case the line rises by 2 as it runs by 5, so the slope is

$$m = \frac{-(-2)}{5} = \frac{2}{5}$$

If $A = 0$, the line is horizontal and has slope 0. If $B = 0$, the line is vertical and we say that the slope is not defined.

EXAMPLE 5   Find the slope of the line $5x - 3y = 8$.
  Solution.   In this case $A = 5$ and $B = -3$, so by (5.3)

$$m = -\frac{A}{B} = \frac{5}{3} \qquad \square$$

EXAMPLE 6   Find the slope of the line through $(-3, 5)$ and $(2, -4)$.
  Solution.   Letting $(x_1, y_1) = (-3, 5)$, $(x_2, y_2) = (2, -4)$ and applying (5.2) gives

$$m = \frac{y_2 - y_1}{x_2 - x_1} = \frac{-4 - 5}{2 - (-3)} = -\frac{9}{5} \qquad \square$$

Slope-Intercept Form

If the equation for a line is in standard form (5.1) with $B \neq 0$, we can solve for y to obtain

$$y = -\frac{A}{B}x + \frac{C}{B}$$

We have seen that $-A/B$ is the slope m. For simplicity, let us rename C/B by the letter b. We then have

$$y = mx + b \qquad\qquad (5.4)$$

Letting $x = 0$ in (5.4), we see that b is the y-intercept of the line. Equation (5.4) is called the **slope-intercept form**.

EXAMPLE 7   Find the slope and y-intercept of the line $7x + 4y = 12$.
   Solution.   Putting the equation into slope-intercept form, we have

$$y = -\frac{7}{4}x + 3$$

Hence, the slope is $-7/4$ and the y-intercept is 3.   □

Point-Slope Form

Suppose we know that a line goes through the point (x_1, y_1) and has slope m. How do we find the equation for the line? Let (x, y) be an arbitrary point on the line and substitute $x = x_2$ and $y = y_2$ into (5.2). We then have

$$y - y_1 = m(x - x_1) \qquad\qquad (5.5)$$

Equation (5.5) is called the **point-slope form** of a line.

EXAMPLE 8   Find an equation for the line through $(-2, -1)$ with slope 3.
   Solution.   Letting $(x_1, y_1) = (-2, -1)$ and applying (5.5) gives

$$y + 1 = 3(x + 2)$$

Although this equation is perfectly adequate for describing the line, we can put it into standard form

$$-3x + y = 5 \qquad\qquad □$$

EXERCISES

In Exercises 1–7, draw pictures to illustrate the concepts.

1. Euclid's Postulate 1   2. Euclid's Postulate 2
3. Euclid's Postulate 3   4. Euclid's Postulate 4
5. Euclid's Postulate 5   6. Example 1
7. Two parallel lines never intersect.
8. If (x_1, y_1) and (x_1, y_2) are points on a line with $y_1 \neq y_2$, show that the line is vertical.
9. Prove that a line is horizontal if and only if its slope is zero.
10. Put the slope-intercept form $y = mx + b$ into standard form.
11. Put the point-slope form $y - y_1 = m(x - x_1)$ into standard form.

Graph the following equations.

12. $x + y = 0$ 13. $x = 5$ 14. $y = -2$
15. $3x + 4y = 6$ 16. $-2x + 3y = -4$ 17. $2x - 4y = 5$

Find the intercepts of the lines given in the indicated exercises.

18. Exercise 12   19. Exercise 13   20. Exercise 14
21. Exercise 15   22. Exercise 16   23. Exercise 17

Find the slopes of the lines given in the indicated exercises.

24. Exercises 12 and 13   25. Exercises 14 and 15
26. Exercises 16 and 17
27. Find the slope of the line through $(2, -3)$ and $(-1, -2)$.
28. Find an equation for the line through $(5, 4)$ with slope 2.
29. Find three points on the line given by the equation $2x - 3y = -5$.
30. Find an equation for the line passing through the origin with slope 3.
31. Find an equation for the line passing through the origin and the point $(1, 1)$.
32. Find an equation for the line passing through the origin that has the same slope as the line $3x - 2y = 1$.
33. A physician starting practice has a library of medical books worth $5000. For tax purposes, its value is assumed to depreciate linearly over a 5-year period. That is, if $V(t)$ is the value after t years, then the graph of $V(t)$ is a line with $V(0) = 5000$ and $V(5) = 0$. Find the equation of this line and the value after 3 years.
34. A new house is bought for $80,000. The owner estimates that the value will increase linearly to $110,000 in 10 years. Find the equation of a line that gives the value of the house over the next 10 years.

5.2 | Euclid's Postulates in the Cartesian Plane

Teacher: "Give me a sentence containing the word geometry."
Student: "A seed was planted and twenty years later it said,
'Gee, I'm a tree.' "

Where do we now stand with Euclid's postulates? The Cartesian plane $\mathbb{R}^2$ defines a plane and points are defined as ordered pairs in $\mathbb{R}^2$. A line is defined as the graph of an equation that has the standard form (5.1). Notice that we can always put the

Photo 5.1
Striking geometrical patterns can be found in architecture (Chrysler Building).

slope-intercept form (5.4) and the point-slope form (5.5) into standard form if we desire. We thus have definitions of points, lines, and planes.

Postulate 1

We can use the point-slope form (5.5) to find the equation of a line passing through two distinct given points (x_1, y_1) and (x_2, y_2). If $x_1 = x_2$, then the line is vertical and has the equation $y = x_1$. If $x_1 \neq x_2$, then the slope m is given by (5.2) and the line has the equation (5.5).

EXAMPLE 1   Find the equation of the line through the points $(-1, 4)$ and $(2, 3)$.
   Solution.   The slope is given by

$$m = \frac{3-4}{2-(-1)} = \frac{-1}{3}$$

Applying (5.5) we have

$$y - 4 = -\frac{1}{3}(x+1) \qquad \qquad \square$$

We conclude that the point-slope form tells us that two points determine one and only one line, so Euclid's Postulate 1 is satisfied on the Cartesian plane.

Postulate 2

What about Postulate 2 that a line extends indefinitely in either direction? If the line is vertical, then it clearly extends indefinitely up and indefinitely down. A similar result holds if the line is horizontal. Otherwise, we can put the equation of the line into slope-intercept form $y = mx + b$ where $m \neq 0$. If $m > 0$, as x becomes arbitrarily large in the positive direction, so does y, and there are points (x_1, y_1) on the line with arbitrarily large positive x- and y-coordinates. Similarly, as x becomes arbitrarily large in the negative direction, so does y, and there are points (x_1, y_1) on the line with arbitrarily large negative x- and y-coordinates. A similar argument holds if $m < 0$. We conclude that Postulate 2 holds on the Cartesian plane.

Postulate 5

As far as Postulate 5 is concerned, we define two lines to be **parallel** if they do not intersect. How do we know if two lines intersect? Well, if two lines intersect, they must have a point in common. To find this point, put the equations for the lines into slope-intercept form, equate their y-coordinates, and find the resulting x-coordinate.

EXAMPLE 2   Find the point of intersection of the lines

$$2x + 3y = -5$$
$$4x - 2y = 3$$

Solution.   Solve the equations for y to obtain

$$y = -\frac{2}{3}x - \frac{5}{3}$$

$$y = 2x - \frac{3}{2}$$

Equating the y-coordinates gives

$$-\frac{2}{3}x - \frac{5}{3} = 2x - \frac{3}{2}$$

Solving for x we have

$$\frac{8}{3}x = \frac{3}{2} - \frac{5}{3}$$

Hence,

$$x = \frac{3}{8}\left(\frac{3}{2} - \frac{5}{3}\right) = \frac{9}{16} - \frac{5}{8} = -\frac{1}{16}$$

Substituting $x = -1/16$ into either of the y equations (say the second) we have

$$y = -\frac{1}{8} - \frac{3}{2} = -\frac{13}{8}$$

Hence, the lines intersect at $(-1/16, -13/8)$. $\square$

EXAMPLE 3   Do the following lines intersect?

$$2x + 3y = -5$$
$$4x + 6y = 6$$

Solution.   Solve the equations for y to obtain

$$y = -\frac{2}{3}x - \frac{5}{3}$$

$$y = -\frac{2}{3}x + 1$$

Equating the y-coordinates gives

$$-\frac{2}{3}x - \frac{5}{3} = -\frac{2}{3}x + 1$$

This gives $-5/3 = 1$ which is impossible, so the lines do not intersect.   □

Notice that the lines in Example 3 have the same slope and this is why they do not intersect. The next theorem shows that we can decide whether two lines intersect by comparing their slopes. The proof of this theorem is given in Exercises 29 through 31 at the end of this section.

Theorem 5.1   Two distinct vertical lines are parallel. Two distinct nonvertical lines are parallel if and only if they have the same slope.

This theorem shows that two distinct lines are parallel if and only if they are both vertical or they both have the same slope. We now show that Euclid's Postulate 5 holds in the Cartesian plane. Suppose ℓ_1 is a line and $P = (x_1, y_1)$ is a point that is not on ℓ_1. If ℓ_1 is vertical, then the vertical line ℓ_2 with equation $x = x_1$ is parallel to ℓ_1 and contains P. If ℓ_1 is not vertical, it has a slope m. Let ℓ_2 be the line whose equation in point-slope form is

$$y - y_1 = m(x - x_1)$$

Then ℓ_2 contains P and has the same slope as ℓ_1, and hence by Theorem 5.1 ℓ_2 is parallel to ℓ_1. We have thus verified Euclid's Postulate 5 in the Cartesian plane.

EXAMPLE 4   Find the equation of a line ℓ containing the point $(5, -1)$ that is parallel to the line $4x + 2y = 7$.

Solution.   The line ℓ has slope -2, so its equation in point-slope form is

$$y + 1 = -2(x - 5)$$ □

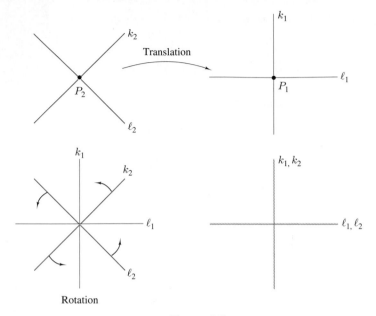

Figure 5.5

Postulate 4

We next examine Euclid's Postulate 4. A right angle is formed at the intersection of two perpendicular lines. Let ℓ_1 and k_1 be perpendicular lines that intersect at point P_1 and let ℓ_2 and k_2 be perpendicular lines that intersect at point P_2. Euclid meant in Postulate 4 that if ℓ_2 and k_2 are translated so that P_2 coincides with P_1, then the translated lines can be rigidly rotated so that ℓ_2 coincides with ℓ_1 and k_2 coincides with k_1. This is illustrated in Figure 5.5.

One can prove algebraically that this translation-rotation procedure can always be performed. However, we shall not do this here because it is clear from Figure 5.5 that this is true. We conclude that Euclid's Postulate 4 holds in the Cartesian plane.

Perpendicular Lines

We have seen that by comparing slopes we can always decide whether two lines are parallel. Can we tell if they are perpendicular by a similar method? The answer is yes. If one of the lines is vertical and the other is horizontal, then they are certainly perpendicular. Now suppose ℓ_1 and ℓ_2 are nonvertical, nonhorizontal lines. If their slopes are m_1 and m_2, respectively, they are perpendicular if and only if $m_1 m_2 = -1$. We shall show in the next section why this is so.

EXAMPLE 5   Show that the following lines are perpendicular.

$$3x + 6y = 11$$
$$4x - 2y = -7$$

Solution.   The slopes of the lines are $m_1 = -1/2$ and $m_2 = 2$. Since $m_1 m_2 = -1$, the lines are perpendicular.   ☐

EXAMPLE 6 Find an equation for the line ℓ that contains $(1, 3)$ and is perpendicular to the line $3x - 4y = 2$.

Solution. Since the given line has slope $3/4$, the slope of ℓ is $m = -4/3$. The point-slope form for the equation of ℓ becomes

$$y - 3 = -\frac{4}{3}(x - 1) \qquad \square$$

We shall show in the next section that Euclid's Postulate 3 also holds in the Cartesian plane. We conclude that the Cartesian plane is a Euclidean geometry. We have not proved Eulcid's postulates. Postulates are accepted properties and cannot be proved. What we have shown is that Euclid's postulates are true in the Cartesian plane. There are many geometrical structures that are Euclidean geometries and the Cartesian plane is just one example. We shall later consider geometries that are non-Euclidean.

EXERCISES

Find an equation for the line through the following pairs of points.

1. $(0, 0)$ and $(0, 1)$
2. $(0, 0)$ and $(1, 0)$
3. $(1, 2)$ and $(3, 2)$
4. $(-1, 2)$ and $(-1, -2)$
5. $(2, 3)$ and $(5, 8)$
6. $(4, 2)$ and $(6, -3)$
7. Let ℓ be a line whose equation is $y = 2x - 3$. Show that there is a point (x_1, y_1) on ℓ with $x_1 \geq 1000$ and $y_1 \geq 1000$.
8. For the line in Exercise 7, show that there is a point (x_1, y_1) on ℓ with $x_1 \leq -1000$ and $y_1 \leq -1000$.
9. Find the point of intersection of the lines $x + y = 1$ and $2x + 3y = 5$.
10. Find the point of intersection of the lines $3x - 4y = 5$ and $2x - 5y = 3$.

Show that the following lines are parallel.

11. $y = 2$ and $y = 3$
12. $x = 1$ and $x = -3$
13. $x + y = 2$ and $2x + 2y = -3$
14. $3x - 4y = 7$ and $-9x + 12y = 5$
15. Find the equation of a line containing the point $(1, 3)$ that is parallel to the line $x + y = 5$
16. Find the equation of a line containing the point $(-5, 7)$ that is parallel to the line $-2x + 3y = 1$.
17. Show that the lines $x + y = 1$ and $x - y = 1$ are perpendicular.
18. Show that the lines $2x + 3y = 5$ and $3x - 2y = 4$ are perpendicular.
19. Find an equation for the line that contains $(0, 0)$ and is perpendicular to the line $3x - y = 2$.
20. Find an equation for the line that contains $(-1, -2)$ and is perpendicular to the line $2x + y = 3$.
21. Find an equation for the line that contains $(1, 2)$ and is perpendicular to the line $x = 1$.
22. If a line ℓ is perpendicular to the line $3x + 4y = 1$, what is the slope of ℓ?
23. If a line ℓ is parallel to the line $3x + 4y = 1$, what is the slope of ℓ?

(Optional Exercises)

A function $f: \mathbb{R}^2 \to \mathbb{R}^2$ of the form $f(x, y) = (x + d, y + e)$ where $d, e \in \mathbb{R}$ are constants is called a **translation**. If ℓ is a line, then $\{f(x, y): (x, y) \in \ell\}$ is called the **translation** of ℓ and is denoted $f(\ell)$.

*24. Let ℓ be the line with equation $2x - y = 1$ and let $f: \mathbb{R}^2 \to \mathbb{R}^2$ be the translation $f(x, y) = (x + 1, y + 3)$. Graph ℓ and graph its translation $f(\ell)$.

*25. Let ℓ be the line with equation $2x + 3y = 5$ and let $f: \mathbb{R}^2 \to \mathbb{R}^2$ be the translation $f(x, y) = (x - 1, y + 1)$. Find the equation for the translation $f(\ell)$.

*26. If ℓ is a line and f is a translation, show that ℓ and $f(\ell)$ have the same slope.

*27. If ℓ_1 and ℓ_2 are parallel and f is a translation, show that $f(\ell_1)$ and $f(\ell_2)$ are parallel.

*28. If ℓ_1 and ℓ_2 are perpendicular and f is a translation, show that $f(\ell_1)$ and $f(\ell_2)$ are perpendicular.

In the following exercises, we shall prove Theorem 5.1.

*29. Show that two distinct vertical lines are parallel.

*30. Let ℓ_1 and ℓ_2 be distinct nonvertical lines with the same slope m. Show that if ℓ_1 and ℓ_2 have a point in common, then they coincide. Conclude that ℓ_1 and ℓ_2 are parallel.

*31. Let ℓ_1 and ℓ_2 be distinct nonvertical lines with different slopes m_1 and m_2, respectively. Show that ℓ_1 and ℓ_2 intersect and hence are not parallel.

5.3 | The Pythagorean Theorem

Mathematicians view themselves as explorers of a unique sort, explorers who seek to discover not just one accidental world into which they happen to be born, but the universal and unalterable truths of all worlds. — DAVID MUMFORD, Twentieth-Century Mathematician

One of the most famous and useful geometric theorems is Pythagoras' theorem. Pythagoras was a philosopher-mathematician who lived in Greece about 500 B.C. and preceded Euclid by 200 years. In Figure 5.6 we illustrate a right triangle. The longest side is the **hypotenuse** and this side has length c. The other two sides are called **legs** and they have lengths a and b.

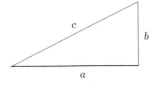

Figure 5.6

| Theorem 5.2 | **(Pythagoras' theorem)** If a and b are the lengths of the legs of a right triangle and c is the length of the hypotenuse, then $a^2 + b^2 = c^2$. |

FOR BETTER OR FOR WORSE   Lynn Johnston

For example, if the legs of a right triangle have lengths 3 and 4 units, what is the length of the hypotenuse? Letting $a = 3$, $b = 4$, and c the length of the hypotenuse, we have

$$c^2 = a^2 + b^2 = 3^2 + 4^2 = 9 + 16 = 25$$

Hence, $c = 5$. The triangle in Figure 5.7 illustrates Pythagoras' theorem for a 3, 4, 5 triangle. Count the number of small squares to verify Pythagoras' theorem for this case. Notice that the horizontal leg, which has length 3, corresponds to 9 small squares and the vertical leg, which has length 4, corresponds to 16 small squares. Finally, the hypotenuse, which has length 5, corresponds to 25 small squares. We can then say, "the square on the hypotenuse equals the sum of the squares on the other two sides."

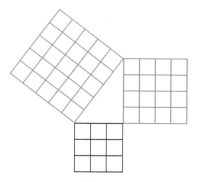

Figure 5.7

EXAMPLE 1   If the length of the hypotenuse of a right triangle is $c = 17$ and the length of one leg is $a = 15$, find the length b of the other leg.

Solution.   Since $a^2 + b^2 = c^2$, we have

$$b^2 = c^2 - a^2 = 17^2 - 15^2 = 289 - 225 = 64$$

Hence, $b = 8$.   □

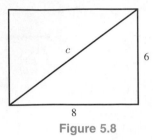

Figure 5.8

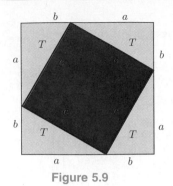

Figure 5.9

EXAMPLE 2　The sides of a rectangle have lengths 6 and 8 units. Find the length c of the diagonal (Figure 5.8).

　　Solution.　Since $c^2 = 8^2 + 6^2 = 64 + 36 = 100$, we have $c = 10$.　　□

We can prove Pythagoras' theorem as follows. Let T be a right triangle whose legs have lengths a and b and whose hypotenuse has length c. Since T is half of a rectangle with sides of length a and b, the area of T is $(1/2)ab$. Draw four copies of T as pictured in Figure 5.9. The area of the larger square is $(a + b)^2$, and the area of the smaller square is c^2. The difference between these two areas is the area of the four triangles, which is $2ab$. We then obtain the equation

$$(a + b)^2 - 2ab = c^2$$

It follows that $a^2 + b^2 = c^2$.

Converse of Pythagoras' Theorem

The converse of Pythagoras' theorem also holds. We now state this converse and the proof will be given in the exercises at the end of this section.

Theorem 5.3　　**(Converse of Pythagoras' theorem)**　If the sides of a triangle have lengths a, b, and c and if $a^2 + b^2 = c^2$, then the triangle is a right triangle.

EXAMPLE 3　Determine whether a triangle whose sides have lengths 15, 36, and 39 is a right triangle.

　　Solution.　If $15^2 + 36^2 = 39^2$, then by Theorem 5.3 the triangular is a right triangle. But this is true since $39^2 = 1521$ and

$$15^2 + 36^2 = 225 + 1296 = 1521$$　　□

There is a systematic method for finding natural numbers that are the lengths of the sides of a right triangle. Let r, s be natural numbers with $r > s$. If we let

$$a = r^2 - s^2, \quad b = 2rs, \quad c = r^2 + s^2$$

Photo 5.2
The design of this building is based
on the structure of a quartz crystal.

then a and b are the lengths of the legs and c is the length of the hypotenuse of a right triangle. To show this, we just need to show that $a^2 + b^2 = c^2$ and apply Theorem 5.3. Performing some algebra gives

$$a^2 + b^2 = (r^2 - s^2)^2 + 4r^2s^2 = r^4 + s^4 - 2r^2s^2 + 4r^2s^2$$
$$= r^4 + s^4 + 2r^2s^2 = (r^2 + s^2)^2 = c^2$$

EXAMPLE 4   Letting $r = 2$ and $s = 1$, we have

$$a = 2^2 - 1^2 = 3, \quad b = 2(2)(1) = 4, \quad c = 2^2 + 1^2 = 5$$

This is the 3, 4, 5 right triangle.

EXAMPLE 5   The lengths of the sides of a right triangle are natural numbers. If one of the legs has length 10, find the lengths of the other two sides.

Solution.   If we let $b = 10$, then $b = 2(5)(1)$. Thus, we can let $r = 5$, $s = 1$, and we have $b = 2rs$. Then

$$a = r^2 - s^2 = 25 - 1 = 24$$

and
$$c = r^2 + s^2 = 25 + 1 = 26 \qquad \square$$

Distance Formula

An important application of Pythagoras' theorem is in finding the distance between two points in the Cartesian plane. Let $P_1 = (x_1, y_1)$ and $P_2 = (x_2, y_2)$ be points in the Cartesian plane and let $d(P_1, P_2)$ be the distance between P_1 and P_2. We shall now prove the **distance formula**

$$d(P_1, P_2) = \sqrt{(x_2 - x_1)^2 + (y_2 - y_1)^2} \qquad (5.6)$$

The situation is illustrated in Figure 5.10. If $x_1 = x_2$ or $y_1 = y_2$, then (5.6) certainly holds (why?). We can therefore assume that $x_1 \neq x_2$ and $y_1 \neq y_2$. If $P_3 = (x_2, y_1)$, then P_1, P_2, and P_3 are the vertices of a right triangle as illustrated in Figure 5.11. We see from Figure 5.11 that $d(P_1, P_3) = x_2 - x_1$ and $d(P_3, P_2) = y_2 - y_1$. (We are assuming that $x_2 > x_1$ and $y_2 > y_1$. Other cases are treated in a similar way.) Applying Pythagoras' theorem gives

$$\begin{aligned}[d(P_1, P_2)]^2 &= [d(P_1, P_3)]^2 + [d(P_3, P_2)]^2 \\ &= (x_2 - x_1)^2 + (y_2 - y_1)^2\end{aligned}$$

Taking square roots, we then obtain (5.6).

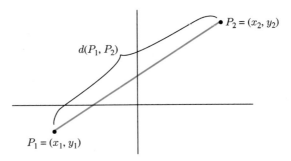

Figure 5.10

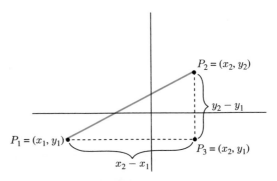

Figure 5.11

EXAMPLE 6 Find the distance between the points $P_1 = (1, 5)$ and $P_2 = (3, 2)$.

　　Solution. Applying (5.6) gives

$$d(P_1, P_2) = \sqrt{(3 - 1)^2 + (2 - 5)^2} = \sqrt{2^2 + 3^2} = \sqrt{13} \qquad \square$$

EXAMPLE 7 Plot the points $A = (-1, -3)$, $B = (6, 1)$, $C = (2, -5)$ and show that the triangle with vertices A, B, and C is a right triangle.

　　Solution. The points are plotted in Figure 5.12. Applying the distance formula gives

$$d(A, B) = \sqrt{7^2 + 4^2} = \sqrt{65}$$
$$d(A, C) = \sqrt{3^2 + 2^2} = \sqrt{13}$$
$$d(B, C) = \sqrt{4^2 + 6^2} = \sqrt{52}$$

Since

$$[d(B, C)]^2 + [d(A, C)]^2 = [d(A, B)]^2$$

by Theorem 5.3, the triangle is a right triangle. $\qquad \square$

Equation of a Circle

We now use the distance formula to show that Euclid's Postulate 3 holds in the Cartesian plane. Let $P_1 = (x_1, y_1)$ be a point in the Cartesian plane and let r be a positive number. Can we find the equation for a circle with center P_1 and radius r? Well, such a circle is the set of all points whose distance from P_1 is r (Figure 5.13). Thus, a point $P = (x, y)$ is on this circle if and only if $d(P, P_1) = r$. Rewriting this as $[d(P, P_1)]^2 = r^2$, the distance formula gives

$$(x - x_1)^2 + (y - y_1)^2 = r^2 \qquad\qquad (5.7)$$

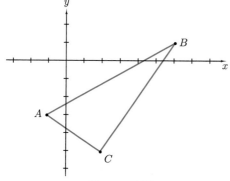

Figure 5.12

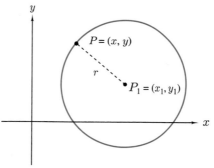

Figure 5.13

[2] M. C. Escher,
Circle Limit IV. (M. C. Escher
Foundation, Baarn, Holland.)

We call (5.7) the **equation of the circle** with center (x_1, y_1) and radius r. The set of points (x, y) that satisfy (5.7) is called the **graph** of (5.7). Since we can graph (5.7), we can always draw a circle with any given center and any given (positive) radius. This verifies Euclid's Postulate 3 in the Cartesian plane.

EXAMPLE 8   Find the equation of the circle with center $(-3, 4)$ and radius 5.
Solution.   Applying (5.7), the equation is $(x + 3)^2 + (y - 4)^2 = 25$.   □

EXERCISES

1. If the length of the hypotenuse of a right triangle is 10 and the length of one leg is 6, find the length of the other leg.

2. If the lengths of the legs of a right triangle are 8 and 15 units, find the length of the hypotenuse.

3. If the lengths of the legs of a right triangle are each 1 unit, find the length of the hypotenuse.

State whether the following numbers are the lengths of the sides of a right triangle.

4. 2, 3, 4   5. $\sqrt{2}, \sqrt{3}, \sqrt{5}$ 6. 1, 1, 2

7. 1, 1, $\sqrt{2}$ 8. 20, 21, 29   9. 16, 30, 34

10. The lengths of the sides of a right triangle are natural numbers. If one leg has length 42, find the lengths of the other two sides.

11. Repeat Exercise 10 for the case that one leg has length 70.

12. Find the distance between $P_1 = (6, -2)$ and $P_2 = (2, 1)$.

13. Find the distance between $P_1 = (-4, -1)$ and $P_2 = (2, 3)$.

14. Show that the triangle with vertices $A = (-3, 4)$, $B = (2, -1)$, $C = (9, 6)$ is a right triangle.

15. Show that the triangle with vertices $A = (7, 2)$, $B = (-4, 0)$, $C = (4, 6)$ is a right triangle.

16. Prove that $d(P_1, P_2) = d(P_2, P_1)$ for any $P_1, P_2 \in \mathbb{R}^2$.

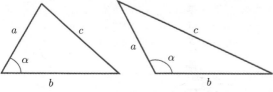 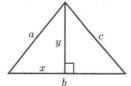

Figure 5.14

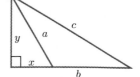

Figure 5.15

17. Prove that $d(P_1, P_2) = 0$ if and only if $P_1 = P_2$.

Find an equation of a circle satisfying the stated conditions.

18. Center $(3, -2)$, radius 4   **19.** Center $(-5, 2)$, radius 5

20. Center $(0, 0)$, contains $(-3, 5)$ **21.** Center $(-4, 6)$, contains $(1, 2)$

22. Draw the circle whose equation is $x^2 + y^2 = 25$.

23. Draw the circle whose equation is $(x + 2)^2 + (y - 5)^2 = 4$.

(Optional Exercises)

In Exercises 24–28 we shall prove Theorem 5.3 (converse of Pythagoras' theorem). Let T be a triangle with sides of length a, b, and c and suppose that $a^2 + b^2 = c^2$. Assume that T is not a right triangle. Then the angle α between the legs must be either smaller or larger than a right angle (Figure 5.14). In both cases, we construct new right triangles with legs of length x and y and hypotenuse of length a as shown in Figure 5.15.

*24. In Case 1, apply Pythagoras' theorem for the two triangles. Then show that
$b^2 - 2bx + a^2 = c^2$.

*25. Conclude from Exercise 24 that $2bx = 0$ and show that this gives a contradiction.

*26. Repeat Exercise 24 for Case 2.   ***27.** Repeat Exercise 25 for Case 2.

*28. Why have we now proved Theorem 5.3?

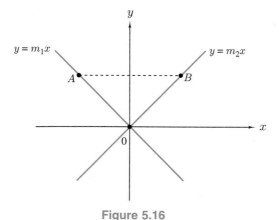

Figure 5.16

In Exercises 29–33 we shall prove that two nonvertical and nonhorizontal lines with slopes m_1 and m_2 are perpendicular if and only if $m_1 m_2 = -1$. For simplicity, we consider the special case of two lines that intersect at the origin 0 as illustrated in Figure 5.16. (The proof for the general case is essentially the same but a little more complicated.)

*29. Show that the lines have the equation $y = m_1 x$ and $y = m_2 x$.

*30. Choose points $A = (x_1, m_1 x_1)$ and $B = (x_2, m_2 x_2)$ where $x_1, x_2 \neq 0$. Show that A is on the line $y = m_1 x$ and B is on the line $y = m_2 x$.

*31. Show that the lines are perpendicular if and only if the triangle with vertices A, B, and 0 is a right triangle.

*32. Show that the lines are perpendicular if and only if $[d(A, 0)]^2 + [d(B, 0)]^2 = [d(A, B)]^2$.

*33. Evaluate the equation in Exercise 32 using the distance formula.

5.4 Perimeter and Area

The universe stands continually open to our gaze, but it cannot be understood unless one first learns to comprehend the language and interpret the characters in which it is written. It is written in the language of mathematics, and its characters are triangles, circles, and other geometric figures, without which it is humanly impossible to understand a single word of it; without these, one is wandering about in a dark labyrinth.

— GALILEO GALILEI (1564–1642), Italian Physicist and Astronomer

---

The **perimeter** of a plane figure is the distance around the figure. In the case of a circle, the perimeter is called the **circumference**.

Suppose we have a rectangle with length L and width W as in Figure 5.17.

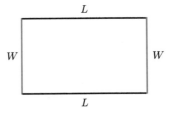

Figure 5.17

Then it is clear that

Perimeter of a Rectangle $P = 2L + 2W$

EXAMPLE 1 Find the perimeter of a rectangle with length 12 inches and width 5 inches.

  Solution. $P = 2(12) + 2(5) = 34$ inches.   □

EXAMPLE 2 Find the perimeter of a right triangle whose legs have length 3 and 4.

  Solution.   The hypotenuse has length 5, so the perimeter is $3 + 4 + 5 = 12$.   □

[3] M. C. Escher,
Smaller and Smaller.
(M. C. Escher Foundation,
Baarn, Holland.)

EXAMPLE 3 I recently took a large box to a UPS (United Parcel Service) station to have it shipped to another state. The box measured 35 inches by 25 inches by 20 inches. The attendant measured the box and informed me that it was too big since their limit is girth plus height equals 130 inches. When I asked her what girth meant, she told me that's the perimeter of the base. Thus, girth plus height is

$$2(35 + 25) + 20 = 140$$

which is 10 inches too large. I then replied, "Why don't we turn the box on its side?" Then girth plus height is

$$2(25 + 20) + 35 = 125$$

"O. K.," she said, "now we can ship it."

If you wrap a string once around a circle and measure this length, you will obtain the circumference c of the circle. If you now divide c by the diameter d of the circle (recall that the diameter is twice the radius r), you will obtain a number that is a little larger than 3. The ancient Greeks proved the amazing fact that the ratio c/d is the same no matter what circle is used. They denoted this important universal constant by the Greek letter π. The value of π is approximately 3.1416. We thus have

> Circumference of a Circle $c = \pi d = 2\pi r$

[4] M. C. Escher, *Moebius Strip I.*
(M. C. Escher Foundation, Baarn,
Holland.)

To compute the area of a plane figure we need a basic unit of area, say a **square inch**. A square inch is the area of a square whose sides have length 1 inch. (A square centimeter, a square foot, and so on, would do just as well.) Suppose we have a rectangle of length 5 inches and width 3 inches. We compute the area of the rectangle by dividing it into squares of area 1 square inch as in Figure 5.18. Since there are 15 squares, we conclude that the area of the rectangle is 15 square inches. We thus have the formula $A = LW$ where A is the area, L the length, and W the width of the rectangle. Of course, this method will not work if L and W are not natural numbers. However, in such a case we can approximate the area with very small squares. We thus make the following definition.

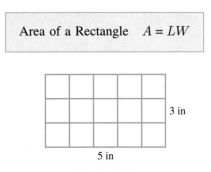

Area of a Rectangle $A = LW$

3 in

5 in

Figure 5.18

Triangle

Since a square is a special case of a rectangle with equal sides, we immediately obtain a formula for the area of a square. If a square has sides of length s, setting $L = W = s$ we obtain the area $A = s^2$. The area formula for a rectangle can be used to find areas for other plane figures. Let T be a right triangle whose legs have length b and h (Figure 5.19). We call b the **base** and h the **height** of T. It does not make any difference which leg corresponds to the base and which leg corresponds to the height. We now double T to form a rectangle of length b and width h (Figure 5.20). Since T is half of a rectangle of area bh, the area of T is $(1/2)bh$.

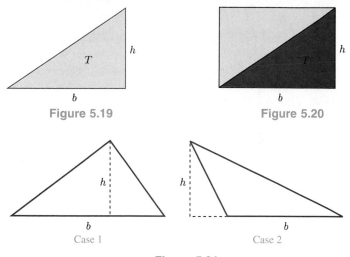

Figure 5.19   Figure 5.20

Case 1   Case 2

Figure 5.21

We next consider an arbitrary triangle T. Choose one of its sides and suppose the length of this side is b. We call this length the **base** of T. Now drop a perpendicular line segment from the vertex opposite to the base side down to the base side (or an extension of the base side). If this line segment has length h, we call h the **height**. From Figure 5.21 we see that there are two cases, one in which the opposite vertex is above the base side, and another in which it is above an extension of the base side.

We first treat Case 1. In this case, T is composed of two right triangles T_1 and T_2 of height h. Suppose the base of T_1 is x. Then the base of T_2 is $b - x$ (Figure 5.22). If the area of T_1 is A_1 and the area of T_2 is A_2, using the area formula for right triangles we obtain the area A of T:

$$A = A_1 + A_2 = \frac{1}{2}xh + \frac{1}{2}(b - x)h = \frac{1}{2}bh$$

Case 2 is treated in a similar way with the same result (Exercise 19). We thus have an area formula for any triangle

> Area of a Triangle $A = \dfrac{1}{2}bh$

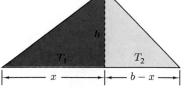

Figure 5.22

Trapezoid

We can now find areas for other plane figures. A **trapezoid** is a four-sided figure having one pair of opposite sides parallel. Figure 5.23 illustrates a typical trapezoid. Besides the usual perpendicular height h, we now have two bases B and b where $B \geq b$. We can find the area A of the trapezoid by dividing it into two triangles as in Figure 5.24. Then A is the sum of the areas of the two triangles:

$$A = \frac{1}{2}Bh + \frac{1}{2}bh = \frac{1}{2}(B+b)h$$

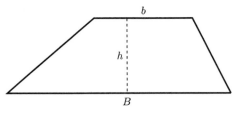

Figure 5.23

Instead of the trapezoid in Figure 5.24, a trapezoid may be like the one in Figure 5.25. We can use a similar method to find the area of the trapezoid in Figure 5.25 and obtain the same result (Exercise 20). We thus have the formula

> Area of a Trapezoid $A = \dfrac{1}{2}(B+b)h$

This formula may be restated as follows. The area of a trapezoid is the average length of the parallel bases times the height.

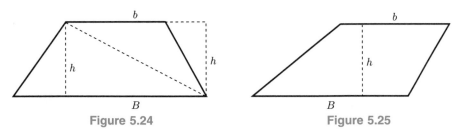

Figure 5.24   **Figure 5.25**

EXAMPLE 4   Find the area of the trapezoid in Figure 5.26 in two ways.
   Solution.

$$A = \frac{1}{2}(B+b)h = \frac{1}{2}(5+3)(2) = 8$$

$$A = 3(2) + \frac{1}{2}(2)(2) = 8 \qquad \square$$

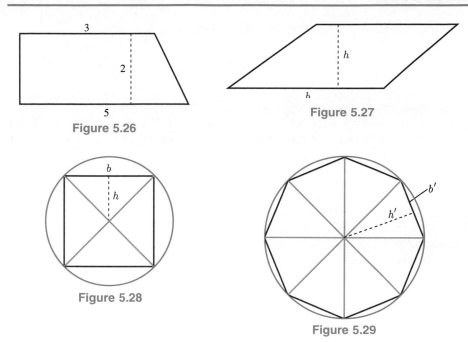

Figure 5.26

Figure 5.27

Figure 5.28

Figure 5.29

Parallelogram

A **parallelogram** is a four-sided figure having both pairs of opposite sides parallel. A typical parallelogram is shown in Figure 5.27

A parallelogram is a special kind of trapezoid in which the two bases are equal. Setting $B = b$ we have

> Area of a Parallelogram $A = bh$

Area of a Circle

We would now like to find the area A enclosed by a circle of radius r. We shall use the method employed by the ancient Greeks. If we divide the circle into four equal parts, we obtain the inscribed square of Figure 5.28. This inscribed square consists of four triangles of base b and height h. Thus the area of the square is

$$A_4 = 4 \left(\frac{1}{2} bh \right)$$

(We could also express this as $A_4 = b^2$, but this is not the form we need for our later calculations.) Of course, A_4 is only a rough approximation to A and is certainly too small.

To get a better approximation, let's divide the circle into eight equal parts. We then have the inscribed octagon of Figure 5.29. This octagon consists of eight triangles of base b' and height h'. Thus the area of the octagon is

$$A_8 = 8 \left(\frac{1}{2} b'h' \right)$$

Now A_8 gives a better approximation to A. Still we are not satisfied, so let's divide the circle into a large number n of equal parts. We then have an inscribed n-sided figure consisting of n triangles with base b_n and height h_n. The area of this n-sided figure is

$$A_n = n \left(\frac{1}{2} b_n h_n \right) = \frac{1}{2} (n b_n) h_n$$

As n gets increasingly large, the area A_n gives a closer and closer approximation to A. We say that A_n **approaches** A **as** n **approaches infinity** and write $A_n \to A$ as $n \to \infty$. Notice that as n gets larger, h_n approaches the radius r. Also, as n gets larger, the perimeter $n b_n$ of the inscribed n-sided figure approaches the circumference $2\pi r$ of the circle. Hence,

$$A_n \to \frac{1}{2} (2\pi r) r = \pi r^2 \qquad \text{as} \quad n \to \infty$$

We have thus arrived at the formula

$$\boxed{\text{Area of a Circle} \quad A = \pi r^2}$$

EXERCISES

1. Find the perimeter of a square with sides of length 3.
2. Find the perimeter of a rectangle with length 10 and width 7.
3. Find the perimeter of a right triangle whose legs have length 8 and 15.
4. Find the circumference of a circle with diameter 4.
5. Find the circumference of a circle with diameter 5.
6. The sides of a hexagon each have length 3. Find its perimeter.
7. Find the area of a square with sides of length 7.
8. Find the area of a rectangle with length 4 and width 3.
9. A right triangle has legs of length 5 and 6. Find its area.
10. Find the area of a triangle with base 9 and height 4.
11. Find the area of a parallelogram with base 10 and height 7.
12. A trapezoid has bases 3 and 5 and height 6. Find its area.
13. Find the area of the trapezoid of Figure 5.30 in two ways.
14. Find the area of a circle with radius 6.
15. Find the area of a circle with diameter 8.
16. Find the area of the figure in Figure 5.31a.
17. Find the area of the figure in Figure 5.31b.

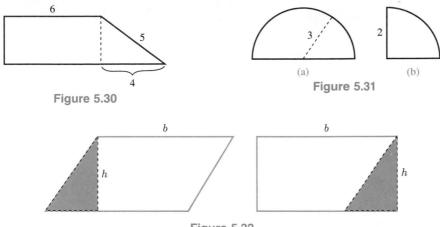

Figure 5.30

Figure 5.31

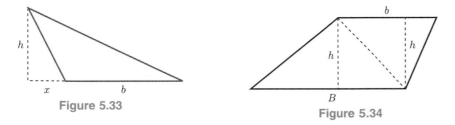

Figure 5.32

18. Show that the area of a parallelogram is $A = bh$, by removing the triangle on the left and attaching it on the right as in Figure 5.32.

19. Show that the area of the triangle in Case 2 of Figure 5.21 is $(1/2)bh$ by subtracting the area of the small right triangle from the area of the large right triangle in Figure 5.33.

Figure 5.33

Figure 5.34

20. Use Figure 5.34 to show that the area of the trapezoid in Figure 5.25 is $(1/2)(B + b)h$.

21. The volume of a rectangular box is the area of its base times its height. What is the volume of a rectangular box three of whose edges have lengths 2, 3, and 4?

22. The volume of a sphere is $(4/3)\pi$ times its radius cubed. What is the volume of a sphere of radius 2?

23. The surface area of a sphere is four times the area of a circle with the same radius. Find the surface area of a sphere with radius 5.

24. The surface area of a box is the sum of the areas of its six sides. Find the surface area of a box three of whose edges have lengths 2, 3, 4.

25. Find the area of a circular ring with inner radius r and outer radius R (see Figure 5.35).

26. The volume of a circular cylinder is the area of its base times its height. Find the volume of a circular cylinder of radius r and height h (see Figure 5.36).

27. The volume of a circular cone is one-third the area of its base times its height. Find the volume of a circular cone whose base has radius r and whose height is h (see Figure 5.37).

28. If a rectangular box measures a by b by c inches, its girth plus height can have three different values. What are these values?

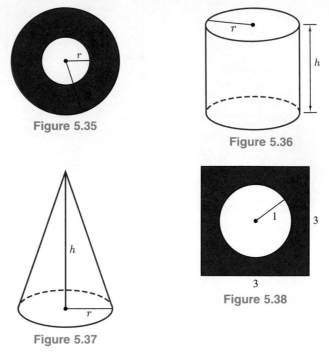

Figure 5.35

Figure 5.36

Figure 5.37

Figure 5.38

29. If the three possible values of girth plus height in Exercise 28 agree, what must be true about a, b, and c?

30. What is the area of the shaded region in Figure 5.38?

5.5 | Areas Under Curves (Optional)

The essence of mathematics lies in its freedom.
— GEORG CANTOR (1845–1981), German Mathematician

The more fundamental or basic is the idea one has learned, almost by definition, the greater will be its breadth of applicability to new problems.
— JEROME BRUNER, Twentieth-Century Educator

In Section 5.4 we found areas for simple plane figures such as a circle, triangle, and trapezoid. We now consider the more complicated problem of finding the area under a curve. Figure 5.39 illustrates a curve that is the graph of the function $f(x) = x^2$ for $0 \leq x \leq 1$. How would you compute the area A between this curve and the x-axis (the colored region in Figure 5.39)?

The ancient Greeks applied a method similar to the one they used to find the area of a circle as discussed in Section 5.4. If we divide the interval $0 \leq x \leq 1$ into equal parts, we can approximate A by rectangular areas. As the number of parts gets larger, the approximations get better. Figure 5.40 illustrates this technique.

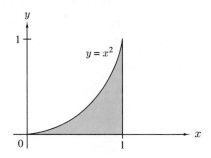

Figure 5.39

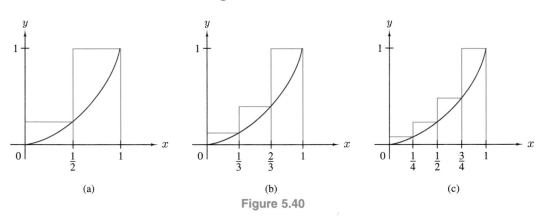

Figure 5.40

The area A_2 in Figure 5.40a is the sum of the areas of the two rectangles

$$A_2 = \frac{1}{2} \left(\frac{1}{2} \right)^2 + \frac{1}{2}$$

In a similar way the areas A_3 and A_4 in Figures 5.40b and 5.40c become

$$A_3 = \frac{1}{3} \left(\frac{1}{3} \right)^2 + \frac{1}{3} \left(\frac{2}{3} \right)^2 + \frac{1}{3}$$

$$A_4 = \frac{1}{4} \left(\frac{1}{4} \right)^2 + \frac{1}{4} \left(\frac{2}{4} \right)^2 + \frac{1}{4} \left(\frac{3}{4} \right)^2 + \frac{1}{4}$$

If we divide the interval $0 \le x \le 1$ into n parts, we obtain the area

$$A_n = \frac{1}{n} \left(\frac{1}{n} \right)^2 + \frac{1}{n} \left(\frac{2}{n} \right)^2 + \frac{1}{n} \left(\frac{3}{n} \right)^2 + \cdots + \frac{1}{n} \left(\frac{n-1}{n} \right)^2 + \frac{1}{n}$$

$$= \frac{1}{n^3} \left[1 + 2^2 + 3^2 + \cdots + (n-1)^2 + n^2 \right]$$

The ancient Greeks were able to simplify this formula for A_n. They were aware of a formula for the sum of the squares of the first n natural numbers,

$$1 + 2^2 + 3^2 + \cdots + n^2 = \frac{n(n+1)(2n+1)}{6} \tag{5.8}$$

[5] M. C. Escher, *Whirlpools.*
(M. C. Escher Foundation, Baarn,
Holland.)

(We have already considered (5.8) in Exercise 19, Section 3.6.) Substituting (5.8) into our expression for A_n gives

$$A_n = \frac{n(n+1)(2n+1)}{6n^3} = \frac{1}{6}\left(\frac{n+1}{n}\right)\left(\frac{2n+1}{n}\right) = \frac{1}{6}\left(1+\frac{1}{n}\right)\left(2+\frac{1}{n}\right)$$

Now as $n \to \infty$, we have $1/n \to 0$, so we see that $A_n \to 1/3$. Thus, the area under the curve $f(x) = x^2$ for $0 \leq x \leq 1$ is $A = 1/3$.

Integration

> The technique of approximating an area under a curve by summing areas of small rectangles is called **integration**.

Methods of integration were developed in the seventeenth century by Newton and Leibniz using their newly discovered calculus. In many cases, calculus gives a much easier way of computing areas than that employed by the ancient Greeks. We shall not treat calculus in detail here (I hope you are not too disappointed), but we shall discuss some of its highlights as far as areas are concerned.

Let $y = f(x)$ be a function whose graph is illustrated in Figure 5.41. This function is considered over the interval $a \leq x \leq b$ and we have labeled the areas A_1, A_2, A_3, A_4. These areas are computed using the approximation by rectangles technique and are positive quantities. The **integral** of f from a to b is denoted $\int_a^b f(x)$ and is defined as

$$\int_a^b f(x) = A_2 + A_4 - A_1 - A_3$$

The reason for the minus signs in front of A_1 and A_3 is that the function f is negative in the corresponding regions. Thus, the integral of f gives the area between the graph of f and the x-axis when f is positive and the negative of this area when f is negative. The symbol $\int$ is the German letter S and represents a "continuous sum." (The sum comes from the fact that we approximate the area by sums of rectangular areas.)

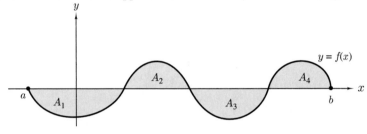

Figure 5.41

For an integral from a to b, the points a and b can be any two points between which the function is defined. In Figure 5.42, $\int_a^b f(x)$ gives the area A of the region between the graph of f and the x-axis from a to b. If f is positive between a and b we have

> The area between the graph of $f(x)$ and the x-axis from a to b is $\int_a^b f(x)$.

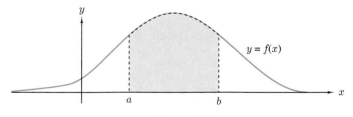

Figure 5.42

The quantity $\int_a^b f(x)$ has various easily verified properties. You should convince yourself by drawing pictures that

$$\int_a^b 0 = 0 \qquad \text{and} \qquad \int_a^b cf(x) = c\int_a^b f(x)$$

for any constant c and

$$\int_a^b [f(x) + g(x)] = \int_a^b f(x) + \int_a^b g(x)$$

In Figure 5.43 we illustrate the fact that if $a < c < b$, then

$$\int_a^b f(x) = \int_a^c f(x) + \int_c^b f(x)$$

Indeed, in this figure we have

$$\int_a^b f(x) = A + B = \int_a^c f(x) + \int_c^b f(x)$$

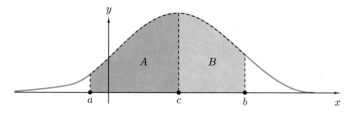

Figure 5.43

Since the area under a single point is 0, we have

$$\int_a^a f(x) = 0$$

For this reason, when we speak of the area under f from a to b, it does not matter whether we are considering the interval $a \le x \le b$ or the interval $a < x < b$ (or $a \le x < b$ or $a < x \le b$).

Integral Formula

For certain functions f, there are formulas for computing $\int_a^b f(x)$. In fact, such formulas are an important branch of calculus called **integral calculus**. We shall not derive these formulas here, but we shall state and use one of the most useful of them. If n is a nonnegative integer, then

$$\int_a^b x^n = \frac{b^{n+1}}{n+1} - \frac{a^{n+1}}{n+1} \qquad\qquad (5.9)$$

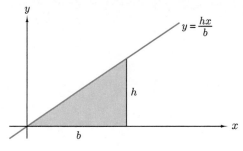

Figure 5.44

Let us look at the special cases $n = 0, 1, 2$.

$$\int_a^b 1 = \int_a^b x^0 = b - a$$

$$\int_a^b x = \int_a^b x^1 = \frac{b^2}{2} - \frac{a^2}{2}$$

$$\int_a^b x^2 = \frac{b^3}{3} - \frac{a^3}{3}$$

EXAMPLE 1 Find the area A under the graph of $f(x) = x^2$ from 0 to 1.

Solution. We have considered this example previously and found the area to be $1/3$. We can use (5.9) to obtain this answer much more easily. In this case $a = 0$ and $b = 1$, so

$$A = \int_0^1 x^2 = \frac{1^3}{3} - \frac{0^3}{3} = \frac{1}{3} \qquad \square$$

EXAMPLE 2 Find the area A of a right triangle with base b and height h.

Solution. Of course, we know the answer is $A = bh/2$, but let us use (5.9) and see if we get the same result. Draw the triangle in the Cartesian plane as in Figure 5.44. The hypotenuse of the triangle lies on the line through the origin with slope h/b. The equation of this line is

$$y = f(x) = \frac{hx}{b}$$

We then obtain

$$A = \int_0^b f(x) = \int_0^b \frac{h}{b} x = \frac{h}{b} \int_0^b x = \frac{h}{b} \left(\frac{b^2}{2} - \frac{0^2}{2} \right) = \frac{1}{2} bh \qquad \square$$

The next example shows that we can find the area between two curves using this method.

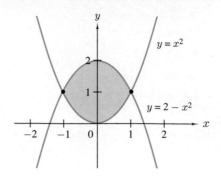

Figure 5.45

EXAMPLE 3   Find the area between the two curves $y = x^2$ and $y = 2 - x^2$.

Solution.   The graphs of the two functions are shown in Figure 5.45. The two graphs intersect when $x^2 = 2 - x^2$ or $x^2 = 1$. This gives the intersection points $x = -1$ and $x = 1$. The area A that we seek is the area under $y = 2 - x^2$ minus the area under $y = x^2$ between -1 and 1. Hence,

$$A = \int_{-1}^{1} (2 - x^2) - \int_{-1}^{1} x^2$$

$$= 2\int_{-1}^{1} 1 - 2\int_{-1}^{1} x^2$$

$$= 2[1 - (-1)] - 2\left[\frac{1^3}{3} - \frac{(-1)^3}{3}\right] = 4 - 2\left(\frac{2}{3}\right) = 2\frac{2}{3} \qquad \square$$

EXERCISES

1. Draw pictures to illustrate that $\displaystyle\int_{a}^{b} 0 = 0$ and $\displaystyle\int_{a}^{a} f(x) = 0$.

2. Draw pictures to illustrate that $\displaystyle\int_{a}^{b} cf(x) = c\int_{a}^{b} f(x)$.

3. Draw pictures to illustrate that $\displaystyle\int_{a}^{b} [f(x) + g(x)] = \int_{a}^{b} f(x) + \int_{a}^{b} g(x)$.

Evaluate the following integrals.

4. $\displaystyle\int_{1}^{2} x^2$

5. $\displaystyle\int_{-1}^{2} x^2$

6. $\displaystyle\int_{1}^{2} x^3$

7. $\displaystyle\int_{2}^{3} x^3$

8. $\displaystyle\int_{-2}^{1} x^3$

9. $\displaystyle\int_{2}^{4} (3x + 2x^2)$

10. $\displaystyle\int_{0}^{1} x^6$

11. $\displaystyle\int_{1}^{2} (x^4 - x^3)$

12. $\displaystyle\int_{-1}^{1} (1 + x + x^2)$

13. If $0 \le f(x) \le g(x)$ for all x, explain why $\int_a^b f(x) \le \int_a^b g(x)$.

14. If $f(x) \ge 0$ for all x and $a \le b \le c$, explain why $\int_a^b f(x) \le \int_a^c f(x)$.

15. Find the area under the graph $f(x) = 1 + x^2$ from 0 to 2.

16. Find the area under the graph $f(x) = x + x^3$ from 1 to 2.

17. What is the area between the curves for the functions $f(x) = x$ and $g(x) = x^2$ for $0 \le x \le 1$?

18. What is the area between the curves for the functions $f(x) = x^2$ and $g(x) = x^3$ for $0 \le x \le 1$?

19. Graph the sequence of functions $f_1(x) = x$, $f_2(x) = x^2$, $f_3(x) = x^3, \ldots$, for $0 \le x \le 1$. Show that the corresponding areas form the sequence $1/2, 1/3, 1/4, \ldots$.

20. Find the area between the curves $y = 4$ and $y = x^2$.

21. Find the area between the curves $y = 16$ and $y = x^4$.

22. Use the method of approximation by rectangles to show that the area under the graph of $f(x) = x$ from 0 to 1 is $1/2$.

Equation (5.9) holds for any $n \in \mathbb{R}$, $n \ne -1$. Evaluate the following integrals.

*23. $\int_1^2 \sqrt{x}$ *24. $\int_0^2 x^{3/2}$ *25. $\int_0^1 x^{5/2}$ *26. $\int_1^2 x^\pi$

5.6 | Non-Euclidean Geometry

Self-similarity is an easily recognized quality. Its images are everywhere in the culture: in the infinitely deep reflection of a person standing between two mirrors, or in the cartoon notion of a fish eating a smaller fish eating a smaller fish eating a smaller fish. Mandelbrot likes to quote Jonathan Swift:

> *"So, Naturalists observe, a Flea*
> *Hath smaller Fleas that on them prey,*
> *And these have smaller Fleas to bite 'em,*
> *And so proceed ad infinitum."*
>
> —JAMES GLEICK, *Chaos: Making a New Science*

Euclidean geometry has many useful applications. It is employed in the design of buildings, parks, streets, furniture, and clothing. For example, consider a simple object like a shoe. There are a huge variety of walking, running, jogging, tennis, basketball, and track shoes. The design of shoes is continually evolving and it requires intricate geometrical analysis. But sometimes we must go beyond Euclidean geometry to non-Euclidean geometry.

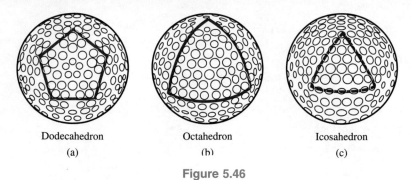

Dodecahedron

(a)

Octahedron

(b)

Icosahedron

(c)

Figure 5.46

Design of Golf Balls

The design of golf balls is probably an application of geometry that has not occurred to you. As you will see, since golf balls are essentially spherical, non-Euclidean geometry must be used for their design. Dunlop Sports Company has been developing and manufacturing golf balls for 400 years. They have recently developed what they claim is the perfect golf ball. It is called the DDH Maxfli. The DDH Maxfli is based on a geometrical figure called a dodecahedron. It incorporates 12 identical spherical pentagons each containing a repeating dimple design (Figure 5.46a). The DDH has 360 dimples of four size variations. Dunlop's older ball, the Titleist, is based on the octahedron configuration and has 336 dimples (Figure 5.46b). Their Top Flight model is based on the icosahedron configuration with 362 dimples (Figure 5.46c).

According to Dunlop, "the flawless symmetry of the DDH makes possible a design free of imperfections, such as seam lines, which disrupt the surface symmetry of other balls during flight." Moreover, "while conducting tests, which took five years to perfect, the Dunlop engineers discovered that, by varying the size and depth of the dimples within the dodecahedron, distances could be significantly increased without sacrificing the already improved accuracy." According to a comparison test, the DDH travels up to 7 yards further than any other ball. Also, using the standard 5-iron machine test, the DDH is 53.2 percent accurate (within 2 yards of target). The next best ball, an octahedron, is only 48.9 percent accurate and the icosahedron, a lowly 27.6 percent. If you would like more information about the dodecahedron, octahedron, and icosahedron, see Section 12.7.

Spherical Geometry

Euclidean geometry gives a description of objects in a simple flat space such as the Cartesian plane $\mathbb{R}^2$. (We can also consider higher dimensional flat spaces such as Cartesian 3-space $\mathbb{R}^3$.) Non-Euclidean geometries are necessary to describe more complicated flat spaces and curved spaces. For example, the earth on which we live is essentially a sphere and a non-Euclidean geometry must be employed to make precise measurements on a sphere. Such a geometry was developed by the German mathematician Georg Riemann (1826–1866). Two points P and Q on a sphere determine a line that gives the shortest distance between P and Q. This shortest distance line is called a **geodesic** and is given by an arc of a great circle on the sphere (Figure 5.47).

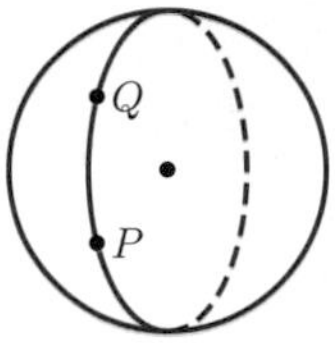

Figure 5.47

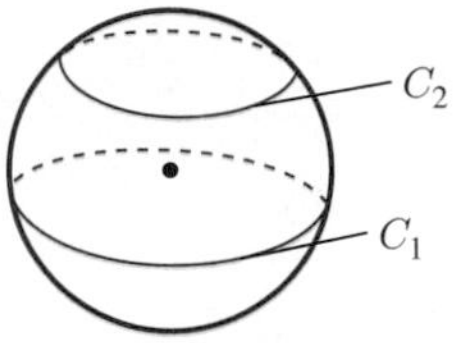

Figure 5.48

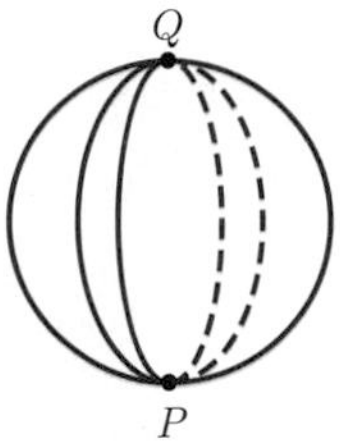

Figure 5.49

A great circle is a circle on the sphere whose center is the center of the sphere. In Figure 5.48 we have two circles C_1 and C_2. Notice that C_1 is a great circle but C_2 is not.

A line in the Riemannian geometry of a sphere continues indefinitely as required by Euclid but does not have infinite length. Some of Euclid's postulates do not hold in Riemannian geometry. Let's consider Postulate 1. The two points P and Q in Figure 5.47 do determine one and only one line. However, if P and Q are at the opposite ends of a diameter of a sphere, say at the north and south poles, they determine many lines (Figure 5.49). Thus, although two points always determine a line, the line may not be unique.

Euclid's fifth postulate also fails in Riemannian geometry. If P is a point that is not on a line ℓ, there is no line through P parallel to ℓ. In fact, any two great circles intersect. This is illustrated in Figure 5.49 where many great circles intersect at the north and south poles.

Hyperbolic Geometry

Another non-Euclidean geometry is the **hyperbolic geometry** which was discovered in the early nineteenth century by Gauss, Lobachevski, and Bolyai. Although they worked independently, their discoveries occurred about the same time. Karl Friedrich Gauss was one of the greatest mathematicians of all times. Nikolai Ivanovich Lobachevski was a Russian mathematician and Janos Bolyai was a Hungarian army officer. The simplest hyperbolic geometry can be represented by a surface formed by revolving a curve called a tractrix about a line as in Figure 5.50.

Tractrix   Hyperbolic geometry

Figure 5.50

[6] M. C. Escher,
Circle Limit III.
(M. C. Escher
Foundation, Baarn,
Holland.)

In a hyperbolic geometry, lines are again represented by geodesics or shortest distance lines. In this geometry, all of Euclid's postulates hold except the fifth. For a hyperbolic geometry, through a point P not on a line ℓ, there are at least two different lines through P that are parallel to ℓ (Figure 5.51).

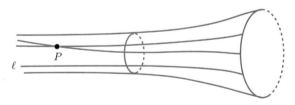

Figure 5.51

Fractal Geometry

A modern non-Euclidean geometry was introduced in 1961 by the IBM mathematician Benoit Mandelbrot. Mandelbrot's geometry is called a **fractal geometry**. Lines have one dimension; planes, spheres, and the hyperbolic geometry we considered have two dimensions; and we live in a three-dimensional world. The reason these new geometries are called fractal is that they have a fractional dimension. For example, one of the fractal geometries has dimension 1.26.

Photo 5.3
Fractal geometry. A computer graphics image entitled *Found in Space*.

Fractals

> A **fractal** is a geometric figure that consists of an identical motif repeating itself on an ever-reduced scale.

Mandelbrot has shown that many everyday objects such as coastlines, snowflakes, clouds, leaves, ferns, and mountain ranges are naturally described by fractals. Ordinary geometric constructions using straight lines and smooth curves and surfaces do not help to describe many of the intricate patterns found in nature. Thus was born a new branch of geometry, fractal geometry.

An example of a fractal is the **tree-fractal** whose motif is the letter T. In Figure 5.52a we have drawn a T in which the vertical line segment (the trunk) has the same length as the two horizontal line segments (the branches). At the end of each branch, we draw another T whose size is one-half the original, and this process (Figures 5.52b, 5.52c, and 5.53) is repeated indefinitely. We arrive at a figure that is **self-similar**. Each vertical branch, however small, can be considered the trunk of a complete tree, a scaled-down copy of the entire figure. We do not have to use a reduction factor of 1/2; we could use a reduction factor of 1/3 or any other fraction.

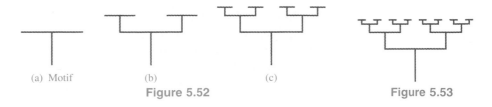

(a) Motif   (b)   (c)

Figure 5.52   **Figure 5.53**

Photo 5.4
Fractal geometry. A computer graphics image entitled *Big Twister*.

Photo 5.5
Fractal geometry. A computer graphics image entitled *Beacon Force*.

In 1915, the Polish mathematician Vaclav Sierpinski invented a fractal that we now call the **Sierpinski sieve**. The **base figure** is an equilateral triangle thought of as a solid object. This triangle is divided into four smaller equilateral triangles, of which the middle one is removed. This figure gives the motif. The process is then continued indefinitely (Figures 5.54 and 5.55).

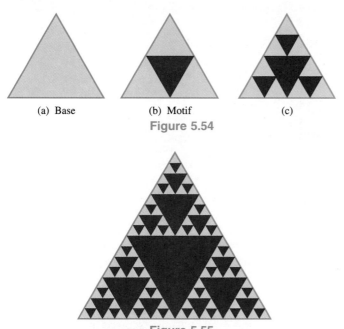

(a) Base   (b) Motif   (c)

Figure 5.54

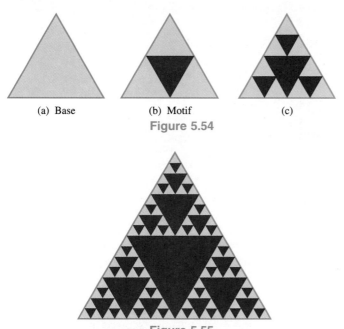

Figure 5.55

In 1904, the mathematician Helge von Koch gave an example of a fractal we now call the **Koch snowflake**. We begin with an equilateral triangle, delete the middle third of each side, and replace it with two sides of an equilateral triangle. We then repeat this process indefinitely (Figures 5.56 and 5.57).

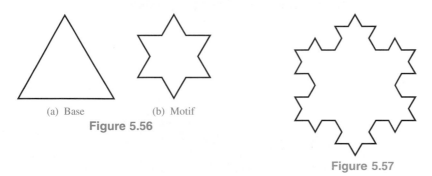

(a) Base   (b) Motif

Figure 5.56

Figure 5.57

Two other fractals, the **Minkowski sausage** and the **Cantor square**, are illustrated in Figures 5.58 and 5.59, and Figures 5.60 and 5.61, respectively.

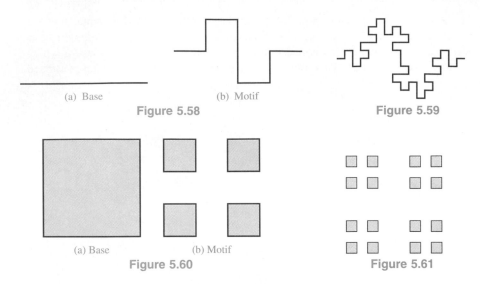

(a) Base   (b) Motif

Figure 5.58   **Figure 5.59**

(a) Base   (b) Motif

Figure 5.60   **Figure 5.61**

There are many important applications of fractals. One example is the study of the human heart. Blood vessels of the heart exhibit fractal-like branching. The large vessels branch into smaller vessels, which in turn branch into even smaller vessels. Each level of vessels in this branching process has a configuration that is similar to that of the next level. This similarity of the various levels is characteristic of fractal behavior. The normal heart rate also exhibits approximate fractal patterns. In one study the heart beat rate of a subject was monitored over a period of 300 minutes. A more detailed analysis of the first 30 minutes exhibited a similarity to the longer time period and an analysis of the first 3 minutes also showed an approximate repetition of the pattern (Figure 5.62).

EXERCISES

1. Discuss longitudes and latitudes on the globe. Are they great circles?

2. Why do airlines want to fly along great circles for long flights?

3. Why does the DDH golf ball have 360 dimples?

4. An octahedral golf ball has eight triangular configurations. Why does it have 336 dimples?

* 5. Why does an icosahedral golf ball have 362 dimples? (*Hint*: See Figure 12.61 and Table 12.2.)

6. In what ways do coastlines, snowflakes, clouds, leaves, ferns, and mountain ranges exhibit fractal behavior?

7. What other objects in nature, besides those mentioned in the text, do you think exhibit fractal behavior?

Continue the following fractals one more step.

8. Figure 5.53   9. Figure 5.55   10. Figure 5.57

11. Figure 5.59   12. Figure 5.61

13. The motif of the **H-fractal** is the letter H and the reduction factor is $1/2$. Continue the fractal construction two steps beyond Figure 5.63.

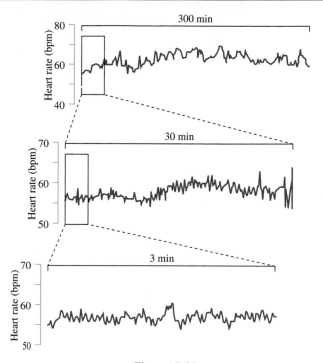

Figure 5.62

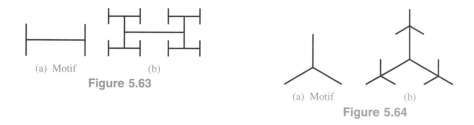

(a) Motif   (b)

Figure 5.63

(a) Motif   (b)

Figure 5.64

14. Draw the first four steps of a tree-fractal with a reduction factor of 1/3.

15. The first two steps of the **ternary tree-fractal** are shown in Figure 5.64. Continue the construction for two more steps.

16. The first two steps of the **Cantor comb fractal** are shown in Figure 5.65. Continue the construction three more steps.

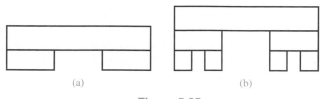

(a)   (b)

Figure 5.65

17. The first three steps in constructing the **Koch fractal** are shown in Figure 5.66. Draw one more step.

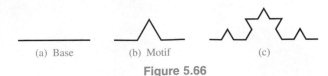

(a) Base   (b) Motif   (c)

Figure 5.66

18. The first three steps in constructing the **cross fractal** are shown in Figures 5.67 and 5.68. Draw one more step.

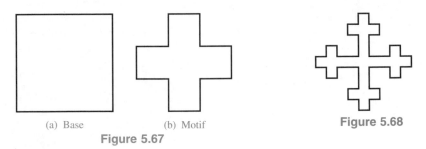

(a) Base   (b) Motif   **Figure 5.68**

Figure 5.67

19. The first three steps in constructing the **box fractal** are shown in Figures 5.69 and 5.70. Draw one more step.

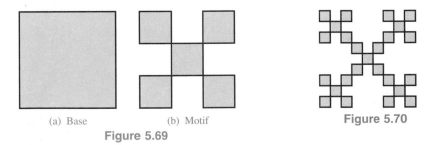

(a) Base   (b) Motif   **Figure 5.70**

Figure 5.69

5.7 Dimension (Optional)

A mathematician is a blind man in a dark room looking for a black hat which isn't there.

—CHARLES DARWIN (1809–1882), British Naturalist

Dimension is an important concept in geometry. Although there are various ways to define dimension, one of the simplest is the following. First consider a line ℓ. Once an origin 0 is established on ℓ, we need only one coordinate x to locate a point P on ℓ (Figure 5.71). We then say that the dimension of ℓ is 1. For a plane, after we establish an origin 0, two coordinates (x, y) are needed to locate a point P, so the dimension of the plane is 2 (Figure 5.72).

Figure 5.71

Figure 5.72

The space in which we live is three-dimensional since three coordinates (x, y, z) are necessary to locate a point (Figure 5.73). A single point P has dimension zero since P itself can be considered an origin and no additional coordinate is needed. A circle has dimension 1 since we can locate a point on a circle by using a single angle α (Figure 5.74). A sphere has dimension 2 since a point P can be located by prescribing two angles, given by the longitude and latitude (Figure 5.75).

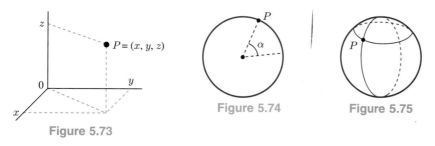

Figure 5.73

Figure 5.74

Figure 5.75

Size

Unfortunately, the previous method of defining dimension is not suitable for a fractal geometry since it is not at all clear what coordinates should be used on a fractal. In fact, as we shall see, the natural dimension for a fractal is not a natural number but a fraction. To define fractal dimensions, we first need the concept of size for a geometric figure. We say that a geometric figure A has half the **size** of a geometric figure B if upon magnification of A by a factor of 2, A is the same as B. For example, the interval $0 \le x \le 1/2$ is half the size of the interval $0 \le x \le 1$, a square of side length $1/2$ is half the size of a square of side length 1, and a cube of side length $1/2$ is half the size of a cube of side length 1 (Figure 5.76).

More generally, let n be a natural number. We say that A **is 1/n the size of** B if upon magnification of A by a factor n, A is the same as B. This concept of size magnification reveals information about the dimension. To magnify a line segment by a factor of 2 we only need to multiply one length by 2, for a square we must multiply two lengths by 2, and for a cube we must multiply three lengths by 2.

To be more precise, suppose we divide a line segment into equal pieces whose size h is $1/n$ the size of the original. We then require $N = n$ pieces (Figure 5.77) so

$$N = \left(\frac{1}{h}\right)^1$$

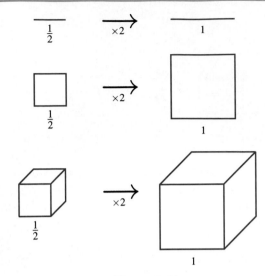

Figure 5.76

If we divide a square into equal pieces whose size h is $1/n$ the size of the original, we require $N = n^2$ pieces (Figure 5.78) so

$$N = \left(\frac{1}{h}\right)^2$$

Finally, if we divide a cube into equal pieces whose size h is $1/n$ the size of the original, we require $N = n^3$ pieces (Figure 5.79) so

$$N = \left(\frac{1}{h}\right)^3$$

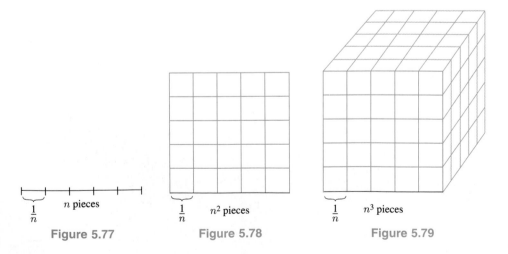

Figure 5.77   Figure 5.78   Figure 5.79

[7] M. C. Escher, *Depth.*
(M. C. Escher Foundation,
Baarn, Holland.)

Dimension Formula

Motivated by the previous discussion, we define the dimension of a geometric figure A as follows. If we divide A into equal pieces of size h compared to A and N pieces are required, then the **dimension** D of A satisfies

$$N = \left(\frac{1}{h}\right)^{D} \tag{5.10}$$

Although (5.10) defines the dimension D, it is difficult to find D using the equation directly. To find D from (5.10) we need to employ logarithms. You will recall that we discussed logarithms in Section 1.6, and it might be helpful for you to review their definition and properties.

Taking logarithms of both sides of (5.10) we obtain

$$\log N = \log \left(\frac{1}{h}\right)^{D} = D \log \left(\frac{1}{h}\right)$$

Hence,

$$D = \frac{\log N}{\log(1/h)} \tag{5.11}$$

We can apply (5.11) to obtain the dimensions of fractals. In constructing a fractal, we begin with a base set B. We then divide B into equal pieces of size h compared to B and then put these pieces back together to form the first stage B_1 of the fractal. We then repeat this procedure on B_1 to form the second stage B_2 of the fractal and continue this process indefinitely. At each stage of the process we can apply (5.11) to obtain the dimension of the fractal.

To illustrate this method, we consider the **Cantor fractal**. To construct this fractal, we begin with an interval and remove its middle third. We then remove the middle thirds of the remaining intervals and continue this process indefinitely (Figure 5.80). The Cantor fractal is certainly larger than a single point, but it is smaller than a line segment. We therefore expect its dimension to be between 0 and 1. In Figure 5.80 we illustrate the base space and the next three stages and tabulate the size h and the number of pieces N. At each stage, the size has the form $h = 1/3^n$ and the number of pieces has the form $N = 2^n$. We conclude that the dimension of the Cantor fractal is

$$D = \frac{\log 2^n}{\log 3^n} = \frac{n \log 2}{n \log 3} = \frac{\log 2}{\log 3} = 0.6309$$

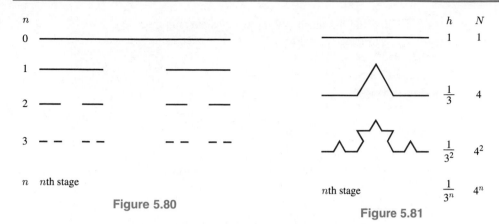

n nth stage

Figure 5.80

Figure 5.81

EXAMPLE 1 Find the dimension of the Koch fractal (Figure 5.66).

Solution. We see from Figure 5.81 that each stage has the form $h = 1/3^n$ and $N = 4^n$. Hence, the dimension is

$$D = \frac{\log 4^n}{\log 3^n} = \frac{n \log 4}{n \log 3} = \frac{\log 4}{\log 3} = 1.2619 \qquad \square$$

EXAMPLE 2 Find the dimension of the Sierpinski sieve (Figure 5.55).

Solution. According to Figure 5.82, each stage has the form $h = 1/2^n$, $N = 3^n$. Hence,

$$D = \frac{\log 3^n}{\log 2^n} = \frac{n \log 3}{n \log 2} = \frac{\log 3}{\log 2} = 1.585 \qquad \square$$

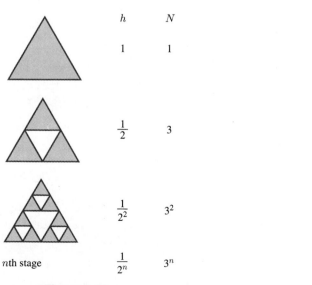

nth stage

Figure 5.82

EXAMPLE 3 Find the dimension of the tree-fractal (Figure 5.53).

Solution.   Referring to Figure 5.52, we construct Table 5.1. This table gives the nth stage $h = 1/2^n$, $N = 2^n$. Hence,

$$D = \frac{\log 2^n}{\log 2^n} = 1$$ □

Table 5.1

Figure	h	N
Base	1	1
1st stage	$1/2$	2
2nd stage	$1/2^2$	2^2
nth stage	$1/2^n$	2^n

EXERCISES

1. What is the dimension of space-time $\mathbb{R}^4 = \{(x, y, z, t): x, y, z, t \in \mathbb{R}\}$? Justify your answer.

2. How would you locate points on the cylinder illustrated in Figure 5.83? What is the dimension of the cylinder?

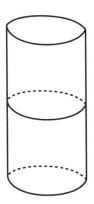

Figure 5.83

3. How would you locate points on the torus in Figure 5.84? What is its dimension?

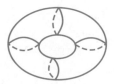

Figure 5.84

4. How would you locate points on the cone in Figure 5.85? What is its dimension?

Figure 5.85

5. What is the relationship between the Cantor fractal of Figure 5.80 and the Cantor comb fractal of Figure 5.65?

6. If a Cantor fractal is constructed by removing the middle half of an interval, what is its dimension?

7. If a Cantor fractal is constructed by removing the middle fifth of an interval, what is its dimension?

Find the dimensions of the following fractals.

8. Koch snowflake (Figure 5.57).

9. Minkowski sausage (Figure 5.59).

10. Cantor square (Figure 5.61).

11. The H-fractal (Figure 5.63).

12. The ternary tree-fractal (Figure 5.64).

13. The box fractal (Figure 5.70).

Chapter 5 Summary of Terms

Euclid's five postulates

lines standard form $Ax + By = C$
  slope-intercept form $y = mx + b$
  point-slope form $(y - y_1) = m(x - x_1)$

Cartesian plane

parallel lines

perpendicular lines

Pythagoras' theorem $a^2 + b^2 = c^2$

converse of Pythagoras' theorem

distance formula $d(P_1, P_2) = \sqrt{(x_2 - x_1)^2 + (y_2 - y_1)^2}$

equation of a circle $(x - x_1)^2 + (y - y_1)^2 = r^2$

circumference of a circle $C = \pi d$

area formulas
rectangle $A = LW$
triangle $A = (1/2)bh$
trapezoid $A = (1/2)(B + b)h$
parallelogram $A = bh$
circle $A = \pi r^2$

integration area formula $A = \displaystyle\int_a^b f(x)$

integration formula $\displaystyle\int_a^b x^n = \frac{b^{n+1}}{n+1} - \frac{a^{n+1}}{n+1}$

non-Euclidean geometry
Riemannian geometry
hyperbolic geometry
fractal geometry

dimension formula $D = \dfrac{\log N}{\log(1/h)}$

Chapter 5 Test

1. Find the intercepts and the slope of the line $5x - 7y = 9$.
2. Find the slope of the line through $(3, 5)$ and $(2, -4)$.
3. Find an equation for the line through $(2, 3)$ that is perpendicular to the line $4x + 3y = 5$.
4. Find the distance between $P_1 = (5, 3)$ and $P_2 = (-4, -2)$.
5. Find an equation for the circle passing through the origin with center $(1, 1)$.
6. Find the area and perimeter of the trapezoid in Figure 5.86.

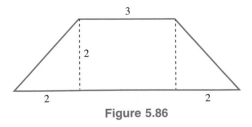

Figure 5.86

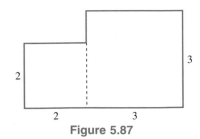

Figure 5.87

7. Find the area and perimeter of the object in Figure 5.87.
8. Find the area of the circle in Exercise 5.

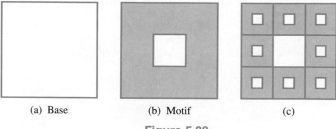

(a) Base   (b) Motif   (c)

Figure 5.88

9. A rectangular box has edges of length 3, 4, and 5 inches. Find three values for its girth plus height.

10. If the square in Figure 5.88a has sides of length 6 inches, find the area of the shaded region in Figure 5.88b.

11. If the square in Figure 5.88a has sides of length 9 inches, find the area of the shaded region in Figure 5.88c.

12. Is there a right triangle with legs of length 4 and 6? If so, what is the length of the hypotenuse?

13. Evaluate the integral $\displaystyle\int_{-1}^{2}(1 - 2x + 3x^2)$.

14. Find the area of the region bounded by the curves $y = x$ and $y = x^2$.

15. A fractal is constructed by dividing a square into nine equal squares and removing the middle one (Figure 5.88). Draw the next stage of the fractal.

16. Are the following statements true or false?
    (a) Every line intersects the x-axis.
    (b) The slope of a vertical line is not defined.
    (c) The point-slope form of a line is $y = mx + b$.
    (d) If $y = mx + b$ is the equation of a line, then b is the x-intercept.
    (e) If two lines are not parallel, then they are perpendicular.
    (f) The lines $y = 2x + 3$ and $y = 2x - 1$ are parallel.
    (g) There is no right triangle with legs of length 2 and 3.
    (h) The magnitude of the perimeter of a square is always less than the magnitude of its area.
    (i) The magnitude of the area of a square is always less than the magnitude of its perimeter.
    (j) The area of a circle with diameter d is $A = \pi d^2/4$.
    (k) The circumference of a circle with radius r is $C = \pi r$.
    (l) The distance from the origin to the point $(1, 1)$ is $\sqrt{2}$.

CONSUMER MATHEMATICS

6

Scientists are increasingly being asked to get out of their ivory towers and many of those who do leave are doing quite well for themselves in the world of business. After all, there's no way you can continue to pursue theory without drying out.

— CARLA LAZZARESCHI, *The Los Angeles Times* (August 29, 1989)

Consumer mathematics is the mathematics of our daily living. Anyone who buys, sells, has a credit card or savings account, borrows money, or pays taxes should know some consumer mathematics. This not only helps us to keep track of our financial activities, it also aids us in saving money. We can save by comparison shopping; selecting the proper bank, credit union, or loan company; or deciding how much down payment to make on a house or car. Consumer mathematics is a tool for survival in a capitalistic society.

6.1 | Interest

My idea then was to get through the course, secure a detail for a few years as assistant professor of mathematics at the Academy, and afterwards obtain a permanent position as professor in some respectable college; but circumstances always did shape my course different from my plans.

— ULYSSES S. GRANT (1822–1885), American President

Many companies and businesses do not produce a product; they merely shuffle currency. For example, you may make a deposit into your savings account. The bank pays you a fee called **interest** for the use of your money. What does the bank do with your funds? Well, it lends them to someone else and charges this other person interest for the privilege of using your money. The interest the bank charges is usually considerably more than it pays you and it pockets the difference. If you buy car insurance, you pay an insurance company an annual fee. The company invests this money and receives interest on it. The company hopes that the amount you pay over a number of years plus the interest is more than it has to pay you if you have an accident. Even if you do have an accident, the insurance company calculates that it will make money overall, since not all of its clients will have accidents.

Simple Interest

The amount of a loan or a deposit is called the **principal** and interest is usually given as a percentage of the principal. This percentage is called the **interest rate**. The interest rate is calculated for a specific time period, for example, daily, weekly, monthly, or annually. Interest computed only on principal is called **simple interest**. Denoting principal by P, interest by I, interest rate by r, and time by t, the simple interest is given by

$$\boxed{\text{Simple Interest} \quad I = Prt} \tag{6.1}$$

EXAMPLE 1   If \$1200 is invested for 4 years at an annual interest rate of 6%, how much simple interest is received?

   Solution.   In this case $P = 1200$, $r = 0.06$, $t = 4$, so applying (6.1) gives

$$I = 1200(0.06)(4) = \$288 \qquad \square$$

EXAMPLE 2   You have invested \$1000 at a quarterly interest rate of 2%. After a period of 2 years, how much simple interest will you receive?

   Solution.   We have $P = 1000$ and $r = 0.02$. Since r is computed quarterly, we must give t in terms of quarters, so $t = 8$. Hence,

$$I = 1000(0.02)(8) = \$160 \qquad \square$$

Compound Interest

Banks usually give a better deal than just simple interest. They calculate interest not just on principal, but on principal plus any previously earned interest. This is called **compound interest**. The period of time for which this interest is credited is called the **compounding period** (or **period**, for short). To see how this works, suppose you deposit \$2000 in a savings account paying an 8% annual interest rate compounded quarterly. The quarterly interest rate is given by

$$r = \frac{0.08}{4} = 2\%$$

Denoting by I_1, I_2, I_3, I_4 the interest at the end of the first, second, third, and fourth quarters, we have

$$I_1 = Prt = 2000(0.02)(1) = \$40$$
$$P + I_1 = \$2040$$
$$I_2 = 2040(0.02)(1) = \$40.80$$
$$P + I_1 + I_2 = \$2080.80$$
$$I_3 = 2080.80(0.02)(1) = \$41.62$$
$$P + I_1 + I_2 + I_3 = \$2122.42$$
$$I_4 = 2122.42(0.02)(1) = \$42.45$$

Thus, at the end of the one year, your account will contain

$$P + I_1 + I_2 + I_3 + I_4 = \$2164.87$$

We thus see that with compound interest, the interest increases with each compounding period. In contrast, simple interest remains the same. If you had earned only simple interest on this last account, after 1 year the interest would be

$$I = Prt = 2000(0.08)(1) = \$160$$

Thus, the compound interest is \$4.87 more than the simple interest for one year.

Compound Amount

We now derive a formula for compound interest. Let P be the principal, r the interest rate per period, and n the number of periods. Denote by $I_1, I_2, \ldots, I_n$ the interest for the first, second,$\ldots$, nth periods, respectively. We then have

$$I_1 = Pr$$

At the end of the first period, the account contains $P + Pr = P(1 + r)$. Hence,

$$I_2 = P(1 + r)r$$

At the end of the second period, the account contains

$$P(1 + r) + P(1 + r)r = P(1 + r)(1 + r) = P(1 + r)^2$$

Continuing, we have

$$I_3 = P(1 + r)^2 r$$

and at the end of the third period, the account contains

$$P(1 + r)^2 + P(1 + r)^2 r = P(1 + r)^2(1 + r) = P(1 + r)^3$$

It is clear that the next terms will be $P(1 + r)^4, P(1 + r)^5, \ldots$, so the **compound amount** A after n periods, with interest rate r per period, becomes

$$\boxed{\text{Compound Amount} \quad A = P(1 + r)^n} \qquad (6.2)$$

In practice, a calculator must be used to compute amounts from (6.2). This is done with the "y^x" key.

EXAMPLE 3 Find the amount accumulated after 1 year if a principal of \$1200 draws interest at an annual rate of 6% compounded quarterly. What if the same interest rate is compounded monthly?

  Solution. In the first case, a period is $1/4$ year, so the interest rate per period is

$$r = \frac{0.06}{4} = 0.015$$

Since $P = 1200$ and $n = 4$, we have from (6.2) that

$$A = 1200(1.015)^4 = \$1273.64$$

In the second case, a period is $1/12$ year, so the interest rate per period is

$$r = \frac{0.06}{12} = 0.005$$

Since $P = 1200$ and $n = 12$, we have

$$A = 1200(1.005)^{12} = \$1274.01 \qquad\qquad \square$$

  We see from Example 3 that compounding monthly gives better interest than compounding quarterly. It is clear that this always happens. With a fixed annual interest rate and principal, the accumulated funds are higher if the compounding period is smaller.

Effective Annual Interest Rate

A bank advertisement offers the following deal on their one-year certificate of deposit. They will pay you an annual rate of 6% compounded monthly with an effective annual interest rate of 6.17%. What does this mean? Clearly, 6.17% sounds better, but how can there be two different interest rates? The **effective annual interest rate** is useful for comparing compound interest schemes. It is defined as the interest rate r' that, when applied at simple interest, gives the same yield after one year as the compounding scheme under consideration.

  Suppose you deposit a principal P at an interest rate of r per period and there are n periods in a year. After 1 year the compound amount in your account will be

$$A = P(1 + r)^n$$

If, on the other hand, you receive only simple interest at the rate r', your account will contain

$$A' = P + Pr' = P(1 + r')$$

In order for these amounts to agree we must have

$$P(1 + r') = P(1 + r)^n$$

or equivalently

$$\boxed{\text{Effective Annual Interest Rate} \quad r' = (1 + r)^n - 1} \qquad (6.3)$$

Equation (6.3) gives the effective annual interest rate r' corresponding to the compound rate r per period when there are n periods per year.

EXAMPLE 4   If a 6% annual interest rate is compounded monthly, what is the effective annual interest rate r' ?
   Solution.   Since the interest rate per period is

$$r = \frac{0.06}{12} = 0.005$$

and there are 12 periods per year, (6.3) gives

$$r' = (1.005)^{12} - 1 = 1.06167 - 1 = 0.06167$$

or $r' = 6.17\%$.   □

Years to Double

A good indication of the profitability of an investment is the number of years it takes to double your money. This number is called the **years to double**. If you invest a principal P and receive simple interest at an annual rate r, then after t years your investment will have grown to an amount

$$A = P + Prt = P(1 + rt)$$

Setting $A = 2P$ we have $2P = P(1 + rt)$, which gives

$$\boxed{\text{Years to Double with Simple Interest} \quad t = \frac{1}{r}} \qquad (6.4)$$

EXAMPLE 5   If you invest $1000 and receive simple interest at an annual rate of 8%, find the years to double.
   Solution.   Applying (6.4) gives

$$t = \frac{1}{0.08} = 12.5 \text{ yr}$$ □

To find the years to double for compound interest, we use (6.2) to obtain $A = P(1+r)^n$. Setting $A = 2P$ we have $2P = P(1+r)^n$ or $2 = (1+r)^n$. Taking logarithms of both sides gives

$$\log 2 = \log(1+r)^n = n \log(1+r)$$

We then have

$$\boxed{\text{Periods to Double with Compound Interest} \quad n = \frac{\log 2}{\log(1+r)}} \qquad (6.5)$$

Remember, in (6.5) n is the number of periods to double and r is the interest rate per period. To find the years to double with compound interest, we must divide n by the number of periods per year.

EXAMPLE 6   If you invest at an annual interest rate of 8% compounded quarterly, find the years to double.

    Solution.   In this case

$$r = \frac{0.08}{4} = 0.02$$

The number of periods to double is, by (6.5),

$$n = \frac{\log 2}{\log(1.02)} = 35$$

The years to double becomes

$$t = \frac{n}{4} = \frac{35}{4} = 8.75 \qquad \qquad \square$$

    Years to double is also a useful measure of inflation. The **annual inflation rate** is the percentage increase of living costs over the previous year. For example, average prices were about 3.5% higher at the beginning of 1991 than they were at the beginning of 1990. This gives an annual inflation rate of 3.5%. For a given annual inflation rate, the years to double gives the number of years it will take for average living costs to double. Put another way, this is the number of years it will take for your present dollar to be worth 50¢.

EXAMPLE 7   Given an annual inflation rate of 4%, find the years to double.

    Solution.   According to (6.5)

$$n = \frac{\log 2}{\log(1.04)} = 17.67 \qquad \qquad \square$$

    If no time is specified for an interest rate, it is always assumed that it is an annual interest rate.

EXERCISES

1. If a principal P is invested at an annual simple interest rate r, show that the amount accumulated after t years is $A = P(1 + rt)$.

Find the simple interest earned for the following deposits at the given annual interest rates.

2. $2000 at 8% for 1 year

3. $5000 at 9% for 2 years

4. $1000 at 11% for 6 months

5. $1500 at 7% for 9 months

6. If $10,000 grows to $15,000 after 5 years, what is the annual simple interest received?

7. If $1000 grows to $1700 at an annual simple interest rate of 6%, how many years have elapsed?

In Exercises 8–10, find the final amount on deposit.

8. $1500 at 9% compounded semiannually for 6 years.

9. $5000 at 8% compounded quarterly for 10 years.

10. $2500 at 10% compounded monthly for 8 years.

11. You have two investment possibilities. You can lend $2000 to Company X with repayment of $4000 in 10 years, or you can lend $2000 to Company Y for 10 years at an annual rate of 8% compounded annually. Assuming the companies are equally good risks, which investment should you choose?

Find the effective annual interest rate for the deposit in the specified exercise.

12. Exercise 8

13. Exercise 9

14. Exercise 10

15. If you invest at an annual simple interest rate of 7%, find the years to double.

16. Repeat Exercise 15 with a rate of 9%.

Find the years to double for the following annual interest rates.

17. 9% compounded semiannually.

18. 7% compounded quarterly.

19. 6% compounded monthly.

20. 10% compounded weekly.

Use the effective annual interest rates to compare the following investments.

21. 5.1% compounded semiannually or 5% compounded quarterly.

22. 6% compounded monthly or 6.2% compounded semiannually.

23. 5.75% compounded quarterly or 5.6% compounded weekly.

For the following annual inflation rates, find the years to double.

24. 3%

25. 3.5%

26. 5%

27. 7%

28. If a is the annual interest rate and n is the number of periods per year, show that the compound amount after t years is

$$A = P \left(1 + \frac{a}{n}\right)^{nt} \qquad (6.6)$$

29. Use (6.6) to solve Exercise 10.

$\boxed{6.2}$ More and More Interest

*The mathematician lives long and lives young; the wings of his soul
do not early drop off, nor do its pores become clogged with earthly
particles blown from the dusty highways of vulgar life.*
— James Sylvester (1814–1897), British Mathematician

Bank X puts up a sign advertising 8% annual interest compounded quarterly. To be competitive, Bank Y advertises 8% annual interest compounded monthly. Not to be outdone, Bank X changes its sign to 8% annual interest compounded weekly. Where is this leading? Will it never end? Finally, a logical conclusion is reached and both banks advertise 8% annual interest compounded continuously! What does this mean? Is each bank compounding every second? Every millisecond? Are the bank's computers furiously computing interest every 1/1000 of a second? More frequent compounding raises the effective interest rate. Does this mean that the effective annual interest rate is going to infinity?

Let's see if more frequent compounding really makes much difference. Table 6.1 compares the compound amount after 1 year for a $1000 deposit in an account giving 8% annual interest. Even if we continue this table to minutes, seconds, milliseconds, etc., we do not get much more than $1083. The difference between annual compounding and compounding hourly is only $3.29 per $1000. Hence the frequency of compounding does not make much difference unless the principal is over $100,000. What makes a greater difference is the annual interest rate and the length of time the principal is in the account. Of course, if the principal is left in the account for a long period of time (say 10 years), then frequent compounding can make a substantial difference.

Table 6.1

Frequency of compounding	Number of periods n	Compound amount $A = 1000(1 + 0.08/n)^n$
Annually	1	$1080.00
Semiannually	2	$1081.60
Quarterly	4	$1082.43
Monthly	12	$1083.00
Weekly	52	$1083.22
Daily	365	$1083.28
Hourly	8760	$1083.29

Continuous Compounding

To understand what compounding continuously means, we need to examine the compound amount formula (6.2). Suppose the annual interest rate is a (in the previous case, $a = 0.08$). If there are n compounding periods per year, then the interest rate per period is $r = a/n$. Starting with principal P, the compound amount after t years is

$$A = P \left(1 + \frac{a}{n}\right)^{nt} \tag{6.7}$$

In the example of the two banks, the compounding becomes more and more frequent so n approaches infinity; that is, n gets arbitrarily large. What happens to A as $n \to \infty$? To get an idea of this, let us look at the simpler quantity $\left(1 + (1/n)\right)^n$ and see what happens for n large (Table 6.2).

Table 6.2

n	$\left(1 + \dfrac{1}{n}\right)^n$
1	2
10	2.5937425
100	2.7048138
1,000	2.7169238
10,000	2.7181459
100,000	2.7182546
1,000,000	2.7182818

We see from Table 6.2 that $\left(1 + (1/n)\right)^n$ approaches the value $2.7182818\ldots$ as $n \to \infty$. This number is a universal constant called the **natural exponential** and is denoted by e. The number e comes up in many applications in the mathematical, physical, business, and social sciences. It is important for studies in population growth and radioactive decay. We shall see in Chapter 8 that it is important in statistics because it is used in the formula for the normal distribution curve. It does not even matter that the n's are natural numbers; they can be any numbers approaching infinity. We can therefore write

$$\left(1 + \frac{1}{x}\right)^x \to e \quad \text{as} \quad x \to \infty \tag{6.8}$$

Now we can write (6.7) as

$$A = P \left(1 + \frac{1}{n/a}\right)^{(n/a)(at)}$$

and letting $x = n/a$ we have

$$A = P \left[\left(1 + \frac{1}{x}\right)^x\right]^{at}$$

As n approaches infinity, so does x. We conclude from (6.8) that as $n \to \infty$, A approaches Pe^{at}. We thus have, with annual interest rate a and t years,

Continuous Compound Amount $A = Pe^{at}$	$\tag{6.9}$

Equation (6.9) is what is meant by **compounding continuously**. The easiest way to compute (6.9) in practice is to use the "e^x" key of a calculator.

EXAMPLE 1 Find the continuous compound amount after 1 year for a principal of $1000 at the annual interest rate of 6%. Compare this with the amount when compounding is weekly.

   Solution.   Applying (6.9) gives the continuous compound amount

$$A = 1000\, e^{0.06} = \$1061.84$$

With weekly compounding we have

$$A = 1000 \left(1 + \frac{0.06}{52} \right)^{52} = \$1061.80 \qquad \square$$

EXAMPLE 2 Compute the value of $100 after 50 years of continuous compounding at annual interest rates of (a) 4%, (b) 5%, (c) 6%.

   Solution.   (a) $at = (0.04)(50) = 2$, so

$$A = 100\, e^2 = \$738.91$$

(b) $at = (0.05)(50) = 2.5$, so

$$A = 100\, e^{2.5} = \$1218.25$$

(c) $at = (0.06)(50) = 3$, so

$$A = 100\, e^3 = \$2008.55 \qquad \square$$

Effective Annual Interest Rate

To compute the effective annual interest rate r' for continuous compounding, we compare (6.9) to the amount from simple interest. For a time $t = 1$ year, we have

$$P(1 + r') = P e^a$$

Hence, the **effective annual interest rate for continuous compounding** at annual interest rate a is

$$\boxed{r' = e^a - 1} \qquad (6.10)$$

EXAMPLE 3 If a 7% annual interest rate is compounded continuously, what is the effective annual interest rate?

   Solution.   Applying (6.10) gives

$$r' = e^{0.07} - 1 = 0.0725 \quad \text{or} \quad r' = 7.25\% \qquad \square$$

EXAMPLE 4 If you invest at an annual interest rate of 8% compounded continuously, find the years to double.

Solution. For a principal P the years to double t satisfies

$$2P = Pe^{0.08t} \quad \text{or} \quad 2 = e^{0.08t}$$

Taking logarithms of both sides gives

$$\log 2 = \log(e^{0.08})^t = t \log e^{0.08}$$

Hence,

$$t = \frac{\log 2}{\log e^{0.08}} = 8.66 \text{ yr} \qquad \square$$

In some situations we would like to know how much to deposit today in order to reach a certain amount A in a given number of years. This deposited sum is called the **present value of A**.

EXAMPLE 5 Janet wants to save to buy a house in 10 years. She figures that she will need $20,000 for a down payment. What lump sum deposited today at 8% compounded quarterly will yield the necessary amount?

Solution. For the compound amount formula, we have $r = 0.08/4 = 0.02$, $n = 40$, $A = 20,000$, so

$$20,000 = P(1.02)^{40}$$

where P is the principal deposited today. In this case, P is the present value of $20,000. Solving for P gives

$$P = \frac{20,000}{(1.02)^{40}} = \$9057.81 \qquad \square$$

Present Value Formulas

Applying (6.2) and (6.9) we can find the present value P for periodic compounding and continuous compounding. In the case of periodic compounding, (6.2) gives for interest rate r per period and n periods

$$\text{Present Value: Periodic Compounding} \quad P = \frac{A}{(1+r)^n} \qquad (6.11)$$

In the case of continuous compounding, (6.9) gives for an annual interest rate a

$$\text{Present Value: Continuous Compounding} \quad P = \frac{A}{e^{at}} \qquad (6.12)$$

EXAMPLE 6   Find the present value of $1000 five years from now if money earns 10% compounded monthly.

   Solution.   In applying (6.11) we have $A = 1000$, $r = 0.10/12 = 0.00833$, $n = 5(12) = 60$. Hence,

$$P = \frac{1000}{(1.00833)^{60}} = \$607.80 \qquad \square$$

EXAMPLE 7   A savings account pays 5% annual interest compounded continuously. What is the present value of $2000 to be paid 5 years from now?

   Solution.   In applying (6.12) we have $A = 2000$, $a = 0.05$, $t = 5$. Hence,

$$P = \frac{2000}{e^{0.25}} = \$1557.60 \qquad \square$$

EXERCISES

Find the continuous compound amount for the following principals and annual interest rates.

1. $2000 at 7% for 5 years
2. $3000 at 6.5% for 10 years
3. $1000 at 5% for 40 years
4. $1000 at 6% for 40 years
5. $1000 at 7% for 40 years.

Find the effective annual interest rate for continuous compounding at the given annual interest rate.

6. 4%   7. 5%   8. 6%   9. 6.5%   10. 7.75%

11. What does $\left(1 + (2/x)\right)^{x}$ approach as $x \to \infty$?
12. What does $\left(1 + (5/x)\right)^{x}$ approach as $x \to \infty$?

Find the present values of the given amounts, years from now, and annual interest rates.

13. $1000, 10 years, 7% compounded quarterly
14. $5000, 5 years, 6% compounded monthly
15. $7000, 6 years, 5% compounded weekly
16. $2000, 7 years, 6% compounded continuously
17. $3000, 10 years, 5.5% compounded continuously
18. Why can we write (6.11) as $P = A(1 + r)^{-n}$?
19. Why can we write (6.12) as $P = Ae^{-at}$?

Find the years to double for the following annual interest rates compounded continuously.

20. 4%   21. 5%   22. 6%   23. 7.5%   24. 9%

25. Show that the years to double for an annual interest rate a compounded continuously is

$$t = \frac{\log 2}{\log e^{a}} \qquad (6.13)$$

*26. Show that (6.13) can be written $t = (\ln 2)/a$ where ln is the logarithm to the base e.
*27. Do Exercise 20 using Exercise 26.

28. Mr. Jones wants to invest for his daughter's college education in 15 years. He calculates that he will need $60,000 at that time. He has an opportunity to earn 9% compounded continuously. How much should he invest?

29. Repeat Example 5 for interest compounded continuously.

30. The State of New York wants to sell bonds worth $1000 each at maturity 10 years from now. The selling price is $525. We could also put our money into certificates of deposit paying 7% compounded annually. Which should we choose?

31. Repeat Exercise 30, except now the bonds sell at $500 and the certificate of deposit 7% interest rate is compounded continuously.

32. Verify the entries in Table 6.1.

33. If an annual interest rate of 9% is compounded every minute, what is the compound amount for a principal of $1000 after 1 year?

34. Repeat Exercise 33 for 10% annual interest rate.

35. Verify the entries in Table 6.2.

36. Compute $\left(1 + (1/n)\right)^n$ for $n = 10,000,000$.

37. Compute $\left(1 + (1/x)\right)^x$ for $x = e^{10}$.

6.3 | Annuities

Do not imagine that mathematics is hard and crabbed, and repulsive to common sense. It is merely the etherealization of common sense.
— LORD KELVIN (1824–1907), British Physicist

Practically every adult participates in some kind of annuity.

> An **annuity** is a sequence of equal periodic payments.

If you make regular deposits of a specified amount to your savings account each month, you are participating in an annuity. Monthly mortgage payments, car loan payments, insurance premiums, and payments into a retirement account are all examples of annuities. If you want to really understand why you are paying $350 a month for your car loan, you need to know the mathematics of annuities.

The payments may be made weekly, monthly, quarterly, annually, or for any fixed period of time. The time between successive payments is called the **payment period** for the annuity. The time from the beginning of the first payment period to the last payment is the **term** of the annuity. In the case of a life insurance policy, if the policy ends after a fixed amount of time, it is called a term policy. Otherwise, it continues until the death of the individual. The **amount** (or **future value**) of an annuity is the sum of all payments and all earned interest.

Photo 6.1

Monthly mortgage payments are an example of an annuity.

Annuity Amount

Suppose you have an annuity of $1000 a year for 4 years at 8% compounded annually. You make your first payment of $1000 at the end of the first year. This $1000 earns interest for the next 3 years, so by the compound amount formula (6.2), it is worth $($1000)(1.08)^3$ at the end of the fourth year. At the end of the second year you pay another $1000 and it earns interest for the next 2 years, so at the end of the fourth year, it is worth $($1000)(1.08)^2$. Your $1000 payment at the end of the third year earns interest for 1 year, so it will be worth $($1000)(1.08)$. The last $1000, paid at the end of the fourth year, earns no interest because the annuity has reached its term and ends. We conclude that the amount A of the annuity is

$$A = 1000(1.08)^3 + 1000(1.08)^2 + 1000(1.08) + 1000$$
$$= 1000 \left[1 + (1.08) + (1.08)^2 + (1.08)^3 \right] = \$4506.11$$

To derive a general formula for the amount A of any annuity, let R be the periodic payment, r the interest rate per payment period, and n the number of payment periods. Just as in the last example, we have

$$A = R(1 + r)^{n-1} + R(1 + r)^{n-2} + \cdots + R(1 + r)^2 + R(1 + r) + R$$
$$= R \left[1 + (1 + r) + (1 + r)^2 + \cdots + (1 + r)^{n-1} \right]$$

But

$$1 + (1 + r) + (1 + r)^2 + \cdots + (1 + r)^{n-1}$$

is a geometric series and we have discussed these in Section 1.2. We saw there that the sum of this series is

$$\frac{1 - (1 + r)^n}{1 - (1 + r)} = \frac{1 - (1 + r)^n}{-r} = \frac{(1 + r)^n - 1}{r}$$

We thus obtain the following.

$$\text{Annuity Amount} \quad A = R\left[\frac{(1+r)^n - 1}{r}\right] \tag{6.14}$$

where A = amount, R = periodic payment, r = rate per period, and n = number of payment periods.

EXAMPLE 1 A company deposits \$1000 each quarter into a fund that pays 8% per year interest compounded quarterly. How much is in the fund (a) at the end of the first year? (b) at the end of the third year? (c) at the end of the sixth year?

Solution. (a) We have $R = 1000$, $r = 0.8/4 = 0.02$, $n = 4$. Applying (6.14) gives

$$A = 1000\left[\frac{(1.02)^4 - 1}{0.02}\right] = \$4121.61$$

(b) In this case $n = 3(4) = 12$ so

$$A = 1000\left[\frac{(1.02)^{12} - 1}{0.02}\right] = \$13,412.09$$

(c) In this case $n = 6(4) = 24$ so

$$A = 1000\left[\frac{(1.02)^{24} - 1}{0.02}\right] = \$30,421.86 \qquad \square$$

EXAMPLE 2 Mrs. Jones has been depositing \$100 each month into a savings account. For the first 10 years the annual interest rate was 6% compounded monthly, but for the last 10 years the annual interest rate has been 9% compounded monthly. What is the total value of this annuity at the end of the 20-year period?

Solution. At the end of the first 10 years the amount of the annuity was

$$A = 100\left[\frac{(1.005)^{120} - 1}{0.005}\right] = \$16,387.94$$

Then over the last 10 years this amount has earned interest at the rate of 9% compounded monthly to accumulate a sum of

$$A_1 = 16,387.94(1.0075)^{120} = \$40,172.70$$

In addition, the monthly deposit of \$100 made over the last 10 years will accumulate, with interest at 9% compounded monthly, to

$$A_2 = 100\left[\frac{(1.0075)^{120} - 1}{0.0075}\right] = \$19,351.42$$

This gives a total for the annuity at the end of the 20-year period of

$$A_1 + A_2 = \$59,524.12$$ □

Suppose you have a financial goal A and want to establish an annuity to reach this goal in a certain time. How do you find your required periodic payments R to reach this goal? If the interest rate per payment period is r and the number of payment periods is n, then we can solve (6.14) for R to obtain

$$R = A \left[\frac{r}{(1 + r)^n - 1} \right] \qquad (6.15)$$

EXAMPLE 3   How much will have to be deposited in a fund at the end of each year at 7% compounded annually to buy a car for \$20,000 in 5 years?
   Solution.   Applying (6.15) gives

$$R = 20,000 \left[\frac{0.07}{(1.07)^5 - 1} \right] = \$3477.81$$ □

EXERCISES

1. Repeat Example 1 with deposits of \$2000 each quarter at 7% per year interest compounded quarterly.

2. What is the result in Example 2 if Mrs. Jones deposits \$200 each month?

Find the future amount of each of the following annuities.

3. \$50 paid monthly for 3 years at 12% compounded monthly.

4. \$100 paid monthly for 5 years at 9% compounded monthly.

5. \$500 paid quarterly for 10 years at 10% compounded quarterly.

6. \$600 paid annually for 25 years at 7% compounded annually.

7. \$300 paid semiannually for 20 years at 8.5% compounded semiannually.

In Exercises 8–12, you are given the amount of an annuity, the term, the payment period, and the annual interest rate, which is compounded per period. Find the required periodic payment.

8. \$5000, 3 years, monthly, 12%   9. \$25,000, 10 years, quarterly, 9%

10. \$50,000, 20 years, semiannually, 8.5%   11. \$30,000, 15 years, annually, 7%

12. \$10,000, 5 years, weekly, 10.4%

13. Parents wish to set up a savings account for their child's education. They plan to deposit \$500 semiannually. The account earns an annual interest of 9% compounded semiannually. To what amount will the account grow in 18 years? How much interest will be earned during the term of the annuity?

14. What amount will have to be invested at the end of each year for 10 years in order to form an annuity of \$300,000 if interest is 9% compounded annually?

15. A company established an annuity to discharge a debt of \$150,000 due in 5 years by making monthly deposits into a fund that pays 10% compounded monthly.
   (a) What is the required size of each deposit?
   (b) How much interest will be earned over the 5-year period?

Photo 6.2
Money deposited with a bank is frequently used to make loans.

6.4 Loans

There is no branch of mathematics, however abstract, which may not someday be applied to the phenomena of the real world.
— NIKOLAI LOBACHEVSKI (1793–1856), Russian Mathematician

In the previous three sections we considered interest paid to you. This was an amount paid to you for the privilege of using your money. But frequently you are on the other side of the table and you are borrowing money from someone else. In this case, you must pay interest for the privilege of using the other party's money.

Suppose you are buying a car and the salesperson tells you that your payment will be $350 per month for 3 years. You might ask how this figure was derived. The answer would probably be that it came from a complicated formula built into the computer and only a mathematical wizard could understand it. Well, there is no reason to accept an answer like that. You are entitled to see the formula and to decide for yourself whether it is fair and reasonable. As we shall see, this formula comes from combining our compound amount formula (6.2) and our annuity amount formula (6.14).

Amortizing

Suppose you apply for a $5000 loan from your local bank. You would like to repay the loan in five equal annual installments over a period of 5 years and the bank's interest charges are 12% compounded annually on the unpaid balance. This method of repayment is called **amortization**.

> **Amortizing** a debt means retiring the debt in a given length of time by equal periodic payments that include compound interest.

Table 6.3

Payment number	Remaining principal at beginning of period	Payment amount	Interest paid (12% of remaining principal)	Amount of payment applied to principal	Remaining principal at end of period
1	5000.00	1387.05	600.00	787.05	4212.95
2	4212.95	1387.05	505.55	881.50	3331.45
3	3331.45	1387.05	399.77	987.28	2344.17
4	2344.17	1387.05	281.30	1105.75	1238.42
5	1238.42	1387.05	146.61	1238.42	0

How much should each annual payment be in order to retire the loan? In this particular case, you will be using the bank's $5000 for 5 years, so if you made no payments during the 5 years, you would owe the bank

$$A = 5000(1.12)^5 = \$8811.71$$

We again call $5000 the **principal**, although in this case the money is the bank's and not yours. To retire this loan, you establish an annuity with the bank. This annuity has an amount $A = \$8811.71$, an interest rate of $r = 12\%$ annually, and $n = 5$ payments. If R is the periodic payment, by the annuity amount formula (6.14) we have

$$5000(1.12)^5 = R \left[\frac{(1.12)^5 - 1}{0.12} \right]$$

Hence,

$$R = \frac{5000(1.12)^5(0.12)}{(1.12)^5 - 1} = \$1387.05$$

Table 6.3 summarizes the amortization schedule for this loan.

Notice that in Table 6.3 the payment amount is divided into two parts, one part to pay the interest on the remaining principal and the other part to reduce the principal. As the payments continue, more of each payment is applied to reducing the principal and less to paying interest. You calculate that you pay the bank a total amount of

$$5(1387.05) = \$6935.25$$

so you pay $1935.25 in interest for the loan. Is this fair and reasonable? We shall see that it is.

First, although the 12% interest is certainly higher than the bank pays its customers for savings accounts, the bank has to make a profit and the 12% is probably competitive with other banks. (You should have done some comparative shopping to make sure of this.) Second, the bank is giving you back the same interest for your annuity. If you had established a separate annuity at another bank, you certainly would not get an interest this high (it might be in the 5 to 7% range).

Amortization Payment

We now derive the general formula for the periodic payments required for amortizing a debt. Let P be the principal of the debt (or loan), R the periodic payment, r the interest rate per period, and n the number of periods. By the compound amount formula (6.2), the total amount owed is $A = P(1 + r)^n$, and by (6.14) the amount of the annuity established to retire the debt is

$$A = R \left[\frac{(1+r)^n - 1}{r} \right]$$

Setting these equal gives

$$P(1+r)^n = R \left[\frac{(1+r)^n - 1}{r} \right]$$

Solving for R and dividing numerator and denominator by $(1 + r)^n$ gives

$$\text{Amortization Payment} \quad R = \frac{Pr}{1 - (1+r)^{-n}} \tag{6.16}$$

where R = periodic payment, P = principal, r = interest rate per period, and n = number of periods.

Remember that in (6.16)

$$(1+r)^{-n} = \frac{1}{(1+r)^n}$$

EXAMPLE 1 You take a car loan of \$15,000, which is to be repaid in 48 equal monthly installments at an annual interest rate of 12% compounded monthly. How much are the monthly payments? How much interest will you pay?

Solution. The monthly payment, computed from (6.16), is

$$R = \frac{15,000(0.01)}{1 - (1.01)^{-48}} = \$395.01$$

The interest you pay will be

$$48(395.01) - 15,000 = \$3960.36 \qquad \square$$

EXAMPLE 2 Mr. and Mrs. Johnson purchase a house for \$120,000. They make a 20% down payment, with the balance amortized by a 30-year mortgage at an annual interest rate of 11% compounded monthly.
(a) What is their monthly mortgage payment?
(b) What is the total amount of interest the Johnsons will pay over the life of the mortgage?

Solution.   (a) Upon subtracting the 20% down payment the principal becomes

$$120,000 - 0.20(120,000) = \$96,000$$

The interest rate per period is

$$r = \frac{0.11}{12} = 0.0091666$$

The number of periods is

$$n = 30(12) = 360$$

Applying (6.16) we have

$$R = \frac{96,000(0.0091666)}{1 - (1.0091666)^{-360}} = \$914.23$$

(b) The total interest is

$$360(914.23) - 96,000 = \$233,122.80 \qquad \square$$

EXERCISES

1. Jill borrows $10,000 at an annual interest of 12% compounded monthly. How much does she owe after 10 years?
2. If you borrow $5000 at an annual interest of 12% compounded quarterly, how much do you owe after 5 years?
3. Verify the entries in Table 6.3.
4. Retabulate Table 6.3 with the same data but for an annual interest rate of 10%.
5. Show that (6.16) is the same as

$$R = \frac{Pr(1 + r)^n}{(1 + r)^n - 1}$$

6. The **present value** of a loan is obtained by solving (6.16) for P. What do you get?

In the following you are given the loan principal, the annual interest rate compounded monthly, and the number of years for the loan. Calculate the monthly payment and the total interest paid.

7. $10,000, 12%, 5 years
8. $20,000, 10%, 10 years
9. $100,000, 11%, 30 years
10. $200,000, 12%, 25 years

Jackie Schneider wants to finance $10,000 on a new car. Determine her monthly payments and the total interest paid if the annual rate of interest is 13% compounded monthly over the following times.

11. 30 months   12. 3 years   13. 4 years   14. 5 years

15. Mr. Smith has a 25-year $100,000 loan that he pays monthly at 13% annual interest compounded monthly. How much does he pay each year?
16. Repeat Exercise 15 for a 30-year loan.
17. Repeat Exercise 15 with 14% interest.

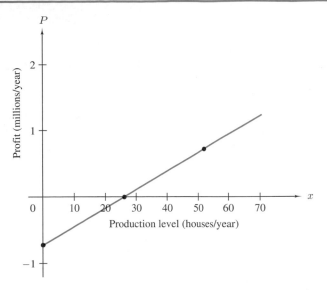

Figure 6.2

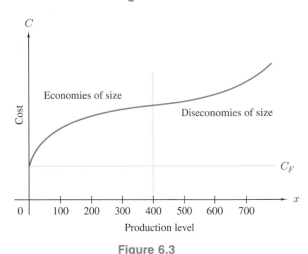

Figure 6.3

EXAMPLE 3   By buying materials in large quantities, the cost per house for the builder in Examples 1 and 2 is $30,000 - 100x$. What is the profit per year if 50 houses are sold?

Solution.   We now have $C_V(x) = (30,000 - 100x)x$. Hence,

$$P(x) = R(x) - C(x) = 60,000x - [750,000 + (30,000 - 100x)x]$$

$$= 30,000x + 100x^2 - 750,000$$

Letting $x = 50$ gives

$$P(50) = 30,000(50) + 100(50)^2 - 750,000 = \$1,000,000 \qquad \square$$

Photo 6.3
Profit, revenue, and cost are important constituents in the economics game.

6.5 Economics

But it should always be required that a mathematical subject not be considered exhausted until it has become intuitively evident.
— FELIX KLEIN (1849–1925), German Mathematician

Besides being consumers, most adults either own a business, work for a profitmaking company, or work for a nonprofit organization such as a religious group, charity, or government. For these people, the economics of business is very important. Even if you are a student or a retired person, economics can have a direct effect on your life. Mathematical economics can be used to describe the prices and interest rates you pay and receive as well as the strength of the prevailing business market. An understanding of economics can help you to make daily and long-term financial decisions.

Profit, Revenue, and Cost

Suppose a plant manufactures a single product, say stoves (the type of product does not matter for this discussion). If you are managing this plant, one of your major decisions is to set the production level, that is, the number of stoves produced each day. How can you do this? One way is to guess. If you guess correctly often enough, your firm will turn a good profit. If not, you may have plenty of time to ponder what went wrong. In that case you might discover that instead of guessing or using trial and error, you should have attempted to analyze the problem. One approach is to study the relationship between the production level x and the profit P.

Profit is the difference between **revenue** R (what comes in) and **cost** C (what goes out).

We thus have

$$P(x) = R(x) - C(x) \qquad (6.17)$$

We usually fix a "short-run period of time," which may be a day, a week, or some other convenient time period. The **production level** x is the number of units produced during this period. A plant can usually be run at different production levels. At one extreme it can be idle ($x = 0$); at the other extreme it can be run at full capacity. We denote the full capacity level by x_m. The set of x values between 0 and x_m is called the **feasible set**. The cost function $C(x)$ is the sum

$$C(x) = C_F + C_V(x)$$

where C_F is the **fixed costs** and $C_V(x)$ the **variable costs**. The fixed costs must be paid even if there is no production at all. These costs include such items as rent, advertising, insurance, interest paid to investors, amortization of plant equipment, and salary of office staff and executives. The variable costs result from actual production such as labor salaries, cost of raw materials, energy, transportation, and so on. Sometimes the variable costs are the same for each unit produced, say A dollars per unit. Then $C_V(x)$ is proportional to x and we have the simple formula

$$C(x) = C_F + Ax \qquad (6.18)$$

EXAMPLE 1 A builder of prefabricated houses has fixed costs of $750,000 per year and variable costs of $30,000x$ where x is the number of units constructed per year. The feasible set is $0 \le x \le 70$.
(a) What is the cost function $C(x)$?
(b) What is the cost of producing 50 houses in a year?
(c) Graph $C(x)$.
  Solution.   (a) Applying (6.18) gives

$$C(x) = C_F + C_V(x) = 750,000 + 30,000x$$

(b) Letting $x = 50$, we have

$$C(50) = 750,000 + 30,000(50) = \$2,250,000$$

(c) The graph of $C(x)$ is shown in Figure 6.1.   □

We next consider revenue. In simple situations, the entire output is sold at a fixed price, say p dollars per unit. Then the revenue function becomes

$$R(x) = px \qquad (6.19)$$

and the profit function is $P(x) = R(x) - C(x) = px - C(x)$.

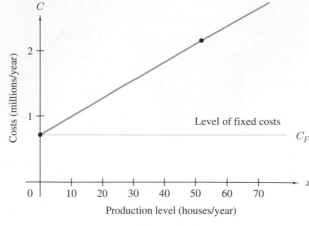

Figure 6.1

EXAMPLE 2 If the builder in Example 1 sells houses at $60,000 each, what is the profit for 50 houses in a year? How many houses must be sold to break even? Graph $P(x)$.
  Solution.   In this case, $p = 60,000$ so

$$P(x) = px - C(x) = 60,000x - (750,000 + 30,000x)$$
$$= 30,000x - 750,000$$

Letting $x = 50$, we have

$$P(50) = 30,000(50) - 750,000 = \$750,000$$

Break-even occurs when $R(x) = C(x)$ or $P(x) = 0$. Solving $30,000x = 750,000$ for x gives $x = 25$ houses. The graph of $P(x)$ is shown in Figure 6.2. Note the negative profit (loss) below $x = 25$.   □

Economies and Diseconomies of Size

In Example 1, the variable costs had the simple form $C_V(x) = Ax$. In this case the cost function $C(x) = C_F + Ax$ was **linear** and the graph was a straight line (Figure 6.1). In complex, large-scale operations, the variable costs are usually not proportional to x. As x increases from $x = 0$, the operation at first tends to become more efficient. Savings may occur by buying raw materials in large quantities, using the workforce more effectively, and introducing labor-saving devices. Such effects, referred to as **economies of size**, reduce the cost per unit. After x increases beyond a certain level, we encounter **diseconomies of size**, which cause the costs per additional unit to rise. These effects might occur if additional employees and office staff must be hired, plant expansion is necessary, and new equipment must be purchased. In such cases, $C(x)$ becomes **nonlinear** and its graph is no longer a straight line. A typical cost function is illustrated in Figure 6.3.

Break-Even Point

The **break-even point** is the production level x at which $P(x) = 0$.

In Example 2 we could find the break-even point easily since we had the linear equation $30,000x = 750,000$, which could be solved for x directly. If $P(x)$ is a nonlinear equation, it is harder to find the break-even point. If $P(x)$ is a **quadratic function**, that is, has the form

$$P(x) = Ax^2 + Bx + C$$

where A, B, C are constants, $A \neq 0$, then we can find the break-even point using the quadratic formula. Recall that the **quadratic formula** states that the zeros (roots) of a quadratic equation $Ax^2 + Bx + C = 0$ are given by

$$x = \frac{-B \pm \sqrt{B^2 - 4AC}}{2A}$$

EXAMPLE 4   Find the break-even point for the profit function in Example 3.
   Solution.   The break-even point is a root of the equation

$$100x^2 + 30,000x - 750,000 = 0$$

We can simplify this equation to $x^2 + 300x - 7500 = 0$. Applying the quadratic formula gives

$$x = \frac{-300 \pm \sqrt{90,000 + 30,000}}{2} = -150 \pm 173.2$$

The two solutions are $x_1 = -323.2$ and $x_2 = 23.2$. The solution $x_1 = -323.2$ is not relevant for our situation since we cannot have a negative production level. Hence, 23.2 is the break-even point. Of course, the builder usually cannot sell a fraction of a house, so we say that the break-even point is about 23 houses. If the builder wants to turn a profit, then at least 24 houses must be constructed and sold. $\square$

EXERCISES

A mail-order supplier operating out of a rented office sells boxes of candy. To sell x boxes per week the following costs are incurred:
   (i)   Office rent: $200 per week   (ii) Mailing costs: $0.90 per box
   (iii) Interest on money invested in inventory: $90 per week
   (iv) Boxes: $5 each   (v) Phone bill: $100 per week

Answer Exercises 1–6 using this information.

1. Which are the fixed costs and which are the variable costs?
2. Find a formula for the weekly costs in terms of x.
3. If each box sells for $10 (including postage), find the cost, revenue, and profit functions.

4. Find the cost, revenue, and profit if 150 boxes are sold in one week.

5. Find the cost, revenue, and profit if 200 boxes are sold in one week.

6. Find the break-even point.

7. Rent-a-Lemon will rent you a car for $15 per day plus 15¢ per mile. Find (a) the cost of renting a car for a day trip of 400 miles and (b) the cost of renting a car for a day and driving x miles.

8. A gasoline station has fixed costs of $1000 per week. The wholesaler supplies the station with fuel at the price $p(x) = 1.10 - 0.00001x$ dollars per gallon if x gallons are ordered. Assuming the station sells fuel for $1.30 per gallon, find the cost, revenue, and profit functions. What is the profit if 6000 gallons is sold for the week?

9. A factory has a linear cost function $C(x) = A + Bx$. Past records show that $C = \$10,000$ when $x = 100$ and $C = \$12,500$ when $x = 150$. From these data find the fixed and variable costs.

10. Rent-a-Lemon offers the following deal on a car rental: $15 per day plus 200 free miles; beyond 200 miles you pay 15¢ per mile. (a) Find the cost $C(x)$ of renting the car for the day and driving x miles if $0 \leq x \leq 200$. (b) Repeat (a) for $x > 200$. (c) What are your costs if you drive 400 miles that day?

11. Graph the cost function $C(x)$ in Exercise 10.

An apartment building contains 60 units and each unit rents for $400 per month. A profit of $65 per month is made for each occupied apartment, while each unoccupied apartment causes a net loss of $35 per month. Letting x be the number of occupied apartments, answer Exercises 12–16.

12. Find the monthly profit function $P(x)$.   13. Find the break-even point.

14. Find the monthly revenue function $R(x)$.   15. Find the monthly cost function $C(x)$.

16. Why are the costs higher for an occupied apartment than for an unoccupied one?

17. A factory has a daily profit function $P(x) = x^2 + 30x - 400$. Find the break-even point.

18. A factory has a daily profit function $P(x) = 2x^2 + 50x - 1500$. Find the break-even point.

Find the roots of the following quadratic equations.

19. $x^2 + 3x - 5 = 0$ 20. $x^2 + 2x + 1 = 0$

21. $2x^2 + 5x + 3 = 0$ 22. $2x^2 - 3x + 1 = 0$

(Optional Exercises)

In Exercises 23–25, we shall prove the quadratic formula. We begin with the quadratic equation

$$Ax^2 + Bx + C = 0, \qquad A \neq 0 \tag{6.20}$$

*23. Show that (6.20) can be written as

$$x^2 + \frac{B}{A}x = -\frac{C}{A} \tag{6.21}$$

*24. By adding $B^2/4A^2$ to both sides of (6.21) show that

$$\left(x + \frac{B}{2A}\right)^2 = \frac{B^2}{4A^2} - \frac{C}{A} \tag{6.22}$$

*25. Show that the quadratic formula is obtained by taking the square root of both sides of (6.22).

**26. What happens in the quadratic formula if $B^2 < 4AC$? Give an example.

Photo 6.3
Profit, revenue, and cost are important constituents in the economics game.

6.5 | Economics

But it should always be required that a mathematical subject not be considered exhausted until it has become intuitively evident.
— FELIX KLEIN (1849–1925), German Mathematician

Besides being consumers, most adults either own a business, work for a profitmaking company, or work for a nonprofit organization such as a religious group, charity, or government. For these people, the economics of business is very important. Even if you are a student or a retired person, economics can have a direct effect on your life. Mathematical economics can be used to describe the prices and interest rates you pay and receive as well as the strength of the prevailing business market. An understanding of economics can help you to make daily and long-term financial decisions.

Profit, Revenue, and Cost

Suppose a plant manufactures a single product, say stoves (the type of product does not matter for this discussion). If you are managing this plant, one of your major decisions is to set the production level, that is, the number of stoves produced each day. How can you do this? One way is to guess. If you guess correctly often enough, your firm will turn a good profit. If not, you may have plenty of time to ponder what went wrong. In that case you might discover that instead of guessing or using trial and error, you should have attempted to analyze the problem. One approach is to study the relationship between the production level x and the profit P.

> **Profit** is the difference between **revenue** R (what comes in) and **cost** C (what goes out).

We thus have

$$P(x) = R(x) - C(x) \tag{6.17}$$

We usually fix a "short-run period of time," which may be a day, a week, or some other convenient time period. The **production level** x is the number of units produced during this period. A plant can usually be run at different production levels. At one extreme it can be idle ($x = 0$); at the other extreme it can be run at full capacity. We denote the full capacity level by x_m. The set of x values between 0 and x_m is called the **feasible set**. The cost function $C(x)$ is the sum

$$C(x) = C_F + C_V(x)$$

where C_F is the **fixed costs** and $C_V(x)$ the **variable costs**. The fixed costs must be paid even if there is no production at all. These costs include such items as rent, advertising, insurance, interest paid to investors, amortization of plant equipment, and salary of office staff and executives. The variable costs result from actual production such as labor salaries, cost of raw materials, energy, transportation, and so on. Sometimes the variable costs are the same for each unit produced, say A dollars per unit. Then $C_V(x)$ is proportional to x and we have the simple formula

$$C(x) = C_F + Ax \tag{6.18}$$

EXAMPLE 1   A builder of prefabricated houses has fixed costs of \$750,000 per year and variable costs of \$30,000x where x is the number of units constructed per year. The feasible set is $0 \le x \le 70$.
(a) What is the cost function $C(x)$?
(b) What is the cost of producing 50 houses in a year?
(c) Graph $C(x)$.
   Solution.   (a) Applying (6.18) gives

$$C(x) = C_F + C_V(x) = 750{,}000 + 30{,}000x$$

(b) Letting $x = 50$, we have

$$C(50) = 750{,}000 + 30{,}000(50) = \$2{,}250{,}000$$

(c) The graph of $C(x)$ is shown in Figure 6.1. □

We next consider revenue. In simple situations, the entire output is sold at a fixed price, say p dollars per unit. Then the revenue function becomes

$$R(x) = px \tag{6.19}$$

and the profit function is $P(x) = R(x) - C(x) = px - C(x)$.

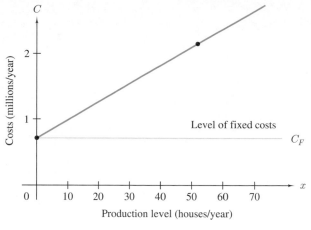

Figure 6.1

EXAMPLE 2   If the builder in Example 1 sells houses at $60,000 each, what is the profit for 50 houses in a year? How many houses must be sold to break even? Graph $P(x)$.

Solution.   In this case, $p = 60,000$ so

$$P(x) = px - C(x) = 60,000x - (750,000 + 30,000x)$$
$$= 30,000x - 750,000$$

Letting $x = 50$, we have

$$P(50) = 30,000(50) - 750,000 = \$750,000$$

Break-even occurs when $R(x) = C(x)$ or $P(x) = 0$. Solving $30,000x = 750,000$ for x gives $x = 25$ houses. The graph of $P(x)$ is shown in Figure 6.2. Note the negative profit (loss) below $x = 25$. □

Economies and Diseconomies of Size

In Example 1, the variable costs had the simple form $C_V(x) = Ax$. In this case the cost function $C(x) = C_F + Ax$ was **linear** and the graph was a straight line (Figure 6.1). In complex, large-scale operations, the variable costs are usually not proportional to x. As x increases from $x = 0$, the operation at first tends to become more efficient. Savings may occur by buying raw materials in large quantities, using the workforce more effectively, and introducing labor-saving devices. Such effects, referred to as **economies of size**, reduce the cost per unit. After x increases beyond a certain level, we encounter **diseconomies of size**, which cause the costs per additional unit to rise. These effects might occur if additional employees and office staff must be hired, plant expansion is necessary, and new equipment must be purchased. In such cases, $C(x)$ becomes **nonlinear** and its graph is no longer a straight line. A typical cost function is illustrated in Figure 6.3.

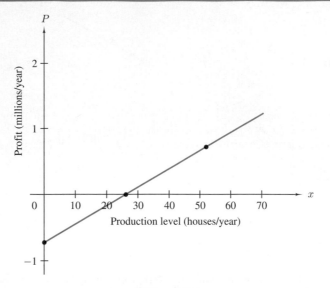

Figure 6.2

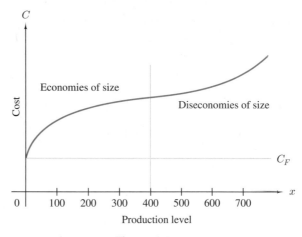

Figure 6.3

EXAMPLE 3   By buying materials in large quantities, the cost per house for the builder in Examples 1 and 2 is $30{,}000 - 100x$. What is the profit per year if 50 houses are sold?

Solution.   We now have $C_V(x) = (30{,}000 - 100x)x$. Hence,

$$P(x) = R(x) - C(x) = 60{,}000x - [750{,}000 + (30{,}000 - 100x)x]$$
$$= 30{,}000x + 100x^2 - 750{,}000$$

Letting $x = 50$ gives

$$P(50) = 30{,}000(50) + 100(50)^2 - 750{,}000 = \$1{,}000{,}000 \qquad \square$$

Photo 6.4
The cost of merchandise and the output of a factory are ruled by the law of supply and demand.

6.6 | Supply and Demand

Richard exerted himself still more. What was his conception of mathematics? A bright network of shining reality spread out infinitely, and one had to feel one's way from knot to knot; yes it was something like that, a complicated web of cosmic weave, like the world itself, which one had to unravel in order to get hold of reality.
— HERMANN BROCH, *The Unknown Quantity*

In Section 6.5 we considered the simple situation in which the entire output is sold at a fixed price p dollars per unit. In this case, the price per unit is constant and the revenue function $R(x) = px$ is linear (Figure 6.4).

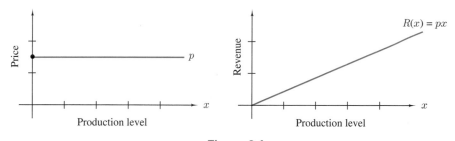

Figure 6.4

Demand Function

Just as the variable costs need not be linear in a realistic situation, so too for revenue. To deal with this, we introduce the demand function.

> The **demand function** $p(x)$ is the price that customers are willing to pay as a function of production level.

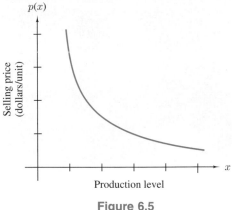

Figure 6.5

Although the price p (dollars per unit) could hypothetically be set at any value, this would not guarantee a maximal profit, or even any profit at all. To arrive at an optimal choice, the producer must test the market to see how demand for the product varies with the asking price. If the price per unit is set too low, the demand will be very great, leaving customers willing to pay but unable to obtain the product. Clearly the manufacturer should then raise the price to increase revenue and profit. Conversely, if the price is too high there will be too few customers and unsold stock will accumulate. For a given production level x, the demand function gives the optimal price $p(x)$ at which exactly x units will be sold, leaving no unsatisfied demand. Frequently, $p(x)$ is found through market surveys or by extrapolating existing sales data.

What is the general relationship between price p and production level x? If p is high, then x must be set low because the demand will be small. On the other hand, if p is low, then x must be set high to supply a greater demand. Put differently, if x is low the price p can be set high because the product is rare, whereas if x is high p must be set low because the market becomes saturated. A typical demand function is illustrated in Figure 6.5 and a typical profit function is shown in Figure 6.6. If $p(x)$ is the demand function, the revenue function becomes

$$R(x) = x\,p(x)$$

We then have the profit function $P(x) = x\,p(x) - C(x)$.

EXAMPLE 1   A small manufacturer has fixed costs of \$1000 per week and variable costs of \$100 per unit produced. Market analysis yields an estimate for the demand function in the feasible set $0 \le x \le 100$ as $p(x) = 300 - 2x$. (a) Find the profit when $x = 30$. (b) If the manufacturer wants to produce 40 units per week, what price should be charged and what is the resulting profit? (c) If the manager decides to set the price at \$200 per unit, how many units can be sold and what is the profit? (d) Find the break-even points.

Solution.   (a) The cost function is $C(x) = C_F + C_V(x) = 1000 + 100x$ and the revenue function is $R(x) = x\,p(x) = (300 - 2x)x$. This gives the profit function $P(x) = R(x) - C(x) = 200x - 2x^2 - 1000$. Setting $x = 30$, we have

$$P(30) = 200(30) - 2(30)^2 - 1000 = \$3200$$

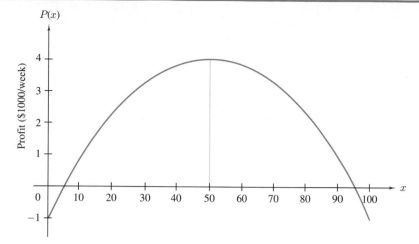

Figure 6.6

(b) Setting $x = 40$, we have

$$p(40) = 300 - 2(40) = \$220$$

$$P(40) = 200(4) - 2(40)^2 - 1000 = \$3800$$

(c) At price \$200, we have $200 = p(x) = 300 - 2x$. Hence, $2x = 100$ or $x = 50$. Setting $x = 50$ gives

$$P(50) = 200(50) - 2(50)^2 - 1000 = \$4000$$

(d) The break-even point occurs when $200x - 2x^2 - 1000 = 0$ or equivalently $x^2 - 100x + 500 = 0$. The quadratic formula gives

$$x = \frac{100 \pm \sqrt{10{,}000 - 2000}}{2} = 50 \pm 44.72$$

The graph of the profit function is shown in Figure 6.6.   □

Supply Function

In Figure 6.5, we illustrated a typical demand function $p(x)$. The function $p(x)$ gives the price that consumers are willing to pay at the production level x. There is a competing function called the supply function.

> The **supply function** $s(x)$ gives the price that manufacturers want to charge at production level x.

If the prevailing price is low, manufacturers are reluctant to supply very many units, but if the price is high, they are eager to increase production. Thus, when s is low, x is low and when s is high so is x. A typical supply function is illustrated in Figure 6.7.

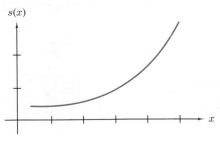

Figure 6.7

Law of Supply and Demand

Let's put the graphs of $p(x)$ and $s(x)$ together as in Figure 6.8. If the prevailing price is p_1, then the demand will have a value x_1 that is much smaller than the amount y_1 that producers are willing to supply and market pressures will force the price down. If the prevailing price is p_2, then the demand x_2 is much larger than the amount y_2 that producers are willing to supply. In this case, the imbalance will force prices up. Eventually an equilibrium will be attained, at which point consumers and producers reach agreement. Equilibrium is determined by the point (x_0, p_0) at which the demand and supply curves cross (Figure 6.9), and this is called the **law of supply and demand**.

> The **equilibrium point** (x_0, p_0) is found by setting $p(x) = s(x)$. We call x_0 the **equilibrium demand** and p_0 the **equilibrium price**.

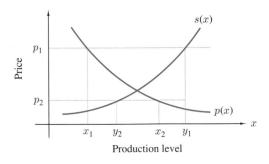

Figure 6.8

EXAMPLE 2   Suppose the supply function for cheese in this country is

$$s(x) = \frac{1}{80}\, x + 1, \qquad 0 \le x \le 180$$

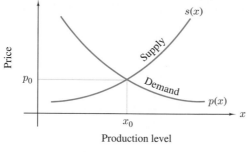

Figure 6.9

where the price s is in dollars per pound and x is in millions of pounds per week while the demand function is

$$p(x) = 3 - \frac{1}{60} x, \qquad 0 \le x \le 180$$

Find the equilibrium price and the number of pounds per week supplied at this price.

Solution. Equilibrium occurs when

$$\frac{1}{80} x + 1 = 3 - \frac{1}{60} x$$

Hence,

$$\left(\frac{1}{80} + \frac{1}{60} \right) x_0 = 2$$

Solving for x_0 gives $x_0 = 68.57$ million pounds. For this production level we have

$$p_0 = s(x_0) = \frac{68.57}{80} + 1 = \$1.86 \text{ per lb}$$

Of course, we could also obtain this result from

$$p_0 = p(x_0) = 3 - \frac{68.57}{60} = \$1.86 \text{ per lb}$$

The situation is graphed in Figure 6.10. □

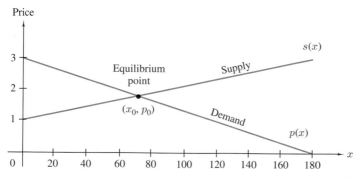

Figure 6.10

EXAMPLE 3 The supply and demand functions for a certain product are

$$s(x) = x^2 + 4x + 200, \qquad 0 \le x \le 20$$
$$p(x) = 500 - 6x - x^2$$

where the price is in dollars per unit and x is the number of units produced per day. Find the equilibrium price and the number of units produced at this price.

Solution. Setting $s(x) = p(x)$ gives

$$x^2 + 4x + 200 = 500 - 6x - x^2$$

Thus, $\qquad\qquad\qquad x^2 + 5x - 150 = 0$

By the quadratic formula,

$$x_0 = \frac{-5 \pm \sqrt{25 + 600}}{2} = -\frac{5}{2} \pm \frac{25}{2}$$

Since a negative x_0 is not relevant to our problem,

$$x_0 = -\frac{5}{2} + \frac{25}{2} = 10 \text{ units per day}$$

Moreover,

$$p_0 = s(10) = (10)^2 + 4(10) + 200 = \$340 \text{ per unit}$$

The graphs are shown in Figure 6.11. ☐

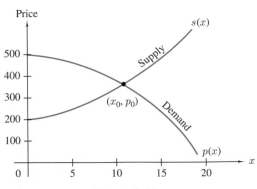

Figure 6.11

EXERCISES

1. Repeat Example 1 with the same data but for the demand function $p(x) = 500 - 3x$.

A department store has found that its weekly sales of dress shirts are related to the price p by the demand function $p(x) = -0.05x + 30$ dollars, where x is the number of shirts they sell in a week. Answer Exercises 2–5.

2. At what price will they sell 100 shirts per week?

3. If they want to sell 200 shirts in a week, what price should they charge?

4. If they charge $22 for a shirt, how many will they sell?

5. If they charge $19 for a shirt, how many will they sell?

6. A gasoline station has found that its demand function is

$$p(x) = 0.80 + \frac{1000}{x} \text{ dollars}$$

where x is the number of gallons sold per day and p is the price per gallon.
   (a) At what price will they sell 2000 gallons per day?
   (b) If they want to sell 5000 gallons per day, what price should they charge?
   (c) If they charge $1.15 per gallon, how many gallons will they sell?

7. A producer has fixed costs $C_F = \$1000$ per week and variable costs $40x$ where x is the production level. The demand function is $p(x) = 55 - 0.01x$. (a) Determine (i) the cost function, (ii) the revenue function, (iii) the profit function. (b) Calculate the profit for $x = 100$, 500, and 1000.

8. The cost function for a gasoline station selling x gallons per week is $C(x) = 1000 + 1.10x$ and the demand function is $p(x) = 1.50 - 0.00002x$ dollars. (a) Find the profit function $P(x)$. (b) What is the profit when the station sells 10,000 gallons per week? (c) At what price did the station have to offer fuel to sell 8000 gallons a week?

In Exercises 9–11 the demand and supply functions are given in dollars per unit. Find the equilibrium price and production level.

9. $p(x) = \dfrac{130 - x}{35}, \quad s(x) = \dfrac{x + 40}{70}$

10. $p(x) = \dfrac{70 - x}{50}, \quad s(x) = \dfrac{x + 100}{80}$

11. $p(x) = \dfrac{245}{3x}, \quad s(x) = \dfrac{x + 30}{60}$

In Exercises 12–17, find both coordinates of the intersection points for the given pairs of curves.

12. $y = x^2 - 2x + 1$ and $y = x$

13. $y = x^2 + x + 1$ and $y = x + 1$

14. $y = x^2 - x - 2$ and $y = 4 + 3x - x^2$

15. $y = x^2 - 9$ and $y = 2x + 7$

16. $y = 500 - 30x - x^2$ and $y = x^2 + 40x$

17. $y = x^2 - 4x + 4$ and $y = 1 + x^2$

18. If the fixed costs are C_F, the variable costs are $C_V(x)$, and the demand function is $p(x)$, what is the profit function $P(x)$?

Chapter 6 Summary of Terms

simple interest $I = Prt$

compound amount $A = P(1 + r)^n$

effective annual interest rate $r' = (1 + r)^n - 1$

years to double with simple interest $t = \dfrac{1}{r}$

periods to double with compound interest $n = \dfrac{\log 2}{\log(1 + r)}$

continuous compound amount $A = Pe^{at}$

effective annual interest rate for continuous compounding $r' = e^a - 1$

present value, periodic compounding $P = A(1 + r)^{-n}$

present value, continuous compounding $P = Ae^{-at}$

annuity amount $A = R\left[\dfrac{(1 + r)^n - 1}{r}\right]$

amortization payment $R = \dfrac{Pr}{1 - (1 + r)^{-n}}$

cost function $C(x) = C_F + C_V(x)$

revenue function $R(x)$

profit function $P(x) = R(x) - C(x)$

break-even point $P(x) = 0$

quadratic equation $Ax^2 + Bx + C = 0$

quadratic formula $x = \dfrac{-B \pm \sqrt{B^2 - 4AC}}{2A}$

demand function $p(x)$

revenue function $P(x) = xp(x)$

supply function $s(x)$

law of supply and demand $p(x) = s(x)$

equilibrium point (x_0, p_0)

Chapter 6 Test

1. If $2000 is deposited at 8.5% compounded monthly, what is the total amount after 5 years?
2. Find the years to double for an annual interest rate of 7.5% compounded monthly.
3. Find the continuous compound amount for a principal of $5000 deposited at an annual interest rate of 8% for 10 years.

4. What is the present value of $8000 at 7.5% compounded quarterly for 10 years?

5. Find the amount of an annuity that has a $200 monthly payment for 10 years at an annual interest rate of 6.5% compounded monthly.

6. What amount must be paid monthly for 12 years for an annuity of $20,000 at 7% compounded monthly?

7. A 10-year, $30,000 loan has an interest rate charge of 13% compounded monthly. What are the monthly payment and the total interest paid?

8. A manufacturer has fixed costs $C_F = \$50,000$ per month. Assume that the cost function $C(x)$ is linear, where x is the monthly production level, and $C(200) = \$150,000$. Find the cost function. What are the costs when $x = 300$?

9. Demand and supply functions are given by

$$p(x) = \frac{-55}{2000}\, x + 120 \qquad \text{and} \qquad s(x) = \frac{35}{2000}\, x + 30$$

   Find the equilibrium production level and price.

10. Find the roots of the equation $3x^2 + 8x - 2 = 0$.

11. A factory has a daily profit function $P(x) = x^2 + 30x - 1000$. Find the break-even point.

12. A manufacturer has a daily profit function $P(x) = -x^2 + 200x - 100$. Graph this function and determine the production level that maximizes the profit.

13. What can you say about the roots of the equation $x^2 + x + 1 = 0$?

14. Are the following statements true or false?
    (a) With the same principal and annual interest, compounding weekly results in more money than compounding monthly.
    (b) The compound amount formula is $A = Pr^n$.
    (c) The effective annual interest rate is always at least as great as the annual interest rate.
    (d) If the annual interest rate is 9%, then the number of periods to double with compound interest is $\log 2$.
    (e) Continuous compounding means compounding every second.
    (f) The present value with periodic compounding is given by $P = A(1 + r)^n$.
    (g) A loan principal is a head of a school for loans.
    (h) Cost is revenue plus profit.
    (i) The quadratic formula gives the zeros of a quadratic equation.
    (j) The law of supply and demand says that when the supply is high so is the demand.
    (k) A lonely person is someone with a lot of loans.

COMPUTER SCIENCE

7

For it is unworthy of excellent men to lose hours like slaves in the labor of calculation which could safely be relegated to anyone else if machines were used.

—GOTTFRIED WILHELM VON LEIBNIZ (1646–1716)
German Mathematician and Philosopher

Throughout history, people have struggled to free themselves from physical and mental drudgery. The creation of the high-speed digital computer is a giant step in human intellectual progress. Although a short time has elapsed since its creation, the computer has had an immense impact on business, industry, and science. Procedures and processes that were unthinkable before the computer are commonplace today. For example, it is generally agreed that it would have been impossible to place someone on the moon without the aid of computers. The design of modern jet aircraft has been accomplished using computers. If you have a high-quality camera, the chances are that the lens was designed by computers. The list of situations in which computer aid is indispensable is almost endless.

Computers are influencing our lives to a greater extent every year and this trend will undoubtedly continue into the future. To get along in our modern society an educated person should know something about computers and preferably have some actual experience with one. People are still required for creative thinking. But computers can perform long, repetitive tasks much quicker and more efficiently. We have long sought the aid of devices to help us carry out long, tedious calculations, and the computer is the latest and most powerful.

Even though computers have enabled us to perform tasks that were previously impossible and have ushered civilization into a new era of technology, they do have certain drawbacks. If you have received an incorrect billing from a large company, you probably realize this. The bill comes from a computer, your letter correcting the error is answered by a computer, as is your next letter, and your next, ... Finally you give up in desperation and pay the incorrect bill. Computers cannot do more than they

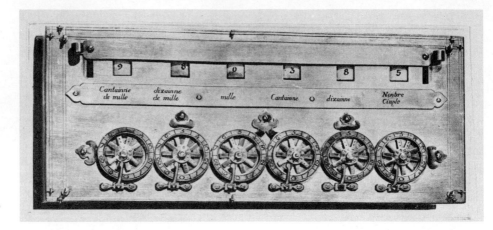

Photo 7.1
Pascal's arith-
metical machine,
1642.

are programmed to do. Most computers cannot correct mistakes, reconsider decisions, make exceptions, or be flexible. In short, they are not human. Work is now being done to make computers more like humans. This can be done by better programming or by improving the computer itself.

7.1 | The History of Computers

When looms weave by themselves, then shall man's slavery end.
— ARISTOTLE (384–322 B.C.), Greek Philosopher

We may say most aptly that the Analytical Engine weaves algebraical patterns just as the Jacquard loom weaves flowers and leaves.
— AUGUSTA ADA, LADY LOVELACE (1801–1880)

The first real step in the development of modern computers was taken in 1642 by Blaise Pascal (Pascal will be mentioned again in a later chapter for his work in probability theory). Pascal was born in 1623 and was a brilliant child with interests in physics, mathematics, and religion. When Pascal was 19, he began developing a machine that was supposed to add long columns of numbers, and he worked for three years on the project. Although the machines he built were unreliable, they employed some of the basic principles used in later mechanical calculators.

The next great advance for calculating machines came, unexpectedly enough, from the weaving industry. In 1725, Basile Bouchon employed a loop of punched paper tape to aid in weaving with a hand loom, and in 1728, M. Falcon used punched cards for the same task. These cards were incorporated into the Jacquard loom and have since come to be called Jacquard cards. Falcon punched holes in the card where he wanted the needles to be directed. To change patterns he had only to change cards. Each card made one line of weave, so that a loop of connected punched cards was used to obtain a complete pattern. This was clearly a giant step for the weaving industry.

Photo 7.2
A great advance for calculating machines came from the weaving industry in 1725.

An employee merely inserted a loop of punched cards, pedaled a foot lever, and the loom automatically wove the fabric.

Jacquard cards gave Charles Babbage (1792–1871) the inspiration that the entire arithmetic function of a calculating machine could be "programmed" in advance. Over a period of 50 years Babbage designed and attempted to build a calculating machine that actually had a "memory" to store information. This machine, called the **analytic engine**, was the forerunner of modern computers.

Babbage was an eccentric Englishman who was years ahead of his time. His analytic engine was a grandiose contraption with thousands of gears, shafts, ratchets, and counters. He was constantly searching for grants and financial aid to help subsidize his project and was frequently successful. However, his many eccentricities often got him into trouble and made him the butt of much ridicule. He was a no-nonsense perfectionist and would not tolerate inaccuracy. Because of this perfectionism he was never able to develop a workable machine.

His desire for precision was not limited to his work but permeated his very existence. After Tennyson wrote "The Vision of Sin," Babbage sent the following note to the poet:

Sir,

In your otherwise beautiful poem there is a verse which reads

"Every moment dies a man,
Every moment one is born."

It must be manifest that if this were true, the population of the world would be at a standstill. In truth the rate of birth is slightly in excess of that of death. I would suggest that in the next edition of your poem you have it read—

"Every moment dies a man,
Every moment 1 and 1/16 is born."

Strictly speaking, this is not correct; the actual figure is so long that I cannot fit it into one line, but I believe that the figure 1 and 1/16 will be sufficiently accurate for poetry.

I am, Sir, yours...

Babbage's analytic engine was all mechanical and was limited by the techniques in gear grinding, the sluggishness of mechanical parts, friction, etc. What was needed was the precision, speed, and accuracy of electronic components, which, of course, were far in the future. Although Babbage never completed the construction of his analytic machine, in 1991 a group of mechanical engineers constructed such a machine. It is powered by turning a large crank and it works! Babbage was right after all.

The first fully electronic digital computer was built in 1946 at the University of Pennsylvania and was called ENIAC (*e*lectronic *n*umerical *i*ntegrator *a*nd *c*alculator). This and other first-generation computers used vacuum tubes, which took up a large amount of room, consumed enormous amounts of electricity, and were not very fast or efficient. The ENIAC filled a 30-by-50-foot room, weighed 30 tons, had 18,000 tubes, and used enough electricity to operate three 150-kilowatt radio stations.

The second-generation computers were developed with the advent of the transistor, about 1952. When the tubes were replaced by transistors, the size and power requirements of the computer were greatly reduced.

Around 1965 the third generation of computers, incorporating microminiaturized circuits, was designed. These **integrated circuits** are formed on tiny chips of silicon much smaller than a thumbnail; in fact, some are barely visible to the naked eye. A small third-generation computer the size of a breadbox could perform calculations much faster and more accurately than the original ENIAC. Programmable pocket computers now have as much computing power as the ENIAC.

Computers today are a huge industry, dominated by IBM (International Business Machines), which has about 60 percent of the market. The computer field is growing at a rate of 15 to 25 percent a year on an international scale, and the total market today is over $500 billion annually. There are now over 200,000 computer installations in the United States. Assuming that each employs about 10 people, this gives a work force of 2 million directly involved in full-time computer operation and use, not to mention those designing, manufacturing, selling, and performing other functions indirectly related to computers. It has been estimated that one worker in three in the

Photo 7.3
Babbage's analytical engine.

United States labor force is associated full- or part-time with computers. No industry in history has grown at so fast a rate.

Although large installations operate on mainframe computers, small businesses, school classrooms, and individuals now use personal computers (microcomputers). These became possible in 1971 with the introduction of the microprocessor, which is a miniaturized central processing unit. There are about 80 million personal computers in use in the United States today.

The fastest and most powerful modern computers are called supercomputers. Such computers are capable of performing more than 20 million instructions per second. The first supercomputer, the Cray-1, was introduced in 1975 by its designer, Seymour Cray. The much faster Cray-2 became available in 1985 and is only one-tenth the size of the Cray-1. Because of their great cost ($5 million to $25 million), supercomputers are used only by large scientific laboratories, research universities, and government centers.

Just as the eighteenth and nineteenth centuries were the years of the industrial revolution in which people developed machines to magnify their physical powers, the twentieth century is the age of the information revolution, a period in which machines are constructed which expand human mental capacity.

How well and to what purpose the information machines will be used in the future rests largely with today's students. For information machines, like energy machines, are tools which people have developed in their conquest of, and cooperation with, nature. Only their users can determine how such machines will affect our lives. For this reason, a study of computer technology is important to us as well as to our society. Only by understanding the potentials and limitations of the information machines can we learn to use them wisely and appreciate their power.

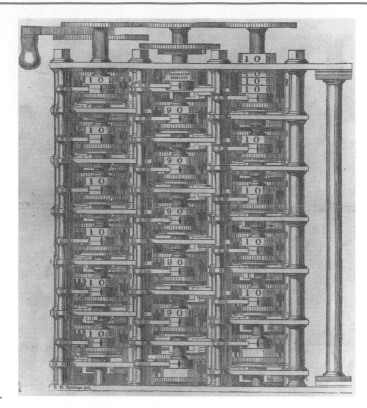

Photo 7.4
Babbage's difference engine.

7.2 | How a Computer Works

Men have become the tools of their tools.
— HENRY THOREAU (1817–1862), American Naturalist and Author

When machines are in league with men, the soul of the alliance must be human, lest its ends become less than human.
— JOHN DIEBOLD,
Twentieth-Century Management and Technology Consultant

Every computing system is made up of five main components: input, storage (or memory), processor (or arithmetic unit), control, and output.

The **input** mechanism is used to enter data and instructions into the machine. This is usually done with a keyboard, a magnetic tape or disk, or an optical scanner. A computer may have many inputs; in fact, they can run into the thousands. Since a computer can work millions of times faster than a person, many people can feed data and instructions into a computer at the same time. The computer can execute these instructions almost simultaneously and feed results and data back to the people through the output unit. This is called **time-sharing**.

The **storage unit** saves or stores information until it is needed for some purpose. The most efficient means known for storing information is using magnetic fields on a

ZIGGY

9·26 © 1989 Universal Press Syndicate Tom Wilson

magnetic disk or magnetic core. On a magnetic core, there are many magnetic "spots" that can be either magnetized one way, which represents 0, or the other way, which represents 1. Similarly, an individual circuit of a computer is either inactive (does not contain electric current), which represents 0, or active (does contain current), which represents 1. Thus the **machine language** of the computer consists of "words" composed of strings of 0s and 1s. All letters, numbers, mathematical symbols, punctuation marks, and instructions are represented or coded by binary numbers. Suppose, for example, that the magnetic spots can be magnetized up ($\uparrow$) or down ($\downarrow$) and that $\uparrow$ represents 1 and $\downarrow$ represents 0. Then the binary word

$$101001101$$

would be represented by

$$\uparrow\downarrow\uparrow\downarrow\downarrow\uparrow\uparrow\downarrow\uparrow$$

This is how information and instructions are stored in the memory unit. The **processor** is the part of the computer in which addition, subtraction, multiplication, and other arithmetic and logical operations are carried out. These operations are performed using electronic circuits.

The **control unit** of the computer directs the flow of data to and from the various components of the computer system. In a regular, rhythmic pattern, the control unit obtains data and instructions, one at a time, from the storage unit and causes the processor to carry out the operations. When data and instructions are fed into the computer through the input unit, control will usually direct them to storage. In solving a problem, the computer is directed by control to start at the first instruction of the program stored at a given location. Once the first instruction has been executed, control usually gets the next instruction from the next higher storage location. This is continued until all instructions are performed.

The last component of a computer is the **output unit**, the means by which the operator obtains the results of the computations or data processing. The output unit may be a magnetic device, a laser printer, a dot-matrix printer, a graphics plotter, or a monitor.

You have seen how binary words or numbers are stored in the memory unit as magnetized spots, but how are these words transmitted through the computer's circuits and from one component to another? Binary words are transmitted by sending a sequence of electric pulses. Computers contain very precise clocks which synchronize the transmission and reception of signals in this form. During each set interval of time, say 1 nanosecond (one-billionth of a second), a pulse, which is approximately a square wave, is either sent or not sent, representing 1 or 0, respectively. In Table 7.1, 0 to 9 are represented by such sequences of square-wave electric pulses.

Table 7.1

Decimal	Binary	Electric pulse
0	0000	
1	0001	
2	0010	
3	0011	
4	0100	
5	0101	
6	0110	
7	0111	
8	1000	
9	1001	

Programs written with binary words are in machine language, the language a computer understands. But such languages are inconvenient and are hard for people to understand. For this reason, most programs are not written in machine language. Other languages, called **algorithmic languages**, have been developed. These languages are more conversational, easier to learn and remember, and closer to our everyday language. These languages also have the advantage of being less prone to mistakes than long strings of 0s and 1s. The algorithmic language is translated into machine language, the natural language of the computer, by a special program called the **compiler**. Over the years many different algorithmic languages have been designed such as FORTRAN, ALGOL, C, COBOL, Pascal, and ADA. One of the simplest languages for the beginner is the BASIC language (*b*eginner's *a*ll-purpose *s*ymbolic *i*nstruction *c*ode), developed at Dartmouth College. You will learn the BASIC language in a later section.

EXERCISES

Explain the function of the following in your own words.

1. Input   2. Output   3. Processor   4. Control

Explain the following terms in your own words.

5. Machine language   6. Algorithmic language

7. Binary word   8. Compiler

Write the following decimal numbers as binary numbers.

9. 10   10. 11   11. 12   12. 13   13. 16   14. 20

Write the following binary numbers as decimal numbers.

15. 10100   16. 10000   17. 1111   18. 11000   19. 100000   20. 11010

Represent the following binary numbers as square-wave electric pulses.

21. 1101   22. 1011   23. 10101   24. 101101

7.3 Software

Thus wise men have been right in taking examples of things which can be investigated with the mind from the field of mathematics, and not one of the ancients who is considered of real importance approaches a difficult problem except by way of a mathematical analogy. That is why Boethius, the greatest scholar among the Romans, said that for a man entirely unversed in mathematics, knowledge of the Divine was unattainable.

— NICHOLAS OF CUSA (1401–1464), Roman Catholic Prelate

The various components of a computer that we discussed in Section 7.2 are examples of computer **hardware**. Thus, the memory, processor, and input and output devices are hardware. There are also peripherals such as modems, communication links, disk libraries, and external storage that are classified as hardware. Although hardware is what makes a computer work, for most practical applications a computer would be useless without software. **Software** consists of specialized programs that permit the user to carry out particular tasks. There are many software packages available for personal computer users. Using a software package is not the same as programming; a software package is a program that has already been created by someone else.

Computers are utilized to perform an immense variety of specialized tasks and each type of task requires its own software. Software packages used for scientific purposes are called **number-crunchers**. For example, a scientist may need the solution of a complicated equation or a large number of equations. There is software designed for **data processing** that involves large amounts of data. Banks and credit card companies use data processing to keep track of customers' accounts. The Internal Revenue Service employs data processing on an immense scale to verify tax returns. Software is also used for **informational retrieval**. For example, airlines use information retrieval to locate information on their reservations, flight times, departures, and destinations. **Pattern recognition** software is designed to recognize and classify various patterns and relationships. For example, in a dollar bill change machine, a

Photo 7.5
Computer software is indispensible for the design and operation of modern rockets.

computer together with an optical scanner must verify that the bill is not counterfeit. Pattern recognition is used at grocery store checkout counters to decipher bar codes on merchandise. Software called **communication packages** is used for transferring information between computers and for electronic mail and fax machines.

We now discuss some common software packages that are available for personal computers. Practically everyone who uses a microcomputer employs a **word processor**. This is a program that allows a user to enter written material into the computer's memory. A word processor has many advantages over the old typewriter. Additions, deletions, and corrections are easy to make. Most word processors have built-in spelling checkers that locate misspelled words. Word processors automatically advance from one line to the next and space letters so that columns of text have perfectly vertical edges. Some common word processors are *MacWrite*, *WordStar*, and *WordPerfect*.

Another useful software package is a **database manager**. This is a program that allows the user to store, sort, and manipulate data in various ways. For example, a small business might want to store its customers' names with information concerning them such as addresses, phone numbers, credit ratings, recent purchases, and frequencies of purchases. The business may want a customer listing that is alphabetical or according to zip code or geographical location. Suppose the business wants to run a sale for its best customers. The database manager would be instructed to locate the customers with frequent purchases or the customers with recent large purchases and print a list.

Our final example is the electronic scratchpad or **spreadsheet**. Two popular spreadsheets are *Excel* and *Lotus 1-2-3*. In a spreadsheet, information is stored in a rectangular array of cells. Usually the array is in numbered rows and lettered columns as in Table 7.2. The user has the ability to address thousands of rows and hundreds of columns if necessary. The **cell** in column B and row 5, for example, is denoted B5.

Table 7.2

	A	B	C	D
1				
2				
3				
4				
5				

The user can enter information into any desired cell. The power of the spreadsheet derives from the fact that arithmetic operations can be performed among cells. For example, suppose we want D2 to be the average of the entries in A2, B2, and C2. We could then enter the following formula into cell D2:

$$\frac{A2 + B2 + C2}{3}$$

Moreover, if the value in any cell is changed, then the values of all cells dependent on this cell are recalculated and displayed.

For example, suppose you have various investments in stocks, bonds, and a savings account and you want to keep track of the dividends and interest. You could set up a spreadsheet as in Table 7.3. The spreadsheet would compute totals for any row or column. It can also perform any desired calculations such as averages, differences, or percentages. If the interest rate of your savings account changes, for example, it would recalculate the values in all the cells. Most spreadsheet packages also have the capability of displaying a summary of the data in the form of a bar graph or line graph. Spreadsheets are not only indispensable for accountants and financial planners, they are being increasingly utilized by the general public as well.

EXERCISES

Explain the following terms in your own words and give examples.

1. Hardware
2. Software
3. Number-crunchers
4. Data processing
5. Information retrieval
6. Pattern recognition
7. Communication packages
8. Word processor
9. Database manager
10. Spreadsheet
11. Draw a spreadsheet with 10 rows and six columns. Label the rows and columns and place an X in the following cells: A3, B7, D9, F8.

Table 7.3

	A	B	C	D	E	F
1	*Month*	*Stock* (IBM)	*Stock* (GM)	*Bonds*	*Savings*	*Totals*
2	Jan.	55.00	75.00	35.00	25.00	190.00
3	Feb.			35.00	25.13	60.13
4	March		75.00	35.00	25.25	135.25
5	April	55.00		35.00	25.37	115.37
6	May		75.00	35.00	25.51	135.51
7	June			35.00	25.63	60.63
8	July	55.00	75.00	35.00	25.76	190.76
9	Aug.			35.00	25.89	60.89
10	Sept.		75.00	35.00	26.01	136.01
11	Oct.	55.00		35.00	26.15	116.15
12	Nov.		75.00	35.00	26.28	136.28
13	Dec.			35.00	26.41	61.41
14	Totals	220.00	450.00	420.00	308.39	1398.38

12. Suppose there is an additional General Electric stock that the user of the spreadsheet in Table 7.3 wants to incorporate. This stock pays $50 quarterly starting in February. Redraw Table 7.3 to include this additional stock.

13. The user of the spreadsheet in Table 7.3 has found that the quarterly dividend for IBM stock is $60 instead of $55. Redraw Table 7.3 to incorporate this change.

14. Set up a spreadsheet for an address book.

15. Set up a spreadsheet for a checking account.

7.4 Flowcharts

A person well-trained in computer science knows how to deal with algorithms: how to construct them, manipulate them, understand them, analyze them. This knowledge prepares him for much more than writing good computer programs; it is a general-purpose mental tool which will be a definite aid to his understanding of other subjects, whether they be chemistry, linguistics or music, etc.

— DONALD KNUTH, Twentieth-Century Computer Scientist

Although a computer has tremendous power, it is up to people to put it to use. We use a computer to perform a certain task, but before we can instruct the computer how to perform the task we must find a systematic step-by-step procedure for the computer to follow. Such a procedure is called an algorithm.

An **algorithm** is a prescribed set of well-defined rules or processes for the solution of a problem in a finite number of steps.

After we have found an algorithm, we must then translate it into a programming language (in this case BASIC) so that the computer will accept it.

You don't have to be a computer scientist to use algorithms. People employ algorithms every day. A recipe in a cookbook is an excellent example of an algorithm. The preparation of a complicated dish is not an easy task, but if the process is broken down into simple steps, any experienced cook can accomplish it. Algorithms are used extensively in sports and entertainment. For example, a football play can be broken down into a succession of basic steps and positions. The number of these basic steps and positions can be very small, but by putting them together in different ways and by allowing for various reactions of the opponents, an endless variety of plays can be devised.

Algorithms executed by a computer can combine billions of elementary steps into a complicated mathematical computation. Also by means of algorithms, a computer can control a manufacturing process, direct a communication system, or coordinate the reservations of an airline as they are received from ticket offices all over the world. Algorithms for such large-scale processes are very complex but they are always built up from simple pieces.

A useful technique for displaying algorithms is to construct a **flowchart**. A computer does only what it is told; it must even be told to start and to stop. A flowchart gives us a method for organizing and displaying the steps of an algorithm so that it can then be put into the form of a program. The flowchart uses geometric figures to express different types of operations and a command or question is written inside each figure.

There are usually five steps involved in solving a problem using a computer:

1. Define the problem.
2. Construct an algorithm for solving the problem.
3. Translate the algorithm into a computer program.
4. Run the program in the computer.
5. Debug (correct) the program or algorithm (or both) if necessary.

The algorithm is usually constructed with the aid of a flowchart. Flowchart 1 shows how the above five steps are accomplished. The "no" alternative after the decision "Does it work?" creates a **loop**, which may have to be traversed many times

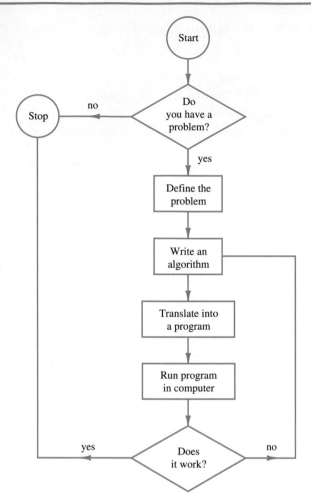

Flowchart 1

until a "yes" alternative is obtained and the procedure stops. The different types of operations in a flowchart are now discussed.

The circle represents starting or stopping the algorithm.

The rectangle represents an instruction, command, or arithmetic type of operation.

Let $N = 2$

Let $C = A + B$

The diamond represents a decision to be made.

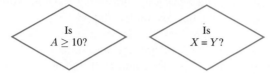

Finally, the pickle (or sausage) represents an input/output operation.

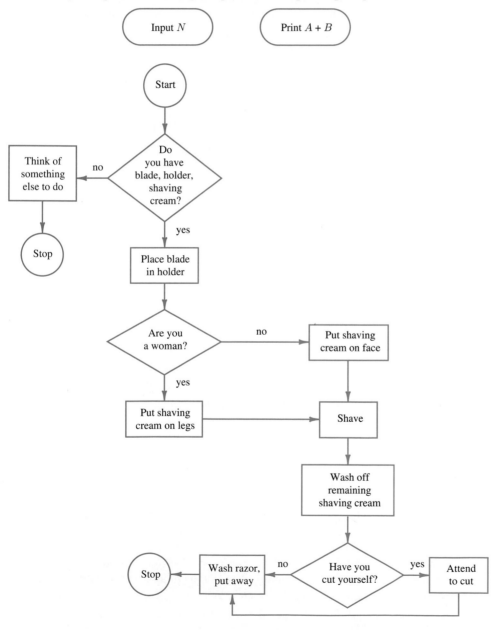

Flowchart 2
Shaving with
a safety
razor

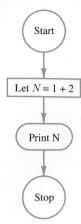

Flowchart 3

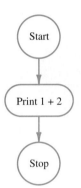

Flowchart 4

First consider a nonmathematical flowchart (Flowchart 2). Such flowcharts give the step-by-step processes that take place in performing tasks of daily life.

Now consider some mathematical examples. Suppose we want to add $1 + 2$. This can be organized according to Flowchart 3. If the computer follows these steps, at step 1 it will start. At step 2 it will perform the operation $1 + 2$ and store the result under the name N. At step 3 it will print the number listed under the name N, namely 3. The computer stops at step 4.

This procedure can also be performed using the simpler Flowchart 4. After starting, the computer will perform the operation $1 + 2$ and then print the result 3. Then the computer will stop.

As a more complicated example, suppose we would like an algorithm for printing the first 100 natural numbers. We could use Flowchart 5.

Table 7.4

Step	What computer does
1	Start
2	Let $N = 0$
3	Let $N = 0 + 1 = 1$
4	Print 1
5	No
3	Let $N = 1 + 1 = 2$
4	Print 2
5	No
3	Let $N = 2 + 1 = 3$
4	Print 3
5	No
⋮	⋮
3	Let $N = 98 + 1 = 99$
4	Print 98
5	No
3	Let $N = 99 + 1 = 100$
4	Print 99
5	Yes
6	Stop

Flowchart 5

1. Start
2. Let $N = 0$
3. Let $N = N + 1$
4. Print N
5. Is $N = 100$?   no / yes
6. Stop

What is meant by: "Let $N = N + 1$"? Surely, we do not mean equation $N = N + 1$ holds, for then $0 = 1$! We really mean the following. The two N's are different; the N on the right side of the equal sign is the **current** value of N, and the N on the left side is the **new** value of N. Thus, after step 2 is first performed, the current value of N is 0. Then, after step 3, the new value of N is its current value, namely 0, plus 1. Thus, after step 3 N has the value 1. The different steps are labeled for reference. Notice that steps 3 to 5 form a loop. This loop is repeatedly traversed until step 5 has a yes answer. When this occurs, the loop is left and the process stops. Table 7.4 shows what the computer does at each step. Such a table is called a **wcd** (what computer does) table.

Now suppose we want to find $1 + 2 + \cdots + 100$ using a computer. Flowchart 6 gives an algorithm for accomplishing this. Notice that at step 4 the new value of N is the current value of N plus 1 and at step 5 the new value of S is the current value of S plus the current value of N. Table 7.5 is a wcd table for this process.

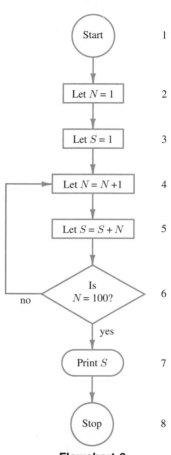

Table 7.5

Step	What computer does
1	Start
2	Let $N = 1$
3	Let $S = 1$
4	Let $N = 1 + 1 = 2$
5	Let $S = 1 + 2 = 3$
6	No
4	Let $N = 2 + 1 = 3$
5	Let $S = 3 + 3 = 6$
6	No
$\vdots$	$\vdots$
4	Let $N = 98 + 1 = 99$
5	Let $S = 4851 + 99 = 4950$
6	No
4	Let $N = 99 + 1 = 100$
5	Let $S = 4950 + 100 = 5050$
6	Yes
7	Print 5050
8	Stop

Flowchart 6

One of the great advantages of flowcharts is that they can sometimes be easily altered to get new results. For example, how can we alter our previous flowchart if we want $1 + 2 + \cdots + 1000$? All we have to do is change step 6 to read: Is $N = 1000$? Now we don't need a computer to find $1 + 2 + \cdots + 100$ since it was shown in Chapter 1 that $1 + 2 + \cdots + n = n(n + 1)/2$. However, we do not have such a handy formula for $1^5 + 2^5 + \cdots + 100^5$, say, and a modification of our previous flowchart would give an algorithm for such sums. Just replace N in step 5 by N^5.

For another example of flowcharts, suppose we want the factors of 10. How do we find them? We divide 10 by an integer, say D, and see if $10/D$ is an integer. If $10/D$ is an integer, then we know that D is a factor; if not, then D is not a factor. A systematic way of doing this would start with $D = 1$ and then increase D by 1 unit

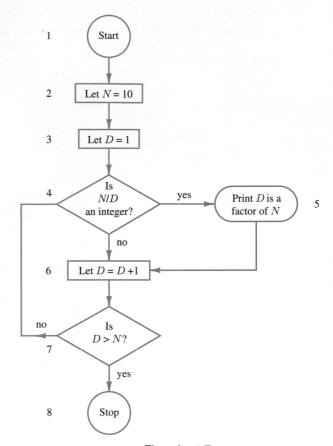

Flowchart 7

until $D = 10$. Flowchart 7 displays this algorithm. Table 7.6 gives the wcd table for this flowchart.

   We do not need a computer to find the factors of 10. But what if we wanted the factors of 55,742? Then a computer would be useful, and all we would have to do is change step 2 to "Let $N = 55742$." If we want the factors of a general N, which is to be decided later, then we can replace step 2 by "Input N." The flowchart that was used for finding the factors of 10 was not very efficient because there are no factors of 10 between 5 and 10 and there was a lot of wasted effort. In fact, for any N there are no factors between $N/2$ and N (why?). Thus we would have a more efficient flowchart if we replace steps 7 and 8 by

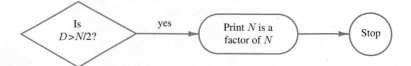

Table 7.6

Step	What computer does	Step	What computer does
1	Start	6	Let $D = 6$
2	Let $N = 10$	7	No
3	Let $D = 1$	4	No
4	Yes	6	Let $D = 7$
5	Print 1 is a factor of 10	7	No
6	Let $D = 2$	4	No
7	No	6	Let $D = 8$
4	Yes	7	No
5	Print 2 is a factor of 10	4	No
6	Let $D = 3$	6	Let Let $D = 9$
7	No	7	No
4	No	4	No
6	Let $D = 4$	6	Let $D = 10$
7	No	7	No
4	No	4	Yes
6	Let $D = 5$	5	Print 10 is a factor of 10
7	No	6	Let $D = 11$
4	Yes	7	Yes
5	Print 5 is a factor of 10	8	Stop

EXERCISES

1. Make a flowchart for sharpening a pencil.

2. Make a flowchart for drinking from a water fountain.

3. Make a flowchart for dressing in the morning. Make a flowchart for undressing at night. How do these two flowcharts compare?

4. Make a flowchart for going out on a date with your girlfriend or boyfriend.

5. Think of some daily activity and make a flowchart that describes it.

6. Make a flowchart for finding (3)(4).

7. Make a flowchart for finding 52!.

8. Make a flowchart for finding the sum of the first 100 even natural numbers.

9. Make a flowchart for finding the sum of the first 100 odd natural numbers.

10. Make a flowchart for finding out whether a natural number is odd or even.

11. Make a flowchart that tells whether a natural number is divisible by 6 or not.

12. Make a flowchart that squares a natural number.

13. What does Flowchart 8 do?

14. How does Flowchart 9 differ from Flowchart 8? Are the results the same?

15. What does Flowchart 10 do?

16. What does Flowchart 11 do?

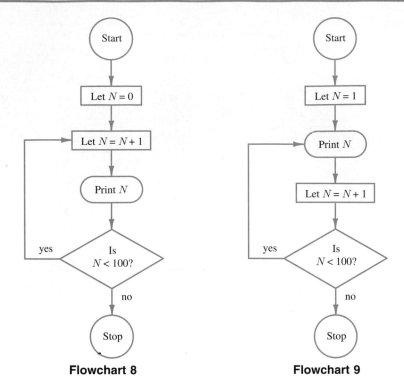

Flowchart 8   **Flowchart 9**

17. What does Flowchart 12 do?
18. Make a wcd table for Flowchart 12.
19. How would the results change if steps 3 and 4 were interchanged in Flowchart 12?
20. How would the results change if steps 4 and 5 were interchanged in Flowchart 12?
21. Make a wcd table (a) for Exercise 19; (b) for Exercise 20.
22. What happens in Flowchart 13?
23. What happens in Flowchart 14?
24. Put the following steps together to make a legitimate flowchart and describe the result.

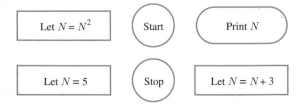

*25. Flowchart 15 tells whether N is prime for $N \geq 3$. Explain why this flowchart works. Make a wcd table for $N = 11$ and $N = 77$.

*26. Flowchart 16 determines which of the numbers A, B, or C is the largest. (a) Explain why this flowchart works. (b) Make a wcd table for $A = 2$, $B = 5$, $C = 7$. (c) Make a wcd table for $A = 5$, $B = 7$, $C = 2$. (d) Make a wcd table for $A = 7$, $B = 2$, $C = 5$. (e) Construct a flowchart that determines which of the numbers A, B, or C is the smallest.

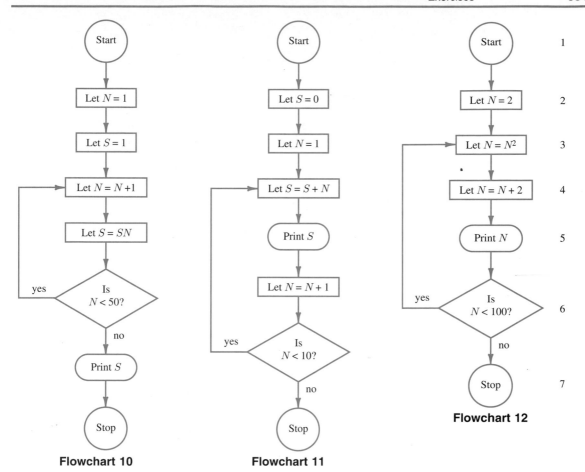

Flowchart 10   **Flowchart 11**

Flowchart 12

*27. (a) The greatest common divisor of two natural numbers A and B is the largest natural number that is a divisor (a factor) of both A and B. Flowchart 17 determines the greatest common divisor of A and B. Explain why this flowchart works. Make a wcd table for $A = 10$, $B = 15$.

(b) The least common multiple of two natural numbers A and B is the smallest natural number that has both A and B as a divisor. Construct a flowchart that determines the least common multiple of A and B.

*28. Flowchart 18 computes the prime factors of a natural number N. Explain why this flowchart works. Make a wcd table for $N = 308$.

*29. Construct a flowchart that determines the primes that are less than or equal to N, where $N \geq 2$. Explain why this flowchart works. Make a wcd table for $N = 20$.

*30. Construct a flowchart that gives the number of factors of N.

*31. A natural number N is **perfect** if N is the sum of its factors other than itself. Construct a flowchart that determines whether N is perfect.

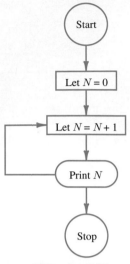

Flowchart 13

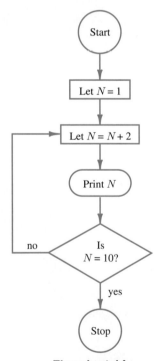

Flowchart 14

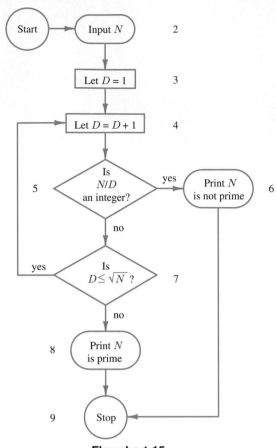

Flowchart 15

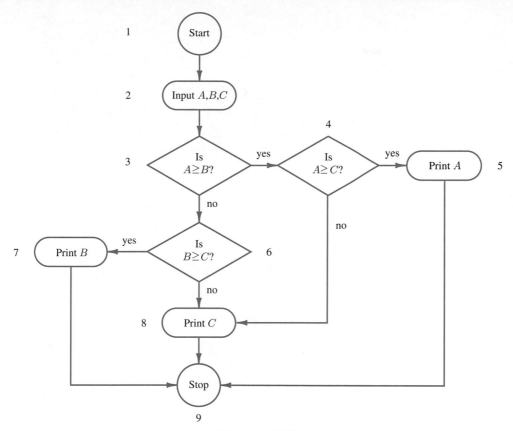

Flowchart 16

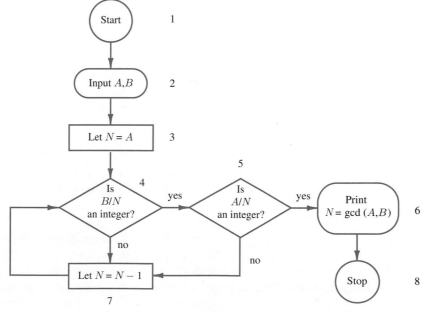

Flowchart 17

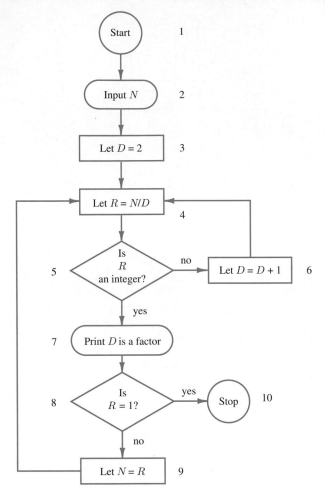

Flowchart 18

7.5 BASIC

The miracle of the appropriateness of the language of mathematics for the formulation of the laws of physics is a wonderful gift which we neither understand nor deserve. We should be grateful for it and hope that it will remain valid in future research and that it will extend, for better or for worse, to our pleasure even though perhaps also to our bafflement, to wide branches of learning.

— E. P. WIGNER, Twentieth-Century Physicist

BASIC is a programming language that is used to communicate with a computer. It is a language used for writing programs and instructions for the computer to execute.

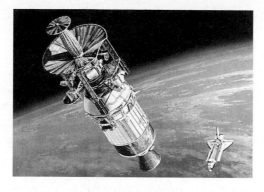

Photo 7.6
Space satellites contain onboard computers for control and communication.

As with any language, you must learn the vocabulary, grammar, and punctuation of BASIC. You will first learn about constants and variables and how to write them in BASIC.

> A **constant** is a number with a definite value.

Constants may appear with or without a decimal point and may or may not have a sign (+ or −). A constant without a sign is assumed to be positive. The only allowable symbols in a BASIC constant are the numeric digits, the decimal point, and the plus or minus sign. For example, 32, −16.1, +0.572, .7689, 51., and 2789 are valid constants in BASIC. The following are not valid BASIC constants: 2,789, 32B, N, N2.

> A **variable** is a quantity that may change values and is represented by a single capital letter or a single capital letter followed by a single digit. (Only capital letters are used in BASIC.)

Some valid BASIC variables are N, B2, X, A1. Thus there are $26 + 26(10) = 286$ possible variables. The following are not valid BASIC variables: BD, 2B, 9, B62.

Constants and variables may be combined with arithmetic operations. Table 7.7 gives the arithmetic operations and the BASIC symbols used to denote them.

Table 7.7

Operation	Symbol
Addition	+
Subtraction	−
Multiplication	*
Division	/
Exponentiation	↑ or ^

For example, the expression $x^3 - 2ax/5$ can be represented in BASIC by

$$X \uparrow 3 - 2 * A * X/5$$

It is important to know the precedence of the various arithmetic operations. For example, when you see the expression $2x^3 + 3$, you know that this does not mean $2(x^3 + 3)$ or $(2x)^3 + 3$ or $2(x + 3)^3$ or $(2x + 3)^3$. You know that in $2x^3 + 3$ the order of precedence is to cube x, then multiply by 2, and finally add 3. The other expressions above are found by grouping with parentheses. In the same way, in BASIC, $2 * X \uparrow 3 + 3$ means first cube X, then multiply by 2, and finally add 3.

Table 7.8 gives the order of precedence for the arithmetic operations. All operations with precedence 1 are done first, then all operations with precedence 2, and finally all operations with precedence 3. If two operations have the same precedence, they are performed from left to right. When parentheses are used, the expressions within the parentheses are evaluated first. Table 7.9 gives some typical mathematical expressions followed by their BASIC representation.

Table 7.9

Mathematical expression	BASIC representation
$2(x^3 + 3)$	$2 * (X \uparrow 3 + 3)$
$(2x)^3 + 3$	$(2 * X) \uparrow 3 + 3$
$2(x + 3)^3$	$2 * (X + 3) \uparrow 3$
$(2x + 3)^3$	$(2 * X + 3) \uparrow 3$
$\left(\dfrac{x + 3}{2}\right)^3$	$((X + 3)/2) \uparrow 3$

Table 7.8

Precedence	Operation
1	$\uparrow$ or $\hat{\ }$
2	$*$ and $/$
3	$+$ and $-$

In BASIC, two arithmetic operations cannot be placed next to each other. Thus we cannot write $X * -A$ but must use $X * (-A)$.

Besides the arithmetic operations there are relational operations. These are given in Table 7.10. For example, the mathematical statement $N \geq T^2$ would have the BASIC representation $N >= T \uparrow 2$.

Table 7.10

Relational operation	BASIC symbol
Is equal to	$=$
Is less than	$<$
Is less than or equal to	$<=$
Is greater than	$>$
Is greater than or equal to	$>=$
Is not equal to	$<>$

LET Statement

Now that you know some of the symbols used in BASIC, how do you write computer instructions? Instructions in BASIC are given by certain specific simple statements. The most fundamental instruction is the LET statement. A LET statement specifies a computation that we want to take place involving constants or variables or both. In BASIC, most instructions are preceded by a **line number**, which is (on most computers) an integer between 1 and 99999. The line number keeps track of the different instructions and indicates the address at which the particular instruction is stored in the computer's memory. Examples of LET statements are

        5 LET X = 3
        10 LET N = 7 + 6
        15 LET A = (2 + M)/(3 + N)
        20 LET N = X ↑ 2 + 4 * X
        30 LET N = N + 1

> Notice that each LET statement has five parts. First is a line number, second the word LET, third a variable, fourth the = symbol, and fifth an **expression**. An expression can be thought of as a rule for computing a value.

An expression may be a single constant, as in line 5, or an arithmetic computation, as in line 10. In statement 15 the expression $(2 + M)/(3 + N)$ has the value 2 plus the current value of the variable M divided by 3 plus the current value of the variable N. In statement 30, the expression N + 1 has the current value of N plus 1.

The LET statement instructs the computer to do two things: (1) evaluate the expression on the right of the = symbol and thus give the expression a value; (2) assign the value of the expression to the variable on the left of the = symbol. The value the variable on the left formerly had is forgotten, and the variable assumes this new value.

PRINT Statement

Once the computer performs its computations, how does it tell you the result? This is accomplished by a PRINT statement. For example, suppose you want the computer to add $1 + 2$. This can be accomplished by the program

        5 PRINT 1 + 2
        10 END

After you type RUN and press the RETURN key, the computer will print

        3

If a variable follows the word PRINT, the computer prints out the current value of the variable. Hence the following program will also instruct the computer to add $1 + 2$.

```
5 LET N = 1 + 2
10 PRINT N
15 END
```

After you type RUN and press the RETURN key, the computer will print

```
3
```

Every BASIC program must conclude with an END statement.

Why have multiples of 5 been used for the line numbers? It is not really necessary. In fact, the line numbers 1, 2, 3 or 10, 20, 30 or any positive integers (≤ 99999) in increasing numeric order could have been used. However, it is useful, especially for more complicated programs, to leave gaps of five or ten between line numbers because you can insert instructions later that you may have forgotten to put in. It is possible to do this because the computer executes instructions in order of line number. Thus the previous program could have read

```
5 LET N = 1 + 2
15 END
10 PRINT N
```

and given the same result. Or suppose you later wished to insert the instruction LET $N = N + 1$ after the instruction LET $N = 1 + 2$. If you had already written the original program, you could add this instruction as

```
5 LET N = 1 + 2
15 END
10 PRINT N
6 LET N = N + 1
```

The PRINT statement can also be used to instruct the computer to print alphabetic or symbolic information. For example, suppose you wish the computer to print PROGRAM NUMBER 1. This would be accomplished by the PRINT statement

```
3 PRINT "PROGRAM NUMBER 1"
```

When this statement is executed, the computer will type the information enclosed in quotation marks. Suppose you type the following program into the computer.

```
3 PRINT "PROGRAM NUMBER 1"
5 LET N = 1 + 2
10 PRINT "THE SUM OF 1 AND 2 IS"; N
15 END
```

After you type RUN and press the RETURN key, the computer will print

```
PROGRAM NUMBER 1
THE SUM OF 1 AND 2 IS 3
```

As another example, suppose you type the following program into the computer.

```
5 LET N = 2
10 PRINT "THE SQUARE OF"; N; "IS"; N ↑ 2
15 END
```

After you type RUN and press the RETURN key, the computer will print

THE SQUARE OF 2 IS 4

You could have written this last program as

```
5 LET N = 2
10 PRINT "IF N = 2 THEN N ↑ 2 ="; N ↑ 2
15 END
```

Upon execution the computer would type

IF N = 2 THEN N ↑ 2 = 4

Notice that a semicolon is used to separate the items you want printed. The semicolon also instructs the computer to stay on the same line with the next printed command. Thus the following program would result in the same printout as the previous one.

```
5 LET N = 2
10 PRINT "IF N = 2 THEN N ↑ 2 =";
15 PRINT N ↑ 2
20 END
```

If you had omitted the semicolon in line 10, the computer would type the PRINT statement with line number 15 on the next line:

IF N = 2 THEN N ↑ 2 =
4

A comma can also be used as a separator in PRINT statements. The only difference is that more space is left. For example, upon executing the program

```
5 LET N = 2
10 PRINT "THE SQUARE OF", N, "IS", N ↑ 2
15 END
```

the computer will print

THE SQUARE OF   2   IS   4

What if a number is too large for the computer to write? For example, some computers cannot work with more than eight significant digits. If the result of a computation is a large number, say 347,265,000,000, then the computer will probably print this as $3.47265E + 11$, which means 3.47265×10^{11}.

GO TO and IF-THEN Statements

Normally, BASIC statements are executed in line-number sequence. However, there are BASIC statements that can change the normal sequence of instruction execution and start a new sequence of execution at another place in the program. Two such BASIC statements are the GO TO and IF-THEN statements. An example of a GO TO statement is

40 GO TO 20

The number following the words GO TO must be a line number in the program, and when this statement is executed the computer will execute next the statement whose line number follows the words GO TO. The computer will then execute statements in line number sequence from this new point until told to change again.

In an IF-THEN statement, if a certain condition is true, then the computer will transfer to a specified line for its next instruction. But if the condition is not true, the computer will go on in normal sequence, executing the next line in sequence following the IF-THEN statement. The following is an example of an IF-THEN statement.

75 IF N > 3 THEN 100

In this statement the computer is instructed to determine whether the current value of the variable N is greater than 3. If it is, the computer is to go to line 100 for its next instruction, and if not, it is to execute the statement following line number 75 in numeric order. The IF-THEN statement corresponds to the diamond-shaped decision procedure used in flowcharts.

BASIC programs corresponding to the flowcharts developed in Section 7.4 are now constructed. The following program instructs the computer to print the first 100 natural numbers (see Flowchart 5).

```
5 LET N = 0
10 LET N = N + 1
15 PRINT N
20 IF N < 100 THEN 10
25 END
```

A BASIC program that computes $1 + 2 + \cdots + 100$ is (see Flowchart 6)

```
10 LET N = 1
20 LET S = 1
30 LET N = N + 1
40 LET S = S + N
50 IF N < 100 THEN 30
60 PRINT S
70 END
```

Suppose we wanted the fifth powers of the first 10 natural numbers. The next BASIC program would do it.

```
10 LET N = 1
20 PRINT N ↑ 5
30 LET N = N + 1
40 IF N <= 10 THEN 20
50 END
```

The BASIC language contains many of the so-called standard functions. For example, SQRT(N) denotes the square root of N. Thus, after executing the following program, the computer will print the number 8.

```
10 PRINT SQRT(64)
20 END
```

Another standard function is INT(N), the **integer part** of the number N. The integer part of a number N is the largest integer that is less than or equal to N. Thus, INT(3.52) = 3, INT(7) = 7, INT(7.801) = 7, INT(−2.32) = −3. Notice that N is an integer if and only if INT(N) = N. Since N $\geq$ INT(N) for any natural number N, it follows that a natural number N is not an integer if and only if N $>$ INT(N).

BASIC has notation for other standard functions, but this will not concern us now.

The next BASIC program instructs the computer to print the factors of 10 and corresponds to Flowchart 7.

```
10 LET N = 10
20 LET D = 1
30 IF N/D > INT(N/D) THEN 50
40 PRINT D; "IS A FACTOR OF"; N
50 LET D = D + 1
60 IF D <= N THEN 30
70 END
```

After you type RUN and press the RETURN key, the computer will type

```
1 IS A FACTOR OF 10
2 IS A FACTOR OF 10
5 IS A FACTOR OF 10
10 IS A FACTOR OF 10
```

INPUT Statement

If you want to compute the factors of many different numbers, instead of changing the program for each of these numbers you can use an INPUT statement. Such a statement allows you to give values to one or more variables. Suppose in the last program you replace line 10 by

```
10 INPUT N
```

After you type RUN and press the RETURN key, the computer will type a question mark. You then type the value of N for which you want the factors. After you press the RETURN key, the computer will run the program with this value of N. For example, suppose you want the factors of 12. After

    ?

you type 12. The resulting printout would appear as follows:

RUN
?12
1 IS A FACTOR OF 12
2 IS A FACTOR OF 12
3 IS A FACTOR OF 12
4 IS A FACTOR OF 12
6 IS A FACTOR OF 12
12 IS A FACTOR OF 12

If you type RUN again, the computer will type another question mark and you can insert a new value of N. This can be repeated as many times as you like.

The next BASIC program instructs the computer to find the prime factors of N and corresponds to Flowchart 18. Notice the GO TO statements in lines 60 and 100.

```
10 INPUT N
20 LET D = 2
30 LET R = N/D
40 IF R = INT(R) THEN 70
50 LET D = D + 1
60 GO TO 30
70 PRINT D; "IS A FACTOR "
80 IF R = 1 THEN 110
90 LET N = R
100 GO TO 30
110 END
```

Suppose you want the prime factors of 308. The printout would appear as follows.

RUN
?308
2 IS A FACTOR
2 IS A FACTOR
7 IS A FACTOR
11 IS A FACTOR

INPUT statements can involve more than one variable. For example, consider a BASIC program that begins with

10 INPUT A, B, C

After you type RUN and press the RETURN key and the computer types a question mark, you can insert any three values as follows:

?5, 62, 147

The computer will then execute the program with these values for A, B, and C.

EXERCISES

1. Write the following expressions using BASIC language notation.
   (a) $7x^4 + 3x^2 - 2x + 4$ (b) $(x - 7)^2 \left[3x - (2/3)\right]^3$
   (c) $\left[(x^2 - 2)^2 + 3\right]^3$ (d) $5/6ax^2$ (e) $x \neq 2$

2. Write the following BASIC expressions using mathematical notation.
   (a) $3 * X \uparrow 2 + 4$ (b) $2 * A * X \uparrow 3 + 7 * Y/4$
   (c) $(3 * X + 2 * (X + 4)) \uparrow 2$ (d) $X <= 5$ (e) $Y <> 2$

3. What different values would the expression $2 * 2 \uparrow 3 + 3$ have if you changed the order of preference?

4. What does the computer do upon execution of the following program?

   5 PRINT 3 + 5
   10 PRINT 3 * 5
   15 PRINT 4/5 + 3 ↑ 2;
   16 PRINT (2 + 7/2) ↑ 2
   20 END

5. What does the computer do upon execution of the following program?

   5 LET N = 5
   10 LET M = 7
   15 PRINT N + M; N * M; N ↑ 2; M ↑ 2
   20 END

6. Identify the following as variables, constants, functions, or none of these.

7	A4	2,176.2	7134.5	4A
INT	X2	SQRT	INTS	BD

7. What will the computer print after executing the following program?

   10 LET A = 3
   20 LET B = 2
   30 LET C = 5
   40 LET X = A + B * C
   50 LET Y = (A + B) ↑ 2
   60 LET Z = (A + 2 * C) ↑ B
   70 PRINT X; Y; Z
   80 END

8. Write a BASIC program that will perform the computation $\left[(6.24)^2 - 24.78\right]^{1/2}$.

9. Write a BASIC program for computing the sum of the squares of the first 100 natural numbers.

10. Write a BASIC program beginning with the statement INPUT N that tells whether N is divisible by 6 or not.

11. Write a BASIC program that tells whether NM is positive, negative, or zero. Begin with the following statements.

```
5 INPUT N
10 INPUT M
```

12. What will the computer print after execution of the following program?

```
10 PRINT "I";
20 GO TO 80
30 PRINT "   ";
40 PRINT "A";
50 GO TO 110
60 PRINT ".";
70 GO TO 140
80 PRINT "   ";
90 PRINT "AM";
100 GO TO 30
110 PRINT "   ";
120 PRINT "COMPUTER";
130 GO TO 60
140 END
```

13. The following two statements can be replaced by one much simpler statement. What is it?

```
50 IF X <= Y THEN 100
60 IF X > Y THEN 100
```

14. Write a BASIC program to print the powers of 2 that are less than 1,000,000.

15. Correct the errors in the following BASIC statements.

```
10 LET A = 2(B + C)
20 GO TO STATEMENT 30
30 V = L * W * H
40 LET A + B = C
50 LET 5 * B = A + C
60 IF A IS LESS THAN B THEN 90
70 IF A = B GO TO LINE 90
80 IF M >= N THEN GO TO 90
90 IF M => N THEN 120
100 A = 3
110 INPUT A,B,C,
120 PRINT OUT N
130 QUIT
140 RUN
LET A = 2
150 IF 10 = N THEN 20
```

16. (a) What will the computer print after executing the following program? (b) How will the printout change if you replace line 50 by 50 PRINT A;? (c) How will the printout change if you change the 20 to 45?

```
10 LET A = 6
20 LET A = A + 1
30 LET B = A + 3
40 LET B = B + 1
50 PRINT A
60 PRINT B
70 END
```

17. What will the computer print after executing the following program?

```
10 INPUT N
20 LET D = 1
30 LET D = D + 1
40 IF D > 10 THEN 80
50 IF N/D > INT(N/D) THEN 30
60 PRINT D; "IS A FACTOR OF"; N
70 GO TO 30
80 END
RUN
?100
```

18. A natural number N is a **perfect square** if $N = M^2$ for some natural number M. Write a BASIC program beginning with 10 INPUT N that tells whether N is a perfect square or not.

19. Natural numbers A, B, C, form a **pythagorean triple** if $A^2 + B^2 = C^2$. Write a BASIC program beginning with 10 INPUT A,B that tells whether A and B are the first two terms of a pythagorean triple.

20. Translate Flowchart 15 into a BASIC program.

21. Translate Flowchart 16 into a BASIC program.

22. Translate Flowchart 17 into a BASIC program.

23. Write a BASIC program that determines the primes that are less than or equal to N, where $N \geq 2$.

*24. Write a BASIC program that instructs the computer to give the number of factors of N.

*25. A natural number N is **perfect** if N is the sum of its factors other than itself. Write a BASIC program that tells whether or not N is perfect.

*26. In Section 1.7 it was shown that if a radioactive material decays at a rate of r percent a day, then its half-life n satisfies $(1 - r)^n = 1/2$. Write a BASIC program to find the half-life of a substance that decays at the rate of 5 percent, 1 percent, and 0.1 percent per day.

*27. In Section 4.6, Exercise 19, friendly numbers were defined. Write a BASIC program that determines whether or not two numbers are friendly. If you have a computer available, use it to show that 1184 and 1210 are friendly and that 17,296 and 18,416 are friendly.

*28. A natural number is **multiplicatively perfect** if it equals the product of its divisors other than itself. Write a BASIC program that determines whether or not N is multiplicatively perfect.

Photo 7.7
Operator in a computer room.

7.6 | More BASIC

Mathematicians are like Frenchmen; whatever you say to them they translate into their own language and forthwith it is something entirely different.
  —Johann Wolfgang von Goethe (1749–1832), German Author

READ and DATA Statements

There are other important and useful BASIC statements. Two of these are the READ and DATA statements. Suppose you have a lot of variables, say, X, N, B1, B2, A, Y, and you want them to have the values 23, 46.5, 72.31, −62.1, 165, and −364, respectively. Instead of using six LET statements you can accomplish your objective with two statements,

    10 READ X, N, B1, B2, A, Y
    20 DATA 23, 46.5, 72.31, −62.1, 165, −364

It does not matter whether there are spaces after the commas of a READ or DATA statement. The DATA statement need not appear immediately after the READ statement but can appear anywhere in the program.

The READ and DATA statements can be used to give the same variable (or variables) different values. For example, if you want the variable N to assume the values 3, 5, 7, 12, −24.6, you can use the statements

    10 READ N
    15 DATA 3, 5, 7, 12, −24.6

The first time the READ statement is executed, N will assume the value 3, the next time the READ statement is executed, N will assume the value 5, etc.

For example, suppose you want 13!, 47!, and 52!. The following program will do it.

```
10 READ M
20 DATA 13, 47, 52
30 LET N = 1
40 LET S = 1
50 LET N = N + 1
60 LET S = S * N
70 IF N < M THEN 50
80 PRINT S
90 IF M = 52 THEN 110
100 GO TO 10
110 END
```

Let us analyze this program. The computer first gives M the value 13. When line 80 is executed, the computer prints the value of 13!. The condition in line 90 is not satisfied, and so line 100 is executed, sending the computer back to line 10. At this point M is given the value 47. When line 80 is executed, the computer now prints the value of 47!. Again, the condition in line 90 is not satisfied, and so the computer goes to line 10 and gives M the value 52. At line 80 the computer prints the value of 52!. Now the condition $M = 52$ is satisfied, and so the computer goes to line 110 and stops. Line 90 was necessary to keep the computer from going back to line 10 a third time and not having any additional data to read.

The next program finds $(x + y)^2$ for pairs of numbers x, y.

```
10 READ X, Y
20 DATA 1, 2, 2, 3, 3, 4, 5, 6
30 PRINT (X + Y) ↑ 2
40 IF N = 5 THEN 60
50 GO TO 10
60 END
```

After you type RUN and press the RETURN key, the computer will type

```
9
25
49
121
```

Suppose you want to add the numbers 1, 2, 3, ..., 10 and keep a running total. The next program will do this.

```
10 LET T = 0
20 READ X
30 LET T = T + X
40 PRINT T;
50 IF X = 10 THEN 80
60 GO TO 20
70 DATA 1, 2, 3, 4, 5, 6, 7, 8, 9, 10
80 END
```

After this program is executed, the computer will type

1   3   6   10   15   21   28   36   45   55

The above computation could also be achieved by the following program.

```
10 LET T = 0
20 LET N = 1
30 LET T = T + N
40 PRINT T;
50 LET N = N + 1
60 IF N <= 10 THEN 30
70 END
```

Now a program to compute your net pay. If you work 40 hours a week at $10 an hour and your deductions are 20 percent, what is your net pay? You are also interested in your net pay if you get a raise to $12 an hour and your deductions go up to 22 percent.

```
10 READ H, R, T
20 DATA 40, 10, .2, 40, 12, .22
30 LET N = H * R * (1 − T)
40 PRINT "YOUR NET PAY IS $"; N
50 IF R = 12 THEN 70
60 GO TO 10
70 END
```

After you type RUN and press the RETURN key, the computer will print

YOUR NET PAY IS $320
YOUR NET PAY IS $374.40

FOR-NEXT Statements

Other important BASIC statements are the FOR-NEXT statements, which appear in pairs in the program. The following are typical FOR-NEXT statements.

```
20 FOR I = 1 TO 10
50 NEXT I
```

When statement 20 is executed the first time, I assumes the value 1. The statements are then executed in order of line number starting with the next statement after 20. When statement 50 is reached, the computer returns to 20, this time giving I the value 2. This is continued until I assumes the value 10. The statements are then executed in line-number order starting with the next statement after 20 and skipping statement 50.

The next program instructs the computer to find the sum of the first 100 natural numbers using FOR-NEXT statements.

```
10 LET A = 0
20 FOR I = 1 TO 100
30 LET A = A + I
40 NEXT I
50 PRINT A
60 END
```

There are other variations of FOR-NEXT statements. For example, in the following statements I assumes the values 2, 4, 6, 8, 10.

```
10 FOR I = 2 TO 10 STEP 2
50 NEXT I
```

Thus, in the above, I goes from 2 to 10 in steps of 2.

In the next statements A goes from 10 to 2 in steps of -2 and hence assumes the values 10, 8, 6, 4, 2.

```
10 FOR A = 10 TO 2 STEP −2
50 NEXT A
```

GOSUB Statement

The last BASIC statement that will be considered is the GOSUB statement. This statement instructs the computer to transfer to a different group of line numbers. When the computer receives a RETURN instruction, it will return to the place where the GOSUB statement was given and then continue in sequence. Follow the steps in the next program and you will understand the GOSUB statement.

```
10 PRINT "THIS PROGRAM";
20 GOSUB 60
30 PRINT "DID YOU DO IT";
40 PRINT "CORRECTLY?"
50 GO TO 90
60 PRINT "ILLUSTRATES THE";
70 PRINT "GOSUB STATEMENT."
80 RETURN
90 END
```

EXERCISES

For Exercises 1–5, what will the computer do upon executing the program?

1.
```
10 READ A, B
20 PRINT A * B
30 DATA 2, 4, 6, 8
40 END
```

2.
```
10 READ A, B
20 PRINT A + B
30 GO TO 10
40 DATA 2, 4, 6, 8
50 END
```

3. 10 READ A, B
   20 PRINT SQRT(A ↑ 2 + B ↑ 2)
   30 IF A = 5 THEN 60
   40 GO TO 10
   50 DATA 3, 4, 5, 12
   60 END

4. 10 READ X, A, B, C
   20 LET Y = A + B − C
   30 LET W = X + Y
   40 LET S = W + Y ↑ 2
   50 LET T = (X + Y) ↑ 3
   60 PRINT W; S; T
   70 DATA 1, 2, 3, −2
   80 END

5. 10 LET N = 0
   20 LET T = 0
   30 READ X
   40 LET N = N + 1
   50 LET T = T + X
   60 IF N = 10 THEN 80
   70 GO TO 30
   80 PRINT T
   90 DATA 1, 2, 3, 4, 5, 6, 7, 8, 9, 10
   100 END

6. The formula for converting degrees Fahrenheit to degrees Celsius is

$$°C = \left(\frac{5}{9}\right)(°F − 32)$$

   Write a BASIC program using READ-DATA statements for converting the following Fahrenheit temperatures to Celsius: 32, 54, −60, 100, and 220.

7. Write a BASIC program using READ-DATA statements that will compute 2^5, 3^5, 4^5, 5^5, 6^5, 7^5, and 8^5.

8. Do Exercise 7 using FOR-NEXT statements

9. What does the following program do?

   10 LET N = 1
   20 FOR I = 1 TO 10
   30 PRINT N
   40 LET N = N + 1
   50 NEXT I
   60 END

10. Write a BASIC program with an INPUT N statement and FOR-NEXT statements that give the sum of the squares of the first N natural numbers.

What do the following programs do?

11. 10 LET S = 0
    20 FOR I = 1 TO 10
    30 LET S = S + 2 ↑ I
    40 NEXT I
    50 PRINT S
    60 END

12. 10 LET N = 1
    20 FOR I = 1 TO 30
    30 LET N = N * I
    40 NEXT I
    50 PRINT N
    60 END

13. Write a BASIC program to ask the value of N and then compute $N + N^2 + N^3 + \cdots + N^{20}$. (*Hint*: See Exercise 11.)

14. The following program can be carried out using a four-line program with FOR-NEXT statements. What is it?

10 LET N = 1
20 PRINT 4 * N ↑ 3 + 5 * N ↑ 2 − 3 * N + 7
30 LET N = N + 1
40 IF N = 11 THEN 60
50 GO TO 20
60 END

15. The program in Exercise 11 can be carried out using a four-line program with READ and DATA statements. What is it?

16. Write a BASIC program using FOR-NEXT statements that computes the sum

$$1 + \frac{1}{2} + \frac{1}{3} + \cdots + \frac{1}{100}$$

17. What will the computer print upon execution of the following BASIC program?

10 FOR N = 1 TO 25
20 LET S = N ↑ 2
30 PRINT S;
40 IF S > 100 THEN 60
50 NEXT N
60 END

18. The following program contains two nested FOR-NEXT statements. What does it do?

10 FOR I = 2 TO 9
20 FOR J = 2 TO 9
30 PRINT I; "*"; J; "="; I * J,
40 NEXT J
50 NEXT I
60 END

19. Write a BASIC program that give the addition table for all natural numbers from 1 to 10. (*Hint*: See Exercise 18.)

20. Given 8 colored marbles of which 2 are red, 3 are blue, and 3 are green, what are the possible combinations that can be made if at least 1 marble of each color is represented in each combination? For example, one such combination is 1 red, 2 blue, 2 green. Explain why the following program gives the answer. Then alter the program so that instead of printing the different kinds of combinations, the total number of combinations is printed.

10 FOR R = 1 TO 2
20 FOR B = 1 TO 3
30 FOR G = 1 TO 3
40 PRINT R; "RED"; B; "BLUE"; G; "GREEN"
50 NEXT R
60 NEXT B
70 NEXT G
80 END

What do the following programs do?

21. 10 LET N = 1
20 FOR I = 1 TO 5
30 LET N = 1 + 2/N
40 NEXT I
50 PRINT N
60 END

22. 10 LET N = 1
20 FOR I = 6 TO 2 STEP −1
30 LET N = 1 + I/N
40 NEXT I
50 PRINT N
60 END

*23. Write a BASIC program to compute the following to any specified length. (*Hint*: See Exercise 21.)

$$1 + \cfrac{1}{1 + \cfrac{1}{1 + \cfrac{1}{1 + \cfrac{1}{1 + \cdots}}}}$$

As this is carried out farther and farther, the numbers are supposed to approach the golden ratio $(\sqrt{5} + 1)/2$. If you have a computer available, run your program for longer and longer lengths (say 1, 5, 10, 20, 25, 26, 27) and see if this happens.

*24. Write a BASIC program using FOR-NEXT statements that tells whether $N^2 - N + 41$ is prime for N running from 1 to 40.

25. What can GOSUB do that GO TO cannot do?

26. What does the following program do?

```
10 PRINT "A";
20 GOSUB 130
30 GOSUB 150
40 GOSUB 170
50 GOSUB 130
60 PRINT "FROM";
70 GOSUB 130
80 GOSUB 150
90 GOSUB 170
100 GOSUB 170
110 PRINT "IPPI";
120 GO TO 190
130 PRINT "   ";
140 RETURN
150 PRINT "M";
160 RETURN
170 PRINT "ISS";
180 RETURN
190 END
```

27. Explain why the following BASIC program computes

$$\frac{N!}{(N - R)!R!}$$

```
10 INPUT N,R
20 LET M = N
30 GOSUB 100
40 LET A = S
50 LET M = R
60 GOSUB 100
70 LET B = S
80 LET M = N − R
90 GOSUB 100
95 LET D = S
96 PRINT A/(B ∗ D)
97 GO TO 160
100 LET S = 1
110 LET T = 1
120 LET T = T + 1
130 LET S = S ∗ T
140 IF T < M THEN 120
150 RETURN
160 END
```

*28. In Section 4.1 the Fibonacci sequence was discussed. Write a BASIC program that will give the first N Fibonacci numbers.

*29. Write a BASIC program that will check your conjecture in Exercise 23 of Section 4.1.

*30. Write a BASIC program that will check your conjecture in Exercise 39 of Section 4.1.

*31. See Exercise 26 of Section 4.1. Write a BASIC program that will compute the first N terms of a Lucas sequence when the first two terms P and Q have been entered.

Chapter 7 Summary of Terms

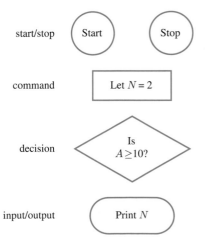

BASIC symbols	equality =	addition +
	subtraction −	multiplication *
	division /	exponentiation ↑ or ^
	grouping ( )	less than <
	greater than >	less than or equal to <=
	greater than or equal to >=	not equal to < >

BASIC statements	LET	END	PRINT	GO TO
	IF-THEN	READ	DATA	GOSUB

Chapter 7 Test

1. What are the five main components of a computer and what are their functions?
2. Why is a machine language necessary?
3. You have probably heard the term "artificial intelligence." What is your interpretation of this term?
4. Draw a diagram of a computer and label its five main components. Give an example of a task that you would like the computer to perform and exhibit on your diagram the necessary steps and the order of execution.
5. Write the following expression using BASIC language notation.

$$\frac{x^2 + 2x + 1}{2x + 1} + \frac{(5x^2 + 1)^2}{3x}$$

6. Suppose you are taking a course that has four exams and a final exam. Each exam counts 15 percent, the final counts 30 percent, and homework counts 10 percent toward your grade. Set up a spreadsheet for these data together with the total number of points on a 100-point scale. Enter the following number of points:

    Exams: 94, 88, 85, 90

    Final: 86

    Homework: 87

7. Make a flowchart for finding 100!.
8. What does Flowchart 19 do?
9. Write a BASIC program for the flowchart in Exercise 8.
10. Write a BASIC program for finding 100!.
11. Write a BASIC program that gives the total number of points on a 100-point scale for the data in Exercise 6.
12. What will be printed when the following program is run?

```
10 READ X, Y, Z
20 DATA 4, 8, 9
30 LET S = X + Y + Z
40 LET A = S/3
50 PRINT A
60 END
```

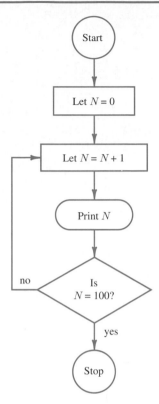

Flowchart 19

13. What will be printed when the following program is run?

    10 LET I = 0
    20 LET I = I + 1
    30 LET Q = I ↑ 3
    40 PRINT I,Q
    50 IF I < 5 THEN 20
    60 END

14. Are the following statements true or false?
    (a) The Cray-1 was one of the first computers.
    (b) A compiler translates machine language into an algorithmic language.
    (c) A spreadsheet is a piece of software that is placed on a bed.
    (d) A flowchart is used to display an algorithm.
    (e) An algorithm is a special kind of logarithm.
    (f) The BASIC symbol for "is not equal to" is <>.
    (g) Every BASIC program must end with a line number and STOP.
    (h) GO TO and GOSUB accomplish the same purpose.
    (i) $(3 * X + 1) \uparrow 2$ means $(3x + 1)^2$ in BASIC.

STATISTICS

8

There are three kinds of lies: lies, damned lies and statistics.
— BENJAMIN DISRAELI (1804–1881), British Politician

The early beginnings of statistics are to be found in census counts during the first century A.D. However, a systematic analysis of data probably began with the work of a London haberdasher, John Graunt (1620–1674). Beginning in 1603, reports of births and deaths in London were published weekly, and by 1623 causes of death were summarized annually in the *Bills of Mortality*. Statistics was born in 1662 when Graunt published his book *Natural and Political Observations upon the Bills of Mortality*. Due to the work of Graunt and his successors, insurance companies were soon established on a profitable and reliable basis.

The concept of statistics is derived from the Latin word *status,* which means condition. This terminology stems from the origin of statistics as giving the status of the state from census counts. Today the word "statistics" is used in a much broader context. Statistics is now a branch of mathematics that deals with the collection, organization, analysis, and interpretation of numerical data. Suppose the presidential election is to be held in one week and we would like to predict the winning candidate. The most reliable way to make such a prediction is to poll every voting-age person in the country. Assuming that there are not very many undecideds and that not very many people will change their minds, this method will produce a prediction that is almost certainly correct. Unfortunately, this method is utterly impractical, so a sample of say 10,000 people are polled. Statisticians have developed techniques for obtaining a representative sample using such parameters as prevailing political opinions and demographics. Obtaining representative samples is part of the problem of data collection and is highly dependent on the specific task at hand. In this chapter we shall not concern ourselves with this specialized problem and shall concentrate on the organization, analysis, and interpretation of data.

Modern society has entered into the information age. We are constantly besieged by huge amounts of information and data. These data come to us not only from the financial, sports, news, and weather sections of newspapers, magazines, and television, but also from the information in science, technology, and business of our daily work.

Photo 8.1
Statistics is frequently used to describe characteristics of large populations.

There is so much information that we rarely see the raw numerical data. We usually see only a summary in some form and it is the job of statistics to make these summaries reliable and useful.

8.1 | Frequency Distributions

In the long run, we are all dead.
— JOHN MAYNARD KEYNES (1883–1946), English Economist

A statistical study begins with a certain **population** or **sample space**. Of course, the word "population" comes from the early use of statistics in census counts or polls, but the term "sample space" is better because it includes more general situations. A sample space would describe not only a population of people, animals, or plants, but also a collection of experimental results in a laboratory or objects such a cars from a factory.

> The elements of a sample space are called **sample points**. Denoting a sample space by S, a **random variable** is a function $X: S \rightarrow D$. The set D is the set of **values** of the random variable X, and D provides the data that we wish to analyze. We call D the **data set** for X.

In statistics we are frequently interested in only the data set D, but we must remember that D always comes from a random variable X. In fact, X gives more information, for X not only gives us D, it also tells us where the values are coming from. Although D is usually a set of numerical values such as the height, weight, or income of people in a certain population, it may also be nonnumerical such as the sex or political preference of people in some sample. Of course, in the latter case we can always convert to a numerical value. For example, we could label Democrats 1 and Republicans 2.

Random Variable

What is random about a random variable? The term "random" is used because the value of a random variable usually cannot be predicted in advance. We may have some idea about the chances or probability that a random variable has a certain value, but we usually cannot predict this value with certainty. For example, suppose we flip a fair coin twice and are interested in whether the second flip is a head (H) or a tail (T). We cannot predict the outcome of the second flip (or the first flip for that matter), but since the coin is fair, we suspect that about half the time the result is a head. We can take our sample space to be the set of all possible outcomes of two flips,

$$S = \{\text{HH, HT, TH, TT}\}$$

The result of the second flip would be the random variable X with values given by

$$X(\text{HH}) = \text{H}, \qquad X(\text{HT}) = \text{T}, \qquad X(\text{TH}) = \text{H}, \qquad X(\text{TT}) = \text{T}$$

In this case, the data set is given by $D = \{\text{H, T, H, T}\}$. Notice that D is not an ordinary set, since we have repeated a value each time it occurs. Since half of the data values are H, if we choose a data value (or a sample point) at random, the probability that we will get an H is $1/2$.

Suppose we are now interested in the number of heads in two flips of a fair coin. This is given by the random variable Y with values

$$Y(\text{HH}) = 2, \qquad Y(\text{HT}) = 1, \qquad Y(\text{TH}) = 1, \qquad Y(\text{TT}) = 0$$

Now the data set is $D = \{2, 1, 1, 0\}$. Since the coin is fair, we suspect that the most likely value for Y is 1. This is true because the value 1 is twice as likely as any other number. However, it is not completely reliable since the value 1 is obtained only half the time. Another description is given by saying that the mean value for D (or X) is 1. This is because

$$\frac{2 + 1 + 1 + 0}{4} = 1$$

As we shall see, statistics is closely related to probability theory, and, in fact, statistics is sometimes called applied probability. However, there is an important difference between the two subjects. In probability theory, we usually deal with an *a priori* probability. That is, we assume that we know the probability beforehand. For example, we might assume that a coin is fair, so that the probability of flipping a head

Table 8.1

Adams	87	Long	67
Baker	78	Mathew	80
Brown	65	Nathan	67
Cantor	75	North	70
Dodds	100	Penn	84
Early	70	Quintana	70
Fine	93	Russell	60
Gore	75	Simpson	74
Hahn	65	Steel	65
Haskins	93	Thomson	84
Jackson	67	Vincent	60
Jones	65	Young	75
Kane	84	Zerbe	98

Table 8.2

Value	Frequency
60	2
65	4
67	3
70	3
74	1
75	3
78	1
80	1
84	3
87	1
93	2
98	1
100	1

on a single toss is $1/2$. But in practice, a coin would have to be perfectly symmetrical in order to be fair. A normal coin might have scratches or notches that would make it slightly biased, or it might be "loaded" by an unscrupulous gambler. How can we tell? Well, we could flip the coin 1000 times and count the number of heads. Suppose we count 550 heads. We would conclude that the coin is biased toward heads and the probability of flipping a head is about 0.55. In statistics, the probability is usually determined experimentally by observing many outcomes. In essence, the probability is derived from a large data set. This is then called an *a posteriori* or **empirical** probability.

Let us now do a little statistics. In my gradebook I have listed the names of the students along with their scores on the last exam. This is tabulated as shown in Table 8.1.

Frequency Distribution

The sample space consists of the students in the class and each student is represented by a sample point. We can define the random variable X that associates with each student that student's test score. As it stands, Table 8.1 does not convey much information. It is just a tabulation of raw data. Right now we are not that interested in the students' names; we are more concerned with analyzing the scores. These are the values of X and they comprise the data set. To organize these values, we first list them in numerical order and tabulate their frequency of occurrence, that is, we construct a **frequency distribution**. Thus, a frequency distribution is a list of the data values together with the number of times each value occurs (Table 8.2).

Histogram

Table 8.2 certainly conveys more information about the data set than Table 8.1. To make the frequency distribution more apparent and visual, we present it in the form of a **bar graph** or **histogram** (Figure 8.1).

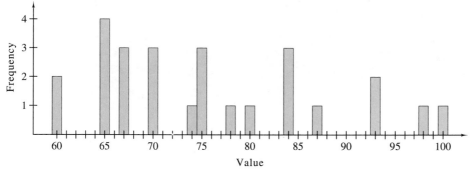

Figure 8.1

Frequency

Figure 8.1 still does not give useful information; it is too detailed. We would like a frequency distribution over a broader scale of values. To do this we divide the scores into **class intervals** of width 5. The lowest class interval is denoted $(55, 60]$. This interval is the set of all data values x such that $55 < x \leq 60$. The next class interval is $(60, 65]$, which is the set of all data values x such that $60 < x \leq 65$. The **frequency** of a class interval is the number of data values that lie in that interval. Table 8.3 gives the frequency distribution for these class intervals.

The histogram corresponding to Table 8.3 is exhibited in Figure 8.2. We have also constructed the **frequency polygon** by joining the midpoints of the tops of the bars with line segments.

The histogram in Figure 8.2 conveys just the right amount of information from the raw data of Table 8.1. Using this information, we can now determine the grades for the exam. These grades would have a frequency distribution something like those listed in Table 8.4.

Table 8.3

Class interval	Frequency
$(55, 60]$	2
$(60, 65]$	4
$(65, 70]$	6
$(70, 75]$	4
$(75, 80]$	2
$(80, 85]$	3
$(85, 90]$	1
$(90, 95]$	2
$(95, 100]$	2

Table 8.4

Class interval	Frequency	Grade
$(55, 60]$	2	F
$(60, 65]$	4	D
$(65, 80]$	12	C
$(80, 90]$	4	B
$(90, 100]$	4	A

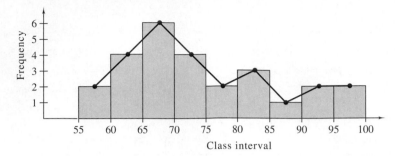

Figure 8.2

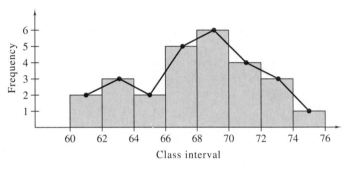

Figure 8.3

Suppose we are now interested in the height of the students in this course. We would then have a random variable Y that associates with each student that student's height (in inches). After polling the class we arrive at the frequency distribution given in Table 8.5. The corresponding histogram together with the freqency polygon are shown in Figure 8.3.

Table 8.5

Class interval	Frequency
$(60, 62]$	2
$(62, 64]$	3
$(64, 66]$	2
$(66, 68]$	5
$(68, 70]$	6
$(70, 72]$	4
$(72, 74]$	3
$(74, 76]$	1

This brings up some interesting questions. Is there a correlation between test scores and height? Roughly speaking, are taller people smarter? Put differently, is there a relationship between the random variable X (or the test score data set) and the random variable Y (or the height data set)? If so, how can we measure the strength of this correlation? In Section 8.7 we shall give methods for answering such questions.

EXERCISES

1. Is the grade distribution in Table 8.4 fair and reasonable? If not, how would you do it?

2. Draw the histogram and frequency polygon for the frequency distribution of Table 8.4.

3. Compute the mean test score from the frequency distribution of Table 8.2.

4. If all you had was the frequency distribution in Table 8.3, how would you compute the mean test score? What would you get?

5. How would you compute the mean grade in Table 8.4?

6. Two students convinced me that I graded their exams wrong. Jones should have gotten a 67 and Kane an 87. Modify Table 8.3 and Figure 8.2 to incorporate these changes.

7. We denote by $[a, b)$ the set of all values x such that $a \leq x < b$. Modify Table 8.3 and Figure 8.2 to incorporate class intervals of width 5 that have this form, starting with $[56, 61)$.

8. We denote by $[a, b]$ the set of all values x such that $a \leq x \leq b$. Can we use class intervals of this form?

9. Let Y be the random variable for two flips of a fair coin considered in the text. Tabulate the frequency distribution and draw the corresponding histogram.

10. Let S be the set of all possible outcomes for three flips of a fair coin and let Y be the random variable that gives the number of heads. List the sample points in S and the value of Y at each sample point. Tabulate the frequency distribution for the data set and draw a histogram.

11. Do Exercise 10 for four flips of a fair coin.

12. The results of the second exam for the students in alphabetical order are given in Table 8.6. Tabulate the frequency distribution as in Tables 8.2 and 8.3 and draw the histogram as in Figure 8.2.

13. Modify Table 8.3 and Figure 8.2 using class intervals of width 6 starting with $(58, 64]$.

14. Tabulate the frequency distribution for the histogram illustrated in Figure 8.4.

15. Modify the histogram in Figure 8.4 so that the class intervals have width 10.

Table 8.6

Adams	90	Long	68
Baker	75	Mathew	82
Brown	68	Nathan	64
Cantor	77	North	73
Dodds	95	Penn	82
Early	75	Quintana	72
Fine	94	Russell	63
Gore	77	Simpson	72
Hahn	63	Steel	70
Haskins	95	Thomson	86
Jackson	68	Vincent	70
Jones	65	Young	77
Kane	82	Zerbe	99

Table 8.7

81	84	85	93	77	71	74	81	88	97
62	77	66	79	82	81	69	55	91	62
71	88	85	91	74	65	79	54	88	93
88	84	91	79	62	75	72	98	93	85
79	82	60	51	93	100	73	69	72	88

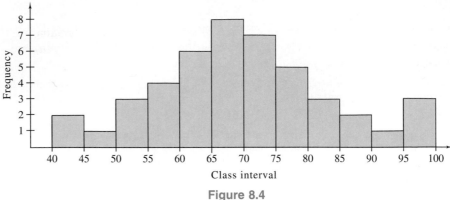

Figure 8.4

16. Table 8.7 shows the high temperatures for the 50 largest cities in the United States on August 24. Tabulate the frequency distribution and draw the histogram with class intervals of width 5.

17. Let S be the sample space $S = \{a, b, c, d, e, f, g, h\}$. Define the random variables X and Y by

$$X(a) = X(e) = X(h) = Y(a) = Y(b) = Y(c) = 0$$
$$X(b) = X(c) = Y(d) = Y(e) = 1$$
$$X(d) = X(f) = Y(f) = 2$$
$$X(g) = Y(g) = Y(h) = 3$$

Define the random variable Z by $Z(s) = X(s)Y(s)$. Give the frequency distributions for X, Y, and Z.

8.2 Central Tendency

It is difficult to understand why statisticians commonly limit their enquiries to averages, and do not revel in more comprehensive views. Their souls seem as dull to the charm of variety as the native of one of our flat English counties, whose retrospect to Switzerland was that, if its mountains could be thrown into its lakes, two nuisances would be got rid of at once.

— SIR FRANCIS GALTON (1822–1911), English Statistician

Suppose we have a large data set and we want to describe it as compactly as possible. Can we describe it with a single number? Of course, a single number cannot completely describe a large amount of data, but it can give an idea of the central tendency of the data, or in some sense it can be the "most representative" of the data set. Such a number is obtained by an averaging process. This section discusses and compares three common types of averages: the **arithmetic mean**, the **median**, and the **mode**.

Summation Notation

We first introduce a helpful notation called **summation notation**. This notation will be employed to make our formulas shorter and easier to apply. If $a_1, a_2, \ldots, a_n$ are numbers, we use the notation

$$\sum_{i=1}^{n} a_i = a_1 + a_2 + \cdots + a_n \tag{8.1}$$

In (8.1), the **index** i takes all integer values from 1 to n and the $\sum$ (capital Greek letter sigma) stands for summation. For example,

$$\sum_{i=1}^{6} a_i = a_1 + a_2 + a_3 + a_4 + a_5 + a_6$$

Another example is

$$\sum_{i=1}^{5} i^2 = 1 + 2^2 + 3^2 + 4^2 + 5^2 = 1 + 4 + 9 + 16 + 25 = 55$$

When we are not concerned with the limits of the index we simply write $\sum a_i$. The index i is called a **dummy variable** and we could just as well use another letter. Thus, for example,

$$\sum_{i=1}^{n} a_i = \sum_{j=1}^{n} a_j$$

If all the a_i's coincide, we write $\sum a$. Notice that

$$\sum_{i=1}^{n} a = na$$

Mean

The **mean** (or **arithmetic mean**), denoted m, for a data set $D = \{x_1, x_2, \ldots, x_n\}$ is defined as

$$m = \frac{x_1 + x_2 + \cdots + x_n}{n}$$

or using summation notation

$$m = \frac{1}{n} \sum_{i=1}^{n} x_i \tag{8.2}$$

Thus, m is defined as the sum of the data values divided by the number of values in the data set.

EXAMPLE 1   Find the mean of the data set $\{1, 7, 3, 4, 6, 8, 1, 2, 4\}$.
   Solution.   In this case we have $n = 9$ and

$$m = \frac{1}{9} \sum_{i=1}^{9} x_i = \frac{1}{9}(1 + 7 + 3 + 4 + 6 + 8 + 1 + 2 + 4) = \frac{36}{9} = 4 \qquad \square$$

As we have seen in Section 8.1, data are sometimes described by a frequency distribution. Suppose the data set is $D = \{x_1, x_2, \ldots, x_n\}$ and the corresponding frequencies are $f_1, f_2, \ldots, f_n$. In this case f_i gives the frequency of x_i, $i = 1, 2, \ldots, n$. We can now calculate the mean according to the formula

$$m = \frac{\sum x_i f_i}{\sum f_i} \tag{8.3}$$

Can you see that (8.2) and (8.3) give the same answer?

EXAMPLE 2   Find the mean for the frequency distribution given in Table 8.8.
   Solution.

$$m = \frac{1 \cdot 2 + 2 \cdot 4 + 3 \cdot 3 + 4 \cdot 1 + 5 \cdot 2}{2 + 4 + 3 + 1 + 2} = \frac{33}{12} = 2.75 \qquad \square$$

Table 8.8

Value	Frequency
1	2
2	4
3	3
4	1
5	2

EXAMPLE 3   Find the mean test score for the frequency distribution in Table 8.2.
   Solution.

$$m = \frac{60 \cdot 2 + 65 \cdot 4 + 67 \cdot 3 + \cdots + 87 + 93 \cdot 2 + 98 + 100}{2 + 4 + 3 + 3 + 1 + 3 + 1 + 1 + 3 + 1 + 2 + 1 + 1} = 75.81 \qquad \square$$

This method for finding the mean works well for frequency distributions when the data are given in terms of single numbers. But what if we have a frequency distribution in terms of class intervals like Table 8.5? We cannot apply (8.3) directly since we do not know what to use for x_i. In this case we replace x_i by the midpoint t_i of the ith class interval, $i = 1, 2, \ldots, n$. This is certainly the reasonable thing to do. The formula for the mean then becomes

$$m = \frac{\sum t_i f_i}{\sum f_i} \tag{8.4}$$

For example, with the frequency distribution in Table 8.5, we have

$$m = \frac{61 \cdot 2 + 63 \cdot 3 + 65 \cdot 2 + 67 \cdot 5 + 69 \cdot 6 + 71 \cdot 4 + 73 \cdot 3 + 75}{2 + 3 + 2 + 5 + 6 + 4 + 3 + 1} = 68$$

There are cases in which (8.3) and (8.4) can both be used. For example, Table 8.2 tabulates a frequency distribution for individual test scores and Table 8.3 tabulates a frequency distribution for the same data in which we combined scores into class intervals. In such a situation, (8.3) and (8.4) can result in different answers. In this case, (8.3) gives the exact mean while (8.4) must be considered an approximation. When there is a large amount of data, this approximation is usually quite good.

EXAMPLE 4   Find the mean test score for the frequency distribution in Table 8.3.
Solution.

$$m = \frac{2(57.5) + 4(62.5) + 6(67.5) + \cdots + 2(92.5) + 2(97.5)}{2 + 4 + 6 + 4 + 2 + 3 + 1 + 2 + 2} = 74.25$$

Notice that this result is somewhat different from that obtained in Example 3.   □

Median

> The **median** of a data set is defined as the value in the "middle" of the list of data values when these are arranged in numerical order.

When there are an odd number of data values, the median is precisely the middle value. However, if there are an even number of values, the median is the mean of the two "middle values." Thus, the median has the property that there are the same number of values above it as below.

EXAMPLE 5   Find the median of the values $5, 7, 3, 8, 1, 2, 7$.
Solution.   Arrange the data in numerical order: $1, 2, 3, 5, 7, 7, 8$. The median is 5 since this is the middle value. There are three values smaller than 5 and three larger than 5.   □

EXAMPLE 6   Find the median of the data set $D = \{4, 3, 6, 7, 2, 8, 2, 1, 3, 4\}$.

Solution.   Listing the values in numerical order gives $1, 2, 2, 3, 3, 4, 4, 6, 7, 8$. The two "middle" values are 3 and 4, so the median is $(3 + 4)/2 = 3.5$. There are five values smaller than 3.5 and five larger.   □

How do we find the median if the data are given in terms of a frequency distribution? The next two examples illustrate the method.

EXAMPLE 7   Compute the median for the frequency distribution of Table 8.9.

Table 8.9

Value	Frequency
3	6
4	2
7	1
8	5
10	3
12	2

Solution.   We could list the data values in numerical order: $3, 3, 3, 3, 3, 3, 4, \ldots$; but this would be tedious. The situation would be worse if there were hundreds or thousands of data values. We therefore need a better method. The number of data values represented is

$$\sum f_i = 6 + 2 + 1 + 5 + 3 + 2 = 19$$

so the tenth value (in numerical order) is the median. Since there are six 3s, two 4s, and one 7, the tenth value is 8. Hence, the median is 8.   □

EXAMPLE 8   Compute the median for the frequency distribution of Table 8.5.

Solution.   The number of data values is

$$\sum f_i = 2 + 3 + 2 + 5 + 6 + 4 + 3 + 1 = 26$$

The two "middle" values are the 13th and 14th values in numerical order. These values are in the class interval $(68, 70]$. We therefore take the median to be 69.   □

Mode

The **mode** of a data set is defined as any value in the set with the highest frequency.

A frequency distribution may have two, three, or more different modes. If a distribution has two modes it is called **bimodal** and if it has three modes it is called **trimodal**.

EXAMPLE 9   Find the mode of the values $2, 3, 7, 1, 5, 3, 7, 8, 3$.

Solution.   The value 3 has the highest frequency, so 3 is the (single) mode.   □

EXAMPLE 10   Find the modes of the values $2, 3, 3, 5, 5, 5, 7, 8, 8, 9, 9, 9, 10$.

Solution.   In this case 5 and 9 both have the highest frequency, so the distribution is bimodal with modes 5 and 9.   □

If the data are given in terms of class intervals, we usually specify **modal classes** or **modal intervals**. These are the intervals with the greatest frequency. For example, for the frequency distribution in Table 8.5, the modal interval is $(68, 70]$.

We have discussed three indicators of central tendency for data sets. Which one of these indicators is the best? Well, it depends on the situation. Consider the data set

$$1, 1, 1, 2, 2, 5, 8, 10, 10, 60$$

The mode is 1, the median is 3.5, and the mean is 10. The mode is not very representative of the data since it is the smallest value of the data set. The mean is only representative of the larger data values. In this case, it appears that the median might be the best indicator of central tendency. The problem with the mean is that it is sensitive to extreme values. For example, if we replace 60 by 160, then the mean shoots up to 20, which is larger than every value but one, while the median and mode remain the same.

Although the mean is responsive to extreme values, in many cases it is the most reliable indicator of central tendency. Let's examine the following data set.

$$1, 1, 1, 1, 1, 2, 6, 7, 9, 11$$

The median is 1.5, which is not very representative, while the mean of 4 indicates the presence of larger values. The mean is usually considered to be the most sensitive of the three indicators. This is because it is responsive to all the values of the data set while the other two indicators are not.

Each of these indicators of central tendency has its advantages and disadvantages. For example, in a certain corporation, 20 people have their birthdates in March. The manager wants to celebrate with a birthday party and would like to choose the best March day. This day should be most representative of the various birthdates. The birthdates are distributed according to Table 8.10. The median is March 9, so if the party is then, nobody is happy. The mean (rounded off to the nearest day) is March 11, so again nobody is happy. However, the mode is March 5, so at least four people will be happy.

EXERCISES

1. Write (8.1) as a sum of n terms for $n = 1, 2, 3, 4, 5$.
2. Show that (8.2) and (8.3) coincide.
3. Find the median and mode for the frequency distribution of (a) Table 8.8, (b) Table 8.2.
4. Find the mean and mode for the data set in (a) Example 5, (b) Example 6, (c) Table 8.9.

Table 8.10

Value	Frequency
2	3
5	4
6	2
8	1
10	3
14	1
15	1
20	3
24	2

5. Find the mean and median for the data set of (a) Example 9, (b) Example 10.

6. Find the mean, median, and mode for the data set in Table 8.6.

7. Find the mean for the frequency distribution of Table 8.10.

8. Find the mean, median, and mode for the histogram in Figure 8.4.

* 9. Find the mean, median, and mode for the data set of Table 8.7.

10. Evaluate the summation $\displaystyle\sum_{i=1}^{5} 3$.

11. If a constant c is added to all the values of a data set, how does this change the median, mean, and mode?

12. If every value of a data set is multiplied by a constant c, how does this change the median, mean, and mode?

13. Let D be the data set $D = \{x_1, x_2, \ldots, x_n\}$ where the data values are listed in numerical order. If n is odd, show that the median is $x_{(n+1)/2}$. If n is even, show that the median is

$$\frac{1}{2}\left(x_{n/2} + x_{n/2+1}\right)$$

14. Evaluate the summation $\displaystyle\sum_{i=1}^{5} \frac{1}{i}$.   15. Evaluate the summation $\displaystyle\sum_{i=1}^{10} i$.

16. Show that the mean and median of two numbers always coincide.

17. Let $D = \{x_1, x_2, x_3\}$ be a data set in which the data values are listed in numerical order. If the mean and median coincide, show that

$$x_3 - x_2 = x_2 - x_1 \tag{8.5}$$

Conversely, if (8.5) holds, show that the mean and median coincide.

18. Let $D = \{x_1, x_2, x_3, x_4\}$ be a data set in which the data values are listed in numerical order. If the mean and median coincide, show that

$$x_1 + x_4 = x_2 + x_3 \tag{8.6}$$

Conversely, if (8.6) holds, show that the mean and median coincide.

19. Give an example of a data set for which the mean, median, and mode all coincide.

*20. Let m_1, m_2, m_3 denote the mean, median, and mode, respectively, of a data set. Show by examples that each of the following six possibilities can occur.

   (a) $m_1 < m_2 < m_3$ (b) $m_1 < m_3 < m_2$ (c) $m_2 < m_1 < m_3$

   (d) $m_2 < m_3 < m_1$ (e) $m_3 < m_1 < m_2$ (f) $m_3 < m_2 < m_1$

21. As n gets large, the summation

$$\sum_{i=1}^{n} \frac{1}{i(i+1)}$$

   is supposed to approach 1. Using a calculator, check to see if this result is reasonable.

22. As n gets large, the summation

$$\sum_{i=1}^{n} \frac{1}{2^i}$$

   is supposed to approach 1. Using a calculator, check to see if this result is reasonable.

23. The following formula holds.

$$\sum_{i=1}^{n} i = \frac{n(n+1)}{2} \tag{8.7}$$

   Using a calculator, check this result for $n = 20$.

24. Let D be the data set $D = \{1, 2, 3, \ldots, n\}$. Use (8.7) to find the mean of D. How does this compare with the median? (See Exercise 13.)

25. (a) If c is a constant, show that

$$\sum_{i=1}^{n} ca_i = c \sum_{i=1}^{n} a_i$$

   (b) Show that

$$\sum_{i=1}^{n} (a_i + b_i) = \sum_{i=1}^{n} a_i + \sum_{i=1}^{n} b_i$$

   (c) Is the following true?

$$\sum_{i=1}^{n} a_i b_i = \sum_{i=1}^{n} a_i \sum_{i=1}^{n} b_i$$

*26. Prove Equation (8.7). *Hint*: Letting

$$S = \sum_{i=1}^{n} i$$

   we have $2S = (1 + 2 + \cdots + n) + (n + \cdots + 2 + 1)$.

*27. Prove that

$$\sum_{i=1}^{n} \frac{1}{i(i+1)} = 1 - \frac{1}{n+1}$$

   Hint: Verify and use the identity

$$\frac{1}{i(i+1)} = \frac{1}{i} - \frac{1}{i+1}$$

*28. Prove that

$$\sum_{i=1}^{n} \frac{1}{2^i} = 1 - \frac{1}{2^n}$$

Hint: Letting

$$S = \sum_{i=1}^{n} \frac{1}{2^i}$$

we have

$$\frac{1}{2} S = \sum_{i=1}^{n} \frac{1}{2^{i+1}}$$

Now subtract.

*29. Schwarz's inequality states that

$$\left(\sum a_i b_i \right)^2 \leq \sum a_i^2 \sum b_i^2 \tag{8.8}$$

To prove (8.8) proceed as follows.

(a) For any constant c show that

$$0 \leq \sum (a_i - cb_i)^2 = \sum a_i^2 - 2c \sum a_i b_i + c^2 \sum b_i^2 \tag{8.9}$$

(b) If $\sum b_i^2 = 0$, then (8.8) holds. If $\sum b_i^2 \neq 0$, let

$$c = \frac{\sum a_i b_i}{\sum b_i^2}$$

and substitute into (8.9).

$\boxed{8.3}$ Deviation

Statistical thinking will one day be as necessary for efficient citizen-ship as the ability to read or write.
　　　　　　　　　—H. G. WELLS (1866–1946), British Novelist

We have seen in Section 8.2 that the mean is a sensitive indicator of the central tendency or average value of a data set (or a random variable). But a single number usually cannot describe a data set very accurately. What we need is an additional number that indicates the degree to which the data are dispersed or spread out. Such a number would give a measure of the variation or deviation of the data about the mean. For example, Table 8.11 tabulates two data sets. They both have mean 6, but the data sets are quite different. In the first set, 6 is fairly representative, while in the second the values are dispersed considerably below and above this value.

Table 8.12

	Janet	George
1st	94	91
2nd	82	94
3rd	83	75
4th	66	76
5th	85	76
Mean	82	82
Median	83	76
Range	28	18

Table 8.11

1st set	2nd set
3	0
5	1
6	6
7	11
9	12

Range

The crudest measure of the dispersion of a data set is its range.

> The **range** of a data set is defined as the difference between the largest and smallest values in the set.

For example, in Table 8.11, the ranges of the two data sets are 6 and 12, respectively. The range is not a very sensitive measure of dispersion and can be misleading. As an illustration, Janet and George had the test scores listed in Table 8.12 for five exams in their English course.

Both students have the same mean score 82, but Janet's median score is 83 and George's median is 76. Comparing the ranges of the scores seems to indicate that George is a more consistent student than Janet. However, this difference is mainly due to Janet's low score on the fourth exam (she had a bad cold that day). If we disregard the fourth exam, Janet's range is only 9 while George's remains at 18. Janet's mean score of 82 is less representative than George's because she was pulled down by one bad day.

The range can be misleading because it depends strongly on the largest and smallest values and these may not be typical values. A more accurate gauge of dispersion is given by the **standard deviation**. It is best to build up to the definition of the standard deviation slowly. Suppose we have a data set $D = \{x_1, x_2, \ldots, x_n\}$ with mean m. Each data value x_i will have a **deviation from the mean** (**deviation** for short) $x_i - m$. This gauges the dispersion from the mean for each individual data value. We would like a gauge of the total deviation from the mean for the whole data set. One way of doing this would be to measure how much the data values deviate from the mean on the average. Unfortunately we cannot define this to be the mean of the values $x_i - m$, $i = 1, 2, \ldots, n$, since this would give

$$\frac{1}{n} \sum_{i=1}^{n} (x_i - m) = \frac{1}{n} \sum_{i=1}^{n} x_i - \frac{1}{n} \sum_{i=1}^{n} m = m - m = 0$$

This provides the useless value 0 no matter what the data set.

Variance

The problem with the previous attempt is that the numbers $x_i - m$ will be both positive and negative and cancel out to zero when summed. In order to obtain a positive quantity, we define the **squared deviation** $(x_i - m)^2$ for the data value x_i. The **variance** (denoted var) of a data set is defined as the mean of the squared deviations. We thus have

$$\text{var} = \frac{1}{n} \sum_{i=1}^{n} (x_i - m)^2 \tag{8.10}$$

Standard Deviation

Since we have squared the deviations $(x_i - m)$ to obtain the variance, the variance can be quite large. To bring it back to a more reasonable value, it is traditional to take its square root.

> We define the **standard deviation** (denoted σ, the Greek lowercase letter sigma) by $\sigma = \sqrt{\text{var}}$.

The standard deviation is thus the square root of the mean of the squared deviations. It is the most common gauge for measuring the spread or dispersion of a data set.

EXAMPLE 1 Compute the standard deviations of the data sets in Table 8.11.
Solution. We first perform the computations in Tables 8.13 and 8.14.

Table 8.13

Value x_i	Deviation $x_i - 6$	Squared deviation $(x_i - 6)^2$
3	−3	9
5	−1	1
6	0	0
7	1	1
9	3	9

Table 8.14

Value x_i	Deviation $x_i - 6$	Squared deviation $(x_i - 6)^2$
0	−6	36
1	−5	25
6	0	0
11	5	25
12	6	36

From the last column of Table 8.13 we have

$$\text{var} = \frac{1}{5}(9 + 1 + 0 + 1 + 9) = 4$$

Hence, $\sigma = \sqrt{\text{var}} = 2$. From the last column of Table 8.14 we have

$$\text{var} = \frac{1}{5}(36 + 25 + 0 + 25 + 369) = 24.404$$

Hence, $\sigma = \sqrt{\text{var}} = 4.94$.   □

EXAMPLE 2   We saw that the range may not be a good indicator of dispersion for Janet's and George's test scores. Is the standard deviation better?

Solution.   Using the mean of 82, we compute σ for Janet and George from Table 8.15.

Table 8.15

Janet	Deviation	Squared deviation	George	Deviation	Squared deviation
94	12	144	91	9	81
82	0	0	94	12	144
83	1	1	75	−7	49
66	−16	256	76	−6	36
85	3	9	76	−6	36

For Janet we have var $= \frac{1}{5}(144 + 0 + 1 + 256 + 9) = 82$ and $\sigma = 9.06$. For George we obtain var $= \frac{1}{5}(81 + 144 + 49 + 36 + 36) = 69.20$ and $\sigma = 8.32$. Thus, their standard deviations are much closer than their ranges.   □

Second Moment

A quantity that is closely related to the variance is the second moment. For a data set $D = \{x_1, x_2, \ldots, x_n\}$, the **second moment** (denoted m_2) is the mean of the squared values. Thus,

$$m_2 = \frac{1}{n}\sum_{i=1}^{n} x_i^2 \tag{8.11}$$

(The mean is sometimes called the **first moment**.) The second moment can be used to compute var and σ. Applying (8.10) and the definition of m we have

$$\text{var} = \frac{1}{n}\sum(x_i - m)^2 = \frac{1}{n}\sum(x_i^2 - 2x_i m + m^2)$$

$$= \frac{1}{n}\sum x_i^2 - \frac{2}{n}m\sum x_i + \frac{1}{n}\sum m^2$$

$$= m_2 - 2m^2 + m^2 = m_2 - m^2$$

The formula

$$\text{var} = m_2 - m^2 \tag{8.12}$$

is frequently easier to apply than (8.10) because we don't have to compute all the subtractions. For example, let us use (8.12) to compute var for the data set in Table 8.13. We have

$$m_2 = \frac{1}{5}(9 + 25 + 36 + 49 + 81) = 40$$

Since $m = 6$ we have var $= m_2 - m^2 = 40 - 36 = 4$.

If data are given in terms of a frequency distribution, then (8.10) becomes

$$\text{var} = \frac{\sum (x_i - m)^2 f_i}{\sum f_i} \tag{8.13}$$

We can again use (8.12), where now the second moment is given by

$$m_2 = \frac{\sum x_i^2 f_i}{\sum f_i} \tag{8.14}$$

EXAMPLE 3 John's grades and corresponding credit hours are shown in Table 8.16. Find John's grade point average and standard deviation.

Table 8.16

Grade	Credit hours
F	3
D	6
C	60
B	45
A	36

Solution. Four grade points correspond to an A, three for a B, two for a C, one for a D, and zero for an F. The grade point average (GPA) is the mean of the grade points where the frequency distribution coincides with the credit hour distribution. Hence,

$$\text{GPA} = m = \frac{\sum x_i f_i}{\sum f_i} = \frac{0 \cdot 3 + 1 \cdot 6 + 2 \cdot 60 + 3 \cdot 45 + 4 \cdot 36}{3 + 6 + 60 + 45 + 36} = \frac{405}{150} = 2.70$$

Applying (8.14) we have

$$m_2 = \frac{\sum x_i^2 f_i}{\sum f_i} = \frac{0 \cdot 3 + 1 \cdot 6 + 4 \cdot 60 + 9 \cdot 45 + 16 \cdot 36}{3 + 6 + 60 + 45 + 36} = \frac{1227}{150} = 8.18$$

Hence, var $= m_2 - m^2 = 8.18 - 7.29 = 0.89$ and $\sigma = \sqrt{\text{var}} = 0.94$. $\qquad\square$

To compute var and m_2 for data given in terms of class intervals, we replace x_i in (8.13) or (8.14) by the midpoint t_i of each class interval. We thus have

$$\text{var} = \frac{\sum (t_i - m)^2 f_i}{\sum f_i} \tag{8.15}$$

and

$$m_2 = \frac{\sum t_i^2 f_i}{\sum f_i} \tag{8.16}$$

EXAMPLE 4 Compute the standard deviation for the frequency distribution in Table 8.5.

Solution.   The mean was found in Section 8.2 to be 68. To apply (8.15) we use Table 8.17.

Table 8.17

t_i	Frequency	$t_i - 68$	$(t_i - 68)^2$
61	2	−7	49
63	3	−5	25
65	2	−3	9
67	5	−1	1
69	6	1	1
71	4	3	9
73	3	5	25
75	1	7	49

Equation (8.15) then gives

$$\text{var} = \frac{49 \cdot 2 + 25 \cdot 3 + 9 \cdot 2 + 5 + 6 + 9 \cdot 4 + 25 \cdot 3 + 49}{26} = 13.92$$

Hence,

$$\sigma = \sqrt{\text{var}} = 3.73$$

We could also use (8.16) and (8.12) to obtain σ. In this case

$$m_2 = \frac{2(61)^2 + 3(63)^2 + 2(65)^2 + 5(67)^2 + 6(69)^2 + 4(71)^2 + 3(73)^2 + (75)^2}{26}$$

$$= 4637.92$$

and $$\text{var} = m_2 - m^2 = 4637.92 - 4624 = 13.92$$

We again obtain $\sigma = 3.73$.   □

In summary, σ measures the spread of the data about its mean and the mean is a reliable representative of the data only when σ is small. For example, Table 8.18 shows the salaries of people in two offices. We can assume that the highest paid person in each office is the boss. The mean salaries in the two offices are

$$m_A = \$52,200 \qquad \text{and} \qquad m_B = \$40,000$$

Table 8.18

Office A	Office B
$150,000	$45,000
36,000	42,000
30,000	40,000
25,000	38,000
20,000	35,000

Should we conclude that the people in Office A are better paid than those in Office B? Not at all. It is clear that the salaries in Office A are skewed by the boss's salary. In Office A the mean is not a good representative while in Office B it is. Let us check the standard deviations. These are

$$\sigma_A = \$49,187 \qquad \text{and} \qquad \sigma_B = \$3406$$

This gives a numerical gauge that the data spread in Office A is much larger than that in Office B.

z Score

Another important method for describing the spread of a data value from the mean is the z score. Suppose a data set has mean m and standard deviation σ. The **z score** of a data value x is defined as

$$z = \frac{x - m}{\sigma} \tag{8.17}$$

Notice that z is just the deviation of x, divided by σ. The z score is a measure of spread in standard units or without units. For example, if x and m are in dollars, then σ is in dollars and the division in (8.17) eliminates units, giving a pure number. Thus, z scores are useful in comparing two data sets with different units.

The z score of the boss in Office A (Table 8.18) is

$$z_A = \frac{150,000 - 52,200}{49,187} = 1.99$$

while the z score of the boss in Office B is

$$z_B = \frac{45,000 - 40,000}{3406} = 1.47$$

This points out the obvious fact that relative to the other workers in their offices, the boss of Office A is more highly paid than the boss of Office B. The situation in the next example is not so obvious.

EXAMPLE 5 Gail scores a 78 on the entrance exam at School A and an 84 at School B. The mean score at A was 72 with standard deviation 12 and the mean score at B was 77 with standard deviation 16. At which school did she receive the best score?

Solution. Gail's z score at School A is

$$z_A = \frac{78 - 72}{12} = 0.50$$

and at School B it is

$$z_A = \frac{84 - 77}{16} = 0.44$$

Since z_A is greater than z_B, the score at School A was superior in comparison with others who took the test. □

We have now defined various concepts for a data set or frequency distribution. The most important of these are the mean and standard deviation. It is worthwhile to remember that these concepts also apply to random variables. For example, the mean and standard deviation of a random variable are just the mean and standard deviation of its set of data values.

EXERCISES

1. If $m = 0$, show that $m_2 = $ var.
2. Verify that σ_A and σ_B given in the text are the correct standard deviations for Table 8.18.
3. Find σ for the data in each of the following.
   (a) Table 8.8   (b) Table 8.2   (c) Section 8.2, Example 1
   (d) Table 8.3   (e) Section 8.2, Example 5   (f) Section 8.2, Example 6
   (g) Table 8.9   (h) Section 8.2, Example 9   (i) Section 8.2, Example 10
   (j) Table 8.10   (k) Table 8.6   (l) Figure 8.4
* 4. Find σ for Table 8.7.
5. Find the z score for each of the values in (a) Table 8.8 and (b) Section 8.2, Example 1.
6. The **third moment** for a frequency distribution is defined by

$$m_3 = \frac{\sum x_i^3 f_i}{\sum f_i}$$

Find m_3 for (a) Table 8.8 and (b) Section 8.2, Example 1.

7. The **mean absolute deviation** for a frequency distribution is defined by

$$\text{mad} = \frac{\sum |x_i - m| f_i}{\sum f_i}$$

(Recall, $|a| = a$ if $a \geq 0$ and $|a| = -a$ if $a < 0$.) Find mad for (a) Table 8.8 and (b) Section 8.2, Example 1.

8. If a constant c is added to every value of a data set, does this change var or σ?

9. If every value of a data set is multiplied by a constant c, how does this change var and σ?

10. If the data set $\{x_1, x_2\}$ has mean 0 and var 1, find x_1 and x_2.

11. If the data set $\{x_1, x_2\}$ has mean 1 and var 1, find x_1 and x_2.

12. If $\sigma = 0$ for a random variable, what can you say about X?

13. Let $D_1 = \{x_1, x_2, \ldots, x_n\}$ and $D_2 = \{x_1^2, x_2^2, \ldots, x_n^2\}$ be two data sets. Show that the mean of D_2 is the second moment of D_1.

14. If X is a random variable, show that the second moment of X is the mean of X^2.

*15. Use Schwarz's inequality to show that mad $\leq \sigma$. (See Exercise 7 of this section and Exercise 29 of Section 8.2.)

*16. Since var $= m_2 - m^2$ we must have $m^2 \leq m_2$. Show that $m^2 \leq m_2$ directly using Schwarz's inequality. (See Exercise 29 of Section 8.2.)

*17. Let X be a random variable with mean m and let M be the random variable defined by $M(s) = m$ for all s in the sample space. Show that the variance of X is the mean of $(X - M)^2$.

8.4 | Continuous Random Variables

If you do mathematics every day, it seems the most natural thing in the world. If you stop to think about what you are doing and what it means, it seems one of the most mysterious.
— P. J. DAVIS AND R. HERSH, Twentieth-Century Mathematicians

Probabilities

There is more to analyzing data than finding their central tendency and standard deviation. We frequently would like to know the probability that the data values lie within certain limits. For example, consider the frequency distribution in Table 8.5. What is the probability that a student's height lies in the interval $(66, 68]$? Since there are five values in this interval and there are a total of 26 students, the chances that a student chosen at random has a height lying within $(66, 68]$ is $5/26 = 0.19$. We say that the **probability** of $(66, 68]$ is 0.19 and write

$$P(66, 68] = 0.19$$

If Y is the random variable that associates students to height, we also write

$$P(66 < Y \leq 68) = 0.19$$

The histogram in Figure 8.3 gives a visual means for describing such probabilities. Can you see that $P(66, 68]$ is the area $A(66, 68)$ of the rectangle under the $(66, 68]$

bar divided by the total area A of the histogram? Indeed, $A(66, 68) = 5 \cdot 2 = 10$ and since $A(60, 62) = 2 \cdot 2$, $A(62, 64) = 2 \cdot 3, \ldots$, we see that $A = 2 \cdot 26 = 52$. Hence,

$$\frac{A(66, 68)}{A} = \frac{10}{52} = 0.19$$

We thus see that probabilities can be computed by comparing areas. However, in the present case, this method only works for probabilities of class intervals. For the continuous random variables that we consider next, this method can be applied to any interval.

The random variables and data sets we have considered up to now were discrete; that is, their values were isolated numbers. However, in many applications, it is important to consider continuous random variables and data sets. These have values that range over a continuous interval, that is, over a set of numbers x satisfying $a \le x \le b$ for some numbers $a < b$. We call such a set a **closed interval** and denote it by $[a, b]$. Of course, since we are only human, any data set that we obtain in practice must be finite, so strictly speaking any statistical random variable or data set is discrete. However, if we have a large number of data values that are very close together, it frequently simplifies matters to imagine that they are continuous. If nothing else, this is often a good approximation. For a continuous random variable X, the probability that X has a value in the closed interval $[a, b]$ is denoted

$$P[a, b] = P(a \le X \le b)$$

Corresponding to a continuous random variable X there is a nonnegative function g called a **density function** such that $P(a \le X \le b)$ is the area under the graph g from a to b.

Square Target

As an example, suppose we throw a dart at a square target whose sides have length 10 inches. Let X denote the random variable that associates the point on the target punctured by the dart with its distance x from the left edge of the target (Figure 8.5). Assume we are not aiming at any particular place and are just throwing the dart at random. The values of X will be numbers in the interval $[0, 10]$. If we throw the dart a few times, then X can be considered to be a discrete random variable. However, if we throw it many times, then its values will begin to "fill" the interval $[0, 10]$, and it is a good approximation to assume that it is continuous.

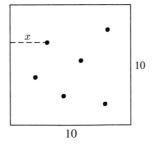

Figure 8.5

Photo 8.2

The points at which a dart strikes a target is an example of a continuous random variable.

Since the dart is randomly thrown, X is just as likely to have one value in $[0, 10]$ as any other value. We define the **density function** $g(x)$ for this distribution to be $g(x) = 1/10$ for all x in $[0, 10]$. Figure 8.6 is the graph of the function $g(x)$. We call $g(x)$ a **uniform** or **constant** density function on $[0, 10]$. If $0 \leq a < b \leq 10$ are arbitrary numbers in $[0, 10]$, then the probability that X has a value in $[a, b]$ is

$$P[a, b] = P(a \leq X \leq b) = \frac{b - a}{10} \tag{8.18}$$

Thus, $P[a, b]$ is the area under the graph of g between a and b.

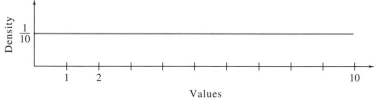

Figure 8.6

Circular Target

Now suppose we throw our dart at a circular target of radius 10 inches. Let X be the random variable that associates the point of the target punctured by the dart with its distance from the center. Again, we are not aiming and the dart can hit the target anywhere. We conclude that the probability that the dart lands in a region of the target is the area of that region divided by the total area 100π of the target. Hence, the probability that X has a value in $[a, b]$ where $0 \leq a < b \leq 10$ is the area of the ring with inner radius a and outer radius b divided by 100π (Figure 8.7). We thus have

$$P[a, b] = P(a \leq X \leq b) = \frac{\pi(b^2 - a^2)}{100\pi} = \frac{1}{100}(b^2 - a^2) \tag{8.19}$$

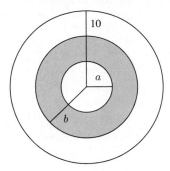

Figure 8.7

What is the density function $g(x)$ for this distribution? It turns out that $g(x) = x/50$ for all x in $[0, 10]$. This function is called a **linear** density function on $[0, 10]$ since its graph is a straight line as pictured in Figure 8.8.

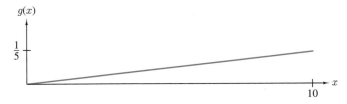

Figure 8.8

To show that $g(x) = x/50$ is actually the density function, we show that the area $A[a, b]$ under the graph of $g(x)$ from a to b coincides with $P[a, b]$. Now $A[a, b]$ is the area of the trapezoid in Figure 8.9. Since the area of such a trapezoid is its base times its mean height (see Exercise 9 or Section 5.3) we have

$$A[a, b] = (b - a) \cdot \frac{1}{2} \left(\frac{b}{50} + \frac{a}{50} \right) = \frac{1}{100} (b^2 - a^2)$$

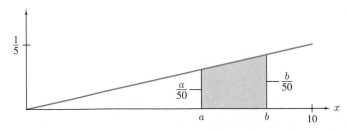

Figure 8.9

EXAMPLE 1 What is the probability that a randomly thrown dart hits the target between 3 and 5 inches from the center?

Solution.

$$P[3, 5] = \frac{1}{100}(5^2 - 3^2) = 0.16 \qquad \square$$

Density Function

> In general, a **density function** is a real-valued function g that is defined for all real numbers x and satisfies the following conditions.
>
> (a) $g(x) \geq 0$ for all x.
> (b) The area under the graph of g is 1.

In short, a density function is a nonnegative function whose graph has unit area. A few words of clarification are in order. First, g must be well-behaved enough that the concept of area under its graph is meaningful. (Such functions are called **integrable**.) Second, the density functions we have previously considered were defined only on finite intervals. However, we can extend them to all real numbers by defining them to be zero outside this interval and nothing is changed.

> We say that a random variable X (or a data set) is **continuous** if there is a density function g such that the probability
>
> $$P[a, b] = P(a \leq X \leq b)$$
>
> of any interval $[a, b]$ is the area under the graph of g from a to b.

It does not matter whether the endpoints are included in the interval since the probability of a single point is always zero for a continuous random variable (or data set). (From now on, anything we say about random variables applies to data sets.) This is because the area under a single point is zero! Of course, the reason for Condition (b) in the definition of density function is that we want $P(-\infty < X < \infty) = 1$.

If X is a continuous random variable, how do we find its mean and standard deviation? That is a good question, and we shall discuss its answer in the next section. A **mode** of X is simply a value x_0 for which $g(x_0)$ is a maximum. The **median** of X is the value x_0 that satisfies

$$P(-\infty < X \leq x_0) = \frac{1}{2} \tag{8.20}$$

Since the total probability is 1, it follows that

$$P(x_0 \leq X < \infty) = \frac{1}{2}$$

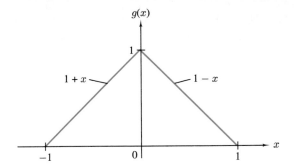

Figure 8.10

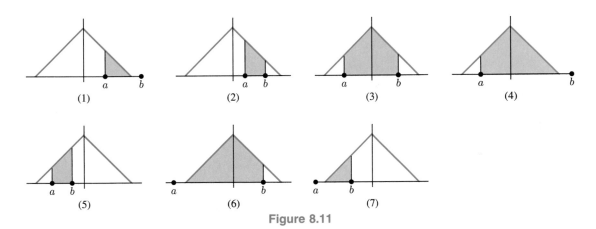

Figure 8.11

Hence, it is just as likely that X has a value below x_0 as above. It is usually easy to find the median of X. For the uniform density of Figure 8.6, it is clear that the median is 5 and every value in [0, 10] is a mode. For the linear density of Figure 8.8, it is clear that the mode is 10. Let us compute the median x_0. In order to satisfy (8.20) we must have

$$\frac{1}{100}(x_0^2 - 0^2) = \frac{1}{2}$$

Hence, $x_0^2 = 50$ so $x_0 = 7.07$.

We close this section with another example. Let g be the function whose graph is shown in Figure 8.10. Since g is nonnegative and the area under its graph is 1, g is a density function. If X is a random variable with this density and $a < b$ are numbers, what is $P(a \le X \le b)$? This depends on where a and b are. The different possible cases are shown in Figure 8.11.

It would be redundant to compute the probabilities for all of these cases. Because of the symmetry of g, any one of the seven cases can be used to compute all the others. In fact, if we find $P(0 \le X \le b)$, we can find the others. If $b \ge 1$, then the area of the triangle gives

$$P(0 \le X \le b) = \frac{1}{2}$$

If $0 \leq b \leq 1$, then we have a trapezoid with base length b and heights 1 and $1 - b$. Hence,

$$P(0 \leq X \leq b) = \frac{b}{2}(1 + 1 - b) = \frac{b}{2}(2 - b) \qquad (8.21)$$

Now consider Case 1 in Figure 8.11. We can write this probability as $P(X \geq a)$. We can see from the picture that

$$P(X \geq a) = \frac{1}{2} - P(0 \leq X \leq a)$$

Applying (8.21) gives

$$P(X \geq a) = \frac{1}{2} - \frac{a}{2}(2 - a) = \frac{1}{2}(1 - 2a + a^2) = \frac{1}{2}(1 - a)^2 \qquad (8.22)$$

Let us consider Case 3 in Figure 8.11. First notice that

$$P(a \leq X \leq b) = P(a \leq X \leq 0) + P(0 \leq X \leq b)$$

By the symmetry of g we have

$$P(a \leq X \leq 0) = P(0 \leq X \leq -a)$$

(remember a is negative so $-a$ is positive). We can now apply (8.21) again to get

$$P(a \leq X \leq b) = P(0 \leq X \leq -a) + P(0 \leq X \leq b)$$

$$= -\frac{a}{2}(2 + a) + \frac{b}{2}(2 - b) \qquad (8.23)$$

$$= \frac{1}{2}\left[b(2 - b) - a(2 + a)\right]$$

EXAMPLE 2 Find $P(X \geq 1/2)$.
  Solution. Applying (8.22) gives

$$P\left(X \geq \frac{1}{2}\right) = \frac{1}{2}\left(1 - \frac{1}{2}\right)^2 = \frac{1}{8} \qquad \square$$

EXAMPLE 3 Find $P(0 \leq X \leq 1/2)$.
  Solution. Applying (8.21) gives

$$P\left(0 \leq X \leq \frac{1}{2}\right) = \frac{1}{4}\left(2 - \frac{1}{2}\right) = \frac{3}{8}$$

Another way of doing this is to use Example 2 to obtain

$$P\left(0 \leq X \leq \frac{1}{2}\right) = \frac{1}{2} - P\left(X \geq \frac{1}{2}\right) = \frac{1}{2} - \frac{1}{8} = \frac{3}{8} \qquad \square$$

EXAMPLE 4 Find $P(-1/3 \leq X \leq 2/3)$.
  Solution. Using (8.23) we have

$$P\left(-\frac{1}{3} \leq X \leq \frac{2}{3}\right) = \frac{1}{2}\left[\frac{2}{3}\left(2 - \frac{2}{3}\right) + \frac{1}{3}\left(2 - \frac{1}{3}\right)\right] = \frac{13}{18} \qquad \square$$

EXERCISES

1. For the frequency distribution in Table 8.5, find the following.
   (a) $P(68, 70]$
   (b) $P(66, 70]$
   (c) The probability that a student's height is greater than 66 inches

2. Why does equation (8.18) hold?

3. For the uniform density of Figure 8.6, find the probability that the dart hits the target more than 7 inches from the left edge.

Exercises 4–6 apply to the circular target example in the text.

4. What is the probability that the dart hits the target within 6 inches from the center?

5. What is the probability that the dart hits the target more than 6 inches from the center?

6. Which of the following is more probable? The dart hits the target within 5 inches from the center. The dart hits the target more than 8 inches from the center.

7. How would (8.19) change if the target had radius c?

8. What is the solution of Example 1 for a target of radius 5 inches?

9. Show that the area of the trapezoid in Figure 8.12 is $(1/2)c(a + b)$ by dividing it into a rectangle and a triangle.

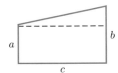

Figure 8.12

10. If the circular target considered in the text has radius c, what is the median of the corresponding linear density?

11. If X is a continuous random variable, verify that

$$P(a \leq X \leq b) = P(a < X \leq b) = P(a \leq X < b) = P(a < X < b)$$

12. If X is a continuous random variable, verify that

$$P(X \geq a) = 1 - P(X \leq a)$$

13. If X is a continuous random variable and $a < b < c$, show that

$$P(a \leq X \leq c) = P(a \leq X \leq b) + P(b \leq X \leq c)$$

14. Let X be a random variable whose density function is given in Figure 8.10. Use (8.21) to find the following.
   (a) $P\left(\frac{1}{4} \leq X \leq \frac{1}{2}\right)$ (b) $P\left(X \geq -\frac{1}{5}\right)$ (c) $P\left(-\frac{3}{4} \leq X \leq -\frac{1}{3}\right)$
   (d) $P\left(X \leq \frac{1}{3}\right)$ (e) $P\left(X \geq -\frac{2}{3}\right)$

15. For the density function in Figure 8.10 it turns out that $\sigma = 0.41$. What is the probability that a value lies within $1\ \sigma$ of 0? Within 2σ of 0?

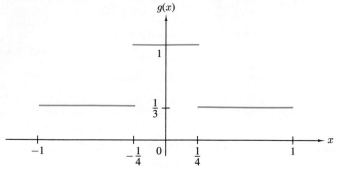

Figure 8.13

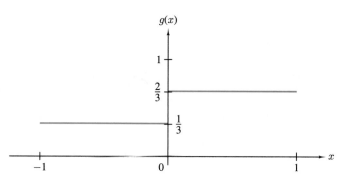

Figure 8.14

Exercises 16–20 apply to Figure 8.11.

16. In Case 2 verify that $P(a \le X \le b) = \frac{1}{2}(b-a)(2-a-b)$.

17. In Case 4 verify that $P(a \le X \le b) = \frac{1}{2}[1-a(2+a)]$.

18. In Case 5 verify that $P(a \le X \le b) = \frac{1}{2}(b-a)(2+a+b)$.

19. In Case 6 verify that $P(a \le X \le b) = \frac{1}{2}[1+b(2-b)]$.

20. In Case 7 verify that $P(a \le X \le b) = \frac{1}{2}(1+b)^2$.

Exercises 21–23 refer to the function g in Figure 8.13 and its corresponding random variable X.

21. Verify that g is a density function.

22. Find the following probabilities.
   (a) $P\left(-\frac{1}{8} \le X \le \frac{1}{4}\right)$ (b) $P\left(-\frac{1}{2} \le X \le \frac{3}{4}\right)$ (c) $P\left(X \ge \frac{1}{8}\right)$
   (d) $P\left(X \le \frac{3}{4}\right)$ (e) $P\left(X \le -\frac{1}{2} \text{ or } X \ge \frac{3}{4}\right)$

23. Find b such that
   (a) $P(0 \le X \le b) = 1/4$ (b) $P(0 \le X \le b) = 1/8$
   (c) $P(0 \le X \le b) = 3/8$ (d) $P(-b \le X \le b) = 5/6$

Exercises 24–28 refer to the function g in Figure 8.14 and its corresponding random variable X.

24. Show that g is a density function and find the median of X.

25. Find $P\left(-\frac{1}{2} \le X \le \frac{1}{2}\right)$.   26. Find $P(X \le -\frac{1}{3})$ and $P(X \ge \frac{1}{3})$.

27. Draw a figure for each of the seven cases analogous to Figure 8.11 and find $P(a \le X \le b)$ for each case.

28. Find b such that
    (a) $P(0 \le X \le b) = \frac{1}{2}$ (b) $P(b \le X \le 0) = \frac{1}{4}$ (c) $P(-b \le X \le b) = \frac{1}{2}$
*29. For the density function in Figure 8.10, let $0 \le c \le 1/2$, and find a b such that
    $P(0 \le X \le b) = c$.

8.5 Area (Optional)

Mathematical rigor is like clothing: in its style it ought to suit the occasion, and it diminishes comfort and restricts freedom of movement if it is either too loose or too tight.
            — G. F. SIMMONS, Twentieth-Century Mathematician

Since areas are so important in the study of continuous random variables, it behooves us to take a closer look. Recall that we have already studied areas under curves in Section 5.5; it might be helpful for you to review that section. We discussed in Section 5.5 that the area between the graph of a function f and the x-axis is given by the integral $\int_{-\infty}^{\infty} f(x)$. We use the infinite limits $-\infty$ and ∞ because the domain of the function may be unbounded. However, in practice the function f usually vanishes outside a finite interval and the infinite limits can be replaced by finite ones.

From our previous discussion, we can now use this concept of area to obtain statistical quantities. Let g be the density function for a random variable X. Then the probability that X has a value between a and b is

$$P(a \le X \le b) = \int_{a}^{b} g(x) \tag{8.24}$$

It can be shown that the mean, second moment, and variance of X are

$$m = \int_{-\infty}^{\infty} xg(x) \tag{8.25}$$

$$m_2 = \int_{-\infty}^{\infty} x^2 g(x) \tag{8.26}$$

$$\text{var} = \int_{-\infty}^{\infty} (x - m)^2 g(x) \tag{8.27}$$

To evaluate these integrals we shall use the following formula from Section 5.5.

$$\int_a^b x^n = \frac{b^{n+1}}{n+1} - \frac{a^{n+1}}{n+1}$$

(8.28)

We now apply these formulas to find the statistical quantities for random variables with certain density functions. Consider the uniform density in Figure 8.6. In this case, $g(x) = 1/10$ for x in $[0, 10]$ and $g(x) = 0$ otherwise. Applying (8.24) and (8.28) we have for $0 \leq a < b \leq 10$

$$P(a \leq X \leq b) = \int_a^b \frac{1}{10} = \frac{1}{10} \int_a^b 1 = \frac{b - a}{10}$$

If a or b lay outside the interval $[0, 10]$, we would get different answers. For example, if $a < 0$ and $0 < b < 10$, then

$$P(a \leq X \leq b) = \int_a^b g(x) = \int_a^0 g(x) + \int_0^b g(x) = \frac{1}{10} \int_0^b 1 = \frac{b}{10}$$

Of course, we do not need (8.28) for these calculations; we already know areas of rectangles. However, (8.28) is useful for finding m and m_2. In fact, from (8.28) we have

$$m = \int_{-\infty}^{\infty} xg(x) = \frac{1}{10} \int_0^{10} x = \frac{1}{10} \left(\frac{10^2}{2} - \frac{0^2}{2} \right) = 5$$

$$m_2 = \int_{-\infty}^{\infty} x^2 g(x) = \frac{1}{10} \int_0^{10} x^2 = \frac{1}{10} \left(\frac{10^3}{3} - \frac{0^3}{3} \right) = 33.33$$

The proof of the formula var $= m_2 - m^2$ is similar to that used for discrete data sets [see work preceding (8.11)]. We then obtain

$$\text{var} = m_2 - m^2 = 33.33 - 25 = 8.33$$

and the standard deviation is

$$\sigma = \sqrt{\text{var}} = 2.89$$

Next, consider the linear density of Figure 8.8. In this case $g(x) = x/50$ for x in $[0, 10]$ and $g(x) = 0$, otherwise. If $0 \leq a < b \leq 10$, we have

$$P(a \leq X \leq b) = \frac{1}{50} \int_a^b x = \frac{1}{50} \left(\frac{b^2}{2} - \frac{a^2}{2} \right) = \frac{1}{100} (b^2 - a^2)$$

a formula we have already derived. For m and m_2 we have

$$m = \int_{-\infty}^{\infty} xg(x) = \frac{1}{50} \int_0^{10} x^2 = \frac{1}{50} \left(\frac{10^3}{3} \right) = 6.67$$

$$m_2 = \int_{-\infty}^{\infty} x^2 g(x) = \frac{1}{50} \int_0^{10} x^3 = \frac{1}{50} \left(\frac{10^4}{4} \right) = 50$$

We then obtain

$$\text{var} = m_2 - m^2 = 50 - 44.44 = 5.66$$

and

$$\sigma = \sqrt{\text{var}} = 2.36$$

We now study the triangular density of Figure 8.10. In this case, the density function g has the form

$$g(x) = \begin{cases} 1 + x & \text{for } -1 \leq x \leq 0 \\ 1 - x & \text{for } 0 \leq x \leq 1 \\ 0 & \text{otherwise} \end{cases}$$

Let us derive the Case 3 formula of Figure 8.11. Because g has different forms to the left of 0 and right of 0, we must break the area into two parts.

$$P(a \leq X \leq b) = \int_a^b g(x) = \int_a^0 g(x) + \int_0^b g(x)$$

$$= \int_a^0 (1+x) + \int_0^b (1-x) = \int_a^0 1 + \int_a^0 x + \int_0^b 1 - \int_0^b x$$

$$= (0 - a) + \left(0 - \frac{a^2}{2}\right) + (b - 0) - \left(\frac{b^2}{2} - 0\right)$$

$$= b - \frac{b^2}{2} - a - \frac{a^2}{2} \tag{8.29}$$

which again we have already derived. Since g is symmetric about the vertical axis, $m = 0$. More precisely, for each value x there is a value $-x$ and these cancel when evalutating (8.25). The second moment is

$$m_2 = \int_{-1}^1 x^2 g(x) = \int_{-1}^0 x^2(1+x) + \int_0^1 x^2(1-x)$$

$$= \int_{-1}^0 x^2 + \int_{-1}^0 x^3 + \int_0^1 x^2 - \int_0^1 x^3 = \frac{1}{3} - \frac{1}{4} + \frac{1}{3} - \frac{1}{4} = \frac{1}{6}$$

Hence,

$$\sigma = \frac{1}{\sqrt{6}} = 0.41$$

We now consider a generalization of this last example. Let $c > 0$ be a constant and define the function

$$g_c(x) = \begin{cases} (c+x)/c^2 & \text{for } -c \leq x \leq 0 \\ (c-x)/c^2 & \text{for } 0 \leq x \leq c \\ 0 & \text{otherwise} \end{cases}$$

The graph of g_c is shown in Figure 8.15. Notice that the area under g_c is 1, so g_c is a density function.

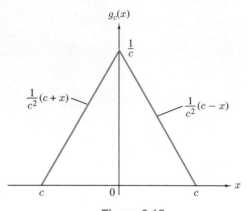

Figure 8.15

As before, $m = 0$ and

$$m_2 = \int_{-c}^{c} x^2 g(X) = \frac{1}{c^2} \int_{-c}^{0} x^2(c+x) + \frac{1}{c^2} \int_{0}^{c} x^2(c-x)$$

$$= \frac{1}{c} \int_{-c}^{0} x^2 + \frac{1}{c^2} \int_{-c}^{0} x^3 + \frac{1}{c} \int_{0}^{c} x^2 - \frac{1}{c^2} \int_{0}^{c} x^3$$

$$= \frac{1}{c} \cdot \frac{c^3}{3} - \frac{1}{c^2} \cdot \frac{c^4}{4} + \frac{1}{c} \cdot \frac{c^3}{3} - \frac{1}{c^2} \cdot \frac{c^4}{4} = \frac{c^2}{6}$$

Hence, $\sigma = c/\sqrt{6} = 0.41c$. By giving c different values we can obtain any desired σ and thus any desired spread. We can thus obtain "thin" or "fat" density functions. Figures 8.16 and 8.17 exhibit g_c for $c = \sqrt{6} = 2.45$ and $c = 1/\sqrt{6} = 0.41$.

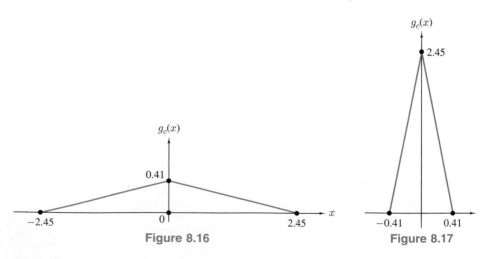

Figure 8.16   **Figure 8.17**

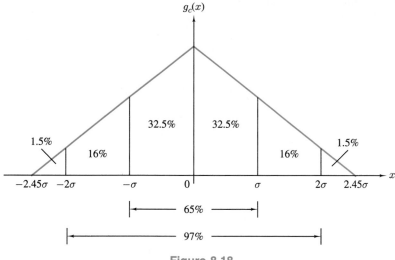

Figure 8.18

We can also translate g_c to the right or left and get any desired mean. For example, $g_c(x - 2)$ has mean 2 and $\sigma = 0.41c$. If we perform the calculation in (8.29) with g replaced by g_c we obtain

$$P(a \leq X \leq b) = \frac{1}{2c} \left[b \left(2 - \frac{b}{c} \right) - a \left(2 + \frac{a}{c} \right) \right] \tag{8.30}$$

whenever $-c \leq a \leq 0 \leq b \leq c$. Let us find the probability that a value lies within one and two standard deviations of the mean $m = 0$. Applying (8.30) gives

$$P(-\sigma \leq X \leq \sigma) = \frac{1}{2c} \left[\sigma \left(2 - \frac{\sigma}{c} \right) + \sigma \left(2 - \frac{\sigma}{c} \right) \right]$$

$$= \frac{\sigma}{c} \left(2 - \frac{\sigma}{c} \right) = \frac{1}{\sqrt{6}} \left(2 - \frac{1}{\sqrt{6}} \right) = 0.65$$

and

$$P(-2\sigma \leq X \leq 2\sigma) = \frac{2\sigma}{c} \left(2 - \frac{2\sigma}{c} \right) = \frac{2}{\sqrt{6}} \left(2 - \frac{2}{\sqrt{6}} \right) = 0.97$$

This shows that 65% of the data values lie within σ of the mean and 97% lie within 2σ of the mean. Of course, 100% lie within $\sqrt{6}\,\sigma = 2.45\sigma$ of the mean. This is illustrated in Figure 8.18.

EXERCISES

1. If $g(-x) = g(x)$ for all x, use (8.25) to show that $m = 0$.

2. If $g(-x) = g(x)$ for all x, show that

$$m_2 = 2 \int_0^\infty x^2 g(x)$$

3. For a continuous random variable show that var $= m_2 - m^2$.

4. Compute $P(a \leq X \leq b)$ using (8.24) for the triangular density of Figure 8.10 in the following cases of Figure 8.11.
   - (a) Case 1
   - (b) Case 2
   - (c) Case 4
   - (d) Case 5
   - (e) Case 6
   - (f) Case 7

5. Show that $m = 0$ for the triangular density of Figure 8.10 by evaluating (8.25).

6. Show that $m = 0$ for g_c (Figure 8.15) by evaluating (8.25).

7. Verify (8.30).

* 8. Graph the function $g_c(x - 2)$ and use (8.25) to show that the mean is 2.

9. Find σ for the density function in Figure 8.13.

10. Find m and σ for the density function in Figure 8.14.

Exercises 11–17 refer to the function g defined by

$$g(x) = \begin{cases} \frac{3}{4}(1 - x^2) & \text{if } -1 \leq x \leq 1 \\ 0 & \text{otherwise} \end{cases}$$

11. Graph g.

12. Show that g is a density function.

13. Evaluate (8.25) to show that $m = 0$.

14. For $-1 \leq a < b \leq 1$, find $P(a \leq X \leq b)$.

15. Find σ.

16. What is the probability that a value lies within σ of 0?

17. What is the probability that a value lies within 2σ of 0?

Exercises 18–24 refer to the function g defined by

$$g(x) = \begin{cases} 6x(1 - x) & \text{if } 0 \leq x \leq 1 \\ 0 & \text{otherwise} \end{cases}$$

18. Graph g.

19. Show that g is a density function.

20. Find m.

21. For $0 \leq a < b \leq 1$, find $P(a \leq X \leq b)$.

22. For $0 \leq a \leq 1$ and $b > 1$, find $P(a \leq X \leq b)$.

23. Find σ.

24. What is the probability that a value lies within σ of $1/2$?

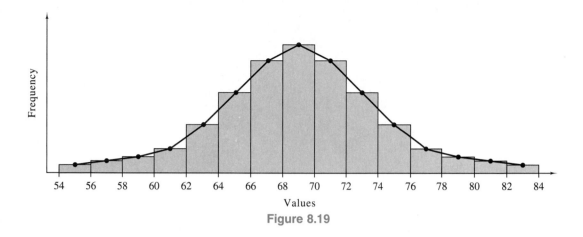

Figure 8.19

8.6 Normal Distributions

Everybody believes in the normal approximation, the experimentalists because they think it is a mathematical theorem, the mathematicians because they think it is an experimental fact.

—G. LIPPMANN (1845–1921), French Physicist

The normal density function (or normal distribution) appears in many applications. In fact, this distribution occurs in various fields of physical science, biology, sociology, psychology, business, and others. Whenever we have a large number of independent trials or experiments, the normal distribution frequently results. The normal density is similar to the triangular density of Figure 8.15 except that it is much smoother. Suppose we request the heights of a random sample of adult males. If we polled 26 adult males, we would obtain a histogram similar to Figure 8.3. However, if we polled a larger sample of say 10,000 we would get a smoother frequency polygon such as in Figure 8.19. If we polled an even larger sample and took smaller class intervals, the histogram (or frequency polygon) would be closely approximated by a constant times the normal density function.

In Figure 8.20 we have illustrated the normal density of mean 0 and standard deviation σ. Similar to the triangular density (Figure 8.18), for the normal density, 68% of the values lie within σ of the mean, 95% within 2σ, and over 99.7% within 3σ. The normal density is perfectly symmetric about the mean, but, unlike the triangular density, it never quite reaches zero although it approaches zero very closely after 3σ from the mean. In fact, the values further than 3σ from the mean are negligible. Like the triangular density (Figures 8.16 and 8.17), this density spreads out for large σ. Moreover, we can translate it to get any mean.

The formula for the normal density of mean m and standard deviation σ is

$$g(x) = \frac{1}{\sigma\sqrt{2\pi}} \, e^{-(x-m)^2/2\sigma^2}$$

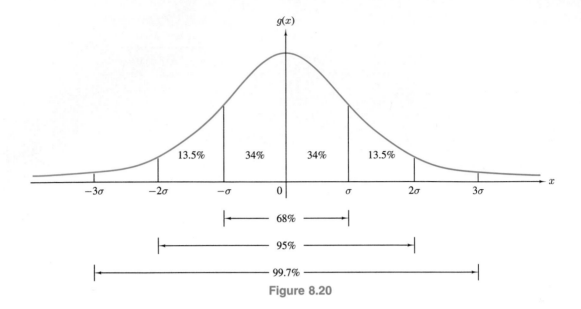

Figure 8.20

where $e = 2.71828$ is called the **natural exponential**. We have shown you this function so you will know how it looks. However, we shall not use this formula to find areas and probabilities. Not even calculus can be used to find exact areas for this function. These must be computed using a table. Such tables list the areas under the normal density between the mean and certain closely spaced values. It would be redundant to make a table for each m and σ, so a standard table is given only for $m = 0$ and $\sigma = 1$. We can convert areas from the standard table to desired areas for arbitrary m and σ using the z score $z = (x - m)/\sigma$. The reason this works is that the z score subtracts m to obtain a zero mean and divides by σ to obtain a unit standard deviation.

Table 8.19 gives the area under the normal density between z score 0 and z. This area is illustrated in Figure 8.21. The reason the table only goes to 3.09 is that the probability of larger values is negligible. Also, negative z values are not given because the 0 mean normal density is symmetric about the vertical axis so the area from 0 to $-z$ is the same as the area from 0 to z.

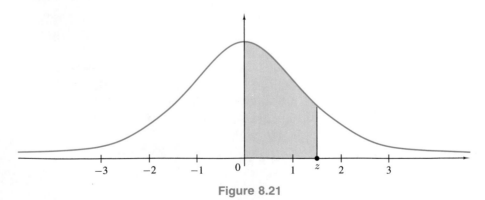

Figure 8.21

Table 8.19

z	0.00	0.01	0.02	0.03	0.04	0.05	0.06	0.07	0.08	0.09
0.0	.0000	.0040	.0080	.0120	.0160	.0199	.0239	.0279	.0319	.0359
0.1	.0398	.0438	.0478	.0517	.0577	.0596	.0636	.0675	.0714	.0753
0.2	.0793	.0832	.0871	.0910	.0948	.0987	.1026	.1064	.1103	.1141
0.3	.1179	.1217	.1255	.1293	.1311	.1368	.1406	.1443	.1480	.1517
0.4	.1554	.1591	.1628	.1664	.1700	.1736	.1772	.1808	.1844	.1879
0.5	.1915	.1950	.1985	.2019	.2054	.2088	.2123	.2157	.2190	.2224
0.6	.2257	.2291	.2324	.2357	.2389	.2422	.2454	.2486	.2517	.2549
0.7	.2580	.2611	.2642	.2673	.2704	.2734	.2764	.2794	.2823	.2852
0.8	.2881	.2910	.2939	.2967	.2995	.3023	.3051	.3078	.3106	.3133
0.9	.3159	.3186	.3212	.3238	.3264	.3289	.3315	.3340	.3365	.3389
1.0	.3413	.3438	.3461	.3485	.3508	.3531	.3554	.3577	.3599	.3621
1.1	.3643	.3665	.3686	.3708	.3729	.3749	.3770	.3790	.3810	.3830
1.2	.3849	.3869	.3888	.3907	.3925	.3944	.3962	.3980	.3997	.4015
1.3	.4032	.4049	.4066	.4082	.4099	.4115	.4131	.4147	.4162	.4177
1.4	.4192	.4207	.4222	.4236	.4251	.4265	.4279	.4292	.4306	.4319
1.5	.4332	.4345	.4357	.4370	.4382	.4394	.4406	.4418	.4429	.4441
1.6	.4452	.4463	.4474	.4484	.4495	.4505	.4515	.4525	.4535	.4545
1.7	.4554	.4564	.4573	.4582	.4591	.4599	.4608	.4616	.4625	.4633
1.8	.4641	.4649	.4656	.4664	.4671	.4678	.4686	.4693	.4699	.4706
1.9	.4713	.4719	.4726	.4732	.4738	.4744	.4750	.4756	.4761	.4767
2.0	.4772	.4778	.4783	.4788	.4793	.4798	.4803	.4808	.4812	.4817
2.1	.4821	.4826	.4830	.4834	.4838	.4842	.4846	.4850	.4854	.4857
2.2	.4861	.4864	.4868	.4871	.4875	.4878	.4881	.4884	.4887	.4890
2.3	.4893	.4896	.4898	.4901	.4904	.4906	.4909	.4911	.4913	.4916
2.4	.4918	.4920	.4922	.4925	.4927	.4929	.4931	.4932	.4934	.4936
2.5	.4938	.4940	.4941	.4943	.4945	.4946	.4948	.4949	.4951	.4952
2.6	.4953	.4955	.4956	.4957	.4959	.4960	.4961	.4962	.4963	.4964
2.7	.4965	.4966	.4967	.4968	.4969	.4970	.4971	.4972	.4973	.4974
2.8	.4974	.4975	.4976	.4977	.4977	.4978	.4979	.4979	.4980	.4981
2.9	.4981	.4982	.4982	.4983	.4984	.4984	.4985	.4985	.4986	.4986
3.0	.4987	.4987	.4987	.4988	.4988	.4989	.4989	.4989	.4990	.4990

We use the notation $P(a \leq z \leq b)$ for the probability that z is between a and b. Table 8.19 then gives the value of $P(0 \leq z \leq b)$ for z score b. As we have mentioned, it does not matter if we include endpoints since single data values have probability zero. So, for example,

$$P(a \leq z \leq b) = P(a < z \leq b) = P(a \leq z < b) = P(a < z < b)$$

We also use the notation $P(z \leq b)$ and $P(z \geq b)$ for the probability that $z \leq b$ and $z \geq b$, respectively. Notice that $P(z \geq 0) = 0.5$ and that $P(-a \leq z \leq 0) = P(0 \leq z \leq a)$.

Suppose we want to find $P(z \leq 1.85)$. We first look up 1.85 in Table 8.19 to obtain

$$P(0 \leq z \leq 1.85) = 0.4678$$

Since the area for $z \leq 0$ is 0.5 (Figure 8.22) we conclude that

$$P(z \leq 1.85) = P(z \leq 0) + P(0 \leq z \leq 1.85)$$
$$= 0.5000 + 0.4678 = 0.9678$$

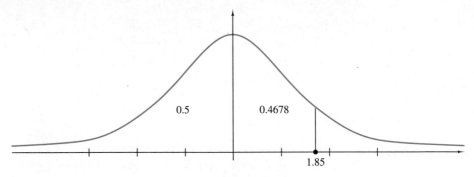

0.5   0.4678

1.85

Figure 8.22

EXAMPLE 1   Find $P(z \geq -0.5)$.
Solution.

$$P(z \geq -0.5) = P(-0.5 \leq z \leq 0) + P(z \geq 0)$$
$$= P(0 \leq z \leq 0.5) + 0.5000$$
$$= 0.1915 + 0.5000 = 0.6915 \qquad \square$$

EXAMPLE 2   Find $P(z \geq 1)$.
Solution.   Since the total area under the curve is 1, we have

$$P(z \geq 1) + P(z \leq 1) = 1$$

Hence,

$$P(z \geq 1) = 1 - P(z \leq 1) = 1 - P(z \leq 0) - P(0 \leq z \leq 1)$$
$$= 1 - 0.5000 - 0.3413 = 0.1587 \qquad \square$$

EXAMPLE 3   Find $P(z \leq -1.5)$.
Solution.

$$P(z \leq -1.5) = P(z \geq 1.5) = 1 -\!\!- P(z \leq 1.5)$$
$$= 1 - P(z \leq 0) - P(0 \leq z \leq 1.5)$$
$$= 1 - 0.5000 - 0.4332 = 0.0668 \qquad \square$$

EXAMPLE 4   Find $P(0.5 \leq z \leq 1.5)$.
Solution.

$$P(0.5 \leq z \leq 1.5) = P(0 \leq z \leq 1.5) - P(0 \leq z \leq 0.5)$$
$$= 0.4332 - 0.1915 = 0.2417 \qquad \square$$

Suppose a normally distributed random variable X has mean 150 and standard deviation 10 and we want to find the probability that $X \leq 165$. To use Table 8.19 we must convert to the z score for the value 165.

$$z = \frac{x - m}{\sigma} = \frac{165 - 150}{10} = 1.5$$

We then have

$$P(X \leq 165) = P(z \leq 1.5) = P(z \leq 0) + P(0 \leq z \leq 1.5)$$
$$= 0.5000 + 0.4332 = 0.9332$$

EXAMPLE 5　The grades on a final examination are known to be normally distributed with $m = 70$ and $\sigma = 12$. What is the probability that a student will make 90 or higher on the final?

Solution.　We first find the z score of 90.

$$z = \frac{90 - 70}{12} = 1.67$$

We then have

$$P(X \geq 90) = P(z \geq 1.67) = 1 - P(z \leq 1.67)$$
$$= 1 - P(z \leq 0) - P(0 \leq z \leq 1.67)$$
$$= 1 - 0.5000 - 0.4525 = 0.0475$$

Thus, 4.75% of the students will score 90 of better on the final. In other words, the probability that a student in the class chosen at random will score 90 or better is 0.0475.　　□

EXAMPLE 6　A brick manufacturer measures the length of each brick as it comes off the assembly line. The measurements are known to be normally distributed with $m = 8$ inches and $\sigma = 0.10$ inches. A brick passes inspection if its length is within the tolerance interval $[7.75, 8.20]$. What percentage of the bricks pass?

Solution.　The z scores for 7.75 and 8.20 are

$$\frac{7.75 - 8}{0.10} = -2.5 \quad \text{and} \quad \frac{8.20 - 8}{0.10} = 2$$

Hence,

$$P(7.75 \leq X \leq 8.20) = P(-2.5 \leq z \leq 2)$$
$$= P(0 \leq z \leq 2.5) + P(0 \leq z \leq 2)$$
$$= 0.4938 + 0.4772 = 0.971$$

Thus, 97.1% of the bricks pass.　　□

EXAMPLE 7　A dry cell battery manufacturer has found that the life of his batteries is normally distributed with $m = 120$ hours and $\sigma = 10$ hours. He wants to advertise that his batteries have a certain guaranteed life and would like no more than 10% of his batteries to fail before this time. How many hours should be guaranteed?

Solution.   We need to find a number b such that

$$P(X \leq b) = 0.10$$

Converting to the z score of b gives

$$z = \frac{b - 120}{10} = \frac{b}{10} - 12$$

Since $b/10 - 12$ is clearly negative, we have

$$P\left(z \leq \frac{b}{10} - 12\right) = P\left(z \geq 12 - \frac{b}{10}\right)$$

$$= P(z \geq 0) - P\left(0 \leq z \leq 12 - \frac{b}{10}\right)$$

$$= 0.5000 - P\left(0 \leq z \leq 12 - \frac{b}{10}\right)$$

Letting

$$P\left(z \leq \frac{b}{10} - 12\right) = 0.10$$

gives

$$P\left(0 \leq z \leq 12 - \frac{b}{10}\right) = 0.5000 - 0.10 = 0.4000$$

According to Table 8.19,

$$12 - \frac{b}{10} \approx 1.28$$

This gives $b \approx 120 - 12.8 \approx 107$

The manufacturer should guarantee a battery life of 107 hours.   □

EXERCISES

1. Find the following probabilities from Table 8.19.
    (a) $P(z \leq 1.7)$ (b) $P(z \leq -0.6)$
    (c) $P(z \leq -1.8)$ (d) $P(z \geq 2.3)$
    (e) $P(z \geq -2.4)$ (f) $P(1.4 \leq z \leq 2.5)$
    (g) $P(-1.2 \leq z \leq -0.4)$ (h) $P(-1.8 \leq z \leq 1.4)$

2. If X is a random variable having a normal distribution with $m = 15$ and $\sigma = 5$, find the probability that X assumes the following values.
    (a) $X \leq 18$ (b) $X \geq 12$ (c) $12 \leq X \leq 18$
    (d) $10 \leq X \leq 22$ (e) $X \geq -1$

3. If X is normally distributed with $m = 2$ and $\sigma = 3$, find (a) $P(X \geq 3)$, (b) $P(0 \leq X \leq 4)$.

4. If X is normally distributed with $m = 1$ and $\sigma = 2$, find a number b for each of the following.

   (a) $P(X \geq b) = 0.1151$ (b) $P(X \geq -b) = 0.8159$

   (c) $P(X \leq b) = 0.9965$ (d) $P(X \leq -b) = 0.2743$

   (e) $P(1 - b \leq X \leq 1 + b) = 0.8664$

5. The manufacturer in Example 6 has another brick with normally distributed length for which $m = 12$ and $\sigma = 0.09$. If the tolerance interval is $[11.80, 12.25]$, what percentage of the bricks pass?

6. The manufacturer in Example 7 has developed a new battery with normally distributed lifetime for which $m = 150$ hours and $\sigma = 8$ hours.

   (a) If he wants the same guarantee arrangement, how many hours should be guaranteed?

   (b) What percentage of these batteries last more than 160 hours?

7. The test grades on a certain test are found to be normally distributed. If grades A or F are given to students whose scores deviate by more than 1.5 standard deviations from the mean, what percentage of students receive grades of B, C, or D?

8. It is found from experience that the number of customers served daily at a supermarket approximates a normal distribution with $m = 650$ and $\sigma = 30$. What percentage of the days will there be more than 700 customers?

9. If a normally distributed random variable has $m = 0$ and $P(0 \leq X \leq 2) = 0.4$, find the approximate value of σ.

10. If a normally distributed random variable has $\sigma = 1$ and $P(X \geq 3) = 0.2$, approximate m.

11. If g is a normal density with $m = 0$ and $g(0) = 1$, find σ.

*12. If g is a normal density with $\sigma = 1$ and $g(0) = e^{-1/2}/\sqrt{2\pi}$, find m.

8.7 | Correlation

Numbers as realities misbehave. However, there is an ancient and innate sense in people that numbers ought not to misbehave. There is something clean and pure in the abstract notion of a number, removed from counting beads, dialects, or clouds; and there ought to be a way of talking about numbers without always having the silliness of reality come in and intrude.

— DOUGLAS R. HOFSTADTER
Gödel, Escher, Bach: An Eternal Golden Braid

So far we have considered one random variable at a time. However, it is frequently important to consider two or more random variables simultaneously. It may happen that two random variables are correlated in the sense that they influence each other and such correlations can be useful for statistical predictions. For example, due to the jet stream and other atmospheric conditions, the weather in Denver is strongly correlated to the weather on the west coast. Because of this correlation, the known west coast weather on Monday can be used to forecast the Denver weather for Tuesday, Wednesday, and Thursday. As another example, political preference is frequently correlated with geographical location, income, education, and other factors. These correlations can be applied to predict the outcome of elections.

Scatter Diagram

Statistical methods for studying correlations were developed by Sir Francis Galton (1822–1911), who was interested in the resemblance between children and their parents. Galton's disciple Karl Pearson (1857–1936) carried out a study in which he measured the heights in inches of 1078 fathers and their sons at maturity, one son per father. He obtained a list of 1078 pairs of heights (x_i, y_i), $i = 1, 2, \ldots, 1078$, where x_i is the father's height and y_i is his son's height. He then plotted these points in a graph called a **scatter diagram** (Figure 8.23). Each dot on the diagram represents a father-son pair. The x-coordinate of the dot measured along the horizontal axis gives the father's height and the y-coordinate measured along the vertical axis gives the son's height.

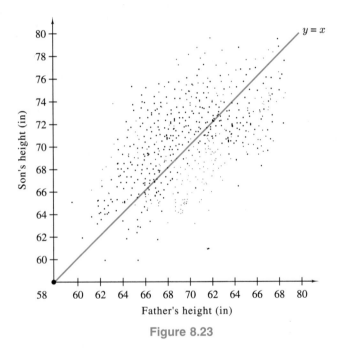

Figure 8.23

Families in which the height of the son was equal to the height of the father would be plotted along the line $y = x$. The scatter diagram confirms the obvious: there is a positive correlation between the heights of fathers and sons. But the scatter diagram says more; it shows how weak the correlation is. This is because there is a considerable spread about the equal-height line $y = x$. Knowing the father's height helps in guessing his son's height, but there is a lot of room for error—reflected in the spread in the y's for any fixed x.

In this situation, we take for our sample space S the set of points in the scatter diagram. That is,

$$S = \{(x_i, y_i): \ i = 1, 2, \ldots, 1078\}$$

We then have two random variables X and Y, where X gives the father's height and Y his son's height. Hence, $X(x_i, y_i) = x_i$ and $Y(x_i, y_i) = y_i$. Since we are interested in the dependence of Y on X, we call X the **independent** random variable and Y the **dependent** random variable.

We now consider the general situation in which we have independent and dependent random variables X and Y, respectively. The statistical quantities that we have already defined give us information about the scatter diagram. In general, the scatter diagram consists of a cloud of points. If m_X and m_Y are the means of X and Y, the approximate center of this cloud is the **mean point** $m_0 = (m_X, m_Y)$ (Figure 8.24).

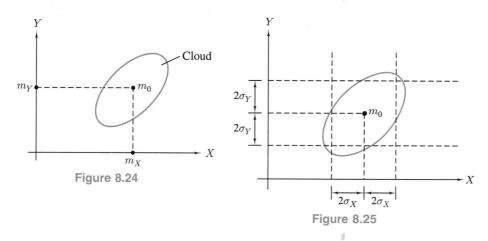

Figure 8.24

Figure 8.25

We can also describe the horizontal and vertical spread of the cloud. If σ_X and σ_Y are the standard deviations of X and Y, most of the points will be within two horizontal σ_X's of m_0 and within two vertical σ_Y's of m_0 (Figure 8.25).

But there is still something missing. We have no information about the association between the two random variables. If the cloud roughly clusters about a straight line we have no information about the spread from this line. For example, a cloud with the same m_0, σ_X, σ_Y as in Figure 8.25 could look like Figure 8.26. This missing information, which measures the spread about a line, will be called the **correlation coefficient**.

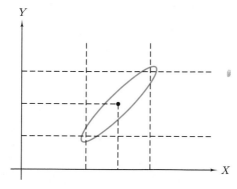

Figure 8.26

The correlation coefficient r measures the strength of the correlation between X and Y and indicates the clustering of sample points about a straight line. It will turn out that r is between -1 and 1 (Exercise 16). When r is close to 1, then the clustering is tight (small spread), and when r is close to 0, the clustering is loose (large spread). If r is negative, then the correlation is negative. In this case, the cloud of sample points slopes down; as X increases, Y actually decreases. For example, in a certain study the correlation coefficient between the IQs of identical twins is 0.95. Their scatter diagram would look like Figure 8.27. In a recent poll the correlation coefficient between income and education for men aged 35 to 44 is 0.60 (Figure 8.28). The correlation coefficient between education and number of children for women in another study is -0.3 (Figure 8.29).

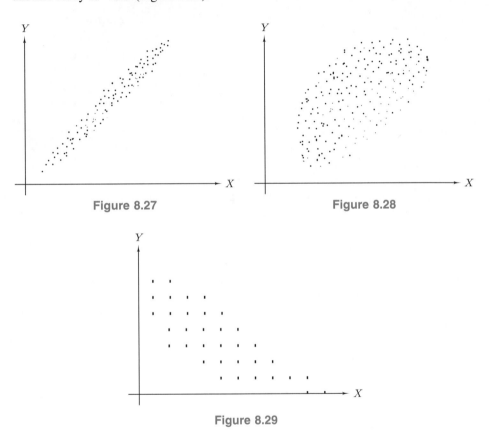

Figure 8.27   Figure 8.28

Figure 8.29

Covariance

We are now ready to define the correlation coefficient. Let X and Y be discrete random variables that are defined on the same sample space and denote their data values by $x_1, \ldots, x_n$ and $y_1, \ldots, y_n$, respectively. By this notation we mean that if x_i is a value of X at some sample point, then y_i is the value of Y at the **same** sample point. Denote the means and standard deviations of X and Y by m_X, m_Y and σ_X, σ_Y, respectively. The **covariance** for X and Y is defined as

$$\text{cov}(X, Y) = \frac{1}{n} \sum_{i=1}^{n} (x_i - m_X)(y_i - m_Y) \tag{8.31}$$

Correlation Coefficient

Notice that the covariance is similar to the variance var_X for a single random variable X. In fact, $\text{cov}(X, X) = \text{var}_X$. We then define the **correlation coefficient** as

$$r = \frac{\text{cov}(X, Y)}{\sigma_X \sigma_Y} \tag{8.32}$$

σ Line

We have previously stated that the closer r is to 1, the more tightly clustered the points of the scatter diagram to a line. What is this line? It is called the **σ line** or **standard deviation line**. As can be roughly seen from Figure 8.25, the σ line is the straight line in the scatter diagram through the mean point $m_0 = (m_X, m_Y)$ with slope σ_Y / σ_X (Figure 8.30). Notice that the point-slope form for the σ line is

$$y - m_Y = \frac{\sigma_Y}{\sigma_X}(x - m_X) \tag{8.33}$$

(Recall our discussion of straight lines in Section 5.1.) This is for positive correlations. When r is negative, the slope of the σ line is reversed; it then becomes $-\sigma_Y / \sigma_X$. We shall not prove these results here, but we shall illustrate them with an example (and you will see other examples in the exercises).

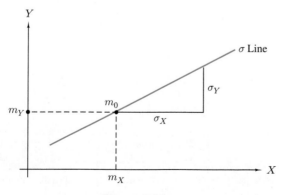

Figure 8.30

Before giving an example, we derive a simpler formula for $\text{cov}(X, Y)$. From (8.31) we have

$$\text{cov}(X, Y) = \frac{1}{n} \sum (x_i - m_X)(y_i - m_Y)$$

$$= \frac{1}{n} \sum x_i y_i - \frac{1}{n} m_X \sum y_i - \frac{1}{n} m_Y \sum x_i + \frac{1}{n} \sum m_X m_Y$$

$$= m_{XY} - m_X m_Y - m_Y m_X + m_X m_Y$$

$$= m_{XY} - m_X m_Y \qquad\qquad (8.34)$$

where

$$m_{XY} = \frac{1}{n} \sum x_i y_i$$

is the mean of the random variable XY. That is, m_{XY} is the mean of the product of the random variables X and Y. [What does (8.34) say when $X = Y$?]

EXAMPLE 1   Six men are asked their heights and weights. The results are shown in Table 8.20. Find the correlation coefficient and plot the scatter diagram and σ line.

Table 8.20

Height X	Weight Y
66	145
67	154
68	170
70	168
71	163
72	172

Solution.   We first compute the quantities in Table 8.21.

Table 8.21

	X	Y	X^2	Y^2	XY
	66	145	4,356	21,025	9,570
	67	154	4,489	23,716	10,318
	68	170	4,624	28,900	11,560
	70	168	4,900	28,724	11,760
	71	163	5,041	26,569	11,573
	72	172	5,184	29,584	12,384
Sums	414	972	28,594	158,018	67,165

We then have

$$m_X = \frac{414}{6} = 69$$

$$m_Y = \frac{972}{6} = 162$$

$$m_{2X} = \frac{28,594}{6} = 4765.67$$

$$m_{2Y} = \frac{158,018}{6} = 26,336.33$$

$$m_{XY} = \frac{67,165}{6} = 11,194.17$$

These give

$$\mathrm{var}_X = m_{2X} - m_X^2 = 4765.67 - 4761 = 4.67$$

$$\mathrm{var}_Y = m_{2Y} - m_Y^2 = 26,336.33 - 26,244 = 92.33$$

$$\mathrm{cov}(X, Y) = m_{XY} - m_X m_Y = 11,194.17 - 11,178 = 16.17$$

$$\sigma_X = \sqrt{\mathrm{var}_X} = 2.16$$

$$\sigma_Y = \sqrt{\mathrm{var}_Y} = 9.61$$

The correlation coefficient becomes

$$r = \frac{\mathrm{cov}(X, Y)}{\sigma_X \sigma_Y} = \frac{16.17}{(2.16)(9.61)} = 0.78$$

This shows a strong positive correlation. To obtain the σ line, we have

$$m_0 = (m_X, m_Y) = (69, 162)$$

and slope

$$s = \frac{\sigma_Y}{\sigma_X} = \frac{9.61}{2.16} = 4.45$$

The point-slope form of the σ line becomes

$$\frac{y - 162}{x - 69} = 4.45$$

which gives the standard form

$$y = 4.45x - 138.05$$

The scatter diagram and σ line are illustrated in Figure 8.31. □

In this section we have discussed linear correlation. This is when the scatter diagram clusters about a straight line. However, there are important cases of nonlinear correlation. For example, the force between two electrically charged particles of different sign decreases in proportion to the reciprocal of the distance squared. A scatter diagram for such a force measurement is illustrated in Figure 8.32. Methods similar to those we have discussed can be used to treat such situations (see Exercise 15).

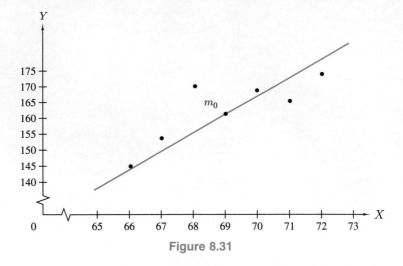

Figure 8.31

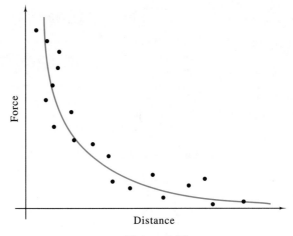

Figure 8.32

EXERCISES

1. Give some examples, that we have not discussed, illustrating (a) positive correlation and (b) negative correlation.

Table 8.22

X	Y
3	1.5
4	2
5	2.5
6	3

2. Show that $\text{cov}(X, X) = \text{var}_X$.

3. If $X = Y$, find r.

4. If $X = Y$, what is the σ line?

5. If $X = -Y$, find r and the σ line.

6. What does (8.34) say if $X = Y$?

7. Another man of height 69 inches and weight 165 pounds is added to the list in Table 8.20. How are the results in Example 1 changed?

8. Let X and Y be random variables defined on the same sample space with values given in Table 8.22. Compute σ_X, σ_Y, $\text{cov}(X, Y)$, and r. Plot the scatter diagram and the σ line. Find a relationship between X and Y and interpret your results.

9. If $Y = cX$ for a constant c, find r and the σ line.

10. If $Y = X + c$ for a constant c, find r and the σ line.

Table 8.23	
X	Y
97	95
99	102
104	107
110	109
121	118

Table 8.24

X	Y
0	25,000
1	23,000
2	32,000
3	35,000
4	33,000
6	44,000

Table 8.25

X	Y
0	7
1	4
3	5
4	2
6	3
8	1

Table 8.26

X	Y
1	1.05
2	0.24
3	0.11
4	0.0627
5	0.0390
6	0.0275

11. If $Y = aX + b$ for a constants a and b, find r and the σ line.

12. A sample of five pairs of identical twins have their IQs tested and the results are listed in Table 8.23. Find r and plot the scatter diagram and σ line.

13. The number of years of education past high school and the annual incomes for six people aged 30 are listed in Table 8.24. Find r and plot the scatter diagram and σ line.

14. The number of years of education past high school and the number of children for six women aged 40 are listed in Table 8.25. Find r and plot the scatter diagram and σ line.

15. The force Y between two charged particles versus their distance apart X are listed in Table 8.26. Plot the scatter diagram for these data. Now let W be the random variable defined by $W = 1/Y^{1/2}$. Find r for X and W and plot the scatter diagram and σ line. How do you interpret these results?

*16. (a) Use the Schwarz inequality (Section 8.2 Exercise 29) to show that

$$|\operatorname{cov}(X, Y)| \le \sigma_X \sigma_Y$$

(b) Apply (a) to show that $-1 \le r \le 1$.

Chapter 8 Summary of Terms

sample points

random variable

data set

histogram

class intervals

frequency distribution

summation notation $\displaystyle\sum_{i=1}^{n} a_i$

mean $\displaystyle m = \frac{1}{n}\sum_{i=1}^{n} x_i$

median

mode

variance $\quad \text{var} = \dfrac{1}{n} \sum\limits_{i=1}^{n} (x_i - m)^2$

standard deviation $\quad \sigma = \sqrt{\text{var}}$

second moment $\quad m_2 = \dfrac{1}{n} \sum\limits_{i=1}^{n} x_i^2$

variance formula $\quad \text{var} = m_2 - m^2$

z score $\quad z = \dfrac{x - m}{\sigma}$

continuous random variables

probability $\quad P(a \le X \le b)$

density function $\quad P(a \le X \le b) = \displaystyle\int_a^b g(x)$

$$m = \int_{-\infty}^{\infty} x g(x)$$

$$m_2 = \int_{-\infty}^{\infty} x^2 g(x)$$

$$\text{var} = \int_{-\infty}^{\infty} (x - m)^2 g(x)$$

normal distribution

scatter diagram

mean point $\quad m_0 = (m_X, m_Y)$

covariance $\quad \text{cov}(X, Y) = \dfrac{1}{n} \sum\limits_{i=1}^{n} (x_i - m_X)(y_i - m_Y)$

correlation coefficient $\quad r = \dfrac{\text{cov}(X, Y)}{\sigma_X \sigma_Y}$

σ line $\quad y - m_Y = \dfrac{\sigma_Y}{\sigma_X} (x - m_X)$

Chapter 8 Test

1. Draw the histogram and frequency polygon for the frequency distribution in Table 8.27.
2. Find the mean for the frequency distribution in Table 8.27.
3. Find the median for the frequency distribution in Table 8.27.
4. Evaluate the summation $\displaystyle\sum_{i=1}^{5} 2^i$.

Table 8.27

Class interval	Frequency
$(0, 5]$	3
$(5, 10]$	5
$(10, 15]$	7
$(15, 20]$	6
$(20, 25]$	8
$(25, 30]$	4
$(30, 35]$	5
$(35, 40]$	2

Table 8.28

Grade	Credit hours
F	4
D	7
C	30
B	42
A	48

Table 8.29

X	Y
0	1
1	3
2	5
3	6
4	8
5	7

5. Is the following always true? $\left[\sum_{i=1}^{n} a_i\right]^2 = \sum_{i=1}^{n} a_i^2.$

6. Find the standard deviation of the data set $D = \{1, 2, 2, 3, 4, 7\}$.

7. Find the mean, median, and mode for the data set in Exercise 6.

8. Discuss the advantages and disadvantages of the mean, median, and mode for describing central tendency.

9. Suppose a length measurement is performed 100 times and results in a data set D. If σ is very small compared to the values in D, what does this say about the measurement?

10. Find the grade point average and standard deviation for the grades in Table 8.28.

11. Let g be the function defined by

$$g(x) = \begin{cases} 4x^3 & \text{if } 0 \leq x \leq 1 \\ 0 & \text{otherwise} \end{cases}$$

    (a) Show that g is a density function.   (b) Find m.
    (c) Find σ.   (d) Find $P\left(\frac{1}{2} \leq X \leq \frac{3}{4}\right)$.

12. If X is normally distributed with $m = 5$ and $\sigma = 0.30$, find $P(4.5 \leq X \leq 5.5)$.

13. Find r and plot the scatter diagram and σ line for the random variables in Table 8.29.

14. If X is a continuous random variable and $a < b < c$, why is $P(a \leq X \leq b) \leq P(a \leq X \leq c)$?

15. Are the following statements true or false?
    (a) The mean of the data set $\{1, 2, 3\}$ is 2.
    (b) A mode of the data set $\{1, 2, 3\}$ is 1.
    (c) $\sum_{i=1}^{5} i = 16.$
    (d) The variance of a data set is always positive.
    (e) The standard deviation is the square root of the variance.
    (f) The mean and the first moment are the same thing.
    (g) The z score is the same for every data value of a data set.
    (h) A random variable is a function with chaotic values.
    (i) A density function is used for measuring intelligence quotients.
    (j) In a normal density, 68% of the values lie within one standard deviation of the mean.

PROBABILITY THEORY

9

It doesn't take a lot of ability
Nor does it require great stability
In order to attain a facility
For understanding probability

How old is mathematics? Nobody knows for certain, although some of the oldest recorded examples of mathematics are the ancient Babylonian clay tablets and the Egyptian papyruses. These were made during the period 2000–600 B.C. and were used mainly for accounting purposes, for example, bills of sale, taxes, and values of possessions. It is also known that the ancient Egyptians used simple geometrical concepts for land surveying. During the sixth, fifth, and fourth centuries B.C. the ancient Greeks developed mathematics to a high degree of sophistication, and there are many recorded examples of their work. However, there is evidence that human use of mathematics goes far back beyond recorded history to prehistoric times. There were Einsteins among the Stone Age and Bronze Age peoples, who lived 10,000 to 40,000 years ago. These men and women were capable of the kind of precise observation and abstract conceptualization that we associate with modern scientists. Instead of rude savages who scarcely could count on their fingers, scientists are finding that prehistoric people often had well-developed concepts of numbers and arithmetic, geometry, and time. They kept track of the orbital motions of sun and moon, sometimes with a precision not known again until the Renaissance. They recorded this knowledge in stone monuments such as Stonehenge that could be used as astronomical observatories and sometimes as computers.

Probability theory, which came much later, began with the study of games of chance. No one really knows how old some of these games are. Dice closely resembling those used today have been found in ancient Greek and Egyptian tombs. The odds in these ancient games of chance were arrived at through experiment and experience. Around 1150 B.C. the Chinese dealt with problems of arrangements similar to our modern theory of permutations and combinations, and about the same time Rabbi Ben Ezra computed that there are 120 possible ways in which two or more

413

of the planetary bodies—sun, moon, Venus, Mercury, Saturn, Mars, and Jupiter—can appear to come close to each other.

Furthermore, ancient Jewish scholars and rabbis of biblical times studied probability theory. Talmudic sources and rabbinical commentaries considered a variety of modern concepts of probability; formulated the arithmetic of probabilities, combinations, and permutations; discussed related logical and philosophical ramifications; and arrived at elementary decision-making guidelines.

9.1 | Introduction

We see ... that the theory of probabilities is at bottom only common sense reduced to calculation; it makes us appreciate with exactitude what reasonable minds feel by a sort of instinct, often without being able to account for it .... It is remarkable that [this] science, which originated in the consideration of games of chance, should have become the most important object of human knowledge.
— PIERRE SIMON DE LAPLACE (1749–1827), French Mathematician

Probability theory in its modern form did not really begin to take shape until 400 years ago. Jerome Cardan (1501–1576), genius, astrologer, philosopher, physician, mathematician, and gambler, was the first to study probability in a systematic fashion. His work *The Book of Games of Chance* was the first book on probability. This book became a gambler's manual and included tips on how to cheat and how to catch others cheating. He correctly computed the odds of various outcomes when two or three dice are tossed and discussed other concepts such as expectations. He actually predicted the day on which he would die (although he had to help fate a little by committing suicide). Chevalier de Mere, a professional gambler, realized that mathematics could be profitably applied to gambling. In 1653 he teamed up with Pascal on problems in dice. Pascal and Fermat developed probability theory into a precise science.

Today probability theory is an indispensable tool. Modern applications of probability theory and its sister science statistics are made in insurance, biology, measurement theory, psychology, economics, manufacturing, social sciences, and politics. You cannot open a newspaper without seeing such applications. On the front page you see the probability of dying of cancer or heart disease from smoking cigarettes and other causes. The weather report gives the probability of precipitation the next day or the next week. The sports page gives the standing of teams, batting averages, and statistics concerning field goals, touchdowns, runs batted in, etc. The business section gives cost-of-living statistics, Dow-Jones industrial averages, New York and American Stock Exchange averages, etc. Today we live in a truly probabilistic world.

Probability theory becomes important when you do not have enough information to describe a situation completely. Using the partial information at your disposal, you cannot predict precisely what will occur, but you can predict the chances that different

Table 9.1

Event	Chances
You will gain back weight lost in a diet.	9 in 10
You will parent a child.	8 in 10
You will qualify for a mortgage.	8 in 10
A criminal will return to jail after release.	8 in 10
A youth will graduate from high school.	75 in 100
A smoker will try to quit.	71 in 100
A wife will outlive her husband.	7 in 10
You will avoid imprisonment if convicted.	6 in 10
A college freshman will graduate.	1 in 2
A child will live past 75.	1 in 2
You will undergo an IRS audit this year.	1 in 100
Your child will be a genius.	1 in 250
A high school player will play in a Super Bowl.	1 in 4,000
The earth will be devastated by a meteorite in your lifetime.	1 in 9,000
A professional golfer will get a hole-in-one.	1 in 15,000
A fan will be hit by a baseball in a major league game.	1 in 300,000
A passenger will be killed in an air crash.	1 in 4.6 million

outcomes have of occurring. You can probably guess who will win the basketball game if you know the records of the two teams. You can get a better grade on the test if you know what the questions are beforehand. The more you know about something, the easier it is to predict what will happen. Almost every important decision you make is based on your feeling that the action you take will produce a favorable result. You know you cannot be sure of the details of the future, but you usually act on the chance that a desirable outcome will take place. You may take certain college courses because you feel that they will increase your chances for success in the business world. When you cross an intersection with a green light, you feel that the chances are good that traffic coming the other way will stop. In the back of your mind you know that an oncoming car can have a brake failure or the driver may not see the light, but you think it unlikely. When you buy a product, such as a television set, you choose the one you do because you feel that the chances are good that you are getting the best buy.

The chance that a certain event will occur is related to the probability or likelihood of that event. When you say that you have one chance in two of graduating college, you mean that the probability is 1/2. Although we try to control our lives as much as possible, we live in a world of chance and probability. Statisticians have computed the chances for events in a huge variety of situations. What are your chances of winning a state lottery? About one in 4 million. This is far less likely than being struck by lightning, which is one in 9100. What are the chances you will be dealt a royal flush in the opening poker hand? About one in 650,000. What are the chances that you will be married? About 75 in 100. Table 9.1 lists various events and the chances that they will occur.

In this chapter probabilistic models are discussed. In applications, these abstract mathematical models serve as tools that enable us to study real phenomena. How these models are applied does not depend on preconceived ideas; using them is a purposeful technique depending on, and changing with, experience. A probabilistic model involves two things:

> (1) Choosing a set to represent the possible outcomes.
>
> (2) Allocating probabilities to these possible outcomes.

9.2 | Probabilistic Models

Life is the art of drawing sufficient conclusions from insufficient premises.   — SAMUEL BUTLER (1612–1680), English Satirist

Before beginning a systematic study of probability theory, consider some simple experiments and how probabilistic models can be used to describe them.

EXAMPLE 1   If you flip a fair (or unbiased) coin, what is the probability that you get a head?

Solution.   You can think of flipping a coin as an experiment. Following the two steps of Section 9.1, you should choose a set to represent the possible outcomes of this experiment and then allocate probabilities to these outcomes. There are two possible outcomes H (heads) and T (tails), so the set of outcomes can be represented by $\Omega = \{H, T\}$. Since the coin is fair (it is not loaded), a head is as likely as a tail, so we allocate a probability $1/2$ to H and $1/2$ to T. We then write $P\{H\} = 1/2$ and $P\{T\} = 1/2$. Notice that $P\{H\} + P\{T\} = 1$, which is interpreted as meaning that the probability of H or T is 1. (This discounts the fact that the coin may land on its edge or disappear before it lands.) It follows that the probability of neither a head nor a tail is zero. This experiment can be represented by the **probability tree** illustrated in Figure 9.1. At start S, the coin is flipped. There are then two possible outcomes H and T, and the probability of each is $1/2$.   □

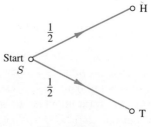

Figure 9.1

POOR ARNOLD'S ALMANAC   Arnold Roth

WHERE CARTOONISTS
GET THEIR IDEAS.

EXAMPLE 2 Suppose we flip a fair (or unbiased) coin two times. We may describe the possible outcomes of this experiment by the collection of ordered pairs HH, HT, TH, TT, which, of course, represent two heads, a head and a tail, a tail and a head, and two tails. These four ordered pairs are called **sample points** for the experiment. They are called sample points because they give samples of the different possible outcomes of the experiment. The **events** for this experiment are sets of sample points. For example, the event A that heads is tossed on the first flip is the set $A = \{$HH, HT$\}$. Hence A is the event consisting of the two sample points in which heads is tossed on the first flip. The event B that heads is tossed on the first flip and tails on the second is the set $B = \{$HT$\}$. Thus B consists of a single sample point. The set of all sample points Ω is called the **sample space**. Thus, in this case, $\Omega = \{$HH, HT, TH, TT$\}$. Now associate probabilities with the different experimental outcomes. Since the coin is fair, we would expect the probability of each sample point to be the same; that is,

$$P\{\text{HH}\} = P\{\text{HT}\} = P\{\text{TH}\} = P\{\text{TT}\} = \frac{1}{4}$$

where $P\{$HH$\}$ denotes the probability that the event $\{$HH$\}$ occurs, etc.

The probabilities can also be read off the probability tree given in Figure 9.2. In the probability tree, the probability of a sample point is obtained by multiplying the probabilities along the corresponding path. At start S, the coin is flipped. There are two possible outcomes, H or T. These give the first branch of the tree. The coin is now flipped again, giving two more branches. The sample point HH corresponds to the path SHH. The probability of moving from S to H is $1/2$ and then the probability of moving from H to H is $1/2$. Multiplying the probabilities along this path, we obtain

$$P\{HH\} = \left(\frac{1}{2}\right)\left(\frac{1}{2}\right) = \frac{1}{4}$$

since $1/2$ and $1/2$ occur along the path SHH.

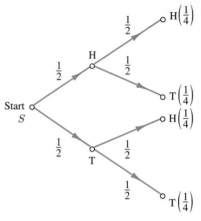

Figure 9.2

The probability that an event occurs should be the sum of the probabilities of the sample points that compose it. Thus the probability that the event A occurs is

$$P(A) = P\{HH, HT\} = P\{HH\} + P\{HT\} = \frac{1}{4} + \frac{1}{4} = \frac{1}{2}$$

EXAMPLE 3   In this example we flip a biased (or loaded) coin three times. The sample space becomes

$$\Omega = \{HHH, HHT, HTH, HTT, THH, THT, TTH, TTT\}$$

If A is the event that a head is tossed on the first flip, then

$$A = \{HHH, HHT, HTH, HTT\}$$

If B is the event that heads is tossed on every flip, then $B = \{HHH\}$. Since the coin is loaded, there will be different probabilities for heads and tails. Suppose after flipping the coin many times we find that a head comes up about one-third of the time and a tail about two-thirds of the time. The probabilities of the sample points can now be obtained from the probability tree in Figure 9.3. For example, let's compute the probability of the sample point THT. This sample point corresponds to the path STHT. Starting with the first flip at S, the probability of obtaining T is $2/3$. The

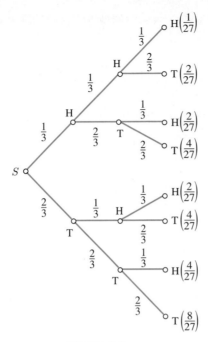

Figure 9.3

coin is flipped a second time and the probability of an H is $1/3$. On the third flip, the probability of T is $2/3$. Multiplying the probabilities along this path gives

$$P\{\text{THT}\} = \left(\frac{2}{3}\right)\left(\frac{1}{3}\right)\left(\frac{2}{3}\right) = \left(\frac{4}{27}\right)$$

In this way we assign the following probabilities to the sample points:

$$P\{\text{HHH}\} = \frac{1}{27}$$

$$P\{\text{HHT}\} = P\{\text{HTH}\} = P\{\text{THH}\} = \frac{2}{27}$$

$$P\{\text{HTT}\} = P\{\text{THT}\} = P\{\text{TTH}\} = \frac{4}{27}$$

$$P\{\text{TTT}\} = \frac{8}{27}$$

Notice that the sum of all the probabilities is 1, so we have

$$P(\Omega) = 1$$

Moreover,

$$P(A) = P\{\text{HHH}\} + P\{\text{HHT}\} + P\{\text{HTH}\} + P\{\text{HTT}\}$$

$$= \frac{1}{27} + \frac{2}{27} + \frac{2}{27} + \frac{4}{27} = \frac{1}{3}$$

EXAMPLE 4 License plates in a certain state have three letters followed by three numerals. The sample space is

$$\Omega = \{AAA111, ABC000, ACZ257, \ldots\}$$

Since there are 26 different letters and 10 different numerals, we have

$$(26)(26)(26)(10)(10)(10) = 17{,}576{,}000$$

sample points. Assuming that all arrangements are equally likely, the probability of each sample point is $1/17{,}576{,}000$. Let A be the event that a certain motorist gets a plate having her initials as the three letters. The possible numbers on her plate range between 000 and 999, so A contains 1000 sample points. Hence,

$$P(A) = \frac{1000}{17{,}576{,}000} = \frac{1}{17{,}576}$$

EXAMPLE 5 Toss a single fair die once. The sample points represent the number on the upward face and are given by a natural number from 1 to 6. The sample space is $\Omega = \{1, 2, 3, 4, 5, 6\}$. Since the die is fair, each sample point is equally likely and has probability $1/6$. Let A be the event "a 1 is tossed." Then $A = \{1\}$ and $P(A) = 1/6$. If B is the event "a 1 is not tossed," then $B = A' = \{2, 3, 4, 5, 6\}$ and $P(B) = 5/6$. (Remember, A' denotes the complement of A.) If C is the event "an even number is tossed," then $C = \{2, 4, 6\}$ and $P(C) = 1/2$. Let D be the event "a 1 or an even number is tossed." Then $D = \{1, 2, 4, 6\} = \{1\} \cup \{2, 4, 6\}$ and

$$P(D) = \frac{2}{3} = P\{1\} + P\{2, 4, 6\} = \frac{1}{6} + \frac{1}{2}$$

EXAMPLE 6 Draw a card at random from a deck. The sample space consists of the possible cards that can be drawn. We let AC, AD, KS, 3H, etc. denote the ace of clubs, ace of diamonds, king of spades, three of hearts, etc. Thus the sample space becomes

$$\Omega = \{AC, AD, AH, AS, 2C, 2D, 2H, 2S, \ldots\}$$

Since each card has the same chance of being picked, the probability of each sample point is $1/52$. If A is the event "an ace is drawn" then

$$P(A) = P\{AC, AD, AH, AS\} = \frac{4}{52} = \frac{1}{13}$$

If B is the event "an honor card is drawn" (an honor card is an ace, king, queen, or jack), then $P(B) = 16/52 = 4/13$. Let C be the event "a 2 is drawn," and let D be the event "a spade is drawn." Then $P(C) = 4/52 = 1/13$ and $P(D) = 13/52 = 1/4$. $C \cup D$ is the event "a 2 or a spade is drawn," and $P(C \cup D) = 16/52 = 4/13$. $C \cap D$ is the event "a 2 and a spade are drawn," and $P(C \cap D) = P\{2S\} = 1/52$.

Table 9.2

Event	Sample points	Probability
A_2	11	1/36
A_3	12, 21	1/18
A_4	22, 13, 31	1/12
A_5	14, 41, 23, 32	1/9
A_6	33, 24, 42, 15, 51	5/36
A_7	16, 61, 25, 52, 34, 43	1/6
A_8	44, 26, 62, 35, 53	5/36
A_9	36, 63, 45, 54	1/9
A_{10}	55, 64, 46	1/12
A_{11}	56, 65	1/18
A_{12}	66	1/36

EXAMPLE 7   Toss a pair of fair dice. The sample points describe the numbers on the two upward faces and are given by a pair of natural numbers from 1 to 6. Since each die has six possibilities, there will be 36 possibilities for the pair. The sample space becomes

$$\Omega = \{11, 12, 13, 14, 15, 16, 21, 22, \ldots, 66\}$$

where 11 represents the sample point "a 1 on each die," 12 is the sample point "a 1 on the first die and a 2 on the second," etc. Since each of these sample points is equally likely, they each have probability 1/36. Let B be the event "a 1 and 2 are tossed." Then $B = \{12, 21\}$, and $P(B) = 1/18$. What is the probability of the event "a 1 and 2 are not tossed"? We notice that this last event is B'. Since B' has 34 sample points, we have $P(B') = 34/36 = 17/18$. In dice games, for example, craps, the important events are those in which certain sums are tossed. Let A_n be the event that the sum of the upward faces of the two dice is n. For example, A_2 is the event "the sum of the upward faces is 2." Thus $A_2 = \{11\}$. In Table 9.2 the sample points in each of these events are listed, and by counting the number of sample points the probabilities can be computed. With this table the probability of other events can be found. For example, the probability of getting a 7 or 11 is

$$P(A_7 \cup A_{11}) = \frac{1}{6} + \frac{1}{18} = \frac{4}{18} = \frac{2}{9}$$

EXERCISES

1. In Example 2, what set represents the event "tails is tossed on the first flip"?

2. In Example 2, what set represents the event "heads is tossed on both flips"?

3. In Example 2, what set represents the event "tails is not tossed on the first flip"? How is this event related to event A of Example 2? How is this event related to the event of Exercise 1?

4. In Example 2, what set represents the event "heads is not tossed on both flips"? How is this event related to the event in Exercise 2?

5. Find the probabilities of the events in Exercises 1 and 2.

6. Find the probabilities of the events in Exercises 3 and 4.

7. In Example 3, what set represents the event "tails is tossed on the first flip"? What is the probability of this event?

8. In Example 3, what set represents the event "heads is tossed on the first flip, and tails is tossed on the second flip"? What is the probability of this event?

9. In Example 3, what set represents the event "two heads are tossed, or two tails are tossed"? What is the probability of this event?

10. In Example 5, what set represents the event "an odd number is tossed"? What is the probability of this event?

11. In Example 2, let C be the event "heads is tossed on the first flip, or tails is tossed on the second flip." What sample points are in C? Show that $A \subseteq C$ and $B \subseteq C$. Find C'. Find $P(C)$ and $P(C')$.

12. In Example 3, suppose heads is tossed three-fourths of the time and tails one-fourth of the time. What probabilities would you now assign to each sample point?

13. In Example 4, what is the probability that an A appears as a letter on a license plate chosen at random?

14. What is the probability that a club or ace or king is drawn from a deck of cards?

15. In Example 7, what is the probability that the sum of the upward faces is a 2 or 12? What is the probability that the sum of the upward faces is less than 7?

16. How can Table 9.3 be used to compute the probabilities of the events A_n in Example 7? Use this table to compute these probabilities.

Table 9.3

+	1	2	3	4	5	6
1	2	3	4	5	6	7
2	3	4	5	6	7	8
3	4	5	6	7	8	9
4	5	6	7	8	9	10
5	6	7	8	9	10	11
6	7	8	9	10	11	12

17. John has bet Harry $1000 that on the next roll of the dice a 7, 11, or **doubles** (same number on both dice) will come up. Would you rather be John or Harry?

18. In Example 4, what is the probability that 123 appears on a license plate?

19. In Example 4, what is the probability that all three letters are the same?

20. In Example 4, what is the probability that all three numbers are the same?

21. What is the probability that a spade or honor card is drawn from a deck of cards?

22. In Example 7, what is the probability that the sum is not 7?

23. Draw the probability tree for four flips of a fair coin.

24. An integer between 1 and 10 (inclusive) is chosen at random. What is the probability that the number is (a) less than 4, (b) greater than 4, (c) odd?

25. An integer between 0 and 99 (inclusive) is chosen at random. What is the probability that (a) precisely one of its digits is a 2, (b) at least one of its digits is a 2, (c) none of its digits is a 2?

26. A fair coin is flipped three times. What is the probability of at least one head? All heads or all tails?

27. An integer between 1 and 15 (inclusive) is selected at random. What is the probability that the integer is (a) divisible by 3, (b) even, (c) odd, (d) the square of an integer?

28. A box contains four red, five white, and six blue balls, and one ball is drawn without looking. What is the probability that it will be (a) red, (b) white, (c) blue, (d) not red, (e) red or white?

29. If Gail has four chances out of 10 of winning a certain game, how likely is she to lose?

30. A fair die is tossed. What is the probability of tossing (a) a 4, (b) less than 4, (c) an odd number?

31. Dolly and Walter each toss a fair die. What is the probability that the tosses are the same?

9.3 | Concepts of Probability Theory

> *As far as the propositions of mathematics refer to reality, they are not certain; and as far as they are certain, they do not refer to reality.*
> — ALBERT EINSTEIN (1879–1955), Physicist

All the probabilistic models in Section 9.2 had certain features in common. They all consisted of a sample space Ω of sample points (or outcomes) and an allocation of probabilities to the sample points. If $a \in \Omega$ is an outcome, we denote the probability that a occurs by $P\{a\}$. Note that in every example in Section 9.2, $P\{a\}$ was always nonnegative and the sum of the probabilities for all the outcomes was 1. These two properties of P are all that are needed to develop a theory of probability. We now state the fundamental concept of the theory.

Fundamental Concept: Given a sample space

$$\Omega = \{a_1, a_2, \ldots, a_n\}$$

we assume that to each sample point a_i there is associated a number $P\{a_i\}$, called the **probability** of a_i, $i = 1, 2, \ldots, n$. $P\{a_i\}$ is a nonnegative real number satisfying

$$P\{a_1\} + P\{a_2\} + \cdots + P\{a_n\} = 1 \tag{9.1}$$

Why do we want $P\{a_i\}$ to be nonnegative? Since $P\{a_i\}$ is interpreted as the likelihood of the outcome a_i, it is natural that this should be nonnegative. Why do we want (9.1) to hold? This is a convention in which we consider 1 to correspond to certainty. We interpret (9.1) as saying that one of the outcomes $a_1, \ldots, a_n$ occurs with certainty. For example, if we flip a fair coin, each of the outcomes H and T has probability $1/2$. We interpret $P\{H\} + P\{T\} = 1$ as meaning that either H or T must certainly occur.

How are the probabilities $P\{a_i\}$ in the fundamental convention determined? This depends upon the particular problem under investigation. Sometimes they must be calculated; other times they must be obtained experimentally; or they may already have been determined by others using sampling, polls, etc., and are simply given to you.

We can draw various conclusions from the fundamental concept. For example, $0 \le P\{a_i\} \le 1$, for every sample point a_i. Indeed, if $P\{a_i\} > 1$ for some i, then the sum in (9.1) would be greater than 1, which is impossible. We can also obtain results about the probabilities of events.

Events

> A subset A of a sample space Ω is called an **event**. If a sample point a is chosen and $a \in A$, we say that A has **occurred**.

If A is an event, we can always form the event A' consisting of all those sample points not contained in A. The event A' occurs if and only if the event A does not occur. If A and B are events, then $A \cap B$ is the event that occurs when both A and B occur and $A \cup B$ is the event that occurs when either A or B (or both) occur. We say that two events A and B are **mutually exclusive** if $A \cap B = \emptyset$. In this case A and B have no sample points in common; if one occurs, the other cannot occur. We can represent events by sets in a Venn diagram just as we did in Sections 1.3 and 1.4.

> We define the **probability** $P(A)$ of an event A as the sum of the probabilities of all the sample points in A.

Probabilistic Formulas

What can we say about the probability of an event? First of all,

$$P(\Omega) = 1 \tag{9.2}$$

This is because Ω consists of all the sample points and the sum of their probabilities is 1. Second, for any event A, we have

$$0 \le P(A) \le 1 \tag{9.3}$$

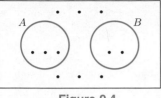

| Figure 9.4 | Figure 9.5 |

This is because the sum of the probabilities of the sample points that happen to be in A cannot be larger than the sum of the probabilities of all the sample points. If A and B are mutually exclusive, they have no common sample points (Figure 9.4). Hence, the sum of the probabilities of the sample points in $A \cup B$ is the sum for those in A plus those in B. Hence,

$$P(A \cup B) = P(A) + P(B) \text{ if } A \cap B = \emptyset \qquad (9.4)$$

If A and B are events, you can see from Figure 9.5 that $A = (A - B) \cup (A \cap B)$ and $(A - B) \cap (A \cap B) = \emptyset$. It follows from (9.4) that

$$P(A) = P(A - B) + P(A \cap B)$$

Hence

$$P(A - B) = P(A) - P(A \cap B) \qquad (9.5)$$

Figure 9.5 also shows that

$$A \cup B = (A - B) \cup (B - A) \cup (A \cap B)$$

and the three sets on the right-hand side are mutually exclusive. It follows from (9.4) and (9.5) that

$$P(A \cup B) = P(A - B) + P(B - A) + P(A \cap B)$$
$$= P(A) - P(A \cap B) + P(B) - P(A \cap B) + P(A \cap B)$$

Hence,

$$P(A \cup B) = P(A) + P(B) - P(A \cap B) \qquad (9.6)$$

We can derive another important equation from (9.5). This equation is

$$P(A') = 1 - P(A) \qquad (9.7)$$

which states that the probability of A' is 1 minus the probability of A. Indeed, from (9.5) and (9.2) we have

$$P(A') = P(\Omega - A) = P(\Omega) - P(A) = 1 - P(A)$$

The above results are now applied to examples. See if you can work these examples yourself before you look at the solutions.

EXAMPLE 1 In an office of 100 secretaries, 65 can type, 60 can take shorthand, and 30 can do both. What is the probability that a secretary can type or take shorthand? What is the probability that a secretary can do neither?

Solution. Let the sample space Ω be the secretaries in the office. Let A be the secretaries who can type and B the secretaries who can take shorthand (Figure 9.6). Assign to each secretary a probability 0.01. The secretaries who can type or take shorthand are those in the set $A \cup B$. Hence

$$P(A \cup B) = P(A) + P(B) - P(A \cap B)$$
$$= 0.65 + 0.6 - 0.3$$
$$= 0.95$$

The secretaries who can do neither are those in the set $A' \cap B'$. Hence

$$P(A' \cap B') = P\left[(A \cup B)'\right]$$
$$= 1 - P(A \cup B)$$
$$= 1 - 0.95$$
$$= 0.05 \qquad \square$$

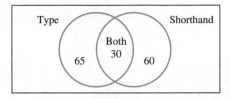

Figure 9.6

EXAMPLE 2   A fair coin is tossed five times. What is the probability that at least one head is obtained? What is the probability that two or more heads are obtained?

Solution.   The sample space Ω is the set of all 5-tuples of H's and T's. Thus

$$\Omega = \{\text{HHHHH, HHHHT}, \ldots, \text{TTTTT}\}$$

Since there are two possibilities for each of the five entries, Ω has $2^5 = 32$ sample points. Each sample point has probability $1/32$. Let A be the event that at least one head is obtained. Then A contains all the sample points except TTTTT, and hence $P(A) = 31/32$. Another way is to consider A'. Since A' is the event that no head is obtained, $P(A') = P\{\text{TTTTT}\} = 1/32$. Hence

$$P(A) = 1 - P(A') = 1 - \frac{1}{32} = \frac{31}{32}$$

Let B be the event that two or more heads are obtained. Then B' is the event that one or no head is obtained, and so

$$B' = \{\text{TTTTT, HTTTT, THTTT, TTHTT, TTTHT, TTTTH}\}$$

Hence

$$P(B') = \frac{6}{32}$$

and

$$P(B) = 1 - P(B') = 1 - \frac{6}{32} = \frac{26}{32} = \frac{13}{16}$$

The reason we calculated $P(B')$ first instead of finding $P(B)$ directly is that B' has fewer sample points than B. $\square$

EXAMPLE 3   A fair die is tossed three times. (a) Find the probability that exactly one 6 is tossed. (b) Find the probability that at least one 6 is tossed.

Solution.   The sample space Ω is the set of triples of natural numbers from 1 to 6. Thus

$$\Omega = \{111, 112, 121, 211, 122, 212, 221, 222, \ldots, 665, 666\}$$

Since there are six possibilities for each of the three entries, Ω has $6^3 = 216$ sample points. Each sample point has probability $1/216$. Let A be the event that exactly one 6 is tossed. To find $P(A)$ count the number of sample points that have exactly one 6. If the 6 occurs in the first position, then there are five possibilities for the second and third positions, so there are 25 sample points with a 6 in the first position but no 6 in the second or third position. Similarly there are 25 sample points with a 6 in the second position but none in the first and third and 25 sample points with a 6 in the third position but none in the first and second. Hence $P(A) = 75/216$. Let B be the event that at least one 6 is tossed. Then B consists of those sample points which have one, two, or three 6s. It was shown that there are 75 sample points with precisely one 6, and there is just one sample point with three 6s. Now consider the sample points with precisely two 6s. If the two 6s are in the first two positions, there are five possibilities for the last position. Similarly if the two 6s are in the last two positions or in the first and last position, then there are five possibilities for the remaining position. Thus there are 15 sample points with precisely two 6s. Finally there are $75 + 15 + 1 = 91$ sample points with at least one 6; therefore $P(B) = 91/216$. This can also be done by considering B'. In fact, B' consists of those sample points which contain no 6s. Hence there are five possibilities for each of the 3 throws, giving $5^3 = 125$ sample points in B'. Thus

$$P(B) = 1 - P(B') = 1 - \frac{125}{216} = \frac{91}{216} \qquad \square$$

Odds

> The ratio of the probability that the event will occur to the probability that the event will not occur is the **odds in favor** of an event. The reciprocal of this ratio is the **odds against** the event.

For example, the odds in favor of tossing a 1 with a fair die are 1 to 5. This is because the probability of tossing a 1 is $1/6$, and the probability of not tossing a 1 is $5/6$; so the ratio is $1/5$. Similarly, the odds against tossing a 1 with a fair die are 5 to 1. These odds can be interpreted as follows: in 6 tosses of a fair die, we can expect that one of the tosses results in a 1 and five of the tosses result in some other number. Hence if A is an event, the odds in favor of A are

$$\frac{P(A)}{P(A')} = \frac{P(A)}{1 - P(A)}$$

[If $P(A) = 1$, the odds in favor of A are not defined.] The odds against A are

$$\frac{P(A')}{P(A)} = \frac{1 - P(A)}{P(A)}$$

[If $P(A) = 0$, the odds against A are not defined.]

Fair Bet

If A is an event, a **fair bet** for the occurrence of A is a wager in which the ratio of the amount bet to the amount of the payoff equals the odds in favor of A.

Can you see why this is called a fair bet? One interpretation is that a fair bet gives you and your opponent equal chances of winning over the long run. In other words, if you continually make fair bets, then you will break even in the long run. For example, suppose the probability that the New York Giants will win their next game is 2/3. Then the odds in favor of a win are 2 (or 2 to 1). If the payoff is $5, then a fair bet would be a wager of $10. If this wager is made for three games then you would expect the Giants to win two of the games and lose one. Thus, you would win $5 for two of the games and lose $10 for one game. In this case you would break even after three games. Of course, if you continue betting, you will be above or below breaking even until you make six bets, at which point it is very likely that you break even again.

See if you can figure out the answers to the next two examples. You might save (or make) yourself some money this way.

EXAMPLE 4   What are the odds in favor of obtaining at least two heads when a fair coin is tossed four times? If you are to receive $10 if at least two heads are tossed, what is a fair bet?

Solution.   There are $2^4 = 16$ sample points in the sample space, each sample point having probability $1/16$. Let A be the event "at least two heads appear." There are 6 sample points with precisely two heads, 4 sample points with precisely three heads, and 1 sample point with four heads. Hence $P(A) = 11/16$ and the odds in favor of A are

$$\left(1 - \frac{11}{16}\right)^{-1}\left(\frac{11}{16}\right) = \frac{11}{5}$$

Therefore, if the payoff is $10, you should bet $22 in a fair bet.   □

EXAMPLE 5   A natural number from 1 to 10 is chosen at random. What is the fair bet that a number from 1 to 4 is chosen if the payoff is $6?

Solution.   The sample space is $\Omega = \{1, 2, \ldots, 10\}$, and each sample point has probability $1/10$. If A is the event "a number from 1 to 4 is chosen," then $P(A) = 4/10$. Hence the odds in favor of A are

$$\left(1 - \frac{4}{10}\right)^{-1}\left(\frac{4}{10}\right) = \frac{4}{6}$$

If the payoff is $6, a fair bet is $4.   □

EXERCISES

1. If $P(A) = 1/3$, what is $P(A')$?

2. If $P(A) = 1/3$, $P(B) = 1/2$, and $P(A \cap B) = 0$, what is $P(A \cup B)$?

3. If $P(A) = 1/3$, $P(B) = 1/2$, and $P(A \cap B) = 1/4$, what is $P(A \cup B)$?

4. A coin is loaded so that heads are twice as probable as tails. What is the probability of a head? A tail?

5. A coin is loaded so that heads is $3/2$ as probable as tails. What is the probability of a head? A tail?

6. Mark, Matt, and Willie take part in a race. Mark and Matt are equally likely to win and each is twice as likely to win as Willie. What is the probability that either Matt or Willie wins?

7. Show that $P(\emptyset) = 0$.

8. Prove (9.4) using (9.6).

9. Show that if $A \cap B = \emptyset$, then $P(A) + P(B) \le 1$.

10. Show that $P(A \cup A') = 1$.

11. Show that $P(A) + P(A') = 1$.

12. Show that $P(A \cap B) = P(A) + P(B) - P(A \cup B)$.

13. Show that $P(A) = 1 - P(A')$.

14. Let A and B be events. If $P(A) = 1/2$, $P(B) = 1/3$, and $P(A \cap B) = 1/4$, find $P(A - B)$, $P(B - A)$, $P(A \cup B)$, $P(A')$, and $P(B')$. Construct a sample space with events A and B so that A, B, and $A \cap B$ will have these probabilities.

15. Prove (9.7) using (9.2) and (9.4). Draw a Venn diagram to illustrate (9.7).

16. In Exercise 14 find $P(A' \cap B')$.

17. In Example 1, find the probabilities of the following.
    (a) A secretary cannot type.
    (b) A secretary cannot take shorthand.
    (c) A secretary either cannot type or cannot take shorthand.
    (d) A secretary can take shorthand but cannot type.
    (e) A secretary can type but cannot take shorthand.

18. In Example 2, find the probability that exactly one head is obtained. Find the probability that exactly two heads are obtained. For a payoff of $54, what is a fair bet that exactly one head is obtained?

19. In Example 3, what is the probability that the sum of the upward faces is greater than or equal to 16? What are the odds in favor of this event?

20. From the numbers 1, 2, 3, 4, 5, first one number is chosen, and then a second selection is made from the remaining four numbers. Assume that all 20 possible results have the same probability. Find the probability that an odd number will be selected   (a) the first time, (b) the second time,   (c) both times.

21. Four men and a woman attend a baroque concert and sit together in a row. If they seat themselves randomly, what is the probability that the woman has men on either side of her?

22. A fair die is tossed. What is the probability that a number that is greater than 4 or even occurs?

23. If someone offered to bet you $1 that if you give him $5 he will give you $100 in exchange, would that be a good bet?

24. Assume that the spinning pointer in Figure 9.7 has an equal probability of stopping on any one of the three numbered regions and does not stop on a line between two regions. Suppose the pointer is spun twice.
    (a) What is the sample space that represents this experiment?
    (b) What is the probability of each sample point?
    (c) What is the probability of the event "the sum of the two numbers at which the pointer stops is 4"?
    (d) What are the odds in favor of the event in part (c)?

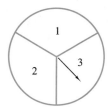

Figure 9.7

25. Let Ω be a sample space, and let $A, B \subseteq \Omega$.
    (a) Prove that $P(A \cup B) \leq P(A) + P(B)$.
    (b) If $A \subseteq B$, prove that $P(A) \leq P(B)$.
26. What is the relationship between (9.4) and (1.3) of Section 1.4?
27. What is the relationship between (9.6) and (1.7) of Section 1.4?
28. If $B \subseteq A$, show that $P(A - B) = P(A) - P(B)$.
29. If $P(A) = 1/2$, what are the odds in favor of A?
30. If $P(A) = 1/3$, what are the odds in favor of A?
31. If the odds in favor of A are $2/3$, what is $P(A)$?
32. If the odds in favor of A are x, what is $P(A)$?
33. If $P(A) = 1/2$ and \$10 is bet that A occurs, what is the payoff for a fair bet?
34. If $P(A) = 1/3$ and \$10 is bet that A occurs, what is the payoff for a fair bet?
35. If $P(A) = 3/4$ and the payoff that A occurs is \$10, what is a fair bet?
*36. If an amount x is bet that A occurs and an amount y is the payoff, show that the bet is fair if $P(A) = x(x + y)^{-1}$.
*37. In a fair bet an amount x is wagered that A occurs and the payoff is an amount y. Show that $P(A) = x(x + y)^{-1}$.
*38. Construct examples to show that the following need not hold.
    (a) If $P(A) = 1$, then $A = \Omega$.
    (b) If $P(A) = 0$, then $A = \emptyset$.
    (c) If $P(A \cup B) = P(A) + P(B)$, then $A \cap B = \emptyset$.
    (d) If $P(A) \leq P(B)$, then $A \subseteq B$.
    (e) If $P(B) = 1 - P(A)$, then $B = A'$.
*39. If $A_1, A_2, \ldots, A_n$ are mutually exclusive events, prove that

$$P\left(\bigcup_{i=1}^{n} A_i\right) = P(A_1) + P(A_2) + \cdots + P(A_n)$$

9.4 | Conditional Probability

Mathematics is the most exact science, and its conclusions are capable of absolute proof. But this is only because mathematics does not attempt to draw absolute conclusions. All mathematical truths are relative, conditional.
— CHARLES PROTEUS STEINMETZ (1865–1923), American Engineer

In this section conditional probability will be discussed. Before seeing the general definition of conditional probability, consider the following two examples.

EXAMPLE 1   A fair coin is flipped twice. Somebody tells us that at least one of the flips is heads. What is the probability that both flips are heads?

Solution.   Let A = {HH} be the event that both flips are heads, and let B = {HT, TH, HH} be the event that at least one flip is heads. In this case we know that event B has occurred. Given that B has occurred, we want the probability of event A. This is the *conditional probability of A given B* and is denoted $P(A \mid B)$. How can we find $P(A \mid B)$? Since B has occurred, we know that the outcome is either HT, TH, or HH. And since HH occurs one-third of the time among these three outcomes, we would expect that $P(A \mid B) = 1/3$. Another way of looking at this is the following. Since B has occurred, we can consider our sample space to be {HT, TH, HH}. Now since the coin is fair, each sample point has probability 1/3, and so again $P(A \mid B) = 1/3$. There is another way to compute $P(A \mid B)$. Notice that

$$1/3 = \frac{1/4}{3/4} = \frac{P(A \cap B)}{P(B)}$$

Thus

$$P(A \mid B) = \frac{P(A \cap B)}{P(B)}$$

This latter formula always holds [assuming $P(B) \neq 0$].   □

Notice that the conditional probability $P(A \mid B)$ is the probability of A given the additional information that B has occurred. This is different from the probability $P(A)$, in which case we have no additional information. For instance, in Example 1, $P(A \mid B) = 1/3$ while $P(A) = 1/4$.

EXAMPLE 2   Suppose a population of 100 people includes 10 color-blind people, 50 females, and 4 color-blind females. Let C, F, and $C \cap F$ be the events that a person chosen at random is color-blind, female, and both color-blind and female, respectively. Then $P(C) = 1/10$, $P(F) = 1/2$, $P(C \cap F) = 1/25$. Suppose we want the probability that a female chosen at random is color-blind or (put differently) the probability that a randomly chosen person is color-blind given that the person is female. In order to compute this probability, we would consider the subpopulation of females and see how many of these people are color-blind. Thus our required probability is 4/50.

$$\frac{4}{50} = \frac{4/100}{50/100} = \frac{P(C \cap F)}{P(F)}$$

We write this probability as $P(C \mid F)$ and call it the conditional probability of C given F.

Definition of Conditional Probability

In general the following definition is made. Let B be an event with positive probability. For an arbitrary event A define the **conditional probability of A given B** to be

$$P(A \mid B) = \frac{P(A \cap B)}{P(B)}$$

If $P(B) = 0$, the conditional probability is not defined. The conditional probability of A given B gives the probability that the event A occurs when we know that the event B has occurred. The idea behind this concept is the following. Suppose we know that B has occurred. This fact gives us additional information. If an arbitrary event A then occurs, we know that the resulting sample point lies in $A \cap B$. Now the probability that A occurs given that B has occurred is the "chance" that a sample point chosen from B lies in $A \cap B$. But this is the "chance" that a sample point lies in $A \cap B$ divided by the "chance" that the sample point lies in B. Thus

$$P(A \mid B) = \frac{P(A \cap B)}{P(B)}$$

Stated differently, since we know that B has occurred, we have replaced the whole sample space Ω by the event B and have replaced the probability function $P(A)$ by the probability function $P(A \cap B)/P(B)$.

Formulas for Conditional Probability

The conditional probability has all the properties of an ordinary probability. For example,

$$P(B \mid B) = 1$$
$$P(A' \mid B) = 1 - P(A \mid B)$$
$$P(A \cup B \mid C) = P(A \mid C) + P(B \mid C) - P(A \cap B \mid C)$$

You will be asked to verify these identities in the exercises.

Notice also that

$$P(A \cap B) = P(B)P(A \mid B) = P(A)P(B \mid A)$$

In words, the probability that A and B both occur equals the probability that A occurs times the probability that B occurs given that A has occurred (or the probability that B occurs times the probability that A occurs given that B has occurred). This can be generalized to three (or more) events A, B, C as follows:

$$P(A \cap B \cap C) = P(A \cap B)P(C \mid A \cap B)$$
$$= P(A)P(B \mid A)P(C \mid A \cap B)$$

Two very useful results involving conditional probabilities are Bayes' rules. Frequently, we know $P(A)$, $P(B)$, and $P(A \mid B)$ but would like to find $P(B \mid A)$. For instance, in Example 1, $P(A) = 1/4$, $P(B) = 3/4$, and $P(A \mid B) = 1/3$. Suppose we want to find $P(B \mid A)$, the conditional probability that at least one flip is heads given that both flips are heads. We can compute $P(B \mid A)$ as follows:

$$P(B \mid A) = \frac{P(B \cap A)}{P(A)} = \frac{P(B)P(A \mid B)}{P(A)}$$

Thus

$$P(B \mid A) = \frac{(3/4)(1/3)}{1/4} = 1$$

With a little thought we could have obtained this result immediately without relying on the formula. Indeed, if we are given that both flips are heads, then we conclude that at least one flip must be heads. Hence, if A occurs, then B certainly occurs, and so $P(B \mid A) = 1$.

In the next example the formula is really needed. Suppose, in Example 2, only the following information is given: $P(C) = 1/10$, $P(F) = 1/2$, and $P(C \mid F) = 2/25$ [$P(C \cap F)$ is not known]. We now want to find $P(F \mid C)$, the conditional probability that a person chosen at random is female given that the person is color-blind. Again, since

$$P(F \mid C) = \frac{P(F \cap C)}{P(C)} = \frac{P(F)P(C \mid F)}{P(C)}$$

we obtain

$$P(F \mid C) = \frac{(1/2)(2/25)}{1/10} = \frac{2}{5}$$

This formula is called **Bayes' first rule**.

Theorem 9.1   **Bayes' first rule**

$$P(B \mid A) = \frac{P(A \mid B)P(B)}{P(A)}$$

Proof:

$$P(B \mid A) = \frac{P(B \cap A)}{P(A)} = \frac{P(B)P(A \mid B)}{P(A)} \qquad \square$$

Bayes' second rule is motivated by the following example. We want to find the probability that a person chosen at random is left-handed. A book on population statistics gives us the following information. In the human population 51 percent are female, 49 percent are male, 20 percent of the females are left-handed, and 15 percent of the males are left-handed. Can we find the percentage of left-handed people from this information? Let M represent the males, F the females, and L the left-handers. The above statistics give $P(F) = 0.51$, $P(M) = 0.49$, $P(L \mid F) = 0.2$, $P(L \mid M) = 0.15$. We would like to find $P(L)$. Intuitively, we might expect that $P(L)$ is the probability of the part of L in F plus the probability of the part of L in M. That is, $P(L) = P(L \cap F) + P(L \cap M)$. From this we would obtain

$$P(L) = P(F)P(L \mid F) + P(M)P(L \mid M)$$
$$= (0.51)(0.2) + (0.49)(0.15) = 0.175$$

It turns out that this is correct and, in fact, this is an example of **Bayes' second rule**.

Theorem 9.2   **Bayes' second rule** Let B_1 and B_2 be mutually exclusive events such that $B_1 \cup B_2 = \Omega$. If $A \subseteq \Omega$, then

$$P(A) = P(B_1)P(A \mid B_1) + P(B_2)P(A \mid B_2)$$

Proof: As illustrated in Figure 9.8,

$$A = A \cap \Omega = A \cap (B_1 \cup B_2) = (A \cap B_1) \cup (A \cap B_2)$$

and $A \cap B_1$ and $A \cap B_2$ are mutually exclusive. Hence, we have

$$P(A) = P(A \cap B_1) + P(A \cap B_2)$$
$$= P(B_1)P(A \mid B_1) + P(B_2)P(A \mid B_2) \qquad \square$$

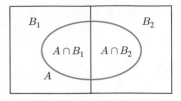

Figure 9.8

Bayes' second rule can be extended from two B's to any finite number of B's. For example, if $B_1 \cup B_2 \cup B_3 = \Omega$ and the B_i's are mutually exclusive, then

$$P(A) = P(B_1)P(A \mid B_1) + P(B_2)P(A \mid B_2) + P(B_3)P(A \mid B_3)$$

Important applications of conditional probabilities are now illustrated by some examples. Before looking at the solutions, see if you can work these examples yourself.

EXAMPLE 3   Two cards are drawn from a deck and are not replaced. (a) If the first card is a spade, what is the probability that the second is a heart? (b) What is the probability that the first card is a spade and the second card is a heart?
   Solution.   (a) Let A be the event "a spade is drawn first," and let B be the event "a heart is drawn second." We are asked to find $P(B \mid A)$. If a spade is drawn first there will be 51 cards left and 13 of these cards are hearts. Hence, $P(B \mid A) = 13/51$. (b) We are now asked to find $P(A \cap B)$. Since there are 13 spades, $P(A) = 13/52 = 1/4$. Hence,

$$P(A \cap B) = P(A)P(B \mid A) = \left(\frac{1}{4}\right)\left(\frac{13}{51}\right) = \frac{13}{204} \qquad \square$$

EXAMPLE 4   *Sampling without replacement.* From a population of n elements labeled $1, 2, \ldots, n$ an ordered random sample is taken. Let i and j be two different elements. Assuming i is the first element drawn, what is the probability that the second element is j? What is the probability of drawing i first and j second?
   Solution.   Let A be the event "i is drawn first," and let B be the event "j is drawn second." Then we want $P(B \mid A)$. Now if i is drawn first there will be $n - 1$ elements left, and so the probability of drawing j from the remaining elements is $1/(n - 1)$. Hence $P(B \mid A) = 1/(n - 1)$. For the second question we want $P(A \cap B)$. But

$$P(A \cap B) = P(A)P(B \mid A) = \frac{1}{n}\left(\frac{1}{n - 1}\right) \qquad \square$$

EXAMPLE 5   *Political polls.* For the purpose of a political poll the United States has been divided into four sections: Northeast, containing three-eighths of the population; South, containing one-eighth; Midwest, containing one-fourth; and West, containing one-fourth. In the poll it is found that in the next election 40 percent of the people in the Northeast say they will vote for Smith, in the South 56 percent will, in the Midwest 48 percent will, and in the West 52 percent will. What is the probability that a person chosen at random will vote for Smith? Assuming a person votes for Smith, what is the probability that that person is from the Northeast?

Solution. Let N, S, M, W be the events that a person chosen at random is from the Northeast, South, Midwest, and West, respectively. Let A be the event that a person chosen at random will vote for Smith. Applying Bayes' second rule gives

$$P(A) = P(N)P(A\,|\,N) + P(S)P(A\,|\,S) + P(M)P(A\,|\,M) + P(W)P(A\,|\,W)$$

$$= \frac{3}{8}(0.4) + \frac{1}{8}(0.56) + \frac{1}{4}(0.48) + \frac{1}{4}(0.52) = 0.47$$

To answer the second question, Bayes' first rule is applied:

$$P(N\,|\,A) = \frac{P(N)P(A\,|\,N)}{P(A)}$$

$$= \frac{(3/8)(0.4)}{0.47} = 0.32 \qquad \square$$

EXAMPLE 6 *Gender distributions.* The size of families in a certain city has been investigated. With the probability that a family has n children denoted by p_n, it was found that $p_0 = 0.2$, $p_1 = 0.1$, $p_2 = 0.3$, $p_3 = 0.2$, $p_4 = 0.1$, $p_5 = 0.07$, and $p_6 = 0.03$. Assume that for any given family size all gender distributions are equally probable; for example, in a family with two children the gender distributions bb, bg, gb, and gg each have the same probability. (a) What is the probability that a family has five children and the gender distribution is bbgbg? (b) What is the probability that a family has no girls? (c) What is the probability that a family has one child if it is known that the family has no girls?

Solution. The sample space for this example is the set of all gender distributions for all possible families. Thus

$$\Omega = \{0, \text{b, g, bg, gb, gg, bb, bbb,} \ldots\}$$

(a) Let A be the event that a family chosen at random has five children and let B be the event $B = \{\text{bbgbg}\}$. Then

$$P(A \cap B) = P(A)P(B\,|\,A)$$

Since there are $2^5 = 32$ possible gender distributions for a family with five children, $P(B\,|\,A) = 1/32$. Hence $P(A \cap B) = 0.07/32 = 0.0022$.

(b) Let C be the event that a family chosen at random has no girls, and let B_n be the event that a family chosen at random has n children, $n = 0, 1, \ldots, 6$. Then applying Bayes' second rule gives

$$P(C) = P(B_0)P(C\,|\,B_0) + P(B_1)P(C\,|\,B_1) + \cdots + P(B_6)P(C\,|\,B_6)$$

$$= 0.2 + \frac{0.1}{2} + \frac{0.3}{2^2} + \frac{0.2}{2^3} + \frac{0.1}{2^4} + \frac{0.07}{2^5} + \frac{0.03}{2^6} = 0.359$$

Photo 9.2
Conditional probability is used to
determine product reliability.

(c) Applying Bayes' first rule, we have

$$P(B_1 \mid C) = \frac{P(B_1)P(C \mid B_1)}{P(C)} = \frac{(0.1)(1/2)}{0.359} = 0.139 \qquad \square$$

EXAMPLE 7 *Machine reliability.* In a bolt factory, machines A, B, and C, respectively, manufacture 25, 35, and 40 percent of the total. Of their output, 5, 4, and 2 percent, respectively, are defective bolts. A bolt is drawn at random from the production batch and is found to be defective. What is the probability that it was manufactured by machine A?

Solution. Let A, B, and C be the events that a bolt is manufactured by machine A, B, C, respectively, and let D be the event that a bolt is defective. Then by Bayes' rules,

$$P(A \mid D) = \frac{P(D \mid A)P(A)}{P(D)} = \frac{P(D \mid A)P(A)}{P(A)P(D \mid A) + P(B)P(D \mid B) + P(C)P(D \mid C)}$$

$$= \frac{(0.05)(0.25)}{(0.25)(0.05) + (0.35)(0.04) + (0.40)(0.02)} = 0.36 \qquad \square$$

EXAMPLE 8 *A prisoner's dilemma.* A prisoner is given the following ultimatum:

> *Before you stand three urns, 100 black balls, and 100 white balls. You may distribute all the balls in the urns in any way you like as long as there is at least one ball in each urn. You will then be blindfolded and the urns will be rearranged. You must point to an urn and then pick a ball from that urn. If you pick a white ball, you are free. If you pick a black ball, you will be executed.*

How should the prisoner distribute the balls to have the greatest probability of freedom, and what is this probability?

Solution. Define the following events:

W = "a white ball is picked"
U_n = "a ball is picked from urn n," $n = 1, 2, 3$

Applying Bayes' second rule gives

$$P(W) = P(U_1)P(W \mid U_1) + P(U_2)P(W \mid U_2) + P(U_3)P(W \mid U_3)$$

$$= \frac{1}{3} P(W \mid U_1) + \frac{1}{3} P(W \mid U_2) + \frac{1}{3} P(W \mid U_3)$$

The prisoner would like to maximize $P(W)$. This can be done by maximizing the three terms on the right of the above equation. The first term can be maximized by making $P(W \mid U_1) = 1$ and hence placing only white balls in the first urn. The second term can be maximized by making $P(W \mid U_2) = 1$ and hence placing only white balls in the second urn. But now all the black balls must go into the third urn. To maximize $P(W \mid U_3)$ the prisoner must put as many white balls as possible into urn 3. Thus the best strategy is one white ball in the first urn, one in the second, and 98 white balls and 100 black balls in the third. Then

$$P(W) = \frac{1}{3} + \frac{1}{3} + \frac{1}{3} \left(\frac{98}{198} \right) \approx \frac{5}{6} \qquad \square$$

Bayes' rules can be very useful in determining the reliability of tests. For example, consider a test for a disease, say a type of cancer. It is common practice to subject the patient to a quite reliable but simple and inexpensive test. If the test comes out positive, the patient is given a more thorough and infallible test (say exploratory surgery) to determine if the disease really exists. If the preliminary test comes out positive, what are the chances that the patient has the disease? This depends upon the reliability of the preliminary test; but even if this first test is very reliable, the chances can be small that the patient is diseased.

Suppose the following data are available. The preliminary test has been administered to many people with the disease, and it is found that 99 percent of them show positive. It is also administered to many people who do not have the disease and 0.1 percent of them show positive. This test is not infallible, but it is certainly very reliable for indicating whether a patient has this disease or not. Finally, it has been found from doctor's records that 0.01 percent of the population have this disease. The important question is: If the patient shows positive on the test, what is the probability $P(D\,|\,+)$ that the patient has the disease?

The probability that a patient tests positive given that the patient has the disease is $P(+\,|\,D) = 0.99$, and the probability that the test is positive given that the patient does not have the disease is $P(+\,|\,D') = 0.001$. Finally, the probability that the patient has the disease is $P(D) = 0.0001$. Using Bayes' rules, $P(D\,|\,+)$ can be computed:

$$P(D\,|\,+) = \frac{P(D)P(+\,|\,D)}{P(+)} = \frac{P(D)P(+\,|\,D)}{P(D)P(+\,|\,D) + P(D')P(+\,|\,D')}$$

$$= \frac{(0.0001)(0.99)}{(0.0001)(0.99) + (0.9999)(0.001)} = \frac{99}{1098} \approx \frac{1}{10}$$

Thus even though the preliminary test indicates that the patient might have the disease, the chances are only 1 out of 10 of really having it.

Consider another example. It is fairly common for astronauts in a space satellite to inform Mission Control that a red warning light has gone on, indicating that a critical part is malfunctioning. Mission Control then assures the anxious television audience that this is probably caused by a faulty warning light and that the critical part is probably all right. How do they know this?

Through tests it has been found that when the part malfunctions, the warning light goes on 99.9 percent of the time. But since electronic components are never perfect, the warning light also goes on 0.01 percent of the time when the part is functioning correctly. This particular part malfunctions 0.0001 percent of the time. Thus the probability that the warning light goes on given that the part is malfunctioning is $P(+\,|\,M) = 0.999$, and the probability that the warning light goes on given that the part is not malfunctioning is $P(+\,|\,M') = 0.0001$. Finally, the probability that the part is malfunctioning is $P(M) = 0.000001$. Hence the probability $P(M\,|\,+)$ that the part is malfunctioning given that the warning light is on can be computed using Bayes' rules.

$$P(M\,|\,+) = \frac{P(M)P(+\,|\,M)}{P(+)}$$

$$= \frac{(0.000001)(0.999)}{(0.000001)(0.999) + (0.999999)(0.0001)} = \frac{999}{109{,}989} \approx \frac{1}{100}$$

The reason behind this is that the critical part itself is much more reliable than the warning light that monitors it.

EXERCISES

1. In Example 1, let C be the event that both flips are tails, and let B be defined as in that example. What is $P(C\,|\,B)$?

2. In Example 1, let D be the event that one flip is heads and the other tails, and let B be defined as in that example. What is $P(D\,|\,B)$?

3. For B, C defined in Exercise 1, find $P(B\,|\,C)$.

4. For B, D defined in Exercise 2, find $P(B \mid D)$.

5. If $P(A) = 1/2$ and $P(B \mid A) = 1/3$, what is $P(A \cap B)$?

6. If $P(A \cap B) = 1/3$ and $P(B) = 1/2$, what is $P(A \mid B)$?

7. If $P(A \mid B) = 1/3$, $P(B) = 1/4$, and $P(A) = 1/2$, what is $P(B \mid A)$?

8. In Example 6, what is the probability of a family in which there is precisely one boy (there can be any number of girls)? Given that a family has precisely one boy, what is the probability that there are three children?

9. In Example 7, what is the probability that a bolt is defective if it is known that it was manufactured by machine A or B?

10. In Example 5, what is the probability that a voter is from the South given that she voted for Smith? From the Midwest? From the West?

11. How would the answer to Example 8 change if there were four urns?

*12. State and prove Bayes' second rule if instead of B_1 and B_2 there were $B_1, B_2, \ldots, B_n$.

13. Show that $P(B \mid B) = 1$.

14. Show that $P(\emptyset \mid B) = 0$.

15. Show that $P(A' \mid B) = 1 - P(A \mid B)$.

*16. Show that $P(A \cup B \mid C) \le P(A \mid C) + P(B \mid C)$.

*17. Show that if $A \cap B = \emptyset$, then $P(A \cup B \mid C) = P(A \mid C) + P(B \mid C)$.

*18. Show that $P(A \cup B \mid C) = P(A \mid C) + P(B \mid C) - P(A \cap B \mid C)$.

19. A fair die is rolled three times. If all the faces are different, what is the probability that one is a 2?

20. Given that a throw of 10 fair dice produced at least one ace, what is the probability of two or more aces?

21. Suppose that 5 men out of 100 and 25 women out of 10,000 are color-blind. A color-blind person is chosen at random. What is the probability that it is a male? (Assume males and females to be in equal numbers.)

22. Suppose that in a population there are m males, f females, and c color-blind people. Suppose 5 percent of the men are color-blind, 2 percent of the men are doctors, 1 percent of the women are doctors, and 3 percent of the color-blind are doctors. Let M, F, C, D be the events that a person chosen at random is male, female, color-blind, or a doctor, respectively. Find $P(M \cap D)$, $P(M \mid D)$, $P(C \mid D)$, and $P(M \mid C)$.

23. Three fair dice are thrown. Consider the following events:
    $A = $ "all faces are the same"
    $B = $ "the sum of the faces is 17"
    $C = $ "at least two of the faces are the same"
    (a) Describe the sample space. How many sample points are there? What is the probability of each sample point?
    (b) Which of the above events imply which other events? Which of the events are mutually exclusive?
    (c) Find $P(A)$, $P(B)$, $P(C)$.
    (d) Find $P(A \mid B)$, $P(A \mid C)$, $P(B \mid A)$, $P(B \mid C)$, $P(C \mid A)$, and $P(C \mid B)$.

24. In a certain school 25 percent of the pupils failed physics, 15 percent failed chemistry, and 10 percent failed both. Select a student at random.
    (a) If the student failed chemistry, what is the probability that she failed physics?
    (b) If the student failed physics, what is the probability that he also failed chemistry?
    (c) What is the probability that the student failed at least one?

25. Suppose two cards are drawn successively from an ordinary deck without replacement. Consider the events

    A = "two hearts are drawn"
    B = "two cards of the same suit are drawn"
    C = "the first card drawn is a heart"

    (a) Describe the sample space.
    (b) How many sample points are there?
    (c) Which of the above events are subsets of which others?
    (d) Find $P(A)$, $P(B)$, $P(C)$.
    (e) Find $P(A \cap B)$, $P(A \cap C)$, and $P(B \cap C)$.
    (f) Find $P(A \mid B)$, $P(B \mid A)$, $P(A \mid C)$, $P(C \mid A)$, $P(B \mid C)$, and $P(C \mid B)$.

26. There are three slot machines each of which normally pays with probability 0.1, but one of the machines is out of order and pays with probability 0.4. We select one of the machines at random and play it once. If we win, what is the probability that we selected the profitable machine? If we lose, what is the probability that we selected the profitable machine?

27. In a class of 100 students, 25 failed physics, 15 failed chemistry, and 10 failed both. A student is selected at random. If the student failed chemistry, what is the probability that the student failed physics?

28. Mike throws a fair die; then Walter throws the same die. What is the probability that Walter will throw a higher number than Mike?

29. Dolly flips one fair coin, and Gail flips two fair coins. Gail wins if she has more heads than Dolly. What is the probability that Gail wins? What is the answer if Gail wins when she has at least as many heads as Dolly?

30. A new couple moves in next door. You are told they have two children. What is the probability that at least one is a boy? You see one of their children playing in the yard and notice that she is a girl. Now what is the probability that at least one is a boy?

31. On the toss of a pair of fair dice, what is the probability that their sum is 8 if you know that neither of the dice came up 3?

32. On the toss of a pair of fair dice, what is the probability that their sum is 6 given that one of the dice came up   (a) even,   (b) odd,   (c) 2 or greater,   (d) 4 or greater?

33. A geologist is using seismographs to test for oil. It is found that if oil is present the test gives a positive result 95 percent of the time and if oil is not present the test gives a positive result 2 percent of the time. Finally, oil is discovered in 1 percent of the cases tested. If the test shows positive, what is the probability that oil is present?

34. In a certain population of people, 25 percent are blue-eyed and 75 percent are brown-eyed. Also 10 percent of the blue-eyed people are left-handed, and 5 percent of the brown-eyed people are left-handed.
    (a) What is the probability that a person chosen at random is blue-eyed and left-handed?
    (b) What is the probability that a person chosen at random is left-handed?
    (c) What is the probability that a person is blue-eyed given that he or she is left-handed?

*35. A fair die is tossed, and a fair coin is flipped the number of times that appears on the die. (For example, if the die shows a 4, the coin is flipped 4 times.)
    (a) How many sample points are there?
    (b) What is the probability of getting three heads given that a 3 appears upward on the die?
    (c) What is the probability of getting only heads?

*36. Suppose there is a human population consisting of mutually exclusive subpopulations or strata $B_1, B_2, \ldots, B_n$. Let p_i be the probability that an individual chosen at random belongs to B_i, $i = 1, 2, \ldots, n$ (so $p_1 + p_2 + \cdots + p_n = 1$). Let A be the event that an individual is left-handed. Suppose the probability that an individual in B_i is left-handed is $P(A \mid B_i) = q_i$, $i = 1, 2, \ldots, n$.

(a) What is the probability that an individual chosen at random is left-handed?

(b) What is the probability that an individual belongs to stratum B_i given that the person is left-handed?

37. An urn contains b black and r red balls. A ball is drawn at random. It is replaced and, moreover, c balls of the color drawn and d balls of the opposite color are added. A new random drawing is made from the urn (now containing $r + b + c + d$ balls), and this procedure is repeated any number of times. The integers c and d may be negative. In particular, choosing $c = -1$ and $d = 0$, we have the model of random drawings without replacement. Give the probabilities of the following events.

(a) A red ball is drawn.

(b) A black ball is drawn second given that a black ball is drawn first.

(c) A black ball is drawn second.

(d) A black ball is drawn in the first and second drawings.

(e) A black ball is drawn in the first three drawings.

9.5 Tree Diagrams*

If we approach the Divine through symbols, then it is most suitable that we use mathematical symbols, these have an indestructible certainty.

— NICHOLAS OF CUSA (1401–1464), Roman Catholic Prelate

A tree diagram, or probability tree, is a useful way of indicating the possible outcomes of a probability experiment. If you know how to use a tree diagram, then you can work problems requiring Bayes' rules without relying on any formulas. Bayes' rules come out automatically in tree diagrams. Some people prefer to use Bayes' rules directly, and others find tree diagrams a very useful pictorial device. You should decide for yourself which method you prefer.

EXAMPLE 1 There are two boxes B_1 and B_2. Box B_1 contains two white balls and one red ball, and box B_2 contains one white and three red balls. A box is picked at random, and one ball is drawn. (a) What is the probability that it is white? (b) If the ball drawn is white, what is the probability that box B_1 was selected?

Solution. We could use Bayes' rules directly to solve these problems, but we shall use tree diagrams. Figure 9.9 is the tree diagram for this experiment. Figure 9.9 shows that there are two choices for the box and then there are two possible colors for the ball that is drawn after the box is selected.

---

*This section is based on unpublished notes by George Springer.

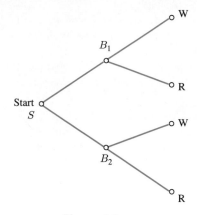

Figure 9.9

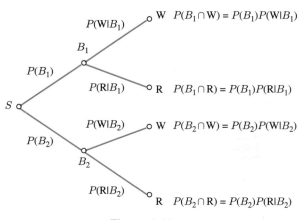

Figure 9.10

Let us label each branch (line segment) of this tree with the probability of selecting that branch. For example, $P(B_1)$ is the probability of selecting branch SB_1, so we write $P(B_1)$ on that segment (see Figure 9.10). If we are at B_1, we have a choice of two paths: B_1W or B_1R. Given that we are at B_1, $P(W \mid B_1)$ denotes the probability of drawing a white ball, and $P(R \mid B_1)$ denotes the probability of drawing a red ball. The tree in Figure 9.10 has all the branches labeled. The path SB_1W, which starts at S, goes to B_1, and then to W, represents the outcome of choosing B_1 and then drawing a white ball. This event is denoted by $B_1 \cap W$. The path SB_1W then has probability $P(B_1 \cap W)$, but we know from the definition of conditional probability that

$$P(B_1 \cap W) = P(B_1)P(W \mid B_1)$$

Thus the path SB_1W has probability equal to the product of the branch probabilities $P(B_1)$ and $P(W \mid B_1)$. In each of the other paths from S to W or R, we get the path probability as the product of the branch probabilities. These are written on the right side of Figure 9.10.

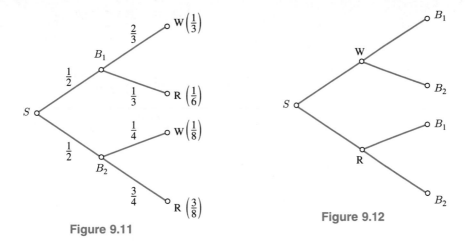

Figure 9.11

Figure 9.12

Let us next fill in the appropriate numbers on the tree. Since a box is chosen at random, $P(B_1) = P(B_2) = 1/2$, and so we write $1/2$ on each branch SB_1 and SB_2. If we have chosen B_1, we have probability $2/3$ of getting a white ball and $P(W \mid B_1) = 2/3$. We then write $2/3$ on the segment B_1W. Similarly, we write $1/3$ on B_1R. Figure 9.11 shows all the branch probabilities filled in.

Now for the path probabilities. For the probability of the path SB_1W, we multiply the two branch probabilities $1/2$ and $2/3$ to get $1/3$. We write $1/3$ after the W to the right of the tree. To the right of the R on the path SB_1R we write $(1/2)(1/3) = 1/6$. Figure 9.11 shows all the path probabilities in the right-hand column.

(a) We can get a white ball in one of two distinct ways corresponding to the paths SB_1W and SB_2W. Thus we have $P(W) = 1/3 + 1/8 = 11/24$, which is the sum of the probabilities attached to these two paths. Notice that this corresponds to the probability statement

$$P(W) = P(W \cap B_1) + P(W \cap B_2) = P(B_1)P(W \mid B_1) + P(B_2)(W \mid B_2)$$

Similarly, we have $P(R) = 1/6 + 3/8 = 13/24$.

(b) In this case, we ask for $P(B_1 \mid W)$, that is, the probability that box B_1 was chosen given that a white ball was drawn. We now make a new tree (Figure 9.12), in which W and R are given as the first two alternatives, and then follow each of these with the two choices B_1 and B_2.

In Figure 9.13 we have labeled each of the branches with the appropriate probabilities. For example, SW is labeled $P(W)$, and SR is labeled $P(R)$. Similarly, WB_1 is labeled $P(B_1 \mid W)$, and path SWB_1 is labeled $P(W \cap B_1) = P(W)P(B_1 \mid W)$.

Let us now see what numbers we can put on this tree. The probabilities $P(W)$ and $P(R)$ were found above [in solving (a)], to be $11/24$ and $13/24$, respectively, and so we write these numbers in Figure 9.14 on SW and SR, respectively. Notice that $P(W \cap B_1) = P(B_1 \cap W)$, and the latter was found at the end of the top path in Figure 9.11. There $P(B_1 \cap W)$ was found to be $1/3$, and so we write $1/3$ at the end of the path SWB_1. Similarly, the path SWB_2 in Figure 9.14 has the same probability as the path SB_2W in Figure 9.11. We label SWB_2 with $1/8$. The other paths are all labeled in the same way by referring back to the corresponding paths in Figure 9.11.

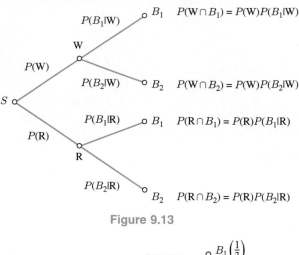

Figure 9.13

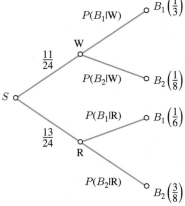

Figure 9.14

Now we want to find $P(B_1 \mid W)$, that is, the probability of the branch WB_1. We saw that the probability of the path SWB_1 is the product of the probabilities of the branches SW and WB_1, and so $(11/24)P(B_1W) = 1/3$. Thus

$$P(B_1 \mid W) = \frac{1/3}{11/24} = \frac{8}{11} \qquad \qquad \square$$

Notice how we found the answer to part (b). We first drew the tree in Figure 9.11 and filled in the known probabilities. From this tree we got $P(W)$ and $P(R)$. Then we reversed the tree, as in Figure 9.14, and used the values of $P(W)$ and $P(R)$ found in Figure 9.11 to fill in the branch probabilities and the path probabilities. We then got the probability of the branch WB_1 by dividing the probability of the path SWB_1 by the probability of the branch SW. We can similarly get $P(B_1 \mid R)$ by dividing $1/6$ by $13/24$; that is,

$$P(B_1 \mid R) = \frac{1/6}{13/24} = \frac{4}{13}$$

Can you find $P(B_2 \mid W)$ and $P(B_2 \mid R)$?

EXERCISES

1. For the experiment posed in this section, find $P(B_2 \mid W)$.

2. For the experiment posed in this section, find $P(B_2 \mid R)$.

3. Find the answers to Example 1 using Bayes' rules.

4. Alter the experiment of this section as follows. Box B_1 contains three white balls and two red balls, and box B_2 contains one white ball and four red balls. Using tree diagrams, solve Example 1.

5. Find $P(B_1 \mid R)$, $P(B_2 \mid W)$, and $P(B_2 \mid R)$ for Exercise 4 using tree diagrams.

6. Solve Exercise 4 using Bayes' rules.

7. Solve Exercise 5 using Bayes' rules.

8. There are three boxes B_1, B_2, and B_3. Box B_1 contains two white balls and one red ball; box B_2 contains one white ball and three red balls; and box B_3 contains one white ball and one red ball. A box is picked at random, and one ball is drawn. Using tree diagrams, answer the following questions.
   (a) What is the probability that the ball is white?
   (b) If the ball drawn is white, what is the probability that box B_1 was selected?

9. Find $P(B_1 \mid R)$, $P(B_2 \mid W)$, and $P(B_2 \mid R)$ in Exercise 8.

10. Solve Exercise 8 using Bayes' rules.

11. Solve Exercise 26 of Section 9.4 using tree diagrams.

12. Solve Exercise 34 of Section 9.4 using tree diagrams.

*13. There are two boxes, B_1 and B_2. Each box has a left and right compartment. In the left compartment of B_1 there are two white balls and three red balls, and in the right compartment of B_1 there is one white ball and one red ball. In the left compartment of B_2 there are two white balls and one red ball, and in the right compartment of B_2 there are three white balls and one red ball. Suppose there is a probability of $1/3$ of picking B_1 and $2/3$ of picking B_2. After a box is picked, a compartment is picked at random and a ball is drawn.
   (a) What is the probability that the ball is white?
   (b) If the ball drawn is red, what is the probability that box B_2 was selected?

*14. Suppose you are playing the following game. A fair die is tossed. If 1, 2, or 3 appears, you win. If 6 appears, you lose. If 4 or 5 appears, you neither win nor lose but call a number between 1 and 6 (inclusive) and toss again. On the next toss, if your number appears, you win, and if it does not, you lose. Use tree diagrams to answer the following.
   (a) What is your probability of winning?
   (b) If you win, what is the probability that you tossed a 4 on the first throw?

9.6 | Independence

How can it be that mathematics, being after all a product of human thought independent of experience, is so admirably adopted to the objects of reality?   — ALBERT EINSTEIN (1879–1955), Physicist

In the examples of Section 9.4 we saw that the conditional probability $P(A \mid B)$ is generally not equal to $P(A)$. This is because the occurrence of B changes the probability of the occurrence of A.

> If $P(A \mid B) = P(A)$, we say that A and B are **independent**.

Thus A and B are independent if the occurrence of B has no effect on the probability that A will occur. Since $P(A \mid B) = P(A \cap B)/P(B)$, we see that A and B are independent if and only if $P(A \cap B) = P(A)P(B)$ [if $P(B) = 0$, then $P(A \mid B)$ is not defined and we use this last identity to define independence]. We see from this last identity that A is independent of B if and only if B is independent of A.

EXAMPLE 1   Suppose a fair coin is tossed twice and A is the event "a head is tossed on the first flip" and B is the event "a head is tossed on the second flip." Then $P(A) = 1/2$ and $P(B) = 1/2$. Since

$$P(A \cap B) = \frac{1}{4} = P(A)P(B)$$

A and B are independent. This agrees with our intuition, since a coin has no memory and does not care what was tossed the first time when the second toss is performed. Thus the result of a toss does not depend on the results of previous tosses.

EXAMPLE 2   Suppose we have a family with two children. Let A be the event "there is at most one girl," and let B be the event "the family has children of both genders." Then $P(A) = 3/4$. Since $P(B) = 1/2$ and $P(A \cap B) = 1/2 \neq P(A)P(B)$, A and B are not independent.

To see that it is not always intuitively obvious whether events are independent, consider the next example.

EXAMPLE 3   Suppose we have a family with three children and consider the events A and B in Example 2. In this case the sample space has eight sample points and the probability of each sample point is $1/8$. Then

$$P(A) = P\{bbg, bgb, gbb, bbb\} = \frac{1}{2}$$

$$P(B) = P\{bbg, bgb, gbb, ggb, gbg, bgg\} = \frac{3}{4}$$

$$P(A \cap B) = P\{bbg, bgb, gbb\} = \frac{3}{8}$$

Hence $$P(A)P(B) = \frac{3}{8} = P(A \cap B)$$

and so A and B are independent.

The next example shows that independence can be used to compute probabilities.

EXAMPLE 4   A biased coin is flipped three times. The probability of a head in a single toss is $1/3$, and the probability of a tail in a single toss is $2/3$. Let H_1, H_2,

H_3, T_1, T_2, T_3 be the event that a head is tossed the first time, a head is tossed the second time, etc. Using the fact that each toss is independent of the others, we can use independence to compute the probabilities of the eight sample points. For example,

$$P\{HHH\} = P(H_1 \cap H_2 \cap H_3)$$
$$= P(H_1)P(H_2 \mid H_1)P(H_3 \mid H_1 \cap H_2)$$
$$= P(H_1)P(H_2)P(H_3) = \left(\frac{1}{3}\right)\left(\frac{1}{3}\right)\left(\frac{1}{3}\right) = \frac{1}{27}$$

$$P\{HTT\} = P(H_1 \cap T_2 \cap T_3)$$
$$= P(H_1)P(T_2 \mid H_1)P(T_3 \mid H_1 \cap T_2)$$
$$= P(H_1)P(T_2)P(T_3) = \left(\frac{1}{3}\right)\left(\frac{2}{3}\right)\left(\frac{2}{3}\right) = \frac{4}{27}$$

Suppose we have three events A, B, C that are pairwise independent. We might think that this implies the independence of such pairs of events as $A \cap B$ and C, but unfortunately this is not necessarily so, as the next example shows.

EXAMPLE 5   Two fair dice are tossed, and three events are defined as follows:

A = "odd face with the first die"
B = "odd face with the second die"
C = "odd face with one die and even face with the other"

Then
$$P(A) = P(B) = P(C) = \frac{1}{2}$$

$$P(A \cap B) = P(A \cap C) = P(B \cap C) = \frac{1}{4}$$

and so A, B, and C are pairwise independent. However,

$$A \cap B \cap C = \emptyset$$

and so
$$P(A \cap B \cap C) \neq P(A \cap B)P(C)$$

Thus $A \cap B$ and C are not independent.

Sometimes when we compute the probability for an experiment, we get results that are contrary to our feeling of what should take place. Such examples remind us that one of the main purposes of probability theory is to handle problems our intuition cannot. Two examples are now examined in which a careful analysis conflicts with intuition.

Colored Cards Example

Suppose you are shown two cards, one that is red on both sides and one that is white on one side and red on the other. Each of the two cards is then placed in a separate

Photo 9.3

What is the probability that two of these people have the same birthday?

envelope. The two envelopes are then mixed, and one of them is placed on a table. The card from inside the envelope is slid out without being lifted, and you observe that the red side is up. What is the probability that the other side is also red? Perhaps you feel that there are two possible outcomes: the other side of card is either red or white. Hence the probability is $1/2$.

But actually the probability is $2/3$. This is because there are three possible outcomes if the red side is up. Either you have the red-white card, or you have the red-red card with one red side up or the other red side up. Out of these three outcomes there are two chances of getting a red below.

Birthday Example

For the second example, suppose you enter a room containing many people. What is the probability that the first person you meet has the same birthday as yours (month and day, not year)? To make things simple, assume that no one is born on February 29 and that being born on a particular day is as likely as being born on any other day. If your birthday is, say, June 4, then the probability that the other person has this birthday is $1/365$. Suppose you walk into a room containing two people; what is the probability that at least two of you have the same birthday? A direct computation is relatively complex, but it is simple to find the probability that none of the three has the same birthday. The probability that the second does not have the same birthday as the first is $364/365$. Since one person's having a certain birthday does not depend on another person's having that birthday, there is independence. Hence

$$P(\text{none share a birthday}) = \frac{364}{365}\left(\frac{363}{365}\right) = 0.991796$$

and therefore

$$P(\text{at least two share a birthday}) = 1 - 0.991796 = 0.008204$$

or about eight times in 1,000 tries. If there are four people in the room, then

$$P(\text{none share a birthday}) = \frac{364}{365}\left(\frac{363}{365}\right)\left(\frac{362}{365}\right) = 0.98364$$

and so

$$P(\text{at least two share a birthday}) = 1 - 0.98364 = 0.01636$$

In general the following formula holds:

$$P(\text{at least two in } n \text{ share a birthday}) = 1 - \frac{(364)(363)\cdots(365 - n + 1)}{365^{n-1}}$$

Working this formula out for even small n is very tedious. Table 9.4 was made with the aid of a computer.

Thus in a group of only 23 people more than half the time there are two sharing a birthday; 97 times out of 100 there are two sharing a birthday in a group of 50 people; while in a group of 90 the odds are 160,000 to 1 that there are two sharing a birthday.

For example, consider the first 36 presidents of the United States. Of this group two share a birthday, Warren Harding and James Polk, November 2. The same calculations that we have made hold just as well with probabilities of deaths. Of the 33 presidents who are dead, Millard Fillmore and William Taft died March 8. John Adams, Thomas Jefferson, and James Monroe died July 4.

Probability Pitfalls

Finally, a word of warning. You can get into trouble blindly applying probability. There are many examples in which you can lie with statistics. A very simple and ludicrous one is now given. Statistically, walking is 92 percent fatal! What is being claimed? It is claimed that 92 percent of the time a person who walks has died. How? Approximately 65 billion people have lived on earth up to today. Of these, approximately 5 billion are still alive or, perhaps better said, are alive today. Therefore 60 out of 65 billion or 92 percent of the people who have walked have died. By the same reasoning, birth is fatal 92 percent of the time since 92 percent of the people that were born have died.

A subtle trap that people sometimes fall into using probability and statistics might be called "significance after the fact." For example, a math student flipped a coin 20 times and got the result THHTTHTTTHHTHTHHTTTH. He then computed that the probability of this event is $1/2^{20} = 1/1,048,576$. He concluded that a miracle had occurred and switched from math to theology. As another example, a psychologist

Table 9.4

Number of people	Probability of event A that at least two have same birthday	Odds in favor of A
5	0.027	1 to 36
10	0.117	3 to 23
15	0.253	1 to 3
18	0.347	53 to 100
20	0.411	70 to 100
21	0.444	80 to 100
22	0.476	91 to 100
23	0.507	103 to 100
24	0.538	116 to 100
25	0.569	132 to 100
27	0.627	168 to 100
30	0.706	240 to 100
35	0.814	438 to 100
40	0.891	817 to 100
50	0.970	32 to 1
60	0.9951	169 to 1
70	0.99916	1190 to 1
80	0.99991	11500 to 1
90	0.99999	160000 to 1

studying extrasensory perception set up an experiment in which he placed 10 envelopes in a row and put a card in some envelopes and nothing in the others. He asked 1,000 subjects to tell which envelopes contained the cards. One of the subjects, Jonathan Fairweather, correctly identified the right envelopes. The psychologist concluded that Fairweather had amazing extrasensory powers and published a paper documenting his remarkable achievement. What is the erroneous reasoning in these two examples?

EXERCISES

1. In Example 1, let C be the event "a tail is tossed on the first flip," and let D be the event "a tail is tossed on the second flip." Are C and D independent? State your reason in physical terms and also using the definition of independence.

2. In Example 2, let C be the event "there is exactly one girl," and let D be the event "there is exactly one boy." Are C and D independent?

3. If the coin in Example 4 is flipped four times, find $P\{HHHH\}$.

4. If the coin in Example 4 is flipped four times, find $P\{HTTT\}$.

5. A fair die is rolled three times. What is the probability that a 4 or less is thrown each time?

6. A fair die is rolled. Let A be the event "1, 2, or 3 is rolled" and let B be the event "3, 4, or 5 is rolled." Are A and B independent?

7. A loaded die is rolled. The probabilities for this die are $P\{1\} = P\{2\} = P\{4\} = P\{5\} = 1/8$ and $P\{3\} = P\{6\} = 1/4$. Are the events defined in Exercise 6 independent in this case?

8. An urn contains 10 balls numbered from 1 to 10. Two balls are drawn at random without replacement. Let A be the event "the first ball is odd," and let B be the event "the second ball is odd." Find $P(A)$, $P(B)$, $P(A \cap B)$. Are A and B independent?

9. A student taking mathematics, sociology, English, and philosophy estimates that his probability of receiving an A is $1/10$, $1/5$, $1/2$, and $3/5$, respectively. Since the four professors involved never talk to each other, the student assumes that the grades can be regarded as independent events. Find the probability that he receives (a) no A's, (b) exactly one A.

10. If $P(A) = 1/3$, $P(B) = 1/4$, $P(A \cup B) = 1/2$, what is $P(A \mid B)$? Are A and B independent?

11. (a) If the event A is independent of itself, find $P(A)$.
    (b) If the event A is independent of A', find $P(A)$.
    (c) If $A \subseteq B$, $P(A) \neq 0$, and A and B are independent, find $P(B)$.

12. A fair coin is flipped three times. Which of the following events are independent of which others? Find their probabilities.
    $A =$ "head on first throw"
    $B =$ "tail on second throw"
    $C =$ "the same side turns up on all three flips"
    $D =$ "at most one head occurs"

13. If A and B are independent and $P(A) = 1/2$, $P(B) = 3/4$, what is $P(A \cup B)$?

14. If $P(A) = 5/6$, $P(B) = 1/3$, and $P(A \cup B) = 8/9$, are A and B independent? No

15. If $P(A \cup B) = P(A)P(B') + P(B)$, show that A and B are independent.

16. If A and B are independent, show that $P(A \cup B) = P(A)P(B') + P(B)$.

17. The Smiths are expecting a baby. Assume that the probability that the baby will be female is $1/2$. Due to their gene makeup, the probability that the baby will have blue eyes is $1/4$. (a) What is the probability that their baby will be female with blue eyes? (b) What is the probability that their baby will be female or have blue eyes? What assumption are you making to work this problem?

18. Consider the toss of a fair die and the events $A = \{1, 2, 3, 4\}$, $B = \{4, 5, 6\}$, $C = \{2, 4, 6\}$. Are these events independent in pairs?

19. Two people, A and B, are practicing archery. The probabilities of their hitting the target are $P(A) = 3/4$, $P(B) = 2/3$. It they shoot at the same time, what is the probability that at least one hits the target?

20. A box contains eight balls numbered 1 to 8. Consider the following draws: $A = \{1, 2, 3, 4\}$, $B = \{2, 4, 6, 8\}$, $C = \{3, 6\}$. Are they independent in pairs? Is $A \cap B$ independent of C?

21. Two cards are drawn from a deck. What is the probability that both cards are aces given that (a) the first card is returned to the deck, (b) it is not returned to the deck?

22. Three consecutive draws are made from a bowl with 4 red, 5 white, and 6 blue balls. What is the probability that the balls will be drawn in the order red, white, blue given that (a) they are replaced, (b) they are not replaced?

23. If the payoff is $10, what is a fair bet that at least two people at a party of 20 have the same birthday?

24. What is the difference between two events being independent and being mutually exclusive? Give an example of two events that are independent but not mutually exclusive. Give an example of two events that are mutually exclusive but not independent.

25. A sports car manufacturer enters two drivers in a race. Let A_1 be the event that driver 1 "shows," that is, is among the first three drivers across the finish line, and let A_2 be the event that driver 2 shows. Assume that the events A_1 and A_2 are independent and that

$P(A_1) = P(A_2) = 0.1$. Compute the following probabilities. (a) Neither of the drivers shows. (b) Both drivers show. (c) Driver 1 shows, given that driver 2 shows.

*26. Prove that if A and B are independent, so are A and B'.

*27. If A, B, C are pairwise independent and $P(A \cap B \cap C) = P(A)P(B)P(C)$, prove that (a) $A \cup B$ and C are independent, (b) $A \cap B$ and C are independent.

*28. Write a BASIC program that gives the probability that at least two people have the same birthday in a group of N people.

9.7 | Probability and Genetics (Optional)

There are things which seem incredible to most men who have not studied mathematics.

— ARCHIMEDES (287–212 B.C.), Greek Scientist

Probability theory was used by the botanist Gregor Mendel in 1850 for his studies of the genetics of plants. One of Mendel's studies concerned the seed color of the pea plant. A single pea plant produces either all green seeds or all yellow seeds. He noticed that there were three types of seeds: pure-line green seeds, pure-line yellow seeds, and hybrid yellow seeds. When planted, the pure-line green seeds grew into plants that upon self-pollination produced only green seeds. The pure-line yellow seeds grew into plants that upon self-pollination produced only yellow seeds, while the hybrid yellow seeds grew into plants that upon self-pollination produced yellow or green seeds (not both). These last yellow and green seeds were pure-line green, hybrid yellow, and pure-line yellow in proportion 1 to 2 to 1; that is, yellow outnumbered green 3 to 1.

Mendel's Explanation

Mendel explained these experimental results as follows. Each egg cell and pollen cell contains two genes, which could be yellow or green characteristic genes (see Figure 9.15). If a YY egg cell unites with a YY pollen cell, the result is a YY seed, which is a pure-line yellow (YY) seed. If a GG egg cell unites with a GG pollen cell, the result is a GG seed, which is a pure-line green (GG) seed. However, if a YG egg cell (there is no difference between YG cells and GY cells) unites with a YG pollen cell, the type of resulting seed is determined purely by chance. In this case, one-fourth of the time we get a pure-line yellow (YY) seed, one-fourth of the time a pure-line green (GG) seed, one-fourth of the time a hybrid yellow (YG) seed in which the Y gene is contributed by the egg cell and the G gene by the pollen cell, and one-fourth of the time a hybrid yellow (YG) seed in which the G gene is contributed by the egg cell and the Y gene by the pollen cell. Pure-line yellow, hybrid yellow, and pure-line green are present in proportion 1 to 2 to 1 and yellow outnumbers green 3 to 1. The probability of a YY seed is $P(YY) = 1/4$, of a YG seed $P(YG) = 1/2$, and of a GG seed $P(GG) = 1/4$. The different possibilities are depicted in Figure 9.16.

In Figure 9.16a, the probability of getting a yellow seed is

$$P(\text{yellow seed}) = 1$$

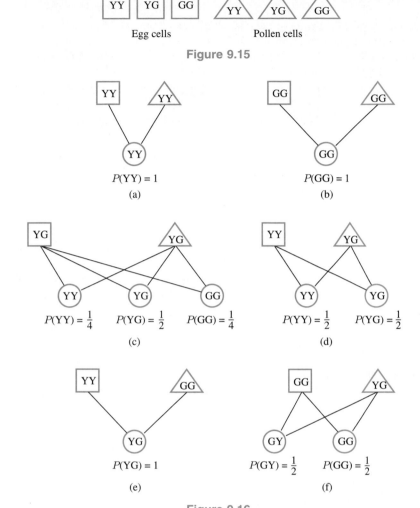

Figure 9.15

Figure 9.16

in Figure 9.16c,

$$P(\text{yellow seed}) = P(YY) + P(YG) = \frac{3}{4}$$

while $$P(\text{green seed}) = P(GG) = \frac{1}{4}$$

In Figure 9.16d,

$$P(\text{yellow seed}) = P(YY) + P(YG) = \frac{1}{2} + \frac{1}{2} = 1$$

In Figure 9.16f,

$$P(\text{yellow seed}) = \frac{1}{2} \quad \text{and} \quad P(\text{green seed}) = \frac{1}{2}$$

Genetic Tables

A useful method for determining the characteristics of the resulting seed after an egg cell and pollen cell unite is by using tables. Suppose a YG egg cell and a YG pollen cell unite, as in Table 9.5. Since there is one YY, one GG, and two YGs (remember, there is no difference between YG and GY since in the resulting seed there is no way to tell whether Y or G came from the egg cell or pollen cell), $P(\text{YY}) = 1/4$, $P(\text{GG}) = 1/4$, and $P(\text{YG}) = 1/2$. As another example suppose a GY egg cell unites with a GG pollen cell. This is described by Table 9.6. Hence in this case $P(\text{GG}) = 1/2$ and $P(\text{YG}) = 1/2$.

Table 9.5

	Y	G
Y	YY	YG
G	GY	GG

Table 9.6

	G	G
G	GG	GG
Y	YG	YG

These methods apply not only to the color characteristics of pea plants but also to any genetic characteristics of all biological systems. *Suppose A, B are the two genes of some genetic characteristic and C, D are the two genes for some other genetic characteristic. Then the genes for the two characteristics combine independently. For example, if an (AB, CD) egg cell unites with an (AB, CD) pollen cell, then the resulting possibilities are found as follows. First find the possible A, B combinations, then the C, D combinations, and finally put these combinations together. This is illustrated in Tables 9.7, 9.8, and 9.9. The probabilities of the different characteristics become

AA, CC = 1/16	AA, DC = 1/8	AA, DD = 1/16
AB, CC = 1/8	AB, CD = 1/4	AB, DD = 1/8
BB, CC = 1/16	BB, CD = 1/8	BB, DD = 1/16

Table 9.7

	A	B
A	AA	AB
B	BA	BB

Table 9.8

	C	D
C	CC	CD
D	DC	DD

---

* Since this book is about mathematics, not genetics, several simplifications have been introduced to avoid long digressions into biological explanations. For example, we use capital and lowercase letters here in a way that differs from standard genetic notation. The student should be aware that in actuality almost all human traits (height, eye color, and the like) are governed not by a simple combination of two genes, as in the model used here, but by complicated interactions of many genes. Laboratory animals are chosen for genetic studies on the basis of the relative simplicity of their genetic makeup and the short time required between generations.

Table 9.9

	CC	DC	CD	DD
AA	(AA, CC)	(AA, DC)	(AA, CD)	(AA, DD)
BA	(BA, CC)	(BA, DC)	(BA, CD)	(BA, DD)
AB	(AB, CC)	(AB, DC)	(AB, CD)	(AB, DD)
BB	(BB, CC)	(BB, DC)	(BB, CD)	(BB, DD)

EXAMPLE 1 Let T represent "tall" and S represent "short."

(a) A man with height genes TT marries a woman with height genes TS. What is the probability that their child is tall? (Tall dominates over short.)

(b) If the man is TS and the woman TS, what is the probability that their child is tall?

(c) If the man is TT and the woman is SS, what is the probability that their child is tall?

(d) A man has a TT father and an SS mother and his wife is SS. What is the probability that their child is tall?

(e) A man has a TS father and a TS mother, and his wife is SS. What is the probability that their child is tall?

Solution.

(a)

	T	S
T	TT	TS
T	TT	TS

$P(\text{tall child}) = P(TT) + P(TS) = 1$

(b)

	T	S
T	TT	TS
S	ST	SS

$P(\text{tall child}) = P(TT) + P(TS) = \frac{1}{4} + \frac{1}{2} = \frac{3}{4}$

(c)

	S	S
T	TS	TS
T	TS	TS

$P(\text{tall child}) = P(TT) + P(TS) = 0 + 1 = 1$

(d) Since we know that the man is TS and his wife is SS, we get

	S	S
T	TT	TS
S	SS	SS

$P(\text{tall child}) = P(TS) = \frac{1}{2}$

(e) In this case we do not know the man's gene content, and so we must use conditional probabilities:

$$P(\text{tall child}) = P(\text{man is TT})P(\text{tall child} \mid \text{man is TT})$$
$$+ P(\text{man is TS})P(\text{tall child} \mid \text{man is TS})$$
$$+ P(\text{man is SS})P(\text{tall child} \mid \text{man is SS})$$
$$= \frac{1}{4}(1) + \frac{1}{2}\left(\frac{1}{2}\right) + \frac{1}{4}(0) = \frac{1}{2} \qquad \square$$

EXAMPLE 2 Let B represent "brown eyes" and b represent "blue eyes," and suppose that brown dominates over blue.
(a) A man has a Bb father and a bb mother and his wife is bb. What is the probability that their child has blue eyes?
(b) What is the probability of a blue-eyed child if the father is Bb, the mother is Bb, and his wife is bb?
(c) What is the probability of a blue-eyed child if the father is Bb, the mother is Bb, and his wife is Bb?
Solution.
(a) $P(\text{blue-eyed child}) = P(\text{man is BB})P(\text{blue-eyed child} \mid \text{man is BB})$

$$+ P(\text{man is Bb})P(\text{blue-eyed child} \mid \text{man is Bb})$$

$$+ P(\text{man is bb})P(\text{blue-eyed child} \mid \text{man is bb})$$

$$= (0)(0) + \frac{1}{2}\left(\frac{1}{2}\right) + \frac{1}{2}(1) = \frac{3}{4}$$

(b) $P(\text{blue-eyed child}) = \frac{1}{4}(0) + \frac{1}{2}\left(\frac{1}{2}\right) + \frac{1}{4}(1) = \frac{1}{2}$

(c) $P(\text{blue-eyed child}) = \frac{1}{4}(0) + \frac{1}{2}\left(\frac{1}{4}\right) + \frac{1}{4}\left(\frac{1}{2}\right) = \frac{1}{4}$ $\qquad \square$

EXAMPLE 3 A (Bb, TS) man marries a (bb, SS) woman. What is the probability that their child will be tall and blue-eyed?
Solution.   This can be done in two ways. Since the height and eye-color genes combine independently, we have

$$P(\text{tall blue-eyed child}) = P(\text{tall child})P(\text{blue-eyed child})$$
$$= \frac{1}{2}\left(\frac{1}{2}\right) = \frac{1}{4}$$

Or we could use Tables 9.10–9.12. Hence

$$P(\text{tall blue-eyed child}) = P(\text{bb, TS}) = \frac{1}{4} \qquad \square$$

Table 9.10

	b	b
B	Bb	Bb
b	bb	bb

Table 9.11

	S	S
T	TS	TS
S	SS	SS

Table 9.12

	TS	TS	SS	SS
Bb	(Bb, TS)	(Bb, TS)	(Bb, SS)	(Bb, SS)
Bb	(Bb, TS)	(Bb, TS)	(Bb, SS)	(Bb, SS)
bb	(bb, TS)	(bb, TS)	(bb, SS)	(bb, SS)
bb	(bb, TS)	(bb, TS)	(bb, SS)	(bb, SS)

Population Genetics

In 1908 Mendel's methods were used by the mathematician G. H. Hardy to study population genetics. His work is illustrated in the following example.

Suppose we have a garden of 750 pea plants 10 percent of which are pure-line yellow seed, 40 percent are pure-line green seed, and the remaining 50 percent are hybrid. We call these plants the parental generation. We thus have the following probabilities for the parental generation: $P_0(YY) = 0.1$, $P_0(GG) = 0.4$, $P_0(YG) = 0.5$. We allow the plants to cross-pollinate at random and would like to predict the distribution of types in the next generation of plants. Let $P_0(Y)$ be the probability that a parental generation egg cell receives a Y gene from a pollen cell. By Bayes' second rule we have

$$P_0(Y) = P(\text{pollinated by YY})P(Y \mid \text{pollinated by YY})$$
$$+ P(\text{pollinated by GG})P(Y \mid \text{pollinated by GG})$$
$$+ P(\text{pollinated by YG})P(Y \mid \text{pollinated by YG})$$
$$= P_0(YY)P(Y \mid YY) + P_0(GG)P(Y \mid GG) + P_0(YG)P(Y \mid YG)$$
$$= (0.1)(1) + (0.4)(0) + (0.5)\left(\frac{1}{2}\right) = 0.35$$

Similarly,

$$P_0(G) = P_0(YY)P(G \mid YY) + P_0(GG)P(G \mid GG) + P_0(YG)P(G \mid YG)$$
$$= (0.1)(0) + (0.4)(1) + (0.5)\left(\frac{1}{2}\right) = 0.65$$

Since the egg cells and pollen cells are completely symmetric, the probability that a parental generation egg cell contributes a Y or G gene upon uniting with a pollen cell is also $P_0(Y) = 0.35$ or $P_0(G) = 0.65$, respectively.

Let $P_1(YY)$ be the probability that a first-generation seed is of type YY, and similarly for $P_1(GG)$, $P_1(YG)$. By independence we have

$$P_1(YY) = P(\text{egg cell receives Y and contributes Y})$$
$$= P(\text{egg cell receives Y})P(\text{egg cell contributes Y})$$
$$= P_0(Y)^2 = 0.35^2 = 0.12$$

Similarly,

$$P_1(GG) = P_0(G)^2 = 0.65^2 = 0.42$$

and $\quad P_1(GY) = P_0(G)P_0(Y) + P_0(Y)P_0(G) = (2)(0.35)(0.65) = 0.46$

Thus the distribution of types of first-generation plants is

$$P_1(YY) = 0.12, \qquad P_1(GG) = 0.42, \qquad P_1(YG) = 0.46$$

Let $P_1(Y)$ be the probability that a first-generation egg or pollen cell contributes a Y gene and similarly for $P_1(G)$. Then

$$P_1(Y) = P_1(YY)P(Y \mid YY) + P_1(GG)P(Y \mid GG) + P_1(YG)P(Y \mid YG)$$
$$= (0.12)(1) + (0.42)(0) + (0.46)\left(\frac{1}{2}\right) = 0.35$$
$$P_1(G) = P_1(YY)P(G \mid YY) + P_1(GG)P(G \mid GG) + P_1(YG)P(G \mid YG)$$
$$= (0.12)(0) + (0.42)(1) + (0.46)\left(\frac{1}{2}\right) = 0.65$$

Since $P_0(Y) = P_1(Y)$ and $P_0(G) = P_1(G)$, the calculation for the distribution of the types of second-generation plants is the same as for the first-generation plants. That is,

$$P_2(YY) = 0.12, \qquad P_2(GG) = 0.42, \qquad P_2(YG) = 0.46$$

This strange result, which we calculated for a specific garden of pea plants, occurs when we follow a single characteristic in any form of organic life. Under random mating the distribution of types of egg or pollen cells is constant in each generation and the distribution of types for the offspring is constant from the first generation on. This explains why national characteristics persist in a population when there is little outside marriage. Thus if 90 percent of a population has brown eyes in a generation after the population has stabilized, then 90 percent of the next generation will have brown eyes.

EXERCISES

1. Suppose long noses are dominant over short noses and the long-nose gene is denoted by L while the short-nose gene is denoted by S. If an LS man marries an LS woman, what is the probability that their child has a short nose?

2. In Exercise 1, if an LS man marries and the probability that their child has a short nose is $1/2$, what is the probability that the man's wife has a short nose?

3. In Exercise 1, if an (LS, Bb) man marries an (LS, Bb) woman, what is the probability that they will have a short-nosed, brown-eyed child?

4. Let T represent tallness and S shortness, and suppose T dominates S. If a TT man marries an SS woman, what is the probability that their child is short?

5. In Exercise 4, if a man is married to an SS woman and their child has a 50 percent chance of being tall, what is the gene structure of the man?

6. If a TS man marries a woman from a certain city where 30 percent of the women are SS, 30 percent are TT, and 40 percent are TS, what is the probability that their child is short?

7. Repeat Exercise 6 with 30, 30, 40 replaced by 20, 25, 55, respectively.

8. Let B represent brown eyes, b represent blue eyes, and suppose B dominates b.

   (a) If a Bb man marries a bb woman, what is the probability that their child has blue eyes?

   (b) A man's father is Bb, his mother is bb, and his wife is Bb. What is the probability that his child has blue eyes?

   (c) In a certain city 30 percent of the women are BB, 20 percent are bb, and 50 percent are Bb. If a Bb man marries a woman from this city, what is the probability that their child has blue eyes?

9. In a population of mice, 40 percent are pure-line white and 60 percent are hybrid black-white. Half of the mice are male, half are female, and the mice mate at random.

   (a) What is the probability that an egg cell receives a black gene? A white gene?

   (b) What percent of the resulting offspring are pure-line black? Pure-line white? Hybrid?

10. Repeat Exercise 9 with 40, 60 replaced by 30, 70, respectively.

11. Suppose we have a garden of pea plants in which 10 percent are pure-line yellow seed, 50 percent are pure-line green seed, and 40 percent are hybrid. The plants cross-pollinate at random. What are the probabilities of getting the different types of plants in the first generation? The second generation?

12. In a population of mice, 30 percent are pure-line black and 70 percent are pure-line white (males and females are distributed equally among both groups). The mice mate at random.

    (a) What is the probability that an egg cell receives a black gene? A white gene?

    (b) What percent of the resulting offspring are pure-line black? Pure-line white? Hybrid?

*13. Let A and B represent two forms of a characteristic, for example, blue and brown eyes. Let $P_0(AA), P_0(AB), P_0(BB)$ be the distribution for the parental generation. Let $P_0(A), P_0(B)$ be the probability that a parental cell contributes gene A or B, respectively. We similarly use $P_1, P_2, \ldots$ for succeeding generations. Under random mating show the following.

   (a) $P_0(A) = P_1(A) = P_2(A) = \cdots$ (b) $P_0(B) = P_1(B) = P_2(B) = \cdots$

   (c) $P_1(AA) = P_2(AA) = \cdots$ (d) $P_1(AB) = P_2(AB) = \cdots$

   (e) $P_1(BB) = P_2(BB) = \cdots$

Chapter 9 Summary of Terms

sample space

sample points and events

probability tree

fundamental convention $P\{a_i\} \geq 0$
$$P\{a_1\} + P\{a_2\} + \cdots + P\{a_n\} = 1$$

probability of event A $P(A) = \sum \{P\{a_i\}: \ a_i \in A\}$

properties of probabilities $P(\Omega) = 1$
$0 \leq P(A) \leq 1$
If $A \cap B = \emptyset$, then $P(A \cup B) = P(A) + P(B)$
$P(A \cup B) = P(A) + P(B) - P(A \cap B)$
$P(A') = 1 - P(A)$

conditional probability $P(A \mid B) = \dfrac{P(A \cap B)}{P(B)}$

Bayes' first rule $P(B \mid A) = \dfrac{P(A \mid B)P(B)}{P(A)}$

Bayes' second rule $P(A) = P(B_1)P(A \mid B_1) + P(B_2)P(A \mid B_2)$
where $B_1 \cap B_2 = \emptyset, \quad B_1 \cup B_2 = \Omega$

tree diagrams

independence $P(A \cap B) = P(A)P(B)$

probability and genetics

Chapter 9 Test

1. If a fair coin is flipped three times, what is the probability of at least two heads?

2. If three fair dice are tossed, what is the probability of the sum 5?

3. If $P(A) = P(B) = 1$, show that $P(A \cap B) = 1$.

4. Let A and B be events. If $P(A) = 1/2$, $P(B) = 1/3$, and $P(A \cap B) = 1/6$, find $P(A - B)$, $P(B - A)$, $P(A \cup B)$, $P(A')$, and $P(B')$.

5. Suppose that 6 men out of 100 and 1 woman out of 100 are color-blind. What is the probability that a person chosen at random is color-blind? If a person chosen at random is color-blind, what is the probability that the person is female? (Assume that males and females are equally likely.)

6. On the toss of a pair of fair dice, what is the probability that their sum is 7 if you know that one of the dice came up 1?

7. Show that A and B are independent if $P(B) = 1 - \dfrac{P(A \cup B - B)}{P(A)}$.

8. If two cards are drawn at random from a deck of cards without replacement, what is the probability that they are both aces?

9. In a population of mice, 20 percent are pure-line white and 80 percent are hybrid black-white. Half of the mice are male, half are female, and the mice mate at random. What percent of the resulting offspring are pure-line white? Pure-line black? Hybrid?

10. If $A \cap B = \emptyset$, show that $P(A \mid A \cup B) = \dfrac{1}{1 + \dfrac{P(B)}{P(A)}}$.

11. You have made a fair bet of $50 that the New York Giants will win their next game and the payoff is $40. What is the probability that they will win?

12. If there is a 70 percent chance that it will rain tomorrow and there is a 50 percent chance that the stock market will fall, what is the probability that it will rain tomorrow and the stock market will fall? What is the probability that it will rain tomorrow or the stock market will fall? What assumption are you making to work this problem?

13. Telephone numbers have three-digit prefixes. A prefix cannot be 911 and the first digit on the left cannot be a 0 or 1. Assume that all allowable prefixes are equally likely. How many allowable prefixes are there? What is the probability that your prefix contains (a) precisely one 1, (b) two 1s, (c) no 1s?

14. Census figures show that 70 percent of the families in the United States have an annual income of less than $40,000 and 30 percent of the children from families in this income bracket graduate from college. Moreover, 40 percent of all the children in this country graduate from college. If you graduate from college, what is the probability that your family's income is less than $40,000?

15. Zip codes are five-digit numbers where each digit is between 0 and 9 inclusive (e.g., 00902). Assume that all zip codes are equally likely. How many zip codes are there? What is the probability that your zip code contains (a) precisely one 2, (b) precisely four 2s, (c) no 2s, (d) at least one 2?

16. Are the following statements true or false?
    (a) $P(A \cup B) = P(A) + P(B)$ for all events A, B.
    (b) $P(A \cap B) = P(A)P(B)$ for all events A, B.
    (c) $P(A) > 0$ for all events $A \neq \emptyset$.
    (d) $P(A \cup B) \leq P(A) + P(B)$ for all events A, B.
    (e) $P(A) < 1$ for all events $A \neq \Omega$.
    (f) $P(A \mid B) \leq P(A)$ for all events A, B.
    (g) $P(A \mid B) \geq P(A)$ for all events A, B.
    (h) $P(A \mid \Omega) = P(A)$ for all events A.
    (i) Ω is independent of every event $A \subseteq \Omega$.
    (j) $\emptyset$ is independent of every event.
    (k) If $P(A) < 1/2$, then $P(A') > 1/2$.
    (l) If x is the odds in favor of A, then $0 \leq x \leq 1$.
    (m) If x is the odds in favor of A and y is the odds against A, then $x = 1/y$.

17. A loaded die gives equal probability to even tosses, each of which is twice as likely as an odd toss. What is the probability that a toss is   (a) even,   (b) odd,   (c) prime,   (d) odd and prime? (The prime numbers are $2, 3, 5, 7, \ldots$.)

18. A fair coin is flipped three times. Consider the events $A =$ "heads on the first flip," $B =$ "heads on the second flip," $C =$ "tails on the last two flips." Are these events independent in pairs?

19. In an airplane, the probability of a defect in the landing gear is 10^{-6} and the probability of a defect in the fuel supply mechanism is also 10^{-6}. If these two are independent, what is the probability of at least one of them occurring?

GAMBLE
IF YOU MUST

10

Gamble if you must, but if you must gamble, know as much about the game as
possible. Ignorance can only increase your losses or decrease your winnings. The
history of gambling is studded with dramatic episodes in which vast fortunes have
ridden on the turn of a card or the roll of a die. In the United States each year
$200 billion changes hands legally—and perhaps another $200 billion illegally—in
gambling.

Despite high stakes, early gamblers played with little idea of the percentages for
or against them. No adequate mathematical analysis of gambling was made until the
year 1654, when both Fermat and Pascal laid the foundations of probability theory.

Are there any hints for winning at gambling? For one thing, in a long game
against the house you are bound to lose. Professional gamblers bet heavily for short
periods. Did you know that there is a school for horse-race betting? This school
advises students to bet on long shots. The head of the school has made a living by
winning about 1 out of 7 long shots. In betting favorites he claims that you win only
1 out of 2.5 times, which is not enough to stay in the black.

In this chapter some of the important games of chance will be analyzed. Similar
methods have also proved useful for solving problems in business and science.

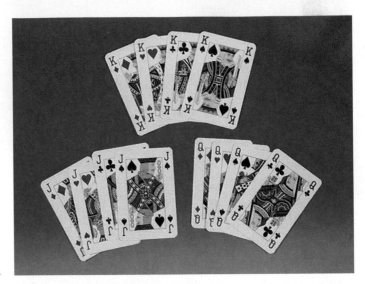

Photo 10.1
The modern cardboard court.

10.1 | History of Playing Cards

card derived from *chartēs* (Greek) and *charta* (Latin) Leaf of
papyrus.   — *Webster's Collegiate Dictionary*, Standard Reference Works

Photo 10.2
The King of
Hearts, sixteenth
century,
Germany.

The history of gambling goes back at least 3000 years. Dice similar to those used today have been found in ancient Egyptian tombs and pyramids. Each game of chance has its own fascinating history and you may enjoy researching at your local library to find the origins of games such as roulette or craps. In this section we shall concentrate on the colorful history of playing cards.

Historians do not agree on the origin of playing cards. There is evidence that they existed at least 2000 years ago in Greece and Rome. It is known that Chinese emperors employed them to entertain their numerous concubines over 1000 years ago. However, the playing cards that we use today had their origin in France and came to us as part of our English heritage.

The most interesting playing cards are those in the Cardboard Court [Kings, Queens, and Knaves (Jacks)]. Each of these cards represents a historical figure. Originally, all of the Court represented French royalty, but British influence has changed some of the characters or their appearance. At various times, portraits of actual royalty were used.

Almost since the beginning of cards, the King of Hearts has represented Charlemagne the Great (King Charles). Charles was King of the Franks in 768 and crowned emperor of the Holy Roman Empire in 800. He is the one king in the pack with an abundance of royal ermine. He still has ermine borders on his robe as he has for centuries. Although Charlemagne had a good stock of wives, none of them has ever been represented as the Queen of Hearts. From early times, the Queen of Hearts has represented Judith of the Bible. Judith is considered as one of the bravest women

Photo 10.3
The King of Spades and Queen of Hearts,
sixteenth century, France.

Photo 10.4
The Knave of Clubs, sixteenth century,
France.

on record. She showed great courage by defeating the fire-eating Assyrian general,
Holofernes.

The Knave of Hearts is the most interesting person in the Cardboard Court. He
represents the fearless soldier La Hire, who died in 1443. Looking closely at the card,
you will notice two distinct features: the feather La Hire holds in his right hand and
the battle-ax behind his head. The battle-ax symbolizes La Hire's love of fighting.
The feather in his right hand is a result of artistic carelessness. Originally the feather
was the flame of a torch that La Hire used to hold in his hand. The torch represented
the torch of love that was supposed to set the heart on fire. Eventually the torch
disappeared and the flame turned into a feather.

The word "spade" comes from the word "sword" and the King of Spades represents King David of the Bible. The sword was King David's emblem. Artists have
frequently portrayed David with an abnormally large sword, the sword of the giant
Goliath, which David claimed after he defeated Goliath.

The Queen of Spades represents the Greek goddess of war and wisdom, Minerva.
She has been on a card for more than three centuries. Minerva is the only queen in
the Court who is allowed to hold a scepter, a wand representing great power.

The Knave of Spades represents Hogier the Dane. He is known as a great soldier
and many songs and poetical romances have been written about him. Most of the
swords he held had a pike-head handle, which accounts for the pike-like object he
presently holds.

Julius Caesar is represented by the King of Diamonds. He is the only king in
the deck who does not hold a sword; instead he holds a great battle-ax. The King of
Diamonds stretches his right hand in a beckoning or bartering way; Julius Caesar was
known as the great barterer. Unlike the other kings in the Cardboard Court, Caesar was
never crowned a reigning monarch although he was a powerful dictator. His passion
to be crowned as emperor was the main reason he was assassinated by a group of his
own senators.

From about 1600, the Queen of Diamonds has represented Rachael from the Bible.
Rachael was the beautiful daughter of Laban, for whom Jacob toiled seven years. The

Photo 10.5
The Knave of
Spades,
sixteenth century,
France.

Knave of Diamonds represents Hector de Maris, a knight of the Round Table. He fought many courageous battles against the Huns and became known as one of the most noble knights in the world.

Alexander the Great is represented by the King of Clubs. Since his appearance as the King of Clubs, Alexander has always been pictured with his orb, the emblem of rule which only a reigning monarch can carry. In our modern deck, the orb is still present, but it floats before the king and the hand that used to grasp it has disappeared.

The Queen of Clubs is Argine. Argine is really not a name but an anagram standing for Regina, which in Latin means queen. Although this queen does not represent a specific historical figure, Argine may have been intended from time to time to represent various queens or perhaps certain rivals to the reigning queen.

The Knave of Clubs is Lancelot, Knight of the Round Table. He was a great archer and during an early period he held a glorified arrow as high as himself or even higher. Unfortunately, our present Knave of Clubs is a miserable depiction of the original cards. The magnificent arrow is now only a staff and the fine feather that used to adorn Lancelot's cap has become a limp leaf.

A surprising fact about playing cards is that in medieval times they were used as calendars. The 52 cards in the deck represented the 52 weeks in a year; the 13 values in each suit represented the 13 lunar months in a year; the four suits represented the four seasons and the 365 spots in a deck represented the 365 days in a year. (In medieval times the spots were arranged differently than they are today.)

Historically, playing cards have been one of the most entertaining inventions of mankind. There must be almost as many decks of playing cards in the world as there are people (about 5.5 billion). The history of playing cards would be of interest to more people if the card makers restored some of the lost details to the modern deck.

10.2 | Averages (Great Expectations)

Mathematics in its widest significance is the development of all types of familiar, necessary, deductive reasoning.
— ALFRED NORTH WHITEHEAD (1861–1947), English Mathematician

The basic concept of probability theory is that of a sample space. For a game of chance the sample points of the sample space represent the possible outcomes of the game. For example, in craps the sample points represent the 36 possible outcomes of tossing two dice; in poker the sample points represent the different possible poker hands. One of the main concerns of this chapter is the average profit (or loss) that a player can expect after playing a game over a reasonably long period of time. For each outcome of the game a player either wins or loses a certain amount (or stays even), depending on the bet. The amount that is won or lost is a function on the sample space, and the average profit (or loss) is the average or expectation of this function. The profit function is an example of a random variable, and you will need to know how to average such random variables.

Photo 10.7
The thrill of the game; a modern gambling casino.

EXAMPLE 1 Suppose a fair coin is flipped twice. You are playing a game in which you win $1 if one or both of the tosses is heads and lose $2 if both tosses are tails. What is your average profit?

Solution. The sample space for this game is

$$\Omega = \{HH, HT, TH, TT\}$$

The probability of each sample point is $1/4$. Your profit depends upon the outcome of the two tosses and is given by a function f, where

$$f(HH) = f(HT) = f(TH) = 1 \text{ and } f(TT) = -2$$

Thus f of a sample point is the amount you win if that sample point occurs as an outcome of the game. Notice that $f(TT) = -2$ means that you lose $2 if TT occurs. Your average profit is the average value $E(f)$ of f. [In Chapter 8, we called $E(f)$ the mean of f. In statistics, the average is usually called the mean.] How is the average of f found? The average of f should be the average of the values of f where each value is weighted according to the probability that that value is obtained. In this case,

$$E(f) = \frac{1}{4} f(HH) + \frac{1}{4} f(HT) + \frac{1}{4} f(TH) + \frac{1}{4} f(TT) = \frac{1}{4} + \frac{1}{4} + \frac{1}{4} + \frac{1}{4}(-2) = \frac{1}{4}$$

Thus your average profit is 25¢. This does not mean that you actually win 25¢ on each play; you win $1 or lose $2. It means that over a large number of plays you will win, on the average, 25¢ per play. □

In Example 1, the profit function f was a random variable. Such a function is called a random variable because its values vary with the outcome and these outcomes are determined only by chance; that is, the outcomes are random. There are other examples of random variables on Ω. For instance, suppose you are interested in the number of heads tossed. Define a function g on Ω where g of a sample point is the number of heads tossed in that outcome. Thus $g(HH) = 2$, $g(HT) = g(TH) = 1$, $g(TT) = 0$. Then the average value $E(g)$ of g becomes

$$E(g) = \frac{1}{4} \, g(HH) + \frac{1}{4} \, g(HT) + \frac{1}{4} \, g(TH) + \frac{1}{4} \, g(TT) = \frac{1}{4} \, (2) + \frac{1}{4} + \frac{1}{4} + 0 = 1$$

This is precisely what should be expected for the average number of heads tossed.

EXAMPLE 2   A loaded coin that comes up heads one-third of the time and tails two-thirds of the time is flipped twice. What is the average number of heads?

Solution.   In this case, $\Omega = \{HH, HT, TH, TT\}$ and the probabilities of the sample points are $P\{HH\} = 1/9$, $P\{HT\} = 2/9$, $P\{TH\} = 2/9$, $P\{TT\} = 4/9$. If f is the random variable giving the number of heads, then $f(HH) = 2$, $f(HT) = 1$, $f(TH) = 1$, $f(TT) = 0$. Hence the average value of f is

$$E(f) = \frac{1}{9} \, f(HH) + \frac{2}{9} \, f(HT) + \frac{2}{9} \, f(TH) + \frac{4}{9} \, f(TT) = \frac{1}{9} \, (2) + \frac{2}{9} + \frac{2}{9} + 0 = \frac{2}{3}$$

Thus, on the average there will be two-thirds head tossed when the coin is flipped twice. Again, this does not mean that two-thirds of a head is actually tossed; either two, one, or no head is tossed. It means that if this coin is tossed twice, many times, then the average number of heads on two tosses is $2/3$. □

EXAMPLE 3   A single fair die is tossed once. What is the average value of the upward face?

Solution.   In this case $\Omega = \{1, 2, 3, 4, 5, 6\}$. If f is the function giving the number on the upward face, then $f(1) = 1$, $f(2) = 2$, $f(3) = 3$, $f(4) = 4$, $f(5) = 5$, $f(6) = 6$. The average value of f becomes

$$E(f) = \frac{1}{6} + \frac{1}{6} \, (2) + \frac{1}{6} \, (3) + \frac{1}{6} \, (4) + \frac{1}{6} \, (5) + \frac{1}{6} \, (6) = 3 \frac{1}{2}$$ □

EXAMPLE 4   Suppose a player at a craps table offers to make you a side wager. He will pay you $3 if a 7 or 11 is tossed, and otherwise you pay him $1. What is your average profit or loss for this wager?

Solution.   You could average over all the 36 possible outcomes, but the following method is easier. Let a denote the outcome that a 7 or 11 is tossed, and let b denote the outcome that a 7 or 11 is not tossed. According to Section 9.2, the probability of a is $P(A_7 \cup A_{11}) = 2/9$. Hence the probability of b is $7/9$. The profit random variable f becomes $f(a) = 3$, and $f(b) = -1$. Hence your average profit is

$$E(f) = \frac{2}{9} \, (3) - \frac{7}{9} = -\frac{1}{9}$$

Thus you would lose about 11¢ on the average for each throw of the dice. □

The general definitions are now given. If $\Omega = \{a_1, \ldots, a_n\}$ is a sample space, a **random variable** is a function $f\colon \Omega \to \mathbb{R}$. Thus a random variable is a real-valued function on the sample space.

The **expectation** or **average value** of f is

$$E(f) = f(a_1)P\{a_1\} + f(a_2)P\{a_2\} + \cdots + f(a_n)P\{a_n\}$$

Hence the expectation of f is the sum of the values of f weighted by the probabilities that these values occur. Random variables will be used mainly to compute the average profit for games of chance. However, you should keep in mind that random variables can be used to describe other important concepts in probability theory. In fact, we saw in Chapter 8 that random variables correspond to many useful statistical quantities.

EXERCISES

1. In Example 1, suppose you win \$3 if two heads are tossed and lose \$1 otherwise. What is your average profit?

2. In Example 2, what is the average number of tails?

3. In Example 3, what is the average of the squares of the numbers appearing on the upward faces?

4. In Example 4, suppose you are paid \$3.50 if a 7 or 11 is tossed and otherwise you lose \$1. What is your average profit?

5. If a natural number from 1 to 10 is chosen at random, what is the average value chosen?

6. If there are five pennies, a nickel, a dime, a quarter, and a half-dollar in a hat, what is the average value of a random selection of a coin from the hat?

7. If there is also a silver dollar in the hat, what is the answer to Exercise 6?

8. In Example 1, let h be the random variable with constant value 2. Compute $E(h)$.

9. In Example 2, f was defined as the random variable giving the number of heads. Let g be the random variable giving the number of tails. Compute $E(f+g)$. How does this compare with $E(f) + E(g)$?

10. Let f and g be random variables on the sample space $\Omega = \{a_1, a_2, \ldots, a_n\}$. Prove that $E(f+g) = E(f) + E(g)$.

11. Define f and g as in Exercise 9. Compute $E(f \cdot g)$. How does this compare with $E(f)E(g)$?

12. In Example 3, if the random variable g is defined by $g(n) = 2n - 6$, what is $E(g)$?

13. Suppose a fair coin is flipped twice. Let f be the random variable giving the number of heads on the first flip, and let g be the random variable giving the number of heads on the second flip. Show that $E(f \cdot g) = E(f)E(g)$.

14. Suppose you play the following game. A pair of fair dice are rolled. If the sum is 6, you win \$2, and if the sum is 9, you win \$5. If neither of these sums comes up, you lose \$1. What is your expected profit per roll?

15. If a pair of fair dice are rolled, what is the average value of the sum?

16. If a fair coin is flipped twice, what is the average of the square of the number of heads?

17. If a fair coin is flipped three times, what is the average of the square of the number of heads?

*18. The **variance** of a random variable f is defined as $E\left[(f - E(f))^2\right]$. Find the variance of the random variable f in Example 1.

*19. Find the variance of the random variable f in Example 2. (See Exercise 18 for the definition of variance.)

10.3 | Gambling Doesn't Pay

It is true that only one out of a hundred wins, but what is that to me?
— FYODOR DOSTOEVSKY (1821–1881), Russian Novelist

If you believe in miracles, head for the keno lounge.
— JIMMY THE GREEK, Contemporary Gambler

The roulette wheel has neither conscience or memory.
— JOSEPH BERTRAND (1822–1900), French Mathematician

In this section some simple games of chance that are played by millions of people all over the world will be analyzed. For each game of chance a sample space is constructed that gives a mathematical model for that game, and the probabilities of the sample points are computed. Different sample spaces can be used to describe a particular game, but it is usually desirable to choose the simplest sample space possible.

Roulette

First consider the game of roulette. In this game a small ivory ball and a horizontal wheel with 38 compartments are set in motion. The compartments are labeled $00, 0, 1, 2, \ldots, 36$, and 18 of these compartments are black and 18 are red. The zero and double-zero compartments are neither red nor black.

Suppose you bet $1 that the ivory ball ends up in compartment 1. According to the rules, if it does you win $35, and if it does not you lose your dollar. What is your expected (or average) profit or loss? A sample space could be constructed with 38 sample points, each representing a different compartment. But it is simpler to construct a sample space with only two points, w and ℓ. If you win, the outcome is w, and if you lose, the outcome is ℓ. In this case a win occurs for only one of the 38 compartments; therefore $P\{w\} = 1/38$ and $P\{\ell\} = 37/38$. If f is the random variable representing your dollar winnings, then $f(w) = 35$ and $f(\ell) = -1$. Hence your expected profit is

$$E(f) = f(w)P\{w\} + f(\ell)P\{\ell\} = (35)\left(\frac{1}{38}\right) + (-1)\left(\frac{37}{38}\right) = -\frac{1}{19} = -0.053$$

Photo 10.8
Placing bets at a roulette
table.

Thus the average profit per play is $-5.3¢$; in other words your average loss is $5.3¢$ per game. This does not mean that you actually lose $5.3¢$; you lose $1 or win $35. It means that over a large number of plays you will lose, on the average, $5.3¢$ per play.

If you bet $1 that the ivory ball will end up in a red compartment and it does, you win $1, while if it does not, you lose your dollar. Is this a better bet? Not at all. Again, let the sample space consist of w and ℓ, where w is the win outcome and ℓ is the lose outcome. In this case, $P\{w\} = 18/38$, $P\{\ell\} = 20/38$ and $f(w) = 1$, $f(\ell) = -1$. Hence the expected profit is

$$E(f) = (1)\left(\frac{18}{38}\right) + (-1)\left(\frac{20}{38}\right) = -0.053$$

which is the same as the previous bet. A way of visualizing your losses is the following. If you start with $100 and bet on red once every minute, you can expect to lose all your money in 1900 minutes, or 31.7 hours.

Numbers

While roulette is the game of the rich, "numbers" is the game of the masses. The averages in the numbers game are pathetic. In this game a player chooses a three-digit number, that is, a number between 000 and 999, and bets $1 that this number will be selected. The number for the day is some well-publicized number that cannot be tampered with, for example, the last three digits of the daily number of stocks bought and sold. If the winning number is the one selected by the player, $700 is returned, for a $699 profit; otherwise the dollar is lost. Let the sample space consist of the two outcomes, win w and lose ℓ. Since the player has one chance out of 1000 to win, $P\{w\} = 1/1000$ and $P\{\ell\} = 999/1000$. The profit random variable is $f(w) = 699$, $f(\ell) = -1$. The average profit per play becomes

$$E(f) = (699)\left(\frac{1}{1000}\right) + (-1)\left(\frac{999}{1000}\right) = \frac{-3}{10}$$

or $-30\cent$. A player betting $1 a day for 10 years can expect to lose $1217.

Chuck-a-Luck

In chuck-a-luck, $1 is bet that one of the six integers $\{1, 2, 3, 4, 5, 6\}$, say 4, is a winner. Three dice in a cage are thoroughly mixed, and the numbers appearing upward are announced when the dice come to rest. The player's dollar is returned plus $1 for each time the player's number (in this case 4) appears; if the player's number does not show, the dollar is lost. A suitable sample space for this situation is $\{a_0, a_1, a_2, a_3\}$, where the outcome a_i occurs if i 4s show. The probability of the event a_0 that no 4 shows is computed first. Now a_0 occurs if 4 does not appear on the first die **and** does not appear on the second **and** does not appear on the third. Since each die is independent of the others and the probability that 4 does not appear on a single die is $5/6$,

$$P\{a_0\} = \left(\frac{5}{6}\right)^3 = \frac{125}{216}$$

The probability of the event a_1 that exactly one 4 appears is computed next. The probability that the first die shows 4 and the other two do not is $(1/6)(5/6)(5/6)$. Since a 4 could appear on the second die but not on the other two or on the third die but not on the other two,

$$P\{a_1\} = (3)\left(\frac{1}{6}\right)\left(\frac{5}{6}\right)\left(\frac{5}{6}\right) = \frac{75}{216}$$

In a similar way,

$$P\{a_2\} = (3)\left(\frac{1}{6}\right)\left(\frac{1}{6}\right)\left(\frac{5}{6}\right) = \frac{15}{216} \qquad \text{and} \qquad P\{a_3\} = \left(\frac{1}{6}\right)^3 = \frac{1}{216}$$

The profit random variable is given by $f(a_0) = -1$, $f(a_1) = 1$, $f(a_2) = 2$, $f(a_3) = 3$. The player's expected profit per play is

$$E(f) = (-1)\left(\frac{125}{216}\right) + (1)\left(\frac{75}{216}\right) + (2)\left(\frac{15}{216}\right) + (3)\left(\frac{1}{216}\right) = -\frac{17}{216} = -0.079$$

or $-7.9\cent$ per play.

Football Pools

Football pools have become a very popular pastime. A typical pool works as follows. A list of 10 football games is printed on a ticket. If one team is thought to be weaker than its opponent by the people running the pool, that team is given enough points to make the game a toss-up. For example, if the Danville Dodgers are a 10-point favorite over the Centralia Zeros, then the Dodgers must win by over 10 points to be a winner in the pool. For $1, a player of the football pool tries to pick any four winners from the 10 games. The player ignores six of the games. If all four selections prove to be winners, $10 is returned for a $9 profit; otherwise the dollar bet is lost. (For simplicity, assume that ties are not allowed.) Let w be the win outcome and ℓ the lose outcome. The player's first selection has a $1/2$ chance of being a winner. Whether the second selection is a winner is independent of the first selection, and so the player's first two selections have a $1/4$ chance of being winners. In this way, $P\{w\} = 1/16$, and $P\{\ell\} = 15/16$. The profit random variable is $f(w) = 9$, $f(\ell) = -1$. The player's expected profit is

$$E(f) = (9)\left(\frac{1}{16}\right) + (-1)\left(\frac{15}{16}\right) = -\frac{6}{16} = -0.375$$

Thus the player loses an average of 37.5¢ for every dollar played. Actually, the situation is even worse. In the above computation, ties are ignored. In practice, if there is a tie, the player loses. No wonder the organizers of such pools make money.

Another way to play this game is to try to pick all 10 winners. The fabulous sum of $150 is returned if the player does choose 10 winners correctly; $20 is returned as an added enticement if nine of the 10 winners are selected. Following a method similar to that in the previous paragraph, the probability of picking 10 winners is $1/2^{10}$, and the probability of picking 9 winners is $10/2^{10}$. This game can be represented by three sample points w_1, w_2, and ℓ, where w_1 occurs if all 10 winners are selected, w_2 occurs if 9 winners are selected, and ℓ occurs if the player loses. Thus $P\{w_1\} = 1/2^{10}$, $P\{w_2\} = 10/2^{10}$, and $P\{\ell\} = 1 - (1/2^{10}) - (10/2^{10})$. The profit random variable is given by $f(w_1) = 149$, $f(w_2) = 19$, $f(\ell) = -1$. The player's expected profit is

$$E(f) = (149)\left(\frac{1}{2^{10}}\right) + (19)\left(\frac{10}{2^{10}}\right) + (-1)\left(1 - \frac{1}{2^{10}} - \frac{10}{2^{10}}\right) = -\frac{337}{512} = -0.658$$

Thus the player loses an average of 65.8¢ per play.

Keno

An unbiased machine selects 20 different numbers from the first 80 positive integers. A simple version of keno is to pay $1 and try to predict one of the selected numbers. If that number is among the 20 selected by the machine, $3.20 is returned for a $2.20 profit. The probability of winning is $P\{w\} = 1/4$, and the probability of losing is $P\{\ell\} = 3/4$. The profit random variable is given by $f(w) = 2.2$, $f(\ell) = -1$. Hence the expected profit is

$$E(f) = (2.2)\left(\frac{1}{4}\right) + (-1)\left(\frac{3}{4}\right) = -0.20 \text{ or } -20¢$$

In a more complicated version of keno a player selects five numbers from the first 80 positive integers. If three of these numbers agree with the machine selection, $3 is returned, for a profit of $2; if four of these numbers agree, $26 is returned, for a profit of $25; and if all five agree, $332 is returned, for a profit of $331. This game can be represented by the sample space $\{a_0, a_1, a_2, a_3, a_4, a_5\}$ where the event a_i occurs if i of the player's numbers are selected by the machine. The probability $P\{a_0\}$ that none of the player's five numbers are selected by the machine is computed first. The probability that the player's first number is not selected is $60/80$. If the first number is not selected, there are 59 unselected numbers left out of 79 numbers. Thus the probability that the player's first two numbers are not selected is $(60/80)(59/79)$. Continuing in this way,

$$P\{a_0\} = \left(\frac{60}{80}\right)\left(\frac{59}{79}\right)\left(\frac{58}{78}\right)\left(\frac{57}{77}\right)\left(\frac{56}{76}\right) = 0.2271842$$

The event a_1 occurs if one of the player's numbers is selected and the other four are not. The probability that the player's first number is selected and the other four are not is

$$\left(\frac{1}{4}\right)\left(\frac{60}{79}\right)\left(\frac{59}{78}\right)\left(\frac{58}{77}\right)\left(\frac{57}{76}\right) = 0.0811372$$

Now the player's second number could be selected and the other four not or the third number could be selected or the fourth or the fifth. Hence

$$P\{a_1\} = (5)(0.811372) = 0.4056861$$

The probability $P\{a_2\}$ is now computed. The probability that the player's first two numbers are selected by the machine but the other three are not is

$$\left(\frac{1}{4}\right)\left(\frac{19}{79}\right)\left(\frac{60}{78}\right)\left(\frac{59}{77}\right)\left(\frac{58}{76}\right) = 0.02704574$$

But the player's first and third or his first and fourth, and so forth numbers could have been selected by the machine. How many such possibilities are there? The number of such possibilities can be counted by listing the following unordered pairs: $\{1, 2\}$, $\{1, 3\}, \{1, 4\}, \{1, 5\}, \{2, 3\}, \{2, 4\}, \{2, 5\}, \{3, 4\}, \{3, 5\}, \{4, 5\}$. (In the next section, these possibilities will be called combinations of five objects taken two at a time, and a formula will be derived for the number of such combinations.) Since there are 10 of these unordered pairs,

$$P\{a_2\} = (10)(0.02704574) = 0.2704574$$

In a similar way,

$$P\{a_3\} = (10)\left(\frac{1}{4}\right)\left(\frac{19}{79}\right)\left(\frac{18}{78}\right)\left(\frac{60}{77}\right)\left(\frac{59}{76}\right) = 0.0839350$$

$$P\{a_4\} = (5)\left(\frac{1}{4}\right)\left(\frac{19}{79}\right)\left(\frac{18}{78}\right)\left(\frac{17}{77}\right)\left(\frac{60}{76}\right) = 0.0120923$$

$$P\{a_5\} = \left(\frac{1}{4}\right)\left(\frac{19}{79}\right)\left(\frac{18}{78}\right)\left(\frac{17}{77}\right)\left(\frac{16}{76}\right) = 0.0006449$$

The profit random variable is given by $f(a_0) = f(a_1) = f(a_2) = -1$, $f(a_3) = 2$, $f(a_4) = 25$, $f(a_5) = 331$. Hence the expected profit becomes

$$E(f) = (-1)(0.2271842 + 0.4056861 + 0.2704574)$$
$$+ (2)(0.0839350) + (25)(0.0120923) + (331)(0.0006449)$$
$$= -0.219$$

The player is better off using the first version of keno since the average loss there is only 20¢ per game, as opposed to an average of 21.9¢ per game in the second version.

Systems

Are there any infallible systems that you can employ that will guarantee a win? That is, are there any winning strategies for gambling in a casino? Unless you have additional information, the answer is no. For example, until about 15 years ago only one deck was used in blackjack and the cards were not shuffled until the entire deck was used. In this case you could remember the previous cards and after about half the deck was dealt you would have a considerable amount of information about the remaining cards. Taking this additional information into account, clever strategies were devised to increase the chances of winning. However, the casinos soon caught on to this and now three decks are usually used and they are shuffled frequently.

A strategy that is continually rediscovered by novice gamblers is called the **martingale method**. To take a simple example, suppose you are playing roulette and you make repeated bets that the ball will settle in a black compartment. Suppose your first bet is \$1. If you win, fine, you make \$1 and you repeat your \$1 bet. If you lose, you double your bet to \$2. If you now win, great, your \$2 win minus your previous \$1 loss leaves you \$1 ahead and you repeat your \$1 bet. But if you again lose on your second bet, you double your bet to \$4. If you now win you have \$4 − \$2 − \$1 = \$1, so again you are ahead \$1 and you repeat your \$1 bet. If you lose, you again double your bet to \$8. This system looks foolproof. Every time you win you are ahead \$1. Even if you have a long string of losses you must eventually get a black to win. To be precise, suppose you have a string of n consecutive losses and win on the $(n+1)$th spin of the wheel. Since you double your bet with each loss, this win gives you 2^n dollars. Now from what we have seen in Section 1.2, your losses add up to

$$1 + 2 + 2^2 + \cdots + 2^{n-1} = 2^n - 1$$

so you are ahead \$1.

What's wrong with this system? First of all, it is very slow. Suppose the wheel is spun once a minute (this is actually a little faster than normal). If you are lucky, you will win about once every 3 minutes. Thus, you make about \$1 every 3 minutes or about \$20 an hour. Not only is this a very boring way to make money, but if you play only an hour, it will not cover your living expenses. Thus you would have to play for about 10 hours to make a decent amount of money. Why not start with a \$2 bet and double every time you lose? Then you will make \$40 an hour. As we shall see, this would increase your risk, which is already considerable.

The martingale method has two main defects. The first is that you have only a limited amount of capital. Suppose you sustain 10 consecutive losses. You have then lost

$$1 + 2 + 2^2 + \cdots + 2^9 = 2^{10} - 1 = \$1023$$

To recoup this loss you must now bet $2^{10} = \$1024$. Thus you must bet \$1024 just to make \$1 and you will need \$2047 (the \$1023 you have already lost plus the \$1024 you need for the current bet) to carry out this method. If you started with about \$1000 you would not have the capital to place your eleventh bet and you would lose all your money. Even if you had unlimited capital you could encounter the second defect. The casino operators know about this method, and for this reason and others as well, they have placed a limit on the amount you can bet. This limit varies with casinos, but it is typically \$1000. Thus the martingale method cannot tolerate 10 consecutive losses. If you start with a \$2 bet and double every time you lose, the system would break down with nine consecutive losses.

We have seen that with 10 consecutive losses you would lose about \$1000 and you could not recoup this loss. How risky is this? Surely, 10 consecutive losses is a very rare event. Let's compute the probability. To lose on a single spin of the wheel the ball must settle in a red, 00, or 0 compartment, so the probability is $P\{\ell\} = 20/38$. Since spins of the wheel are independent, the probability of two consecutive losses is $\left(P\{\ell\}\right)^2 = (20/38)^2$. Continuing, we see that the probability of 10 consecutive losses is

$$\left(P\{\ell\}\right)^{10} = \left(\frac{20}{38}\right)^{10} = 0.0016310 \approx \frac{1}{613}$$

Thus, there is about one chance in 613 for 10 consecutive losses. Roughly speaking, 10 consecutive losses will occur about once in every 613 wheel spins. But if you play for 10 hours, you will see about 600 wheel spins. Hence, there is a good chance that 10 consecutive losses will occur during this time period. If you double \$2 bets the system breaks down with probability

$$\left(P\{\ell\}\right)^9 = \left(\frac{20}{38}\right)^9 = 0.0030990 \approx \frac{1}{323}$$

which is almost twice as likely as before. In summary, the martingale method might work if you are lucky, but the risk is so high that it is not worth it.

There are systems that are marginally successful for roulette, but they are quite subtle and require a careful study of the wheel. No roulette wheel (or any other gambling apparatus) is perfect. There are always slight dents or notches in the compartment slots and the tolerances are never exact so the openings between slots vary slightly. Hence, the probability for a number is not exactly $1/38$ and the probability of a color is not exactly $18/38$. If you observe a particular roulette wheel over a long time period and write down the numbers that occur, you will find that some numbers occur slightly more often than others. If you now bet on these numbers, you will have gained a small advantage. Unfortunately, this advantage is usually not enough to overcome the natural advantage of the casino.

A better system relies on carefully watching the croupier (the wheel operator). The croupier is supposed to spin the wheel at different speeds and toss the ball in different ways at each play of the game. However, toward the end of a long work shift, the croupier usually gets tired and starts spinning the wheel and tossing the ball with a certain regularity. When this happens, by first noting the beginning compartment before the spin, you can make a rough prediction about the pie-shaped area in which the ball will settle. For example, if you can predict that the ball will settle in a pie-shaped area that is $1/6$ of the wheel then you can bet on the six or seven numbers in this area. Of course, this will not work every time, but if it only works one-fourth of the time you will make a good profit. For example, suppose you bet $1 on each of seven numbers every play of the game and the system works one-fourth of the time. (Of course, the seven numbers will be different with each play, depending on your prediction.) After four spins you have bet $28 and one of your numbers wins. Hence, you make a profit of $35 - $27 = $8. Unlike the martingale method, in which your risk increases with larger bets, your risk remains the same but your expected profit increases. The disadvantage of this system is that a sharp croupier can easily foil it with the flip of a wrist.

EXERCISES

1. In European roulette, there are 37 compartments, $0, 1, 2, \ldots, 36$. Thus European roulette is the same as the roulette discussed in this section except that there is no double zero.
   (a) If a player bets $1 on red, what is the expected profit?
   (b) If a player bets $1 on 3, what is the expected profit?

2. How would the profit of roulette change if there were no zero or double zero?

3. What is a player's expected profit in the numbers game (a) if $800 is returned for a winning number? (b) If $900 is returned? (c) If $1000 is returned?

4. Suppose the numbers game is changed so that a player tries to choose a winning four-digit number. If $8000 is now returned on a winning $1 bet, what is the expected profit?

5. Suppose the numbers game is changed so that a player chooses two three-digit numbers (which may be the same) and bets $1 that both these numbers are selected. If $500,000 is returned for two winning numbers, what is the expected profit?

6. In chuck-a-luck, why is $P\{a_2\} = 15/216$ and $P\{a_3\} = 1/216$?

7. Suppose chuck-a-luck is played exactly the same except that $10 is returned if the player's number appears three times. Now what is the expected profit?

8. If chuck-a-luck were played with four dice, what would the expected profit be?

9. (a) Show that the probability of picking 10 winners in a football pool is $1/2^{10}$ and the probability of picking precisely nine winners is $10/2^{10}$.
   (b) What is the probability of picking precisely eight winners?

10. In keno, show that the following hold.
    (a) $P\{a_3\} = (10)(1/4)(19/79)(18/78)(60/77)(59/76)$
    (b) $P\{a_4\} = (5)(1/4)(19/79)(18/78)(17/77)(60/76)$
    (c) $P\{a_5\} = (1/4)(19/79)(18/78)(17/77)(16/76)$

11. Suppose a player bets $2 in roulette: $1 that the ball ends up in compartment 1 and $1 that it ends up in 2. What is the expected profit?

12. Suppose there were 36 compartments in roulette labeled $00, 0, 1, 2, \ldots, 34$ and the payoffs were the same as before. What would the expected profit be for a bet on 1? A bet on red?

13. Suppose you make repeated bets on black in roulette. What is the probability of five consecutive losses?

14. What is the probability that neither red nor black will occur on three consecutive roulette spins?

15. You have decided that on a certain roulette wheel the number 7 occurs about once in every 35 spins.
    (a) If you make repeated bets of $1 on number 7, what is your expected profit?
    (b) If the wheel is spun once every minute, about how long will it take to make $20?

16. When does the martingale method break down for doubled $1 bets in roulette if the maximum bet is $500?

17. Repeat Exercise 16 for doubled $2 bets.

18. Repeat Exercise 16 for a maximum bet of $2000.

*19. Discuss the following system. You begin with a $1 bet on black in roulette and you play as in the martingale method except that you double your bet after every win.

*20. Invent your own system for roulette (it must, of course, be legal). Discuss its advantages and disadvantages.

*21. Consider the following game. An unbiased machine selects 10 different integers from among the first 20 positive integers. A player bets $1 and chooses 10 of the first 20 positive integers. If the player's 10 choices agree with those of the machine, $1000 is returned; if 9 of the choices agree, $100 is returned; and otherwise the $1 is lost. What is the player's expected profit?

10.4 Permutations and Combinations

Though this be madness, yet there is method in 't.
— WILLIAM SHAKESPEARE (1564–1616), English Dramatist

In this section more tools are developed to help compute probabilities. But first a definition.

> A **permutation** of a set of objects is a selection in a specific order of a certain number of objects in the set.

For example, suppose you have two jars labeled A and B and you want to know the number of ways you can place these jars in a row on a shelf. You could place jar A first to get AB, or you could place jar B first to get BA. We then say that the number of permutations of *two objects taken two at a time* is 2. What if you have three jars A, B, and C that you want to place in a row? You could arrange the jars in the following six ways:

$$ABC, ACB, BAC, BCA, CAB, CBA$$

We then say that the number of permutations of *three objects taken three at a time* is 6. You could compute the number 6 without actually arranging jars by the following

Photo 10.9
The luck of the throw.

reasoning. Jar A can be placed first and now there are two ways to place the other two jars. Or jar B can be placed first and there are two ways to place the other two jars. Finally, we could place jar C first with two ways of placing the other two jars. This gives a total of six different arrangements of the jars.

What if you had four jars A, B, C, D but you wanted to place only two of them in a row on a shelf? The number of possible arrangements is the number of permutations of *four objects taken two at a time*. What is this number? Let's first find this number by displaying the possible arrangements:

$$AB, AC, AD, BA, BC, BD, CA, CB, CD, DA, DB, DC$$

You thus see that the answer is 12. Let's now try to think this through. There are four possibilities for the first jar. Once you have placed the first jar, there are now three possibilities for the second jar. This gives a total of $(4)(3) = 12$ possible arrangements.

EXAMPLE 1   How many three-letter initials are there for people's names in which each letter is different?

Solution.   In this case the set of objects is {A, B, C, . . . , Z}, and selections (or arrangements) are made of these letters taken three at a time. For example, ABC is one possible selection and this is considered to be different from the selections ACB, BCA, CAB. Each selection is called a permutation of 26 objects taken three at a time. Now the first initial can be chosen in 26 different ways. For each of these ways there are 25 possibilities for the second initial. For each of the first two initials there are 24 possibilities for the third initial. Thus, the answer is $(26)(25)(24) = 15,600$. Notice that we can write this number as $(26)(26 - 1)(26 - 2)$.   □

EXAMPLE 2   How many four-digit house numbers are there in which each digit is different and not zero?

Solution.   We seek the number of permutations of nine objects $\{1, 2, . . . , 9\}$ taken four at a time. For example, some possibilities are 1234, 2951, 7346. Reasoning as before, this number is $(9)(8)(7)(6) = 3024$. This number can also be written as $(9)(9 - 1)(9 - 2)(9 - 3)$.   □

In general, the following theorem holds.

Theorem 10.1   The number of permutations of n objects taken r at a time is

$$_nP_r = n(n-1)(n-2)\cdots(n-r+1)$$

Proof: There are n choices for the first object. For each of these n choices there are $n-1$ choices for the second. For each of the $n(n-1)$ first two choices there are $n-2$ choices for the third, etc. Since r objects are taken at a time, continue this r times to get

$$n(n-1)(n-2)\cdots[n-(r-1)] = n(n-1)(n-2)\cdots(n-r+1) \qquad \square$$

Notice the shorthand notation $_nP_r$ used in Theorem 10.1 for the cumbersome expression $n(n-1)(n-2)\cdots(n-r+1)$. The easiest way to remember $_nP_r$ is to start with n and keep multiplying the previous number by 1 less until there are r numbers being multiplied. Thus $_3P_1 = 3$, $_3P_2 = (3)(2)$, $_7P_3 = (7)(6)(5)$, $_{10}P_4 = (10)(9)(8)(7)$.

EXAMPLE 3   How many permutations are there of five letters from the alphabet?
   Solution. $_{26}P_5 = (26)(25)(24)(23)(22) = 7,893,600$ $\qquad \square$

EXAMPLE 4   Five cities are to be visited on a vacation trip. If each city is to be visited only once, how many ways can the trip be planned?
   Solution. $_5P_5 = (5)(4)(3)(2)(1) = 120$ $\qquad \square$

EXAMPLE 5   A company plans to run a contest among its 20 leading salespeople. If a prize of \$100 is given to the winner and \$50 to the runner-up, how many ways can the contest turn out?
   Solution. $_{20}P_2 = (20)(19) = 380$ $\qquad \square$

Since expressions like $(5)(4)(3)(2)(1)$ appear often, the factorial notation $5!$ is frequently used. (Remember, by convention, $0! = 1$.) Thus, for example,

$$_7P_3 = (7)(6)(5) = \frac{7!}{4!} = \frac{7!}{(7-3)!}$$

$$_{26}P_5 = (26)(25)(24)(23)(22) = \frac{26!}{21!} = \frac{26!}{(26-5)!}$$

In general,

$$\boxed{\text{Permutation Formula} \quad _nP_r = \frac{n!}{(n-r)!}}$$

We now consider combinations. With a combination of objects we are interested in a certain number of objects while with a permutation the order of the objects is also considered.

A **combination** of a set of objects is a selection of a certain number of objects in which order is disregarded.

When permutations and combinations are considered, it is always assumed that each object is distinct and can be used only once. In general, each combination of objects in a set corresponds to a number of permutations.

EXAMPLE 6 How many two-element subsets are there of a three-element set?

Solution. Suppose the three-element set is $A = \{a, b, c\}$. In describing a two-element subset of A, the order of the two elements is immaterial. For example, $\{a, b\}$ and $\{b, a\}$ describe the same subset. Thus we would like to find the number of combinations of three objects taken two at a time. We can display the two-element subsets as follows:

$$\{a, b\}, \quad \{a, c\}, \quad \{b, c\}$$

Thus, the answer is three. □

Notice that each combination in Example 6 corresponds to two permutations. Thus, $\{a, b\}$ corresponds to the permutations ab and ba, $\{a, c\}$ corresponds to ac and ca, while $\{b, c\}$ corresponds to bc and cb. These give all the permutations of a, b, c taken two at a time.

Figure 10.1 depicts the combinations of five objects taken three at a time. Each of these combinations corresponds to $3! = 6$ permutations (see Figure 10.2). Thus if $_5C_3$ denotes the number of combinations of five objects taken three at a time, then

$$_5P_3 = (3!)(_5C_3) \qquad \text{or} \qquad _5C_3 = \frac{_5P_3}{3!} = \frac{5!}{3!2!}$$

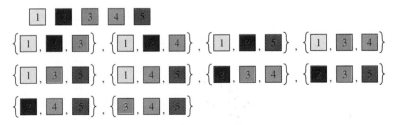

Figure 10.1

In general, any combination containing r objects can be arranged into $_rP_r = r!$ permutations of these r objects. Thus each combination of n objects taken r at a time represents $r!$ permutations of the original set of n objects taken r at a time. Hence if the number of combinations of n objects taken r at a time is $_nC_r$, then $_nP_r = (r!)(_nC_r)$. Thus the following theorem has been proved.

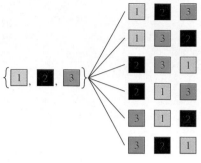

Figure 10.2

Theorem 10.2   (Combination formula) The number of combinations of n objects taken r at a time is

$$_nC_r = \frac{n!}{r!(n-r)!}$$

For example,

$$_7C_3 = \frac{7!}{3!4!} = \frac{(7)(6)(5)}{(3)(2)(1)} = 35$$

$$_{26}C_5 = \frac{26!}{5!21!} = \frac{(26)(25)(24)(23)(22)}{(5)(4)(3)(2)} = 65{,}780$$

EXAMPLE 7 How many three-person committees can be selected from a group of 18 people?

Solution.   Since the order of the people in the committees is not relevant, the solution is the number of combinations of 18 people taken three at a time. This is

$$_{18}C_3 = \frac{18!}{3!15!} = \frac{(18)(17)(16)}{(3)(2)} = 816 \qquad \qquad \square$$

EXAMPLE 8 How many five-card poker hands are possible?

Solution.   The order in which the cards are dealt in a hand is not relevant, so the number of combinations of 52 cards taken five at a time is required. This is

$$_{52}C_5 = \frac{52!}{5!47!} = 2{,}598{,}960 \qquad \qquad \square$$

EXAMPLE 9 How many bridge hands are possible?

Solution.   Since a bridge hand consists of 13 cards, the answer is

$$_{52}C_{13} = 635{,}013{,}559{,}600$$

This explains why bridge books and columns are so popular. Since there are so many possible bridge hands, the probability that one will be dealt the same hand twice in a lifetime is astronomically low. $\qquad \square$

EXAMPLE 10   Two persons are chosen at random from a group of 20 married couples.
(a) What is the probability that the pair of persons chosen are married to each other?
(b) What is that probability if one male and one female are chosen?

Solution.   (a) The total number of possible pairs is $_{40}C_2$, and the number of married couples is 20; therefore the probability in question is

$$\frac{20}{_{40}C_2} = \frac{(20)(2!)(38!)}{40!} = \frac{1}{39}$$

(b) There are $_{20}C_2$ pairs of females and $_{20}C_2$ pairs of males, and so the number of pairs consisting of one male and one female is

$$_{40}C_2 - (2)(_{20}C_2) = 400$$

This could also be found as follows. Since each of the 20 males can be paired with each of 20 females, the total number of male-female pairs is $(20)(20) = 400$. Since the number of married pairs is 20, the probability in question is $20/400 = 1/20$.   □

EXERCISES

1. Evaluate $_6C_3$, $_7C_4$, $_6P_3$, and $_7P_4$.

2. Verify the number in Example 8.

3. Verify the number in Example 9.

4. What are the answers in Example 10 if there are 30 married couples?

5. Evaluate $_{50}P_1$, $_{50}P_2$, $_{50}P_3$.

6. Evaluate $_{50}C_1$, $_{50}C_2$, $_{50}C_3$

7. First, second, third, and fourth prizes are to be awarded in a track meet in which 16 runners are entered. In how many different ways can the prizes be awarded?

8. A Scrabble player with seven different letters in his rack decides to test all possible five-letter permutations of his letters before making his next play. If he tests one permutation each second, how long will it take before he is ready to play?

9. In an Olympic hockey schedule, the United States plays six games. How many ways can the results end in two wins, three losses, and one tie?

10. Find the number of executive committees of president, vice president, secretary, and treasurer that are possible in a club with 12 members.

11. How many choices does a student have if he can choose 7 out of 10 questions on a quiz?

12. A quiz consists of five true-false questions. How many ways can you answer the five questions, and what is the probability that you will get at least two correct by simply guessing?

13. There are five teams in a football league. Assuming there are no identical records at the end of the season, how many possible rankings are there for the teams of this league?

14. Two different natural numbers from 1 through 9 are picked at random. What is the probability that they are both odd? Both even? What are the answers if the two numbers can be the same?

15. (a) How many possible three-letter initials can a person have?
    (b) What if no two initials are the same?
    (c) What if exactly two of the initials are the same?
    (d) What is the probability that a person's first initial is A if it is known that no two of her initials are the same?

16. Solve the equation $(n+2)! = 90n!$ for n.

17. Solve the equation $_nP_4 = 1680$ for n.

18. Show that $_nC_r = {_nC_{n-r}}$.

19. Let $S = \{1, 2, 3, 4\}$. List the combinations of elements of S taken one at a time; two at a time; three at a time; four at a time.

20. Let $S = \{1, 2, 3\}$. List the permutations of elements of S taken one at a time; two at a time; three at a time.

21. Notice that $(a+b)^2 = a^2 + 2ab + b^2 = {_2C_0}a^2 + {_2C_1}ab + {_2C_2}b^2$. Do the analogous calculations for $(a+b)^3$. What do you think the result would be for $(a+b)^4$?

*22. Solve the equation $(5)(_nP_3) = (2)(_{n-1}P_4)$ for n.

*23. Solve the equation $(15)(_nP_1) + {_nP_2} = {_nP_3}$ for n.

*24. Write a BASIC computer program that will compute $_NP_R$. Begin with 10 INPUT N, R.

*25. Write a BASIC computer program that will compute $_NC_R$. Begin with 10 INPUT N, R.

26. From a set of two men and four women, three people are chosen at random. What is the probability that among the three (a) exactly one is a man, (b) all are women, (c) at least two are women?

27. If $_nP_r = {_mP_s}$, then necessarily does $n = m$ and $r = s$?

*28. If $u \geq v + w$, prove $(_uP_v)(_{u-v}P_w) = {_uP_{v+w}}$.

10.5 | Lottery, Poker, and Bridge

In play there are two pleasures for your choosing. The one is winning, and the other is losing.

—LORD BYRON (1788–1824), English Poet

Lottery

The history of lotteries goes back to ancient times. In the biblical period, they were used to select people to perform certain tasks. Today they are a legalized form of gambling that is very popular in many states. The money collected from lotteries is considered a voluntary tax and many people think it is a pleasant way to increase a state's revenue. As pleasant and tempting as it may appear, you should play the lottery only if you have money to spare. Your chances of coming out ahead in a lottery are extremely small.

Each state that has a lottery employs its own variation of the game. We shall consider a lottery that is fairly typical although it differs in certain details from other variations. The player chooses six different integers between 1 and 42 inclusively and marks them on a ticket. The ticket is turned in together with a dollar. If the chosen

Photo 10.10
Your play, poker face.

integers agree with the six integers of the next random drawing, the play wins the grand prize. (There are two random drawings a week.) If no player chooses the correct six integers, then the grand prize increases according to a formula depending on the number of tickets sold. Players then try to win this larger grand prize at the next lottery drawing. If more than one player chooses the six winning numbers, they divide the grand prize equally. The grand prize begins at $1 million and sometimes builds to $10 million or more before it is won. Of course, a player can buy any desired number of lottery tickets.

Why did the lottery officials decide to use integers from 1 to 42? As we shall see, this results in a probability of $1/5,245,786$ that a single ticket will win. If several million tickets are sold, then there is a reasonable chance that someone will win the grand prize. There is also a good chance that nobody will win, so the grand prize increases in value and makes the game more exciting.

Besides the grand prize, lesser prizes are also awarded in the lottery. If you match any five of the six winning numbers, you win $2000 (above your $1 entry fee). If you match any four numbers, you win $50 and if you match any three numbers, you win $3.

Let's compute the various probabilities. Our sample space consists of all the combinations of 42 numbers taken six at a time. Hence, the number of sample points is

$$_{42}C_6 = \frac{42!}{6!\,36!} = 5,245,786$$

and the probability that a chosen combination wins the grand prize is

$$P(W_1) = \frac{1}{5,245,786}$$

Let W_2 be the event that a chosen combination matches any five of the six winning numbers (but not all six). There are six chances out of 42 that the first number in the combination gives a match, resulting in the probability 6/42. Given that the first number is a match, there are now five chances out of 41 that the second number is a match, so the probability is 5/41. Continuing, we have probabilities 4/40, 3/39, and 2/38 for the third, fourth, and fifth numbers, respectively. Since the sixth number does not give a match, this results in the probability 36/37. Multiplying these probabilities gives the probability that the first five numbers in the chosen combination match winning numbers. Since we are interested in matching any five numbers, we must multiply by $_6C_5 = 6$ to obtain $P(W_2)$. We conclude that the probability that a chosen combination wins second prize is

$$P(W_2) = \left(\frac{6}{42}\right)\left(\frac{5}{41}\right)\left(\frac{4}{40}\right)\left(\frac{3}{39}\right)\left(\frac{2}{38}\right)\left(\frac{36}{37}\right)\,_6C_5 \approx \frac{1}{24,286}$$

In a similar way, the probabilities of winning the $50 third prize and the $3 fourth prize are

$$P(W_3) = \left(\frac{6}{42}\right)\left(\frac{5}{41}\right)\left(\frac{4}{40}\right)\left(\frac{3}{39}\right)\left(\frac{36}{38}\right)\left(\frac{35}{37}\right)\,_6C_4 \approx \frac{1}{555}$$

and

$$P(W_4) = \left(\frac{6}{42}\right)\left(\frac{5}{41}\right)\left(\frac{4}{40}\right)\left(\frac{36}{39}\right)\left(\frac{35}{38}\right)\left(\frac{34}{37}\right)\,_6C_3 \approx \frac{1}{37}$$

The probability of winning one of the four prizes becomes

$$P(W) = P(W_1) + P(W_2) + P(W_3) + P(W_4) \approx \frac{1}{35}$$

and the probability of losing is

$$P(L) = 1 - P(W) = \frac{34}{35}$$

Suppose the grand prize is $1,000,000 and let us assume that we do not have multiple winners for the grand prize. (For the second prize, all winners gain $2000 and similarly for the third and fourth prizes.) Your average profit for each lottery ticket becomes

$$\frac{1,000,000}{5,245,786} + \frac{2000}{24,286} + \frac{50}{555} + \frac{3}{37} - \frac{34}{35} \approx -52.7¢$$

Thus, you lose about 53¢ on the average every time you purchase a lottery ticket. It does not help to buy more than one ticket. If you buy two tickets, then your average loss is about $1.05.

Of course, your average profit will go up if the grand prize is larger and in this case it might be a good idea to buy more than one ticket. However, very large grand prizes, say $10 million or more, cause much excitement. In this case, a lot more tickets are sold and there is a good chance for multiple winners. To carry this to the extreme, suppose the grand prize is $10 million. Why not play 5,245,786 different numbers and be a sure winner?

Poker

Poker and bridge are two of the most popular card games. Huge fortunes have transferred hands as a result of poker games, while bridge is usually played for the challenge and fame. In this section, the game of poker is considered first, and the probabilities of the different poker hands are computed. Combinations play an important part in these computations; for example, there are $_{52}C_5 = 2{,}598{,}960$ different hands. There are different variations of poker, but in practically all of them five or more cards are dealt and the players choose their five best cards. For simplicity, only five-card poker hands will be considered here. There are 10 main different types of hands. These are now listed in descending order of their value, and their probabilities are computed.

1. **Royal flush** Ace, king, queen, jack, ten of the same suit. Since there are four possible royal flushes, one for each suit, the probability is $4/2{,}598{,}960 \approx 1/650{,}000$. What are the chances that a poker player will hold a royal flush in a lifetime? A professional player who plays every night, seven days a week plays about 50,000 hands a year. This player will hold a royal flush on the average of once every 13 years. The amateur will be lucky to see a royal flush, let alone hold one once in a lifetime.

2. **Straight flush** Five cards in a sequence in the same suit but not a royal flush. There are nine possible spade straight flushes since they can begin with $1, 2, \ldots, 9$. Since there are four suits, the probability is $36/2{,}598{,}960 \approx 1/70{,}000$. The professional player might hold one straight flush a year.

3. **Four of a kind** Let us compute the probability of getting four aces. In such a hand there are four aces plus an odd card. There are 48 possibilities for this odd card. Hence there are 48 different hands with four aces. Since there are 13 different card values, the probability of four of a kind is $(13)(48)/2{,}598{,}960 \approx 1/5000$.

4. **Full house** Three of a kind together with a pair. The number of hands with two 2s and three 3s is $(_4C_2)(_4C_3) = (6)(4) = 24$. Thus the number of full houses consisting of 2s and 3s is 48 (we could have three 2s and two 3s). Since there are $_{13}C_2$ possible pairs of card values, the probability of a full house becomes $(48)(_{13}C_2)/2{,}598{,}960 \approx 1/700$.

5. **Flush** Five cards in a single suit but not a straight. We first compute the number of spade flushes. Since a spade flush consists of a combination of 13 spades taken five at a time, there are $_{13}C_5$ spade flushes (this includes the straights and royal flushes, which we will subtract later). Multiplying by 4 and subtracting straights and royal flushes, we get

$$\frac{(4)(_{13}C_5) - 40}{2{,}598{,}960} = \frac{5108}{2{,}598{,}960} \approx \frac{1}{500}$$

We could compute the number of hands with five spades without using combinations as follows. The probability of getting the first spade is 13/52. The probability of getting the second spade given that we have drawn the first is $12/51$, etc. Hence the probability of five spades is $(13/52)(12/51)(11/50)(10/49)(9/48)$.

6. **Straight** Five cards in a sequence not all the same suit. Consider the straights beginning with 2. Since there are four different 2s, four different 3s, etc., we have 4^5 different straights beginning with 2 (this includes the straight flushes, which we subtract later). When we multiply by 10 (a straight could begin with $1, 2, 3, \ldots, 10$) and subtract the number of straight and royal flushes, the probability becomes

$$\frac{(10)(4^5) - 40}{2,598,960} = \frac{10,200}{2,598,960} \approx \frac{1}{250}$$

Many beginning poker players have the feeling that a straight is harder to get than a flush, but we now see that a flush is twice as hard to get.

7. **Three of a kind** We first compute the number of hands with three 2s. There are $_4C_3$ different ways of getting three 2s and $_{48}C_2$ possibilities for the other two cards. Since there are 13 possible card values, and since we have to subtract the number of full houses, the probability becomes

$$\frac{(13)(_4C_3)(_{48}C_2) - 3744}{2,598,960} = \frac{54,912}{2,598,960} \approx \frac{1}{48}$$

8. **Two pairs** We first compute the number of hands with two 2s and two 3s. There are $_4C_2$ possible pairs of 2s, $_4C_2$ possible pairs of 3s, and 44 possibilities for the remaining fifth card. Since the number of possible pairs of card values is $_{13}C_2$, the probability becomes

$$\frac{(_{13}C_2)(_4C_2)(_4C_2)(44)}{2,598,960} = \frac{123,552}{2,598,960} \approx \frac{1}{21}$$

9. **One pair** We first compute the number of hands with two 2s. There are $_4C_2$ possible pairs of 2s, there are now 48 possibilities for the third card (this card cannot be a 2), 44 possibilities for the fourth card (this card cannot be a 2 or have the same value as the third card), and 40 possibilities for the fifth card. Since the order of these last three cards is irrelevant, we must divide by 3!. Finally we multiply by 13 since there are 13 different possible card values. The probability thus becomes

$$\frac{(_4C_2)(48)(44)(40)(13)}{(3!)(2,598,960)} = \frac{1,098,240}{2,598,960} \approx \frac{1}{2.5}$$

10. **Nothing of interest** This is the hand that the author usually gets and, as we shall see, is the hand most often dealt. To find the probability we add the probabilities in items 1 through 9 above and subtract the result from 1. This gives the probability

$$\frac{1,302,540}{2,598,960} \approx \frac{1}{2}$$

If you get this hand, you are well advised to drop out (unless you are a very good bluff) since, as we have seen in item 9, the chances are very good that somebody else will have a pair.

A few words about strategy in poker. As we have seen, the probability of getting a straight flush, say, is exceedingly low. But what if you are playing five-card stud (four cards face up and one face down) and after all the cards are dealt you hold a full house and one of your opponents shows the 3, 4, 5, 6 of spades? This opponent bets the limit. Should you stay in or fold? If you stay in because the chances of a straight flush are only one in 70,000, you are not reasoning correctly. In fact, now that you know that your opponent has the 3, 4, 5, 6 of spades, the opponent's chances of having a straight flush are considerably higher. In particular, with your knowledge of your own five cards, your opponent's four upward cards, and the other cards that were dealt up, you know about 20 or more cards. If none of these cards is a 2 or 7 of spades, your opponent's chances of having a straight flush have increased to 1 in 15. You should now make your decision accordingly.

Bridge

Some bridge problems are now considered.

EXAMPLE 1 Find the probability of drawing a bridge hand in which every card is a 9 or lower (ace is high in bridge).

Solution.   There are $(8)(4) = 32$ cards that are 9 or lower. Hence the probability is $_{32}C_{13}/_{52}C_{13} \approx 0.000547$. Thus if you draw such a hand, you certainly are justified in complaining about bad luck. $\square$

EXAMPLE 2 In a bridge game we refer to the four players as South, West, North, and East. If South wins the contract, then North becomes dummy and places her cards face up on the table. Suppose that South observes that he and his partner (North) hold seven spades. Then West and East must jointly hold the remaining six spades. Find the probability that these six spades are distributed 4 and 2 between West and East.

Solution.   Suppose West has four spades. There are $_6C_4$ ways that these four spades can be distributed in West's hand. The remaining nine cards in West's hand can be selected from 20 nonspades in $_{20}C_9$ different ways. We could also find East holding four spades. The probability thus becomes

$$\frac{(2)(_6C_4)(_{20}C_9)}{_{26}C_{13}} = \frac{78}{161} \approx 0.4845 \qquad \square$$

Other results of this type are exhibited in Table 10.1, which gives the distribution of cards in a single suit.

Table 10.1

Held jointly by S and N	Held jointly by W and E	Distribution between W and E	Probability
7	6	3–3	$286/805 \approx 0.3553$
		4–2	$78/161 \approx 0.4845$
		5–1	$117/805 \approx 0.1453$
		6–0	$12/805 \approx 0.0149$
8	5	3–2	$78/115 \approx 0.6783$
		4–1	$13/46 \approx 0.2826$
		5–0	$9/230 \approx 0.0391$
9	4	2–2	$234/575 \approx 0.4070$
		3–1	$286/575 \approx 0.4974$
		4–0	$11/115 \approx 0.0975$
10	3	2–1	$39/50 \approx 0.78$
		3–0	$11/50 \approx 0.22$
11	2	1–1	$13/25 \approx 0.52$
		2–0	$12/25 \approx 0.48$

EXERCISES

1. How would you answer the question posed at the end of the lottery discussion?
2. Verify that $_{42}C_6 = 5{,}245{,}786$.   3. Find $_{41}C_6$.
4. Verify the equation for $P(W_3)$.   5. Verify the equation for $P(W_4)$.

Assuming we have no multiple winners, find the average profit for the following lottery grand prizes.

6. \$2,000,000   7. \$3,000,000   8. \$4,000,000

9. Suppose the lottery is the same as discussed except that now integers between 1 and 40 inclusive are chosen. Find (a) $P(W_1)$,   (b) $P(W_2)$,   (c) $P(W_3)$,   (d) $P(W_4)$,   (e) the average profit when the grand prize is \$1,000,000.

10. Repeat Exercise 9 with 40 replaced by 38.

11. A card is drawn from an ordinary deck. The card is replaced, and a second card is drawn.
    (a) What is the probability that both cards are aces?
    (b) Given that the first card is an ace, what is the probability that the second card is an ace?

12. Two cards are dealt from an ordinary deck.
    (a) How many possible hands are there?
    (b) What is the probability that a pair is dealt?
    (c) What is the probability that both cards are the same suit?

For the distributions in Exercises 13–17, verify that the probabilities in Table 10.1 are correct.

13. S and N hold 11, W and E hold 2.   14. S and N hold 10, W and E hold 3.
15. S and N hold 9, W and E hold 4.   16. S and N hold 8, W and E hold 5.
17. S and N hold 7, W and E hold 6.

18. Three cards are dealt from a deck of 52 cards.
    (a) How many different hands are there?
    (b) What is the probability of three of a kind?
    (c) What is the probability of a pair?
    (d) What is the probability of three of the same suit?

19. Four cards are dealt from a deck. What is the probability that all four are
    (a) the same suit,   (b) four of a kind,   (c) three of a kind?

20. In a five-card poker hand consider the events A = "three aces are dealt" and B = "four aces are dealt." Find $P(B \mid A)$.

21. What is the probability of being dealt the following bridge hands? (a) All the same suit. (b) Twelve of the same suit. (c) All honor cards (jacks, queens, kings, aces).

22. If two cards are selected from an ordinary deck, which is more probable, that both the cards are spades or that one of the cards is a black ace?

23. What is the probability of being dealt a poker hand containing the ace of spades?

24. In poker, what is the probability of being dealt the following hand? AS, KD, 2H, 3H, 7C

25. A poker hand is dealt. What is the probability that it contains exactly (a) one ace,   (b) two aces,   (c) three aces,   (d) four aces,   (e) five aces?

26. What is the probability of being dealt a poker hand consisting only of black cards?

27. What is the probability of being dealt a poker hand that contains exactly one face card? Exactly two face cards? All face cards?

10.6 Craps

> *In order to seek truth it is necessary once in the course of our life to doubt as far as possible all things.*
> — RENÉ DESCARTES (1596–1650), French Mathematician and Philosopher

Craps is the most popular game of chance in the world's casinos. One reason is that among all the casino games craps gives the best odds. In this section the probability of winning at craps is computed. Conditional probability is the basic tool used in this computation.

The rules of craps are as follows. Two dice are tossed. The sum of the two numbers that appear upward is, of course, an integer between 2 and 12. If 2, 3, or 12 is the sum, the player loses. If a 7 or 11 is the sum, the player wins. If the sum is any other integer (the only possibilities are 4, 5, 6, 8, 9, and 10), then the player neither wins nor loses. Instead, this sum is designated as the player's **point**. The player then continues to toss the dice until either the point or a 7 appears, whichever comes first. If the point appears, the player wins; if a 7 appears, the player loses.

The simplest sample space that can be used to describe this game is

$$\{a_2, a_3, \ldots, a_{12}\}$$

where a_i occurs if the sum is i, $i = 2, 3, \ldots, 12$. The probabilities of these events are

Photo 10.11
Players having a fling at craps.

computed as in Example 6 in Section 9.2 and have the following values:

$$P\{a_2\} = P\{a_{12}\} = 1/36 \qquad P\{a_3\} = P\{a_{11}\} = 1/18$$
$$P\{a_4\} = P\{a_{10}\} = 1/12 \qquad P\{a_5\} = P\{a_9\} = 1/9$$
$$P\{a_6\} = P\{a_8\} = 5/36 \qquad P\{a_7\} = 1/6$$

Let w be the event that occurs if the player wins and ℓ the event that occurs if the player loses. The probability that the player wins on the first toss is

$$P(w) = P(a_7 \cup a_{11}) = P\{a_7\} + P\{a_{11}\} = \frac{1}{6} + \frac{1}{18} = \frac{2}{9}$$

The probability that the player loses on the first toss is

$$P(\ell) = P\{a_2 \cup a_3 \cup a_{12}\} = P\{a_2\} + P\{a_3\} + P\{a_{12}\} = \frac{1}{36} + \frac{1}{18} + \frac{1}{36} = \frac{1}{9}$$

Figure 10.3 shows a probability tree that describes the first toss. The upper two circles end the game. However, if any of the six lower circles is reached, the process is continued to the next stage. That is, the dice are tossed again until either a 7 is the sum or the point is the sum.

Now consider what happens if the first toss of the dice results in the sum of 5. This is the fourth circle from the top in Figure 10.3. The dice are tossed additional times until a 5 or a 7 is the sum. All other results are ignored. If the sum is 5, the player wins, and if the sum is 7, the player loses. Since all other results can be ignored, the problem is equivalent to looking at the four ways of tossing a 5 and the six ways of tossing a 7. Thus out of the ten relevant outcomes, four give a 5 and six give a 7. Hence the probability of winning given that the point is 5 becomes $P(w \mid a_5) = 4/10 = 2/5$, and the probability of losing given that the point is 5

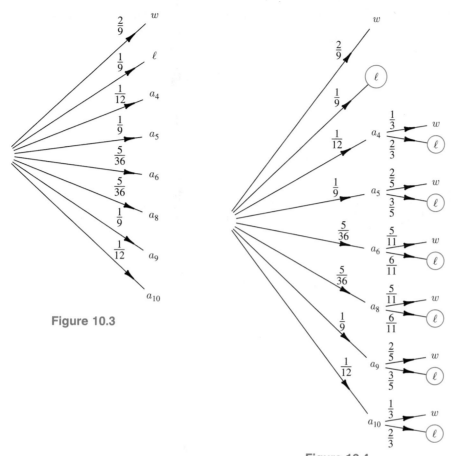

Figure 10.3

Figure 10.4

becomes $P(\ell \mid a_5) = 6/10 = 3/5$. In a similar way the following probabilities are found:

$$P(w \mid a_9) = \frac{3}{5}, \qquad P(w \mid a_4) = P(w \mid a_{10}) = \frac{1}{3}, \qquad P(w \mid a_6) = P(w \mid a_8) = \frac{5}{11}$$

The complete probability tree can now be depicted as in Figure 10.4.

There are seven paths that lead to win. Each of these paths represents a mutually exclusive event. For example, the topmost path represents the event "either a 7 or 11 occurs on the first toss." The bottommost path represents the event "a 10 on the first toss and a 10 is tossed before a 7 is tossed on succeeding tosses." Since these two events cannot both occur, they are mutually exclusive.

Since all the winning paths are mutually exclusive, the probability of winning is the sum of the probabilities of the seven winning paths. Thus the probability of winning a game of craps is

$$\frac{2}{9} + \frac{1}{36} + \frac{2}{45} + \frac{25}{396} + \frac{25}{396} + \frac{2}{45} + \frac{1}{36} = \frac{244}{495} = 0.49292929$$

This probability is slightly less than $1/2$. On the average, a player should expect to win 244 and lose 251 games of craps out of every 495 games.

EXERCISES

1. Prove that $P(w \mid a_4) = 1/3$ and $P(w \mid a_6) = 5/11$.
2. What is the probability that a player's point is 4 and the player wins?
3. What is the probability that the player's point is 6 given that the player wins?
4. What is the probability that a craps game ends with one toss?
5. What is the probability that a craps game ends with exactly two tosses?
6. What is the probability that a craps player wins with exactly two tosses in one game?
7. What is the probability that a player's point is 6 and the player wins with exactly two tosses?
8. What is the probability that a player's point is 6 and the player wins with exactly three tosses?
9. Suppose a dice game is played exactly like craps except that after a player tosses a point the player loses if the point is tossed later and wins if a 7 is tossed later. Construct the complete probability tree for this game and compute the probability of winning.
*10. Alter the game of craps as follows. On the first toss a player wins with a 6 or 10, loses with a 2, 3, or 12, and throws a point with a 4, 5, 7, 8, 9, or 11. If a point is thrown, the player wins if the point is later thrown and loses if a 6 is later thrown. Construct the complete probability tree for this game and compute the probability of winning.

$\boxed{10.7}$ Gambler's Ruin (Optional)

> *Do not try to know everything if you wish to know anything.*
> — DEMOCRITUS OF ABDERA (470 B.C.), Greek Philosopher

The gambler's ruin problem is formulated as follows. A gambler and an opponent are playing a game. On each play of the game either the gambler or the opponent wins $1. We shall consider the simple case in which each has the probability $1/2$ of winning on each play. The game is continued until either the gambler or the opponent is ruined, that is, loses all of his or her money. Suppose the gambler starts with n dollars and the opponent starts with $a - n$ dollars ($a \geq n$), so that their combined capital is a dollars. The sample space consists of sequences of W's and L's where a W indicates a win and an L a loss for the gambler. A typical sample point is WLLWWWLL.... The sequence terminates if the gambler or the opponent is ruined, that is, when the gambler's capital is either 0 or a. The problem is to find the probability that the gambler will be ruined.

Random Walks

This problem can also be interpreted physically. Suppose a particle starts at an initial position n on the real line and moves at regular time intervals a unit in the positive

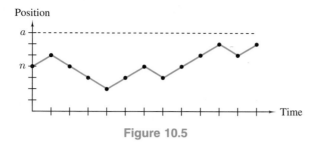

Figure 10.5

direction with probability $1/2$ or a unit in the negative direction with probability $1/2$. The process terminates when the particle reaches either 0 or a for the first time. A typical particle path is shown in Figure 10.5. Such a process is sometimes called a **random walk**. The position of the particle corresponds to the gambler's capital at a given instant.

　Random walks are used to describe one-dimensional diffusion processes in which a gas (or disease, bacteria, etc.) diffuses through a medium. They also describe Brownian motion, in which a particle is subjected to a great number of molecular collisions that impart a random motion to it. This can also be done in two or three dimensions. In two dimensions the problem can be rephrased as follows. If a person leaves a tavern and walks randomly either forward, backward, to the left, or to the right, what is the probability that the person will arrive home before he or she is back at the tavern again?

　Before solving the gambler's ruin problem, we must first formulate it mathematically. Suppose the gambler starts with n dollars and the opponent starts with $a - n$ dollars ($a \geq n$). Let q_n be the probability of the gambler's ultimate ruin. What information do we have concerning q_n? First, if the gambler starts with no dollars, he will certainly be ruined (he already is); so $q_0 = 1$. Furthermore, if the gambler starts with a dollars, he certainly will not be ruined; so $q_a = 0$. These are called **boundary conditions** for the problem.

Gambler's Ruin Formula

Now suppose $0 < n < a$. After the first play of the game either the gambler loses (event L_1) and has $n - 1$ dollars or wins (event W_1) and has $n + 1$ dollars. From Bayes' second rule, the probability that the gambler is eventually ruined equals the probability of event L_1 times the conditional probability of ruin given L_1 plus the probability of event W_1 times the conditional probability of ruin given W_1. That is,

$$P(\text{ruin}) = P(L_1)P(\text{ruin} \mid L_1) + P(W_1)P(\text{ruin} \mid W_1)$$

This is the same thing as the difference equation

$$q_n = \frac{1}{2} q_{n-1} + \frac{1}{2} q_{n+1}, \qquad 0 < n < a$$

　The gambler's ruin problem has thus been reduced to the following mathematical problem. Solve the above difference equation for q_n subject to the boundary conditions.

Table 10.2

Probability of ruin q_n	Initial capital of gambler n	Initial capital of opponent $a - n$	Total capital a
1/2	10	10	20
2/3	10	20	30
1/3	20	10	30
10/11	100	1,000	1,100
1,000/1,001	1,000	1,000,000	1,001,000

Difference equations were discussed in Section 1.7, and you might review them at this point. How can we solve the above difference equation? First of all, $q_0 = 1$. If we let $n = 1$ in the difference equation, we obtain

$$q_1 = \frac{1}{2} q_0 + \frac{1}{2} q_2 \quad \text{or} \quad q_2 = 2q_1 - 1$$

Letting $n = 2$, we have $q_2 = (1/2)q_1 + (1/2)q_3$, and substituting the above value of q_2 into this equation gives $q_3 = 3q_1 - 2$. Continuing in this way, we find

$$q_4 = 4q_1 - 3$$
$$q_5 = 5q_1 - 4$$
$$\vdots \quad \vdots$$
$$q_n = nq_1 - (n - 1)$$
$$\vdots \quad \vdots$$
$$q_a = aq_1 - (a - 1)$$

Substituting the boundary condition $q_a = 0$ into this last equation gives $0 = aq_1 - (a-1)$; so

$$q_1 = \frac{a - 1}{a} = 1 - \frac{1}{a}$$

We thus find that

$$q_n = n \left(1 - \frac{1}{a}\right) - (n - 1) = 1 - \frac{n}{a}$$

which solves the difference equation subject to the boundary conditions. This is the *only* solution of the difference equation subject to the boundary conditions (why?). This solves the gambler's ruin problem.

Let us substitute some values for n and a into the solution $q_n = 1 - n/a$. For example, suppose the gambler began with \$10 and the opponent began with \$10. Then $n = 10$ and $a = 20$; so $q_{10} = 1 - 1/2 = 1/2$. Just as expected, if the gambler and the opponent start with the same capital, then the gambler has a probability of $1/2$ of being ruined. Table 10.2 gives q_n for some different values of n and a.

The last row in Table 10.2 corresponds to the case of an individual playing against a casino in a fair game. The individual starts with a capital of $1000 and the casino with $1,000,000. The probability that the individual loses everything is $1000/1001$. In other words, the individual has only one chance in 1001 to break the casino and win $1,000,000 in a fair game. If the game is weighted toward the casino (which is always the case), the probability of ruin is even greater.

The moral of the story is that if you are going to play a long game against an equally skilled player, make sure your opponent is poorer. Also this shows that if you are playing against a very rich opponent (say a casino), you will eventually lose in a long game and so it is better to play high stakes for a short period.

Suppose a gambler whose initial capital is $1000 is not interested in breaking the casino, which is highly unlikely, but only interested in making $10. In this case we can think of the opponent as having $10. If the game is fair, the probability that the gambler will win $10 before being ruined becomes

$$1 - q_{1000} = 1000/1010 = 0.990$$

Thus if the gambler is not greedy, the chance of making $10 is excellent. What are the gambler's chances of making $100, $1000? These are $1000/1100 = 0.909$ and $1000/2000 = 1/2$. If the game is weighted toward the casino, then these probabilities will go down.

EXERCISES

1. If the gambler has an initial capital of $10,000 and the opponent has an initial capital of $15,000, what is the probability of the gambler's eventual ruin?

2. Repeat Exercise 1 with $10,000 and $15,000 replaced by $100 and $120, respectively.

3. If the gambler starts with $100 and the probability of his eventual ruin is $3/4$, what is his opponent's initial capital?

4. Repeat Exercise 3 with $3/4$ replaced by $5/6$.

5. If the probability of the gambler's eventual ruin is $2/5$ and his opponent starts with $1000, what is the gambler's initial capital?

6. Repeat Exercise 5 with $2/5$ replaced by $3/5$.

7. A gambler with $100 playing against a rich opponent decides to quit after he wins $50. What is the probability that the gambler will be successful?

8. Prove that $q_{a-n} + q_n = 1$. Why does this show that the gambler is either eventually ruined or wins everything so that the game does not go on forever but eventually ends?

9. Show that if f_n and g_n are solutions of the difference equation $q_n = \frac{1}{2} q_{n+1} + \frac{1}{2} q_{n-1}$, then so is $cf_n + dg_n$, where c and d are any constants.

*10. Show that $f_n = c + dn$ is a solution of the difference equation $q_n = \frac{1}{2} q_{n+1} + \frac{1}{2} q_{n-1}$ for any constants c and d. Find the values of c and d so that f_n also satisfies the boundary conditions of this section. Now compare f_n with the solution obtained in this section.

11. Show that the difference equation becomes $q_n = pq_{n+1} + (1 - p)q_{n-1}$ if the gambler has probability p of winning on a single play.

*12. Show that

$$f_n = c + d \left(\frac{1-p}{p} \right)^n$$

satisfies the difference equation in Exercise 11. Find the values of c and d so that f_n also satisfies the boundary conditions.

Chapter 10 Summary of Terms

sample space

random variable

average value (expectation, mean)

$$E(f) = f(a_1)P\{a_1\} + f(a_2)P\{a_2\} + \cdots + f(a_n)P\{a_n\}$$

roulette

numbers

chuck-a-luck

football pools

keno

martingale method

permutations $\quad {}_nP_r = \dfrac{n!}{(n-r)!}$

combinations $\quad {}_nC_r = \dfrac{n!}{r!(n-r)!}$

lottery

poker

bridge

craps

gambler's ruin $\quad q_n = 1 - \dfrac{n}{a}$

Chapter 10 Test

1. A loaded coin comes up heads one-fourth of the time and tails three-fourths of the time. If the coin is flipped twice, what is the average number of heads?

2. Let $\Omega = \{a_1, a_2, a_3, a_4\}$ be a sample space and let $f\colon \Omega \to \mathbb{R}$ be the random variable defined by $f(a_i) = i^2 - 8$, $i = 1, 2, 3, 4$. If each sample point has the same probability, find $E(f)$ and $E(f^2)$.

3. What is a player's expected profit in the numbers game if \$850 is returned for a winning number?

4. If you bet $1 on red and $2 on black in roulette, what is your expected profit?

5. Prove that $_nC_r = \left(1 - \dfrac{r}{n+1}\right) {_{n+1}C_r}$.

6. If a set S has 10 elements, how many subsets does S have with 5 elements?

7. Four cards are dealt from a deck. What is the probability that there are two pairs?

8. Four cards are dealt from a deck. What is the probability of a four-card straight?

9. In the gambler's ruin problem, a gambler with initial capital $100 plays against a rich opponent and decides that he will quit after he wins $10. What is the probability that the gambler is successful?

10. A small lottery is played as follows. You choose six different integers between 1 and 10 inclusive. If your numbers match the six numbers in a later random selection, you win $100. Otherwise, you lose a dollar. What is your average profit if there is only one winner?

11. How many four-person committees can be selected from a group of 20 people?

12. $_nC_r$ is sometimes called "n choose r." Why is this?

13. If a random variable has only nonnegative values, show that its average value is nonnegative.

14. Are the following statements true or false?
    (a) The average profit is always positive.
    (b) If you've been losing for a long time, you're due to start winning so it's a good idea to bet high.
    (c) You have a better chance to win at craps than at roulette.
    (d) $_2C_1 = 2$.
    (e) $_2P_1 = 2$.
    (f) $_nP_r/{_nC_r} = r!$
    (g) $_nP_r \leq {_nC_r}$ for all n and r.
    (h) In poker, a straight is better than a flush.
    (i) The Queen of Hearts represents Charlemagne's wife.
    (j) The difference between a permutation and a combination is that in a combination the order of the objects is immaterial.

15. What is the probability of five consecutive even numbers in roulette?

16. How many different 10-digit numbers, such as 8,730,152,649, can be written using all 10 digits? Numbers starting with zero are excluded.

MATRICES

11

The goal of mathematics is to think the thinkable.
— C. J. KEYSER, Twentieth-Century Mathematician

Although matrices have been useful in science and engineering for a long time, they have recently become important in the mathematical analysis of social and economic problems. A matrix gives a means for organizing a large array of data. Then using matrix algebra, these data arrays can be treated as single entities and be subjected to various mathematical operations. This gives an economical and efficient way of manipulating and analyzing data. The idea of matrices goes back to the English mathematician Arthur Cayley (1812–1895) who first introduced them in 1858. The recent widespread use of matrices stems from the advent of high-speed computers that can perform efficient calculations on large quantities of numbers.

11.1 Definition of a Matrix

matrix derived from Latin *matr-*, *mater*, womb. Something within which something else originates or develops.
— *Webster's New Collegiate Dictionary*

A **matrix** is a rectangular array of real numbers. An example of a matrix is

$$M = \begin{bmatrix} 1 & 3 \\ -2 & 7 \\ 5 & 1 \end{bmatrix}$$

A matrix is not just a collection of numbers; the locations of the numbers in the matrix are important. This is because we want to keep track of each individual number and perhaps store it and retrieve it. Each horizontal line of numbers is a **row** and each vertical line of numbers is a **column** of the matrix. The matrix M has three rows and two columns. For this reason, we say that M has **order** 3×2 (read "3 by 2"). Each number appearing in a matrix is an **entry** of the matrix and we can locate an entry by specifying its row and column. For example, in the matrix M the entry in the first row and second column is 3. Notice that the entries in the first row, first column and third row, second column are both 1.

In general, a matrix is a rectangular array of numbers arranged in m rows and n columns. We represent a matrix by a capital letter as follows.

$$A = \begin{bmatrix} a_{11} & a_{12} & \cdots & a_{1n} \\ a_{21} & a_{22} & \cdots & a_{2n} \\ \vdots & & \ddots & \\ a_{m1} & a_{m2} & \cdots & a_{mn} \end{bmatrix}$$

The entries of a matrix are designated using the lower case of the same letter that symbolizes the matrix itself. For the matrix A, the entry a_{ij} is the number in the ith row and jth column, and we call a_{ij} the **ijth entry**. The **order** (or **dimension**) of A is $m \times n$ and we say that A is an $m \times n$ matrix. To designate the order of a matrix, we sometimes use the notation

$$A_{m \times n} \qquad \text{or} \qquad \left[a_{ij} \right]_{m \times n}$$

EXAMPLE 1   The following are examples of matrices.

$$A = \begin{bmatrix} 3 & 5 & -2 \\ 1 & 7 & 6 \end{bmatrix} \qquad \begin{array}{l} A \text{ is a } 2 \times 3 \text{ matrix.} \\ a_{12} = 5, \quad a_{23} = 6 \end{array}$$

$$B = \begin{bmatrix} 2 & -1 & 3 \\ 0 & 4 & -3 \\ -1 & 5 & 7 \end{bmatrix} \qquad \begin{array}{l} B \text{ is a } 3 \times 3 \text{ matrix.} \\ b_{22} = 4, \quad b_{12} = -1, \quad b_{32} = 5 \end{array}$$

EXAMPLE 2   A television set manufacturer has two factories F1 and F2. Sets are produced in these factories and shipped to three warehouses W1, W2, and W3 (Figure 11.1). The sets are distributed from the warehouses to wholesale and retail outlets. The shipping costs to the various warehouses are tabulated in Table 11.1.

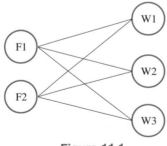

Figure 11.1

Table 11.1

From	To	Cost per item
F1	W1	$3.00
F1	W2	$2.75
F1	W3	$3.10
F2	W1	$2.60
F2	W2	$2.80
F2	W3	$2.40

The cost schedule can be displayed in a 2×3 matrix using factories as the row labels and warehouses as the column labels,

$$C = \begin{array}{c} \begin{array}{ccc} \text{W1} & \text{W2} & \text{W3} \end{array} \\ \begin{bmatrix} 3.00 & 2.75 & 3.10 \\ 2.60 & 2.80 & 2.40 \end{bmatrix} \begin{array}{c} \text{F1} \\ \text{F2} \end{array} \end{array}$$

Notice that c_{ij} is the cost per item shipped from factory i to warehouse j, $i = 1, 2$, $j = 1, 2, 3$. For example, $c_{23} = 2.40$ is the cost for shipping a set from factory F2 to warehouse W3.

Special Types

We next consider special types of matrices. An $m \times 1$ matrix has just one column and is called a **column vector**, while a $1 \times n$ matrix has just one row and is called a **row vector**.

EXAMPLE 3   The 3×1 matrix V is a column vector. The 1×4 matrix W is a row vector. The 1×1 matrix X is both a row vector and a column vector. Such a matrix is really no different than a number and we usually identify it with that number.

$$V = \begin{bmatrix} -2 \\ 5 \\ 1 \end{bmatrix} \qquad W = \begin{bmatrix} 5 & 7 & 0 & -3 \end{bmatrix} \qquad X = \begin{bmatrix} 4 \end{bmatrix}$$

A matrix with the same number of rows as columns is called a **square matrix**. Square matrices have order $n \times n$ for some natural number n. The **diagonal** of a square matrix A consists of entries of the form $a_{11}, a_{22}, \ldots, a_{nn}$. A **diagonal matrix** is a square matrix whose off-diagonal entries are zero.

EXAMPLE 4   The matrix A is a 3×3 square matrix whose diagonal consists of the entries $5, 4, 3$. The matrix B is a 4×4 diagonal matrix.

$$A = \begin{bmatrix} 5 & 2 & 0 \\ -2 & 4 & 1 \\ 0 & -1 & 3 \end{bmatrix} \qquad B = \begin{bmatrix} 3 & 0 & 0 & 0 \\ 0 & 1 & 0 & 0 \\ 0 & 0 & -2 & 0 \\ 0 & 0 & 0 & 8 \end{bmatrix}$$

A diagonal matrix whose diagonal consists of all 1s is called an **identity matrix**. The $n \times n$ identity matrix is denoted by I_n or sometimes simply as I. As we shall see, in matrix algebra I_n acts in much the same way as the number 1 in regular algebra. A **zero matrix** is any matrix that has zero for every entry. We denote the $m \times n$ zero matrix by $O_{m \times n}$ or sometimes simply as O. In matrix algebra, a zero matrix acts like the number zero in regular algebra. Some examples of identity and zero matrices are

$$I_2 = \begin{bmatrix} 1 & 0 \\ 0 & 1 \end{bmatrix} \qquad I_3 = \begin{bmatrix} 1 & 0 & 0 \\ 0 & 1 & 0 \\ 0 & 0 & 1 \end{bmatrix} \qquad I_4 = \begin{bmatrix} 1 & 0 & 0 & 0 \\ 0 & 1 & 0 & 0 \\ 0 & 0 & 1 & 0 \\ 0 & 0 & 0 & 1 \end{bmatrix}$$

$$O_{2 \times 3} = \begin{bmatrix} 0 & 0 & 0 \\ 0 & 0 & 0 \end{bmatrix} \qquad O_{3 \times 4} = \begin{bmatrix} 0 & 0 & 0 & 0 \\ 0 & 0 & 0 & 0 \\ 0 & 0 & 0 & 0 \end{bmatrix}$$

Transpose

If A is an $m \times n$ matrix, the **transpose** A^t of A is the $n \times m$ matrix whose rows are the columns in A (in the same order) and whose columns are the rows in A (in the same order). Denoting the entries of A^t by a_{ij}^t, this is the same as requiring that $a_{ij}^t = a_{ji}$. Notice that the transpose of a row vector is a column vector and the transpose of a column vector is a row vector.

EXAMPLE 5 Examples of transposes are

$$A = \begin{bmatrix} 2 & 3 \\ -1 & 7 \\ 0 & 6 \end{bmatrix} \qquad A^t = \begin{bmatrix} 2 & -1 & 0 \\ 3 & 7 & 6 \end{bmatrix}$$

$$B = \begin{bmatrix} 1 & 0 & 3 \\ 2 & 1 & 5 \\ 7 & 6 & 8 \end{bmatrix} \qquad B^t = \begin{bmatrix} 1 & 2 & 7 \\ 0 & 1 & 6 \\ 3 & 5 & 8 \end{bmatrix}$$

$$V = \begin{bmatrix} 2 & 0 & -1 \end{bmatrix} \qquad V^t = \begin{bmatrix} 2 \\ 0 \\ -1 \end{bmatrix}$$

Two matrices A and B are said to be **equal** if they have the same order and if $a_{ij} = b_{ij}$ for every pair of subscripts ij. For example, if

$$A = \begin{bmatrix} 1 & 2 \\ 3 & 4 \end{bmatrix} \qquad \text{and} \qquad B = \begin{bmatrix} 1 & 3 \\ 2 & 4 \end{bmatrix}$$

then $A \neq B$ even though they contain the same set of numbers. If A equals B we write $A = B$.

The next example illustrates how a matrix can be described by a formula.

EXAMPLE 6 Write the 3×3 matrix in which

$$a_{ij} = \begin{cases} 0 & \text{if } i + j \text{ is even} \\ 1 & \text{if } i + j \text{ is odd} \end{cases}$$

Solution. In Table 11.2 we list the values of i and j, whether $i + j$ is even or odd, and the corresponding value of a_{ij}. The matrix A becomes

$$A = \begin{bmatrix} 0 & 1 & 0 \\ 1 & 0 & 1 \\ 0 & 1 & 0 \end{bmatrix} \qquad \qquad \square$$

Table 11.2

i	j	$i+j$	Even/odd	a_{ij}
1	1	2	even	0
1	2	3	odd	1
1	3	4	even	0
2	1	3	odd	1
2	2	4	even	0
2	3	5	odd	1
3	1	4	even	0
3	2	5	odd	1
3	3	6	even	0

EXERCISES

1. Give the order of each matrix in Example 5.

2. If the order of A is $m \times n$, show that the order of A^t is $n \times m$.

3. For the matrix

$$A = \begin{bmatrix} 5 & 0 & -1 & 2 \\ \frac{2}{3} & 1 & 6 & -2 \\ 7 & 3 & -3 & 8 \end{bmatrix}$$

   give the elements (a) a_{12}, (b) a_{33}, (c) a_{23}, (d) a_{32}, (e) a_{11}, (f) a_{34}.

4. For matrix A of Exercise 3, give the row-column location of the numbers (a) 1, (b) -3, (c) 2/3, (d) 2, (e) -1, (f) -2.

5. For matrix A of Exercise 3, what is A^t?

6. Give the transposes of the matrices in Example 4.

7. Show that the transpose of a row vector is a column vector and the transpose of a column vector is a row vector.

8. If I is an identity matrix, show that $I^t = I$.

9. If D is a diagonal matrix, show that $D^t = D$.

For Exercises 10–13, write the 4×4 matrix in which a_{ij} has the stated value.

10. $a_{ij} = i + j$

11. $a_{ij} = j/i$

12. $a_{ij} = \begin{cases} 0 & \text{if } i = j \\ 1 & \text{if } i < j \\ -1 & \text{if } i > j \end{cases}$

13. $a_{ij} = \begin{cases} 0 & \text{if } i = j \\ i & \text{if } i > j \\ j & \text{if } i < j \end{cases}$

14. If the last column of matrix A of Exercise 3 is removed, what is the diagonal of the new matrix?

15. If the first column of matrix A of Exercise 3 is removed, what is the diagonal of the new matrix?

16. What is the diagonal of the matrix in Exercise 10?

17. What is the diagonal of the matrix in Exercise 11?

18. Write out the matrix I_5.

19. Write out the matrix $O_{4 \times 5}$.

20. Prove that $(A^t)^t = A$ for any matrix A.

21. How many entries does an $m \times n$ matrix have?

22. How many entries are there in the diagonal of an $n \times n$ matrix?

23. Represent the cost schedule of Example 2 as a matrix using factories as the column labels and warehouses as the row labels. How is this matrix related to C of Example 2?

$\boxed{11.2}$ Addition, Subtraction, Scalar Multiplication

The science of figures, to a certain degree, is not only indispensably requisite in every walk of civilized life; but the investigation of mathematical truths accustoms the mind to method and correctness of reasoning, and is an employment peculiarly worthy of rational beings. In a clouded state of existence, where so many things appear precarious to the bewildered researcher, it is here that the rational faculties find a firm foundation to rest upon. From the high ground of mathematical and philosophical demonstrations, we are insensibly led to far nobler speculations and sublimer meditations.

— GEORGE WASHINGTON (1732–1799), American President
—Letter dated June 20, 1788

We now begin the study of matrix algebra by describing how matrices can be added, subtracted, and multiplied by numbers. Numbers are sometimes called **scalars** and multiplying a matrix by a number is called **scalar multiplication**.

Addition

Two matrices of the same order are added by adding their corresponding entries. Thus, if A and B are $m \times n$ matrices, then their **sum** C is the $m \times n$ matrix with $c_{ij} = a_{ij} + b_{ij}$, and we write $C = A + B$.

If two matrices do not have the same order, they cannot be added.

EXAMPLE 1   The following are matrix sums.

(a) $\begin{bmatrix} 2 & 5 & 6 \end{bmatrix} + \begin{bmatrix} 3 & 0 & -4 \end{bmatrix} = \begin{bmatrix} 2+3 & 5+0 & 6-4 \end{bmatrix} = \begin{bmatrix} 5 & 5 & 2 \end{bmatrix}$

(b) $\begin{bmatrix} 1 & 0 & 3 \\ 0 & -2 & -1 \\ 5 & 7 & 2 \end{bmatrix} + \begin{bmatrix} 2 & 1 & 4 \\ 3 & 5 & 6 \\ 0 & -1 & 2 \end{bmatrix} = \begin{bmatrix} 1+2 & 0+1 & 3+4 \\ 0+3 & -2+5 & -1+6 \\ 5+0 & 7-1 & 2+2 \end{bmatrix} = \begin{bmatrix} 3 & 1 & 7 \\ 3 & 3 & 5 \\ 5 & 6 & 4 \end{bmatrix}$

(c) $\begin{bmatrix} 1 & 0 \\ 2 & 3 \end{bmatrix} + \begin{bmatrix} 2 & 4 & 5 \\ 1 & 6 & 7 \end{bmatrix}$

The sum of (c) is not defined because the two matrices have different orders.

EXAMPLE 2   A concession stand sells hamburgers, hot dogs, and sodas. In the morning they sold 12 hamburgers, 15 hot dogs, and 24 sodas. In the afternoon they sold 24 hamburgers, 30 hot dogs, and 46 sodas. Representing the morning and afternoon sales by row vectors

$$M = \begin{bmatrix} 12 & 15 & 24 \end{bmatrix} \qquad A = \begin{bmatrix} 24 & 30 & 46 \end{bmatrix}$$

the total sales according to type become

$$S = M + A = \begin{bmatrix} 36 & 45 & 70 \end{bmatrix}$$

Properties of Addition

The operation of addition has three basic properties.

(1) $A + O = A$ (O is the additive identity.)
(2) $A + B = B + A$ (Commutative law of addition.)
(3) $(A + B) + C = A + (B + C)$ (Associative law of addition.)

Of course, the matrices in Properties 1, 2, and 3 are assumed to have the same order. The proofs of these properties are quite simple. For example, to prove Property 2 we have

$$A + B = \begin{bmatrix} a_{ij} \end{bmatrix} + \begin{bmatrix} b_{ij} \end{bmatrix} = \begin{bmatrix} a_{ij} + b_{ij} \end{bmatrix} = \begin{bmatrix} b_{ij} + a_{ij} \end{bmatrix}$$
$$= \begin{bmatrix} b_{ij} \end{bmatrix} + \begin{bmatrix} a_{ij} \end{bmatrix} = B + A$$

Scalar Multiplication

> For **scalar multiplication**, we can multiply any matrix A by a number k. This is done by multiplying each entry of A by k. In symbols this gives
>
> $$kA = \begin{bmatrix} ka_{ij} \end{bmatrix}$$

The definition of Ak is the same as for kA; that is, $Ak = kA$. For example,

$$3 \begin{bmatrix} 3 & 4 & 7 & 0 \end{bmatrix} = \begin{bmatrix} 9 & 12 & 21 & 0 \end{bmatrix}$$

$$\frac{1}{2} \begin{bmatrix} 8 & 6 \\ 4 & 1 \\ 2 & 3 \end{bmatrix} = \begin{bmatrix} 4 & 3 \\ 2 & \frac{1}{2} \\ 1 & \frac{3}{2} \end{bmatrix}$$

We can also use scalar multiplication to factor a number out of a matrix. For example,

$$\begin{bmatrix} \frac{3}{4} & \frac{1}{2} & -2 \\ \frac{3}{2} & \frac{5}{4} & \frac{1}{2} \end{bmatrix} = \frac{1}{4} \begin{bmatrix} 3 & 2 & -8 \\ 6 & 5 & 2 \end{bmatrix}$$

Properties of Scalar Multiplication

The operation of scalar multiplication has the following properties. If c and d are real numbers and A and B are matrices of the same order, then

(1) $(c+d)A = cA + dA$.
(2) $c(A + B) = cA + cB$.
(3) $c(dA) = (cd)A$.

The proof of Property 3, for example, is

$$c(dA) = c\left[da_{ij}\right] = \left[(cd)a_{ij}\right] = (cd)\left[a_{ij}\right] = (cd)A$$

The **negative** $-A$ of a matrix A is defined as $-A = (-1)A$. Since $-\left[a_{ij}\right] = \left[-a_{ij}\right]$, each entry of $-A$ is the negative of the corresponding entry of A. If O is the zero matrix with the same order as A, then it is clear that

$$A + (-A) = O$$
which we write as $$A - A = O$$

Subtraction

If A and B have the same order, we define the **subtraction** of B from A as

$$A - B = A + (-B)$$

The entries of $A - B$ are the entries of A minus the corresponding entries of B. In symbols, we have
$$A - B = \left[a_{ij}\right] + \left[-b_{ij}\right] = \left[a_{ij} - b_{ij}\right]$$

EXAMPLE 3   Find $A - B$ if

$$A = \begin{bmatrix} 1 & 4 & 6 \\ 3 & -2 & 0 \end{bmatrix} \quad \text{and} \quad B = \begin{bmatrix} 2 & 3 & 4 \\ -4 & -1 & 2 \end{bmatrix}$$

Solution.   We have

$$A - B = \begin{bmatrix} -1 & 1 & 2 \\ 7 & -1 & -2 \end{bmatrix} \qquad \square$$

Matrix Equations

We can solve certain simple matrix equations just like we solve regular algebraic equations. Suppose A and B are known matrices of the same order and we want to find the unknown matrix X that satisfies

$$2X + A = B$$

We first subtract A from both sides,

$$2X + A - A = B - A$$

Hence,
$$2X = B - A$$

We then multiply both sides by 1/2, giving

$$X = \frac{1}{2}(B - A)$$

EXAMPLE 4   Solve the matrix equation $5X + A = 4(X + B)$ where

$$A = \begin{bmatrix} 3 & 4 & 0 \\ 1 & 5 & 2 \\ 0 & 1 & 6 \end{bmatrix} \quad \text{and} \quad B = \begin{bmatrix} 0 & 1 & 0 \\ 3 & 0 & 2 \\ 1 & 2 & 3 \end{bmatrix}$$

Solution.   We have

$$5X + A = 4X + 4B$$
$$X + A = 4B$$
$$X = 4B - A$$

Hence,

$$X = 4 \begin{bmatrix} 0 & 1 & 0 \\ 3 & 0 & 2 \\ 1 & 2 & 3 \end{bmatrix} - \begin{bmatrix} 3 & 4 & 0 \\ 1 & 5 & 2 \\ 0 & 1 & 6 \end{bmatrix}$$

$$= \begin{bmatrix} 0 & 4 & 0 \\ 12 & 0 & 8 \\ 4 & 8 & 12 \end{bmatrix} - \begin{bmatrix} 3 & 4 & 0 \\ 1 & 5 & 2 \\ 0 & 1 & 6 \end{bmatrix} = \begin{bmatrix} -3 & 0 & 0 \\ 11 & -5 & 6 \\ 4 & 7 & 6 \end{bmatrix} \qquad \square$$

EXERCISES

Given the matrices

$$A = \begin{bmatrix} 2 & 0 & -3 \\ 4 & 3 & 1 \\ -2 & 2 & 0 \end{bmatrix} \quad B = \begin{bmatrix} -1 & 5 & 7 \\ 6 & -3 & -2 \\ 0 & 1 & 4 \end{bmatrix} \quad C = \begin{bmatrix} 3 & 2 & -1 \\ 8 & 0 & 5 \\ -4 & 6 & 3 \end{bmatrix}$$

determine the results of the following operations.

1. $A + B$ 2. $A + C$ 3. $B + C$
4. $A - B$ 5. $B - A$ 6. $B - C$
7. $C - A$ 8. $3A$ 9. $\frac{1}{2}B$
10. $2C$ 11. $A - (B + C)$ 12. $(A - B) - C$
13. $3A - 2(B - C)$ 14. $2(A - 3B + 4C)$

Write each of the following matrices as the scalar multiplication of a fraction and a matrix of integers.

15. $\begin{bmatrix} \frac{2}{3} & -\frac{7}{18} \\ \frac{5}{9} & -\frac{1}{3} \\ \frac{5}{6} & -\frac{1}{9} \end{bmatrix}$ 16. $\begin{bmatrix} \frac{3}{4} & \frac{7}{12} \\ \frac{2}{3} & -\frac{5}{6} \end{bmatrix}$ 17. $\begin{bmatrix} 0 & \frac{2}{3} & -\frac{4}{5} \\ \frac{1}{3} & -3 & \frac{2}{15} \end{bmatrix}$ 18. $\begin{bmatrix} \frac{7}{2} & -\frac{5}{4} & \frac{3}{8} \end{bmatrix}$

Given the matrices

$$A = \begin{bmatrix} 0 & 3 \\ 4 & -2 \end{bmatrix} \qquad B = \begin{bmatrix} 1 & -\frac{1}{2} \\ 3 & 0 \end{bmatrix} \qquad C = \begin{bmatrix} 1 & 0 \\ 1 & -2 \end{bmatrix}$$

solve each of the following matrix equations for the matrix X.

19. $3X = A$

20. $X + A = C$

21. $X - B = 0$

22. $X + \frac{1}{3}A = B$

23. $X + B = A + C$

24. $2(X - A) = 3(X + B) + C$

25. Show that for 1×1 matrices, the operations of addition, subtraction, and scalar multiplication reduce to the usual operations for numbers.

26. Prove that Property 1 of addition holds.

27. Prove that Property 3 of addition holds.

28. Prove that Property 1 of scalar multiplication holds.

29. Prove that Property 2 of scalar multiplication holds.

30. If A and B are matrices of the same order, prove that $(A + B)^t = A^t + B^t$.

31. If k is a number, prove that $(kA)^t = kA^t$.

11.3 Matrix Multiplication

The classic puzzle form, of course, is mathematics. It is so good for converting troubles into manageable problems that we can justify studying the puzzle forms of mathematics on their own. I have never heard of a recommendation to an educational authority that mathematics be cut from the curriculum because most of it is, for any practical purposes, useless.

—JEROME BRUNER, *In Search of Mind: Essays in Autobiography*

Addition for two matrices of the same order was defined by simply adding corresponding entries of the matrices. You might suspect that to multiply A times B, A and B should have the same order and we should multiply corresponding entries. But it turns out that this is not a useful definition of multiplication. For various applications, some of which we shall discuss, we need a different definition of multiplication.

Vector Multiplication

We begin with the multiplication of a row vector with a column vector.

> If A is a $1 \times n$ row vector and B is an $n \times 1$ column vector, their **product** AB is the number defined by
>
> $$AB = \begin{bmatrix} a_1 & a_2 & \cdots & a_n \end{bmatrix} \begin{bmatrix} b_1 \\ b_2 \\ \vdots \\ b_n \end{bmatrix} = a_1 b_1 + a_2 b_2 + \cdots + a_n b_n$$

We call this operation **vector multiplication**. (It is also called an **inner product** or a **scalar product**, but we shall not use these terms.) Notice that this product is defined only when A and B have the same number of entries.

EXAMPLE 1   Find the product AB when

$$A = \begin{bmatrix} 2 & 0 & 3 & 1 \end{bmatrix} \qquad B = \begin{bmatrix} 4 \\ -5 \\ 1 \\ -6 \end{bmatrix}$$

Solution.   By definition we have

$$AB = 2(4) + 0(-5) + 3(1) + 1(-6) = 8 + 0 + 3 - 6 = 5 \qquad \square$$

EXAMPLE 2   During a one-hour period a concession stand sold 10 hot dogs, 15 hamburgers, 12 ice creams, and 20 sodas. They charge $2.50 for a hot dog, $3.00 for a hamburger, $2.00 for an ice cream, and $1.00 for a soda. Use vector multiplication to find the total amount received.

Solution.   We define a price row vector P and a column quantity vector Q as follows:

$$P = \begin{bmatrix} 2.5 & 3 & 2 & 1 \end{bmatrix} \qquad Q = \begin{bmatrix} 10 \\ 15 \\ 12 \\ 20 \end{bmatrix}$$

The total amount received becomes

$$PQ = 2.5(10) + 3(15) + 2(12) + 1(20) = \$114 \qquad \square$$

EXAMPLE 3   In physics a vector with three entries is used to describe a quantity in three-dimensional space. For example, the column vector

$$V = \begin{bmatrix} 1 \\ 3 \\ 4 \end{bmatrix}$$

gives the location of a point that is 1 unit from the origin in the x direction, 3 units in the y direction, and 4 units in the z direction (Figure 11.2). To find the work done when displacing an object from the origin to a point D, one multiplies the force row vector F times the displacement column vector D.

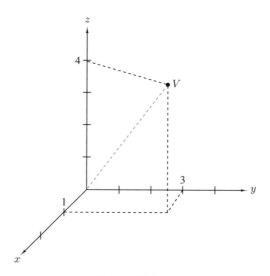

Figure 11.2

For example, if a force whose x, y, and z components are 2, 4, and 1 pounds, respectively, displaces an object distances of 3, 2, 5 feet in the x, y, z directions, respectively, the resulting work is

$$W = FD = \begin{bmatrix} 2 & 4 & 1 \end{bmatrix} \begin{bmatrix} 3 \\ 2 \\ 5 \end{bmatrix} = 6 + 8 + 5 = 19 \text{ ft} \cdot \text{lb}$$

We shall use vector multiplication to define multiplication for matrices. But first we must define when it is possible to multiply two matrices.

We can multiply A times B only if the number of columns of A equals the number of rows of B.

Thus, the product of A and B is defined only if the order of A has the form $m \times p$ and the order of B has the form $p \times n$. As we shall show, the order of their product AB is then $m \times n$ (Figure 11.3).

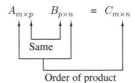

Figure 11.3

Notice that this relationship between order already holds for vector multiplication. If A is a $1 \times n$ row vector and B is an $n \times 1$ column vector, then AB is a 1×1 matrix which is the same as a number, $A_{1 \times n} B_{n \times 1} = C_{1 \times 1}$.

Now suppose A is an $m \times p$ matrix. Then each of the rows of A can be thought of as a $1 \times p$ row vector, and since there are m rows there are m of these row vectors, $R_1, R_2, \ldots, R_m$ (Figure 11.4). If B is a $p \times n$ matrix, each of the columns of B can be thought of as a $p \times 1$ column vector, and there are n of these column vectors, $C_1, C_2, \ldots, C_n$ (Figure 11.4).

$$A_{m \times p} = \begin{bmatrix} \longleftarrow & R_1 & \longrightarrow \\ \longleftarrow & R_2 & \longrightarrow \\ & \vdots & \\ \longleftarrow & R_m & \longrightarrow \end{bmatrix} \qquad B_{p \times n} = \begin{bmatrix} \uparrow & \uparrow & & \uparrow \\ C_1 & C_2 & \cdots & C_n \\ \downarrow & \downarrow & & \downarrow \end{bmatrix}$$

Figure 11.4

Matrix Multiplication

The **product** $C = AB$ of A and B is the $m \times n$ matrix whose ijth entry is the product of the ith row of A and the jth column of B. That is,

$$c_{ij} = R_i C_j$$

We thus have

$$C = \begin{bmatrix} R_1C_1 & R_1C_2 & \cdots & R_1C_n \\ R_2C_1 & R_2C_2 & \cdots & R_2C_n \\ \vdots & & & \vdots \\ R_mC_1 & R_mC_2 & \cdots & R_mC_n \end{bmatrix}$$

Notice that C does indeed have order $m \times n$.

EXAMPLE 4   Find AB for the matrices

$$A = \begin{bmatrix} 2 & 1 & 4 \\ 3 & -2 & 0 \end{bmatrix} \quad \text{and} \quad B = \begin{bmatrix} 5 & 2 & 6 \\ 1 & 0 & 4 \\ 0 & 2 & 1 \end{bmatrix}$$

Solution.   Notice that A has order 2×3 and B has order 3×3, so the product $C = AB$ is defined and has order 2×3. By definition, we have

$$AB = \begin{bmatrix} \begin{bmatrix} 2 & 1 & 4 \end{bmatrix} \begin{bmatrix} 5 \\ 1 \\ 0 \end{bmatrix} & \begin{bmatrix} 2 & 1 & 4 \end{bmatrix} \begin{bmatrix} 2 \\ 0 \\ 2 \end{bmatrix} & \begin{bmatrix} 2 & 1 & 4 \end{bmatrix} \begin{bmatrix} 6 \\ 4 \\ 1 \end{bmatrix} \\ \begin{bmatrix} 3 & -2 & 0 \end{bmatrix} \begin{bmatrix} 5 \\ 1 \\ 0 \end{bmatrix} & \begin{bmatrix} 3 & -2 & 0 \end{bmatrix} \begin{bmatrix} 2 \\ 0 \\ 2 \end{bmatrix} & \begin{bmatrix} 3 & -2 & 0 \end{bmatrix} \begin{bmatrix} 6 \\ 4 \\ 1 \end{bmatrix} \end{bmatrix}$$

$$= \begin{bmatrix} 10+1+0 & 4+0+8 & 12+4+4 \\ 15-2+0 & 6+0+0 & 18-8+0 \end{bmatrix} = \begin{bmatrix} 11 & 12 & 20 \\ 13 & 6 & 10 \end{bmatrix} \qquad \square$$

EXAMPLE 5   Find AB for the matrices

$$A = \begin{bmatrix} 3 \\ 2 \\ 1 \end{bmatrix} \quad \text{and} \quad B = \begin{bmatrix} -2 & 5 \end{bmatrix}$$

Solution.   Notice that since A has order 3×1 and B has order 1×2, the product AB is defined and has order 3×2.

$$AB = \begin{bmatrix} 3 \\ 2 \\ 1 \end{bmatrix} \begin{bmatrix} -2 & 5 \end{bmatrix} = \begin{bmatrix} 3(-2) & 3(5) \\ 2(-2) & 2(5) \\ 1(-2) & 1(5) \end{bmatrix} = \begin{bmatrix} -6 & 15 \\ -4 & 10 \\ -2 & 5 \end{bmatrix} \qquad \square$$

EXERCISES

In Exercises 1–6, suppose A has the first of the given orders and B has the second. State whether AB is defined and if so, give the order of AB.

1. $3 \times 1, \quad 3 \times 2$ **2.** $1 \times 3, \quad 3 \times 2$ **3.** $4 \times 1, \quad 1 \times 5$

4. $2 \times 5, \quad 3 \times 4$ **5.** $7 \times 6, \quad 6 \times 3$ **6.** $5 \times 2, \quad 5 \times 3$

For Exercises 7–14, perform the indicated matrix multiplications.

7. $\begin{bmatrix} 1 & 0 & -1 & 2 \end{bmatrix} \begin{bmatrix} 2 \\ 3 \\ -2 \\ 4 \end{bmatrix}$

8. $\begin{bmatrix} \frac{1}{2} & 3 & -1 \end{bmatrix} \left(\begin{bmatrix} 2 & \frac{1}{3} & 1 \end{bmatrix} \right)^t$

9. $\begin{bmatrix} \frac{1}{3} & \frac{1}{2} & \frac{1}{4} \end{bmatrix} \begin{bmatrix} 1 \\ 2 \\ 3 \end{bmatrix}$

10. $\begin{bmatrix} 1 & 5 \\ 2 & 0 \\ -1 & 4 \end{bmatrix} \begin{bmatrix} 3 & 0 & 1 & -4 \\ 1 & 2 & 5 & -1 \end{bmatrix}$

11. $\begin{bmatrix} 2 & 1 & -5 \\ -1 & 0 & 4 \\ 1 & 3 & 2 \end{bmatrix} \begin{bmatrix} 1 & 0 & 4 \\ 2 & 1 & 3 \\ -1 & 3 & -2 \end{bmatrix}$

12. $\begin{bmatrix} 1 & 0 & 0 \\ 0 & 1 & 0 \\ 0 & 0 & 1 \end{bmatrix} \begin{bmatrix} 5 & 6 & -7 \\ -8 & 9 & 4 \\ -3 & 2 & 1 \end{bmatrix}$

13. $\begin{bmatrix} 3 & 0 & 0 \\ 0 & 5 & 0 \\ 0 & 0 & -1 \end{bmatrix} \begin{bmatrix} 6 & 0 & 0 \\ 0 & -1 & 0 \\ 0 & 0 & 7 \end{bmatrix}$

14. $\begin{bmatrix} 1 & 5 \\ -2 & 3 \end{bmatrix} \left(\begin{bmatrix} 2 & 4 \\ -2 & 1 \end{bmatrix} \begin{bmatrix} 1 & 0 \\ 6 & 5 \end{bmatrix} \right)$

15. In Example 2, suppose 20 hot dogs, 12 hamburgers, 15 ice creams, and 25 sodas were sold. Use vector multiplication to find the total amount received.

16. In Example 2, suppose the same number of items were sold but hot dogs cost $2.00, hamburgers $2.50, ice creams $1.50, and sodas $0.75. Use vector multiplication to find the total amont received.

17. Find the work done for a force vector, $F = \begin{bmatrix} 3 & 5 & 7 \end{bmatrix}$ and a displacement vector,

$D = \begin{bmatrix} 2 & 1 & 3 \end{bmatrix}^t.$

18. Find the work done for a force vector, $F = \begin{bmatrix} 3 & 5 & 2 \end{bmatrix}$ and a displacement vector,

$D = \begin{bmatrix} 1 & -2 & 4 \end{bmatrix}^t.$

19. If O is a zero matrix and OA is defined, prove that $OA = O$.

20. If I is an identity matrix and A is a square matrix of the same order, prove that $IA = A$.

21. If A and B are diagonal matrices of the same order, prove that AB is a diagonal matrix. What are the entries of AB?

*__22.__ If A is an $m \times p$ matrix and B is a $p \times n$ matrix and $C = AB$, prove that

$$c_{ij} = \sum_{k=1}^{p} a_{ik} b_{kj}, \qquad i = 1, \ldots, m; \quad j = 1, \ldots, n$$

*__23.__ If AB is defined, prove that $B^t A^t$ is defined and $B^t A^t = (AB)^t$.

Photo 11.1 Matrices have been applied to airline scheduling problems.

11.4 Uses of Matrix Multiplication

When we cannot use the compass of mathematics or the torch of experience it is certain that we cannot take a single step forward.
—FRANCOIS VOLTAIRE (1694–1778), French Philosopher

Every new *body of discovery is mathematical in form, because there is no other guidance we can have.*
—CHARLES DARWIN (1809–1882), British Naturalist

An important property of matrix multiplication is that if a column vector is multiplied from the left by a matrix then another column vector is obtained. To check this, let A be an $m \times n$ matrix and let V be an $n \times 1$ column vector. It follows that AV is an $m \times 1$ column vector. Similarly, if a row vector is multiplied from the right by a matrix, then another row vector is obtained. Indeed, if A is an $m \times n$ matrix and W is a $1 \times m$ row vector, then WA is a $1 \times n$ row vector.

EXAMPLE 1 An airline company has flights from a hub city to cities C1, C2, C3, and C4. Its approximate costs per flight in thousands of dollars are given by the following matrix.

$$A = \begin{array}{c} \\ \\ \\ \\ \end{array} \begin{array}{cccc} \text{C1} & \text{C2} & \text{C3} & \text{C4} \\ \left[\begin{array}{cccc} 3 & 5 & 4 & 2 \\ 2 & 3 & 3 & 1 \\ 1 & 3 & 2 & 1 \end{array}\right] & & & \end{array} \begin{array}{l} \text{Fuel} \\ \text{Personnel} \\ \text{Overhead} \end{array}$$

Thus, its flights to city C1 have approximate fuel costs of $3000, personnel costs of $2000, and overhead $1000. The flights to C2 have approximate fuel costs of $5000, and so on. Of course, in practice the company has more accurate figures than these, but we are using simple numbers for illustration. During a one-week period, the number of scheduled flights is organized in a column vector as follows

$$V = \begin{bmatrix} 10 \\ 5 \\ 7 \\ 15 \end{bmatrix} \begin{matrix} \text{C1} \\ \text{C2} \\ \text{C3} \\ \text{C4} \end{matrix}$$

The total costs for fuel, personnel, and overhead per week can be computed by matrix multiplication:

$$C = AV = \begin{bmatrix} 3 & 5 & 4 & 2 \\ 2 & 3 & 3 & 1 \\ 1 & 3 & 2 & 1 \end{bmatrix} \begin{bmatrix} 10 \\ 5 \\ 7 \\ 15 \end{bmatrix} = \begin{bmatrix} 113 \\ 71 \\ 54 \end{bmatrix} \begin{matrix} \text{Fuel} \\ \text{Personnel} \\ \text{Overhead} \end{matrix}$$

Notice that this computation gives the total fuel costs as follows.

$3000 fuel for C1 $\times$ 10 flights = $30,000
$5000 fuel for C2 $\times$ 5 flights =   25,000
$4000 fuel for C3 $\times$ 7 flights =   28,000
$2000 fuel for C4 $\times$ 15 flights =   30,000
                                   $\overline{\$113,000}$

Similar computations yield personnel costs of $71,000 and overhead costs of $54,000. Note the efficiency of the matrix method. In practice, a large airline would have flights to dozens of cities and its costs would be broken down into dozens of different categories. The various costs are stored as a matrix in a computer memory bank. The flight schedules for a particular week can be entered into the computer. The computer would then perform the matrix multiplication and output the total costs in each category.

EXAMPLE 2   The Moon Calculator Company manufactures three calculator models, I, II, and III. Each calculator is composed of subassemblies a, b, and c, the number of which are given by the matrix S.

$$S = \begin{matrix} & \text{I} & \text{II} & \text{III} \\ & \begin{bmatrix} 2 & 1 & 4 \\ 1 & 3 & 2 \\ 3 & 1 & 2 \end{bmatrix} & & \end{matrix} \begin{matrix} a \\ b \\ c \end{matrix}$$

Thus, Model I requires two a subassemblies, one b subassembly, and three c subassemblies, and so on. Each subassembly is made from components 1, 2, 3, and 4, the number of which are tabulated in the matrix

$$C = \begin{array}{c} \\ \\ \\ \\ \end{array} \begin{matrix} a & b & c \\ \begin{bmatrix} 1 & 2 & 4 \\ 3 & 1 & 2 \\ 2 & 3 & 1 \\ 1 & 2 & 3 \end{bmatrix} & \begin{matrix} 1 \\ 2 \\ 3 \\ 4 \end{matrix} \end{matrix}$$

The matrix $A = CS$, obtained by multiplying the component matrix C times the subassembly matrix S, gives the number of each type of component required in the construction of each model.

$$A = CS = \begin{bmatrix} 1 & 2 & 4 \\ 3 & 1 & 2 \\ 2 & 3 & 1 \\ 1 & 2 & 3 \end{bmatrix} \begin{bmatrix} 2 & 1 & 4 \\ 1 & 3 & 2 \\ 3 & 1 & 2 \end{bmatrix} = \begin{matrix} \begin{matrix} \text{I} & \text{II} & \text{III} \end{matrix} \\ \begin{bmatrix} 16 & 11 & 16 \\ 13 & 8 & 18 \\ 10 & 12 & 16 \\ 13 & 10 & 14 \end{bmatrix} \begin{matrix} 1 \\ 2 \\ 3 \\ 4 \end{matrix} \end{matrix}$$

Note that the matrices must be multiplied in the correct order. In fact, we cannot even form the product SC since S has order 3×3 and C has order 4×3.

Let's check to see that A gives the right amounts. The product of the first row of C with the first column of S is

$$1(2) + 2(1) + 4(3) = 16$$

or 16 type 1 components in Model I. Now Model I contains two type a subassemblies each containing one type 1 component, one type b subassembly each containing two type 1 components, and three type c subassemblies each containing four type 1 components. Thus, there are 16 type 1 components in Model I. A similar computation works for the other cases, but you can see that the matrix method is much simpler and more automatic.

The company wants to manufacture 20 Model I, 10 Model II, and 5 Model III calculators per day and needs to know the number of components of each type required per day. This can be found by multiplying A times the quantity vector

$$Q = \begin{bmatrix} 20 \\ 10 \\ 5 \end{bmatrix}$$

We thus have

$$AQ = \begin{bmatrix} 16 & 11 & 16 \\ 13 & 8 & 18 \\ 10 & 12 & 16 \\ 13 & 10 & 14 \end{bmatrix} \begin{bmatrix} 20 \\ 10 \\ 5 \end{bmatrix} = \begin{matrix} \begin{bmatrix} 510 \\ 430 \\ 400 \\ 430 \end{bmatrix} & \begin{matrix} 1 \\ 2 \\ 3 \\ 4 \end{matrix} \end{matrix}$$

The company pays \$1.00, \$2.00, \$0.50, and \$3.00 for components of type 1, 2, 3, and 4, respectively. Upon forming the cost vector $V = [\,1 \quad 2 \quad 0.5 \quad 3\,]$, their component costs for each model type can be expressed as

$$VA = \begin{bmatrix} 1 & 2 & 0.5 & 3 \end{bmatrix} \begin{bmatrix} 16 & 11 & 16 \\ 13 & 8 & 18 \\ 10 & 12 & 16 \\ 13 & 10 & 14 \end{bmatrix} = \begin{matrix} \text{I} & \text{II} & \text{III} \\ \begin{bmatrix} 86 & 63 & 102 \end{bmatrix} \end{matrix}$$

Thus, the components cost \$86, \$63, and \$102 for models of type I, II, and III, respectively. To find the company's total component costs per day we have

$$VAQ = \begin{bmatrix} 1 & 2 & 0.5 & 3 \end{bmatrix} \begin{bmatrix} 510 \\ 430 \\ 400 \\ 430 \end{bmatrix} = \$2860$$

This could also be computed from

$$\begin{bmatrix} 86 & 63 & 102 \end{bmatrix} \begin{bmatrix} 20 \\ 10 \\ 5 \end{bmatrix} = \$2860$$

EXERCISES

Let A be the matrix

$$A = \begin{bmatrix} 1 & 0 & -1 & 2 \\ 2 & 1 & 3 & 4 \\ -3 & 2 & 1 & 0 \end{bmatrix} \tag{11.1}$$

In Exercises 1–3 multiply the given column vectors by A on the left and in Exercises 4–6, multiply the given row vectors by A on the right.

1. $\begin{bmatrix} 2 \\ 1 \\ 3 \\ 1 \end{bmatrix}$
2. $\begin{bmatrix} -2 \\ 3 \\ 0 \\ -1 \end{bmatrix}$
3. $\begin{bmatrix} 1 \\ 2 \\ -3 \\ 1 \end{bmatrix}$

4. $\begin{bmatrix} 2 & 0 & 1 \end{bmatrix}$
5. $\begin{bmatrix} -1 & 2 & -2 \end{bmatrix}$
6. $\begin{bmatrix} 1 & 0 & 4 \end{bmatrix}$

7. Find $(AV)W$ and $A(VW)$ where A is matrix (11.1), V is the column vector in Exercise 1, and W is the row vector in Exercise 4.

8. Find $(AV)W$ and $A(VW)$ where A is matrix (11.1), V is the column vector in Exercise 2, and W is the row vector in Exercise 5.

9. Find $(AV)W$ and $A(VW)$ where A is matrix (11.1), V is the column vector in Exercise 3, and W is the row vector in Exercise 6.

10. In Example 1, find the total costs for fuel, personnel, and overhead for a flight schedule

$$V = \begin{bmatrix} 20 \\ 10 \\ 9 \\ 18 \end{bmatrix}$$

11. In Example 2, find the company's total component costs for the quantity vector

$$Q = \begin{bmatrix} 25 \\ 15 \\ 10 \end{bmatrix}$$

12. A manufacturer produces two products, I and II. One unit of product I requires four parts of type A and three parts of type B. One unit of product II requires two type A parts, five type B parts, and three type C parts. Display this information in a 2×3 parts matrix P. The company has received an order for 300 units of product I and 450 units of product II. Set up a row vector Q displaying this information. Using matrix multiplication, find the number of parts of each type required to fill the order.

13. If A is matrix (11.1) and V, W are the vectors in Exercises 1 and 2, show that $A(V + W) = AV + AW$.

14. If A is matrix (11.1) and V, W are the vectors in Exercises 4 and 5, show that $(V + W)A = VA + WA$.

15. If A is matrix (11.1) and V is the vector in Exercise 1, show that $(AV)^t = V^t A^t$.

16. If A is matrix (11.1) and W is the vector in Exercise 4, show that $(WA)^t = A^t W^t$.

11.5 Inverses

In most sciences one generation tears down what another has built and what one has established another undoes. In mathematics alone each generation adds a new story to the old structures.
— HERMANN HANKEL (1839–1873), German Mathematician

Properties of Matrix Multiplication

We first consider the algebraic properties of matrix multiplication. If A, B, and C are matrices, then whenever the products are defined we have

(1) $A(BC) = (AB)C$ (Associative law)
(2) $A(B + C) = AB + AC$ (Distributive law)

These two laws are also basic properties for products of real numbers. However, there are properties that hold for the product of real numbers that do not hold for matrix products. For example, if a and b are two real numbers, we know that the commutative law of multiplication, $ab = ba$, holds. This law does not apply to matrix multiplication; if AB and BA are both defined, they need not be equal.

EXAMPLE 1   For the matrices

$$A = \begin{bmatrix} 1 & 0 \\ 0 & 0 \end{bmatrix} \quad \text{and} \quad B = \begin{bmatrix} 0 & 1 \\ 0 & 0 \end{bmatrix}$$

we have

$$AB = \begin{bmatrix} 0 & 1 \\ 0 & 0 \end{bmatrix} \quad \text{and} \quad BA = \begin{bmatrix} 0 & 0 \\ 0 & 0 \end{bmatrix}$$

Commuting Matrices

This last example also shows that the product of two nonzero matrices can be the zero matrix. Matrices A and B are said to **commute** if $AB = BA$. Any two diagonal matrices of the same order commute. For example,

$$\begin{bmatrix} 1 & 0 \\ 0 & 2 \end{bmatrix} \begin{bmatrix} -2 & 0 \\ 0 & 3 \end{bmatrix} = \begin{bmatrix} -2 & 0 \\ 0 & 6 \end{bmatrix} = \begin{bmatrix} -2 & 0 \\ 0 & 3 \end{bmatrix} \begin{bmatrix} 1 & 0 \\ 0 & 2 \end{bmatrix}$$

If A is an $m \times n$ matrix and $I = I_n$ is an identity matrix, then $AI = A$. Similarly, if $I = I_m$, then $IA = A$. This property is the reason I is called an identity matrix.

EXAMPLE 2   For the matrices

$$A = \begin{bmatrix} \frac{1}{2} & \frac{1}{2} \\ 1 & 0 \end{bmatrix} \quad \text{and} \quad B = \begin{bmatrix} 0 & 1 \\ 2 & -1 \end{bmatrix}$$

we have

$$AB = \begin{bmatrix} \frac{1}{2} & \frac{1}{2} \\ 1 & 0 \end{bmatrix} \begin{bmatrix} 0 & 1 \\ 2 & -1 \end{bmatrix} = \begin{bmatrix} 1 & 0 \\ 0 & 1 \end{bmatrix} = I$$

and

$$BA = \begin{bmatrix} 0 & 1 \\ 2 & -1 \end{bmatrix} \begin{bmatrix} \frac{1}{2} & \frac{1}{2} \\ 1 & 0 \end{bmatrix} = \begin{bmatrix} 1 & 0 \\ 0 & 1 \end{bmatrix} = I$$

This gives an example of two nondiagonal matrices that commute. Since $AB = BA = I$, we say that B is the **inverse** of A and write $B = A^{-1}$.

> For a square matrix A, the **inverse** A^{-1} of A is the square matrix that satisfies
> $$AA^{-1} = A^{-1}A = I$$

Invertible Matrices

Note that for a matrix to have an inverse it must be a square matrix. But even if A is a square matrix, A may not have an inverse. For example, a square zero matrix O cannot have an inverse since $OB = BO = O$ for any square matrix B of the same order. If a matrix has an inverse it is said to be **invertible** or **nonsingular** and if it does not have an inverse it is called **singular**. If A is invertible, then A has only one inverse; that is, if A has an inverse, then that inverse is unique. Let's prove this result.

Suppose A has two inverses, B and C. Since B is an inverse of A we have

$$AB = I$$

Multiplying this equation on the left by C gives

$$(CA)B = C(AB) = C$$

Since C is also an inverse of A we have $CA = I$. Hence, $B = C$, so all inverses of A are equal.

We have seen that in order to check whether B is the inverse of A we must verify that $AB = I$ and $BA = I$. However, there is an important theorem of matrix theory that states that if one of these equations holds, then so does the other. For this reason we only have to verify one of these equations.

Row Operations

We now give a method for finding the inverse of a matrix, if it exists. This method uses operations on the rows of a matrix called **row operations**. There are three types of row operations.

Type 1. Interchange two rows.

Type 2. Multiply a row by a nonzero number.

Type 3. Add a multiple of a row to another row.

EXAMPLE 3   Let M be the matrix

$$M = \begin{bmatrix} 1 & 0 & 2 & 3 \\ 3 & -1 & 4 & 5 \\ 0 & 3 & -2 & 4 \end{bmatrix}$$

Denote the rows of M by R_1, R_2, R_3. An example of a Type 1 operation is to interchange the first two rows, which we denote by $R_1 \leftrightarrow R_2$. Performing this operation on M gives

$$\begin{bmatrix} 1 & 0 & 2 & 3 \\ 3 & -1 & 4 & 5 \\ 0 & 3 & -2 & 4 \end{bmatrix} \xrightarrow{R_1 \leftrightarrow R_2} \begin{bmatrix} 3 & -1 & 4 & 5 \\ 1 & 0 & 2 & 3 \\ 0 & 3 & -2 & 4 \end{bmatrix}$$

An example of a Type 2 operation is to multiply the third row by 2; we denote this operation by $2R_3$. Performing this operation on M gives

$$\begin{bmatrix} 1 & 0 & 2 & 3 \\ 3 & -1 & 4 & 5 \\ 0 & 3 & -2 & 4 \end{bmatrix} \xrightarrow{\;2R_3\;} \begin{bmatrix} 1 & 0 & 2 & 3 \\ 3 & -1 & 4 & 5 \\ 0 & 6 & -4 & 8 \end{bmatrix}$$

An example of a Type 3 operation is to multiply the first row by 3 and then add the result to the third row. We denote this operation by $3R_1 + R_3$. Performing this operation on M gives

$$\begin{bmatrix} 1 & 0 & 2 & 3 \\ 3 & -1 & 4 & 5 \\ 0 & 3 & -2 & 4 \end{bmatrix} \xrightarrow{\;3R_1 + R_3\;} \begin{bmatrix} 1 & 0 & 2 & 3 \\ 3 & -1 & 4 & 5 \\ 3 & 3 & 4 & 13 \end{bmatrix}$$

Augmented Matrix

> If A is a square matrix, the **augmented matrix** is the matrix $[A \mid I]$ where I is the identity matrix with the same order as A.

The augmented matrix is still a matrix; it just has twice as many columns as A and the vertical bar is only used to separate A from I. For example, the augmented matrix for

$$A = \begin{bmatrix} 1 & 1 \\ 1 & -1 \end{bmatrix}$$

is the matrix

$$[A \mid I] = \begin{bmatrix} 1 & 1 & 1 & 0 \\ 1 & -1 & 0 & 1 \end{bmatrix}$$

To find the inverse of a matrix A we perform row operations on the augmented matrix until the matrix to the left of the bar, namely A, is transformed into I. Then the matrix to the right of the bar, namely I, is transformed into A^{-1}. In symbols,

$$[A \mid I] \to [I \mid A^{-1}]$$

Let's illustrate this method for the particular matrix A just considered.

$$\begin{bmatrix} 1 & 1 & 1 & 0 \\ 1 & -1 & 0 & 1 \end{bmatrix} \xrightarrow{-R_1 + R_2} \begin{bmatrix} 1 & 1 & 1 & 0 \\ 0 & -2 & -1 & 1 \end{bmatrix} \xrightarrow{-\frac{1}{2}R_2} \begin{bmatrix} 1 & 1 & 1 & 0 \\ 0 & 1 & \frac{1}{2} & -\frac{1}{2} \end{bmatrix}$$

$$\xrightarrow{-R_2 + R_1} \begin{bmatrix} 1 & 0 & \frac{1}{2} & \frac{1}{2} \\ 0 & 1 & \frac{1}{2} & -\frac{1}{2} \end{bmatrix}$$

We then obtain

$$A^{-1} = \begin{bmatrix} \frac{1}{2} & \frac{1}{2} \\ \frac{1}{2} & -\frac{1}{2} \end{bmatrix}$$

Let's check to see if this is really the inverse of A.

$$\begin{bmatrix} 1 & 1 \\ 1 & -1 \end{bmatrix} \begin{bmatrix} \frac{1}{2} & \frac{1}{2} \\ \frac{1}{2} & -\frac{1}{2} \end{bmatrix} = \begin{bmatrix} \frac{1}{2} + \frac{1}{2} & \frac{1}{2} - \frac{1}{2} \\ \frac{1}{2} - \frac{1}{2} & \frac{1}{2} + \frac{1}{2} \end{bmatrix} = \begin{bmatrix} 1 & 0 \\ 0 & 1 \end{bmatrix}$$

You must admit that this simple method is quite amazing. Although the explanation for why it works is not complicated, it is a little long. If you are really interested, you might consult a text on matrices [(Friedberg, 1989), (Jacob, 1990), (Schneider, 1987)].

Procedure for Finding Inverses

The following steps summarize the procedure for finding the inverse of a matrix A.

(1) Form the augmented matrix $\left[A \mid I \right]$.

(2) Use row operations to get 1 in row 1, column 1.

(3) Use row operations to get 0s in the rest of column 1.

(4) Use row operations to get 1 in row 2, column 2.

(5) Use row operations to get 0s in the rest of column 2.

(6) Continue until A is transformed into I.

(7) The matrix to the right of the bar is A^{-1}.

(8) If any of these steps is not possible, A has no inverse.

EXAMPLE 4   Find the inverse, if it exists, of the matrix

$$A = \begin{bmatrix} 1 & 2 \\ 2 & 4 \end{bmatrix}$$

Solution.   Carrying out the above procedure gives

$$\begin{bmatrix} 1 & 2 & | & 1 & 0 \\ 2 & 4 & | & 0 & 1 \end{bmatrix} \xrightarrow{-2R_1 + R_2} \begin{bmatrix} 1 & 2 & | & 1 & 0 \\ 0 & 0 & | & -2 & 1 \end{bmatrix}$$

We cannot use row operations to get 1 in row 2, column 2 and still maintain the zero in row 2, column 1. Hence A does not have an inverse, so A is singular.   □

EXAMPLE 5 Find the inverse, if it exists, of the matrix

$$A = \begin{bmatrix} 0 & 1 & 2 \\ 1 & 2 & 3 \\ 1 & 3 & 4 \end{bmatrix}$$

Solution. Carrying out the above procedure gives

$$\begin{bmatrix} 0 & 1 & 2 & | & 1 & 0 & 0 \\ 1 & 2 & 3 & | & 0 & 1 & 0 \\ 1 & 3 & 4 & | & 0 & 0 & 1 \end{bmatrix} \xrightarrow{R_1 \leftrightarrow R_2} \begin{bmatrix} 1 & 2 & 3 & | & 0 & 1 & 0 \\ 0 & 1 & 2 & | & 1 & 0 & 0 \\ 1 & 3 & 4 & | & 0 & 0 & 1 \end{bmatrix}$$

$$\xrightarrow{-R_1 + R_3} \begin{bmatrix} 1 & 2 & 3 & | & 0 & 1 & 0 \\ 0 & 1 & 2 & | & 1 & 0 & 0 \\ 0 & 1 & 1 & | & 0 & -1 & 1 \end{bmatrix}$$

$$\xrightarrow{-2R_2 + R_1} \begin{bmatrix} 1 & 0 & -1 & | & -2 & 1 & 0 \\ 0 & 1 & 2 & | & 1 & 0 & 0 \\ 0 & 1 & 1 & | & 0 & -1 & 1 \end{bmatrix}$$

$$\xrightarrow{-R_2 + R_3} \begin{bmatrix} 1 & 0 & -1 & | & -2 & 1 & 0 \\ 0 & 1 & 2 & | & 1 & 0 & 0 \\ 0 & 0 & -1 & | & -1 & -1 & 1 \end{bmatrix}$$

$$\xrightarrow{-R_3} \begin{bmatrix} 1 & 0 & -1 & | & -2 & 1 & 0 \\ 0 & 1 & 2 & | & 1 & 0 & 0 \\ 0 & 0 & 1 & | & 1 & 1 & -1 \end{bmatrix}$$

$$\xrightarrow{R_3 + R_1} \begin{bmatrix} 1 & 0 & 0 & | & -1 & 2 & -1 \\ 0 & 1 & 2 & | & 1 & 0 & 0 \\ 0 & 0 & 1 & | & 1 & 1 & -1 \end{bmatrix}$$

$$\xrightarrow{-2R_3 + R_2} \begin{bmatrix} 1 & 0 & 0 & | & -1 & 2 & -1 \\ 0 & 1 & 0 & | & -1 & -2 & 2 \\ 0 & 0 & 1 & | & 1 & 1 & -1 \end{bmatrix}$$

We then obtain

$$A^{-1} = \begin{bmatrix} -1 & 2 & -1 \\ -1 & -2 & 2 \\ 1 & 1 & -1 \end{bmatrix}$$

You should check to make sure that this is really the inverse of A. This is done by multiplying this matrix times A and verifying that the result is I. $\quad\square$

EXERCISES

For Exercises 1–8, use the following matrices.

$$A = \begin{bmatrix} 1 & 2 \\ 3 & 4 \end{bmatrix} \qquad B = \begin{bmatrix} 2 & -1 \\ -1 & 3 \end{bmatrix} \qquad C = \begin{bmatrix} 3 & 2 \\ -1 & 1 \end{bmatrix}$$

1. Show that the associative law $A(BC) = (AB)C$ holds.
2. Show that the associative law $B(AC) = (BA)C$ holds.
3. Show that the distributive law $A(B + C) = AB + AC$ holds.
4. Show that the distributive law $B(A + C) = BA + BC$ holds.
5. Show that the distributive law $(A + B)C = AC + BC$ holds.
6. Do A and B commute?
7. Do A and C commute?
8. Do B and C commute?

In Exercises 9–18, perform the indicated row operations on the following matrix.

$$M = \begin{bmatrix} 2 & 1 & -2 & 3 \\ 1 & 0 & -4 & 4 \\ 3 & 5 & -1 & 2 \end{bmatrix}$$

9. $R_1 \leftrightarrow R_2$
10. $R_2 \leftrightarrow R_3$
11. $3R_1$
12. $2R_2$
13. $(-1)R_3$
14. $(1/2)R_1$
15. $-R_1 + R_2$
16. $3R_2 + R_3$
17. $(1/2)R_3 + R_1$
18. $-2R_2 + R_1$
19. In Example 5, show that the matrix obtained is the inverse of A.

Find the inverse of the matrices in Exercises 20–27, if possible. Check your answers using matrix multiplication.

20. $\begin{bmatrix} 4 & 5 \\ -2 & 6 \end{bmatrix}$

21. $\begin{bmatrix} 1 & 2 \\ 3 & 7 \end{bmatrix}$

22. $\begin{bmatrix} 2 & -1 \\ 3 & 4 \end{bmatrix}$

23. $\begin{bmatrix} 0 & 2 \\ 1 & -3 \end{bmatrix}$

24. $\begin{bmatrix} 2 & 1 & 0 \\ 1 & 0 & 2 \\ 3 & 1 & 1 \end{bmatrix}$

25. $\begin{bmatrix} 4 & 3 & 1 \\ 8 & 6 & 2 \\ 1 & 0 & 5 \end{bmatrix}$

*26. $\begin{bmatrix} 3 & 2 & 0 & 1 \\ 0 & 1 & 4 & 5 \\ 1 & -1 & 2 & 0 \\ 1 & 0 & 1 & 3 \end{bmatrix}$

*27. $\begin{bmatrix} 1 & 0 & 2 & 0 \\ 0 & 2 & 4 & 3 \\ 1 & -1 & 0 & 5 \\ 0 & 1 & 1 & 8 \end{bmatrix}$

28. Prove that any two diagonal matrices of the same order commute.
29. Let A be an $m \times n$ matrix. If AB and BA are both defined, prove that B is an $n \times m$ matrix.
30. If A and B commute, prove that A and B are both square matrices of the same order.

Find the inverse of the following matrices by inspection.

31. $A = \begin{bmatrix} 3 & 0 \\ 0 & 1 \end{bmatrix}$

32. $A = \begin{bmatrix} 1 & 0 & 0 \\ 0 & 2 & 0 \\ 0 & 0 & 3 \end{bmatrix}$

33. If I is an identity matrix, what is its inverse?

34. If $AB = 0$ and A is invertible, prove that $B = 0$.

35. If $AB = 0$ and B is invertible, prove that $A = 0$.

36. If $AB = 0$, prove that either A or B is singular.

37. If D is a diagonal matrix with nonzero entries on its diagonal, prove that D is invertible. What are the entries of D^{-1}?

*38. If D is an invertible diagonal matrix, prove that its diagonal entries are nonzero.

39. If A is invertible, prove that $(A^{-1})^{-1} = A$.

40. If A is invertible and $k \neq 0$, prove that $(kA)^{-1} = \frac{1}{k}A^{-1}$.

41. If A and B are invertible and have the same order, prove that $(AB)^{-1} = B^{-1}A^{-1}$.

42. If A is invertible, prove that $(A^t)^{-1} = (A^{-1})^t$. (You may use Exercise 23 of Section 11.3.)

**43. Prove that the associative law of multiplication holds.

**44. Prove that the distributive law of multiplication holds.

11.6 Applications

Human ingenuity cannot concoct a cipher that human ingenuity cannot resolve.

—Edgar Allan Poe (1809–1849), American Poet and Writer

Systems of Linear Equations

One of the most important applications of matrices is in finding the solution of a system of linear equations. A simple example of a system of linear equations is

$$\begin{aligned} x + y &= 6 \\ x - y &= 4 \end{aligned} \qquad (11.2)$$

This is a system of two linear equations in two unknowns x and y. The **solution** for this system is a pair of values for x, y that satisfy these equations. Now we do not really need matrices to solve this system because it is so simple. In fact, all we have to do is add the equations to obtain $2x = 10$. We conclude that $x = 5$. Substituting this value of x into the first equation gives $y = 1$. If we substitute $x = 5$ into the second equation we again obtain $y = 1$, which we must, since these values satisfy both equations.

Even though it is not necessary, let us use matrices to solve (11.2) as an illustration of the method. We can rewrite (11.2) in matrix form as

$$\begin{bmatrix} 1 & 1 \\ 1 & -1 \end{bmatrix} \begin{bmatrix} x \\ y \end{bmatrix} = \begin{bmatrix} 6 \\ 4 \end{bmatrix} \qquad (11.3)$$

The 2×2 matrix A on the left is called the **coefficient matrix**. But we have shown in Section 5 that the inverse of the coefficient matrix A is

$$A^{-1} = \begin{bmatrix} \frac{1}{2} & \frac{1}{2} \\ \frac{1}{2} & -\frac{1}{2} \end{bmatrix}$$

If we multiply (11.3) on the left by A^{-1} we obtain

$$\begin{bmatrix} x \\ y \end{bmatrix} = \begin{bmatrix} \frac{1}{2} & \frac{1}{2} \\ \frac{1}{2} & -\frac{1}{2} \end{bmatrix} \begin{bmatrix} 6 \\ 4 \end{bmatrix} = \begin{bmatrix} 5 \\ 1 \end{bmatrix}$$

Thus, we again obtain $x = 5$, $y = 1$.

EXAMPLE 1   Solve the following system of linear equations.

$$y + 2z = 3$$
$$x + 2y + 3z = -2$$
$$x + 3y + 4z = 4$$

Solution.   We can rewrite the system of equations in matrix form as

$$\begin{bmatrix} 0 & 1 & 2 \\ 1 & 2 & 3 \\ 1 & 3 & 4 \end{bmatrix} \begin{bmatrix} x \\ y \\ z \end{bmatrix} = \begin{bmatrix} 3 \\ -2 \\ 4 \end{bmatrix} \qquad (11.4)$$

In Example 5 of Section 11.5, we found the inverse of the coefficient matrix A to be

$$A^{-1} = \begin{bmatrix} -1 & 2 & -1 \\ -1 & -2 & 2 \\ 1 & 1 & -1 \end{bmatrix}$$

Multiplying (11.4) on the left by A^{-1} gives

$$\begin{bmatrix} x \\ y \\ z \end{bmatrix} = \begin{bmatrix} -1 & 2 & -1 \\ -1 & -2 & 2 \\ 1 & 1 & -1 \end{bmatrix} \begin{bmatrix} 3 \\ -2 \\ 4 \end{bmatrix} = \begin{bmatrix} -11 \\ 9 \\ -3 \end{bmatrix}$$

Hence, $x = -11$, $y = 9$, $z = -3$ is the solution.   □

EXAMPLE 2   Solve the following system of linear equations.

$$y + 2z = 5$$
$$x + 2y + 3z = 0$$
$$x + 3y + 4z = 1$$

Solution. Notice that this system of equations is the same as in Example 1 except for the numbers on the right-hand side. One of the advantages of the matrix method is that once the inverse of the coefficient matrix is known, it can be used no matter what numbers appear on the right-hand side. Thus the inverse can be stored in the memory bank of a computer. If the numbers on the right are then entered into the computer, the computer can perform the matrix multiplication and output the solution. In this case we have

$$\begin{bmatrix} x \\ y \\ z \end{bmatrix} = \begin{bmatrix} -1 & 2 & -1 \\ -1 & -2 & 2 \\ 1 & 1 & -1 \end{bmatrix} \begin{bmatrix} 5 \\ 0 \\ 1 \end{bmatrix} = \begin{bmatrix} -6 \\ -3 \\ 4 \end{bmatrix}$$

Hence, $x = -6$, $y = -3$, $z = 4$. $\qquad\qquad\qquad\qquad\qquad\qquad\square$

For simplicity, we shall consider only systems of linear equations that have the same number of equations as unknowns. In this case the coefficient matrix is a square matrix. A system of linear equations need not have a unique solution or any solution at all. For example, consider the system of equations

$$x + 2y = 1$$
$$2x + 4y = 4$$

Dividing the second equation by 2 gives

$$x + 2y = 1$$
$$x + 2y = 2$$

Since we cannot have a number that equals both 1 and 2, there is no solution to this system. Another way to see this is to examine the coefficient matrix

$$A = \begin{bmatrix} 1 & 2 \\ 2 & 4 \end{bmatrix}$$

We have seen in Example 4 of Section 5 that A is singular.

Another example is the system of linear equations

$$x + 2y = 1$$
$$2x + 4y = 2$$

Dividing the second equation by 2 gives the first equation, so we really have only one equation in two unknowns. There are many solutions of this equation; for example, $x = 1$, $y = 0$; $x = 2$, $y = -1/2$; $x = 0$, $y = 1/2$; and so on. Again, the coefficient matrix is singular. In general, there is a theorem in matrix theory that states that a system of n linear equations in n unknowns has a unique solution if and only if the coefficient matrix is nonsingular. Thus, the matrix method of solution that we have considered works whenever there is a unique solution.

EXAMPLE 3   Solve the following system of linear equations.

$$x + y + z = a$$

$$y + z = b$$

$$z = c$$

Solution.   In this example, the constants a, b, and c are unspecified, so if we find a solution we can then substitute any values we want for a, b, and c. In this way we can find solutions for all possible a, b, c values simultaneously. We now give two ways of solving this system of equations. The first is the method of elimination. Substituting c for z in the second equation gives $y = b - c$. Substituting b for $y + z$ in the first equation gives $x = a - b$. We thus have the solution $x = a - b$, $y = b - c$, $z = c$.

The second method uses matrices. We find the inverse of the coefficient matrix as follows.

$$\begin{bmatrix} 1 & 1 & 1 & | & 1 & 0 & 0 \\ 0 & 1 & 1 & | & 0 & 1 & 0 \\ 0 & 0 & 1 & | & 0 & 0 & 1 \end{bmatrix} \xrightarrow{-R_2 + R_1} \begin{bmatrix} 1 & 0 & 0 & | & 1 & -1 & 0 \\ 0 & 1 & 1 & | & 0 & 1 & 0 \\ 0 & 0 & 1 & | & 0 & 0 & 1 \end{bmatrix}$$

$$\xrightarrow{-R_3 + R_2} \begin{bmatrix} 1 & 0 & 0 & | & 1 & -1 & 0 \\ 0 & 1 & 0 & | & 0 & 1 & -1 \\ 0 & 0 & 1 & | & 0 & 0 & 1 \end{bmatrix}$$

Hence,

$$\begin{bmatrix} x \\ y \\ z \end{bmatrix} = \begin{bmatrix} 1 & -1 & 0 \\ 0 & 1 & -1 \\ 0 & 0 & 1 \end{bmatrix} \begin{bmatrix} a \\ b \\ c \end{bmatrix} = \begin{bmatrix} a - b \\ b - c \\ c \end{bmatrix}$$

so again $x = a - b$, $y = b - c$, $z = c$. □

Secret Codes

Another application of matrices is in cryptography or secret codes. Putting messages into code or ciphers dates back to the ancient Greeks. Secret codes have been used for thousands of years to maintain secure communications. Only those who knew the secret "key" could decipher such messages. Today codes are very important for industrial, governmental, legal, diplomatic, and financial transactions. An example is your credit card number. You would not want an unauthorized person to obtain this number. Matrices and their inverses have been used successfully to encode and decode secret messages.

Suppose we want to send a message written in English. For simplicity, we shall ignore punctuation. We now number the alphabet as shown in Table 11.3, where we use 27 to correspond to a space between words. If we had wanted to include punctuation marks, numbers, and other symbols we would just continue this list.

Table 11.3

Letter	a	b	c	d	e	f	g	h	i	j	k	l	m	n
Number	1	2	3	4	5	6	7	8	9	10	11	12	13	14
Letter	o	p	q	r	s	t	u	v	w	x	y	z	space	
Number	15	16	17	18	19	20	21	22	23	24	25	26	27	

Now choose a square matrix with known inverse such as

$$A = \begin{bmatrix} 0 & 1 & 2 \\ 1 & 2 & 3 \\ 1 & 3 & 4 \end{bmatrix} \qquad A^{-1} = \begin{bmatrix} -1 & 2 & -1 \\ -1 & -2 & 2 \\ 1 & 1 & -1 \end{bmatrix} \qquad (11.5)$$

We have selected this matrix for illustration. In practice, a much larger matrix would be used so as to make it difficult to break the code. Suppose we want to send the message "lunch today." This message is coded as

$$\begin{array}{ccccccccccc} L & U & N & C & H & - & T & O & D & A & Y \\ 12 & 21 & 14 & 3 & 8 & 27 & 20 & 15 & 4 & 1 & 25 \end{array}$$

We next separate the code numbers into groups of three and write these as column vectors.

$$V_1 = \begin{bmatrix} 12 \\ 21 \\ 14 \end{bmatrix} \qquad V_2 = \begin{bmatrix} 3 \\ 8 \\ 27 \end{bmatrix} \qquad V_3 = \begin{bmatrix} 20 \\ 15 \\ 4 \end{bmatrix} \qquad V_4 = \begin{bmatrix} 1 \\ 25 \\ 27 \end{bmatrix}$$

Notice that we added a space at the end to fill in the vector V_4.

Each of the column vectors is multiplied on the left by A giving

$$W_1 = AV_1 = \begin{bmatrix} 49 \\ 96 \\ 131 \end{bmatrix} \qquad W_2 = AV_2 = \begin{bmatrix} 62 \\ 100 \\ 135 \end{bmatrix}$$

$$W_3 = AV_3 = \begin{bmatrix} 23 \\ 62 \\ 81 \end{bmatrix} \qquad W_4 = AV_4 = \begin{bmatrix} 79 \\ 132 \\ 184 \end{bmatrix}$$

The message is transmitted as

$$49, 96, 131, 62, 100, 135, 23, 62, 81, 79, 132, 184$$

This is called the **encoded message**. The receiver, in possession of A^{-1}, decodes the message by separating the encoded numbers into column vectors and multiplying by A^{-1}.

$$A^{-1}W_1 = A^{-1} \begin{bmatrix} 49 \\ 96 \\ 131 \end{bmatrix} = \begin{bmatrix} 12 \\ 21 \\ 14 \end{bmatrix} = \begin{bmatrix} L \\ U \\ N \end{bmatrix}$$

$$A^{-1}W_2 = A^{-1} \begin{bmatrix} 62 \\ 100 \\ 133 \end{bmatrix} = \begin{bmatrix} 3 \\ 8 \\ 27 \end{bmatrix} = \begin{bmatrix} C \\ H \\ - \end{bmatrix}$$

and so on.

EXERCISES

1. Solve the system of linear equations $x + y = 2$, $x - y = 1$.
2. Solve the system of linear equations $x + y = 10$, $x - y = 5$.

Using the work in Example 1, solve the following systems of linear equations.

3. $\quad y + 2z = 0$
   $x + 2y + 3z = 1$
   $x + 3y + 4z = 2$

4. $\quad y + 2z = -5$
   $x + 2y + 3z = 7$
   $x + 3y + 4z = -3$

5. Show that the following system of equations has no solution.

$$2x - 3y = 1$$
$$-6x + 9y = 2$$

6. Show that the following system of equations has many solutions.

$$2x - 3y = 1$$
$$-6x + 9y = -3$$

Using the work in Example 3, solve the following systems of equations.

7. $x + y + z = 1$
   $\quad\; y + z = 2$
   $\quad\qquad z = 3$

8. $x + y + z = -3$
   $\quad\; y + z = 7$
   $\quad\qquad z = 5$

9. Verify that the vectors W_1, W_2, W_3, and W_4 of the text are correct.
10. Using matrix A in (11.5), encode the message "strike now."
11. Using matrix A in (11.5), encode the message "now or never."

In Exercises 12–17, use the matrix method to solve the systems of linear equations.

12. $\quad x + y = 3$
    $2x + y = 4$

13. $\quad x + 3y = -3$
    $2x + 5y = 16$

14. $\quad x + y = 1$
    $4x + 8y = 7$

15. $\quad 2x + y = 10$
    $-5x + 2y = 20$

16. $x + 5y - 2z = 5$
    $3x - y + z = 4$
    $4x + y - z = 3$

*17. $\quad 2x + 6y - z = 18$
     $2y + 3z = 9$
     $3x - 5y + 8z = 4$

In Exercises 18–22, use the code matrix

$$A = \begin{bmatrix} 3 & 3 & 6 \\ 0 & 1 & 2 \\ 5 & 3 & 2 \end{bmatrix}$$

18. Encode the message "lunch today."
19. Encode the message "strike now."
20. Encode the message "now or never."
21. Find the inverse of A.
22. Decode the message 177, 45, 145, 243, 67, 191, 192, 43, 158, 165, 51, 73.
23. Using the matrix method, solve the system of linear equations

$$w + 3y = 1$$
$$2w - x + z = 2$$
$$3w + 2y = -1$$
$$x + z = 3$$

24. Using the matrix method, solve the system of linear equations

$$w + 3y = 5$$
$$2w - x + z = -3$$
$$3w + 2y = 1$$
$$x + z = 6$$

Chapter 11 Summary of Terms

matrix terms   row
               column
               entry
               order

special matrices   square
                   diagonal
                   row vector
                   column vector
                   identity matrix
                   zero matrix

operations of matrices   transpose A^t
                         addition $A + B$
                         subtraction $A - B$
                         scalar multiplication kA
                         multiplication AB

matrix inverse $AA^{-1} = A^{-1}A = I$

singular and nonsingular matrices

augmented matrix $[A \mid I]$

row operations   (1) Interchange two rows.
                     (2) Multiply a row by a nonzero number.
                     (3) Add a multiple of a row to another row.

finding the inverse $[A \mid I] \rightarrow [I \mid A^{-1}]$

systems of linear equations

cryptography

Chapter 11 Test

1. For the following matrix A find   (a) a_{13},   (b) a_{22},   (c) the order of A,   (d) $-A$,   (e) the ij for which $a_{ij} = 5$, (f) the order of A^t.

$$A = \begin{bmatrix} 1 & -3 & 4 \\ 5 & 0 & 1 \end{bmatrix}$$

2. If an $n \times n$ matrix A has the following entries, show that $A = I$.

$$a_{ij} = \begin{cases} 1 & \text{if } i = j \\ 0 & \text{if } i \neq j \end{cases}$$

3. Find $A + B$, $A - B$, $B - A$, A^t, and $3B$ for the matrices

$$A = \begin{bmatrix} 2 & 1 & 4 \\ -2 & 0 & 2 \\ 1 & 5 & -7 \end{bmatrix} \quad \text{and} \quad B = \begin{bmatrix} -1 & 0 & 8 \\ 7 & -3 & 2 \\ 3 & 4 & -5 \end{bmatrix}$$

4. For the matrices in Exercise 3, solve the following matrix equation for X.

$$3(X + A) = 2(X - B) + 4A$$

5. Find AB and BA for the matrices in Exercise 3.

6. If $A = \begin{bmatrix} 1 & 2 & 3 \end{bmatrix}$ and $B = \begin{bmatrix} -2 & 1 & -1 \end{bmatrix}$, find AB^t and $A^t B$.

7. If V is a row vector, prove that $VV^t \geq 0$ and $VV^t = 0$ only if $V = 0$.

8. For matrix A in Exercise 3, perform the following row operations.
    (a) $R_1 \leftrightarrow R_3$ (b) $(1/3)R_2$ (c) $-R_2 + R_3$

9. (a) What do you get if you multiply an arbitrary 3×3 matrix on the left by the following matrix?

$$M = \begin{bmatrix} 1 & 0 & 0 \\ 0 & 0 & 0 \\ 0 & 0 & 0 \end{bmatrix}$$

   (b) What if you multiply on the right with M?

10. Find the inverse of the following matrix.

$$\begin{bmatrix} 1 & 0 & 2 \\ 3 & 1 & 0 \\ 0 & 4 & 5 \end{bmatrix}$$

11. Show that the following matrix is singular.

$$\begin{bmatrix} 1 & 2 & 3 \\ -2 & 0 & -1 \\ -1 & 2 & 2 \end{bmatrix}$$

12. Solve the following system of linear equations.

$$x + 2z = 4$$
$$3x + y = 6$$
$$4y + 5z = 1$$

13. The **trace** of a square matrix is the sum of its diagonal entries. Find the traces of matrices A and B of Exercise 3.

14. The **determinant**, det A, of a 2×2 matrix A is defined as det $A = a_{11}a_{22} - a_{12}a_{21}$. Find the determinant of the matrix

$$A = \begin{bmatrix} 3 & -4 \\ 2 & -1 \end{bmatrix}$$

15. Are the following statements true or false?
    (a) For a 3×3 matrix A, the entry a_{34} does not make sense.
    (b) An $m \times 1$ matrix is a row vector.
    (c) A $1 \times n$ matrix is a column vector.
    (d) Every diagonal matrix is a square matrix.
    (e) Every square matrix is a diagonal matrix.
    (f) If A has order $n \times r$ and B has order $r \times s$, then AB has order $n \times s$.
    (g) An identity matrix commutes with every matrix with the same order.
    (h) If $A = A^t$, then A is a square matrix.
    (i) If A is a square matrix, then $A = A^t$.
    (j) Every system of linear equations has at least one solution.
    (k) A system of linear equations cannot have more than one solution.

GRAPH THEORY

12

Mathematics must take its motivation from concrete specific substance and aim again at some layer of "reality." The flight into abstraction must be something more than a mere escape; start from the ground and reentry are both indispensable, even if the same pilot cannot handle all phases of the trajectory.
— RICHARD COURANT, Twentieth-Century Mathematician

The mathematician, carried along on his flood of symbols, dealing apparently with purely formal truths, may still reach results of endless importance for our description of the physical universe.
— KARL PEARSON (1857–1936), English Mathematician

Recently the students and teachers of a grade school decided to try an experiment. They would go through the entire day without doing or using anything that involved mathematics. They had to cover the clocks because clocks have numbers, they could not read books because the pages and chapters were numbered, and they found that even their discussions were severely limited. They couldn't engage in sports since these used scores, and they couldn't even play on the playground since all their games involved boundaries, which is a mathematical concept. They discovered that they couldn't eat lunch since the ovens had electric circuits requiring mathematics for their design and they decided that they shouldn't even be inside the school building because it was constructed using blueprints, dimensions, geometry, and other mathematical concepts. The entire situation quickly became hopeless, and they had to dismiss school for the rest of the day.

The importance and influence of mathematics in our daily lives is overwhelming. Carry out your own experiment and see how far you can get without using mathematics. It is well known that mathematics is indispensable in the physical sciences such as physics, chemistry, operations research, computer technology, and engineering. But graph theory has applications in such areas as communication sciences, genetics, psychology, sociology, economics, anthropology, and linguistics.

12.1 | Introduction

What does a millimeter of pizazz more or less matter when we are
discussing the nth degree of wonderful?   —THOMAS DISCH
         "Jerome Robbins' Broadway in Circles," *The Nation* (May 1989)

Königsberg Bridge Problem

Leonhard Euler (1701–1783) is the father of graph theory. He solved the Königsberg
Bridge problem in 1736, using graph theory. The city of Königsberg in Prussia (now
Kaliningrad in Russia) consisted of two islands linked to each other and to the banks
of the Pregel River by seven bridges (see Figure 12.1).

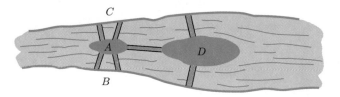

Figure 12.1

On warm summer evenings the citizens of Königsberg would stroll through the
city and cross over some of the bridges. For a really long stroll, one would cross all
seven bridges. But this posed an interesting problem. Can one begin at any of the four
land areas, walk across each bridge exactly once, and return to the starting point?

Intuitively, a graph is a finite set of points called **vertices** and a set of lines called
edges joining some (or none or all) of the vertices (a precise definition is given
in the next section). Two vertices that are joined by an edge are called **adjacent**,
and two edges meeting at a common vertex are called **adjacent**.

Euler replaced each land area by a vertex and each bridge by an edge joining the
corresponding vertices, obtaining the graph (really, a multigraph) in Figure 12.2. Notice that there are four vertices and seven edges in this graph. Vertices A and B are
adjacent because they are joined by edges; similarly, A and C, A and D, C and D,
and B and D are adjacent. The vertices B and C are not adjacent because there is
no edge joining them. The edge AD is adjacent to the edge BD because they meet
at the common vertex D. What other edges are adjacent?

Euler showed that for a stroll across each bridge exactly once and back to the
starting point to be possible, each vertex must be the meeting point of an even number
of edges. Since this is not the case, the stroll is impossible.

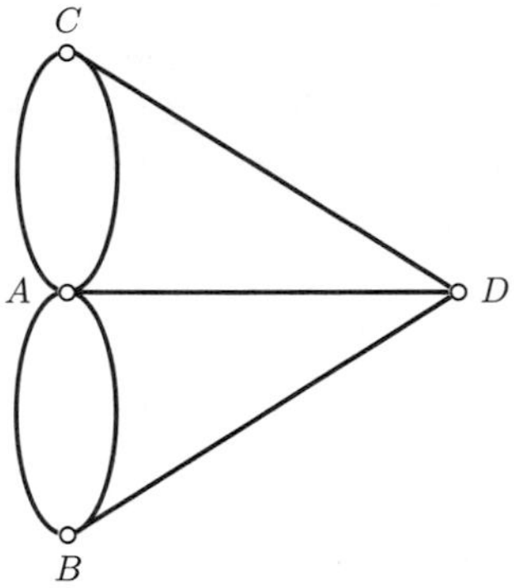

Figure 12.2 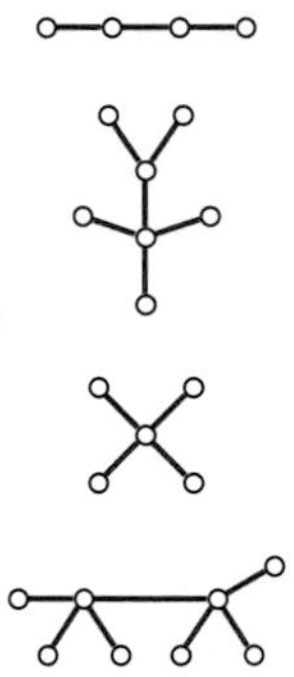   Figure 12.3   Figure 12.4

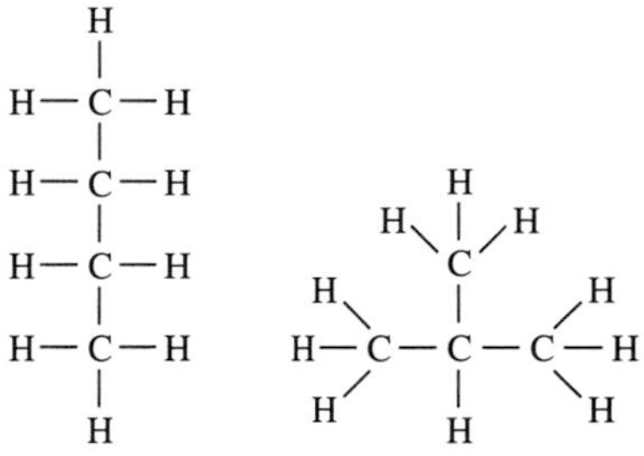

Trees

Kirchhoff developed the theory of trees in 1847 to solve systems of linear equations that give the current in each branch and around each circuit of an electric network. A **tree** may be defined to be a graph in which any two vertices are connected by a sequence of adjacent edges and the number of vertices is 1 more than the number of edges. Some examples of trees are given in Figure 12.3 (the last one is known for its bark). Kirchhoff was able to replace each electric network by a tree, which greatly simplified the computations.

In 1857 Cayley used trees in organic chemistry to enumerate the isomers of the saturated hydrocarbons C_nH_{2n+2} with a given number n of carbon atoms. This reduced to the number of trees with $3n + 2$ vertices in which each vertex had degree 1 or 4 (the degree of a vertex is the number of edges incident to it); see Figure 12.4.

Around the World

Graph theory is also very useful for solving puzzles. In some cases these are more than just amusing exercises, and they lead to deep and important results. An example is a game invented by the famous mathematician Sir William Hamilton in 1859. This game uses a regular dodecahedron whose 20 vertices are labeled with the names of important cities. The player is challenged to travel "around the world" by finding a closed circuit along the edges that passes through each vertex exactly once (not all the edges need be used). Hamilton sold his idea to a maker of games for 25 guineas, which was a clever move because the game was never a financial success. Mathematically, however, the game has led to many important investigations and questions. In graph-theoretic terms, the edges and vertices of the dodecahedron are represented by the graph in Figure 12.5. Try your hand at it; can you go around the world?

Four-Color Problem

Another problem which started out as an amusing puzzle and was discovered by mathematicians about 1850 has become one of the most famous problems in mathematics.

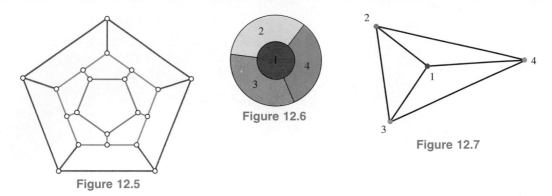

Figure 12.6

Figure 12.7

Figure 12.5

This is the four-color problem. In graph-theoretic terms, the four-color problem can be stated: Is it possible to color the vertices of a planar graph with four or fewer colors so that adjacent vertices have different colors? This has turned out to be a profound problem that engaged the attention of many of the most brilliant mathematical minds in the world for over a century until it was finally solved in the affirmative by Kenneth Appel and Wolfgang Haken in 1976. Figure 12.6 illustrates a map with four countries labeled 1, 2, 3, and 4. Since two countries with a common boundary must be colored differently, this map requires four colors. This map is represented by the graph in Figure 12.7. The countries are represented by vertices also labeled 1, 2, 3, and 4. Two vertices are adjacent if the corresponding countries have a common boundary. Since each vertex is adjacent to the other three vertices, we again see that four colors are required.

Since its early beginnings, many applications of graph theory have been found not only in mathematics and the physical sciences but in the humanities and life sciences as well. Some of these applications will be considered later in this chapter.

Puzzles

Before beginning a systematic study of graph theory, we will use graphs to solve some simple puzzles. First consider Exercise 21 in Section 1.2. This problem can be considered to be a one-person game. In this game you have certain allowable moves that you continue to make until you win (or lose). Pretend you have a board with a river and two banks drawn on it, and you also have wooden models of a man, a goat, a wolf, a cabbage, and a boat. Denote a move by a circle around the letter M and no more than one of the letters W, G, C, which represent placing the corresponding objects into the boat and transporting them to the opposite shore. An allowable move is one in which the wolf and goat or the goat and cabbage are not left together in the absence of the man. What is your first move? The only possible first move is (MG). What should the second move be? There are two possible second moves: the man could turn around at the opposite shore and come back (MG), or the man could leave the goat on the opposite shore and come back alone (M). The first of these alternatives would be silly, because you would then be back where you started, so even though this is a possible move, you ignore it. Now the man comes back alone and is allowed to pick up either the wolf (MW) or the cabbage (MC), so there are two possibilities for the third move, ignoring the silly move in which the man goes back alone (M). Continuing in this way, you obtain the graph in Figure 12.8. After the last move all have been transferred across the river and the game has been won. Notice that there are two possible paths for winning the game.

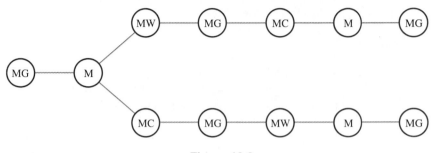

Figure 12.8

Exercise 29 in Section 1.2 is a little more complicated because in this case there are more allowable moves. Denote the three men by A, B, C and their wives by a, b, c, respectively. Two letters inside a circle mean that these two people cross in a rowboat. The allowable first moves are

Starting with the first move, (ab), three winning strategies are given (see Figure 12.9). There are others that are not pictured.

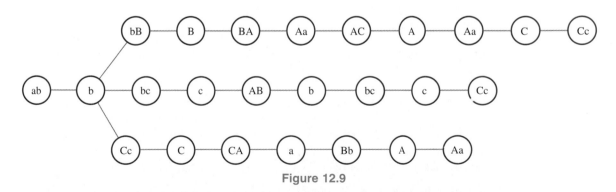

Figure 12.9

An interesting problem that can be solved using graphs is the following. Show that at any party with six people there are three mutual acquaintances or three mutual strangers. That is, there must be at least three people who know each other or at least three people who do not know each other. The situation in this problem may be represented by a graph G with six vertices representing the people at the party

and edges joining those vertices that represent acquainted people. The problem is to demonstrate that G has three mutually adjacent vertices or three mutually nonadjacent vertices. For example, in Figure 12.10 four possible graphs with six vertices are depicted. In graphs G_1 and G_2 there are three mutual acquaintances, but there are not three mutual strangers. In graph G_4 there are three mutual acquaintances and three mutual strangers, while in graph G_3 there are not three mutual acquaintances but there are three mutual strangers.

Complement

> The **complement** $\overline{G}$ of a graph G is defined to have the same vertices as G, but vertices in $\overline{G}$ are adjacent if and only if they are nonadjacent in G.

For example, the complements of the graphs in Figure 12.10 are given in Figure 12.11.

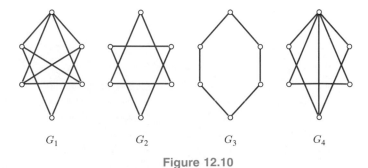

G_1 $\qquad$ G_2 $\qquad$ G_3 $\qquad$ G_4

Figure 12.10

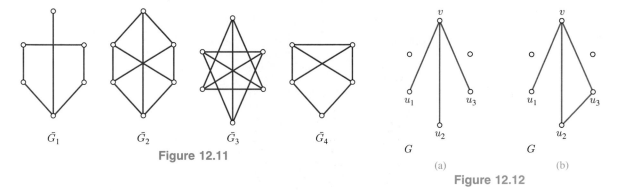

Figure 12.11

Figure 12.12

Note that there are three mutual acquaintances if and only if the corresponding graph has a triangle and that there are three mutual strangers if and only if the complement of the corresponding graph has a triangle. For example, G_1 and G_2 have triangles while $\overline{G}_1$ and $\overline{G}_2$ do not; G_3 does not have triangles; $\overline{G}_3$ does; and G_4 and $\overline{G}_4$ both have triangles. Thus the problem is solved by the following graph-theoretical theorem.

Theorem 12.1   For any graph G with six vertices, G or $\overline{G}$ contains a triangle.

Proof: Let v be a vertex of G. Now v is adjacent to three vertices u_1, u_2, u_3 of G or $\overline{G}$ since if v is not adjacent to at least three vertices of G it is adjacent to at least three vertices of $\overline{G}$. Suppose u_1, u_2, u_3 are vertices of G (Figure 12.12a). If any two of the vertices u_1, u_2, u_3 are adjacent, then they are two vertices of a triangle whose third vertex is v (Figure 12.12b). If no two of the vertices u_1, u_2, u_3 are adjacent in G, then they are vertices of a triangle in $\overline{G}$. If u_1, u_2, u_3 are vertices of $\overline{G}$, the proof is similar.   □

EXERCISES

In Exercises 1–6 draw maps of countries (imaginary or real) in which countries with a common boundary are colored with different colors.

1. Draw a map that requires two colors.
2. Draw a map that requires three colors.
3. Draw a map different from the one in the text that requires four colors.
4. Draw a map with rectangular countries that requires four colors.
5. The graph corresponding to a map is constructed as follows. Let a vertex correspond to each country, and if two countries have a boundary in common, make the corresponding vertices adjacent. Draw a map different from the one in the text requiring four colors and construct the corresponding graph.
6. Try to draw a map requiring five colors. Can you see why such a map is so hard to construct?
7. Methane is the hydrocarbon CH_4. Draw the tree for methane.

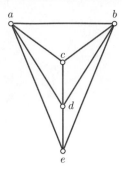

Figure 12.13

8. Ethane is the hydrocarbon C_2H_6. Draw the tree for ethane.

9. Propane is the hydrocarbon C_3H_8. Draw the tree for propane.

10. In Exercise 29 of Section 1.2, start with the opening move (Aa) and find a winning strategy.

11. Solve Exercise 18 of Section 1.2, using graphs.

12. In a party with three people, show that there are either two mutual acquaintances or three mutual strangers. Also show that there are either two strangers or three mutual acquaintances.

13. In a party with five people everyone knows everyone else. What is the graph for this party? The complement?

14. In a party with six people, must there be either four mutual acquaintances or four mutual strangers?

15. Find the complements of the graphs in Figure 12.3.

16. Find the complements of the graphs in Figure 12.12.

17. What is the smallest number of colors that can be used to color the vertices of the graphs in Figure 12.10 so that adjacent vertices are colored differently?

18. Color the vertices of the graph in Figure 12.13 so that adjacent vertices are colored differently. What is the smallest number of colors that can be used?

19. Solve the around-the-world problem.

20. Find the complement of the graph in Figure 12.13.

21. Find some of the isomers of C_5H_{12}.

22. Draw three different trees with six vertices.

23. Let G be a graph with n vertices, and let v be a vertex of G. Show that the number of edges incident to v in G plus the number of edges incident to v in $\overline{G}$ is $n - 1$.

*24. My wife and I attended a party where there were two other married couples. Various handshakes took place. No one shook hands with himself or herself or with his or her spouse, and no one shook hands with the same person more than once. After all the handshakes were over, I asked each person, including my wife, how many hands he or she had shaken. Each gave a different answer. How many hands did my wife shake? What is the answer if there were three other married couples? Four other married couples?

12.2 Graphs and Multigraphs

The whole of mathematics consists in the organization of a series of aids to the imagination in the process of reasoning.
— ALFRED NORTH WHITEHEAD (1861–1947), English Mathematician

Definition of a Graph

In this section some of the basic definitions that will be used throughout this chapter are given. First a rigorous definition of a graph.

> A **graph** is a finite nonempty set V together with a collection E of subsets of V each consisting of two elements. The elements of V are called **vertices**, and the elements of E are called **edges**.

For example, if $V = \{a, b, c, d\}$ and $E = \{\{a, b\}, \{a, d\}, \{b, c\}, \{b, d\}\}$, then $G = (V, E)$ is a graph.

Diagram of a Graph

A graph is diagramed as follows. Draw the vertices as small circles, and if two vertices belong to an edge, join the corresponding circles with a line. The graph in the previous paragraph is diagramed in Figure 12.14.

In the diagram of a graph, the position of the vertices is immaterial as long as the edges are correctly represented by lines. For example, Figure 12.15 is also a diagram for this same graph. The lines representing edges need not be straight lines. Thus Figure 12.16 is also a diagram of this graph.

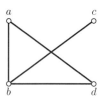

Figure 12.14

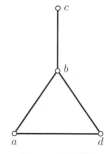

Figure 12.15

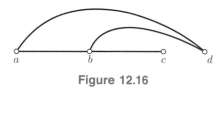

Figure 12.16

Figures 12.3 to 12.13 are all diagrams of graphs. For example, Figure 12.13 is the diagram of the graph (V, E), where $V = \{a, b, c, d, e\}$ and

$$E = \{\{a, b\}, \{a, c\}, \{a, d\}, \{a, e\}, \{b, c\}, \{b, d\}, \{b, e\}, \{c, d\}, \{d, e\}\}$$

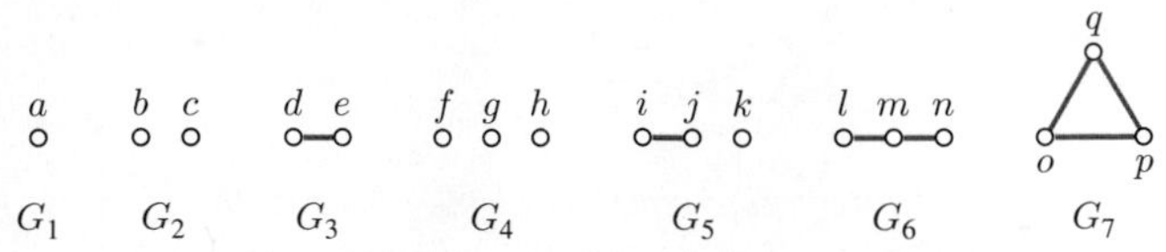

Figure 12.17

In Figure 12.17 all the graphs with one, two, or three vertices are diagramed. Table 12.1 gives the vertex set and the edge set of each of the graphs diagramed in Figure 12.17.

Table 12.1

Graph	Vertex set	Edge set
G_1	$\{a\}$	$\emptyset$
G_2	$\{b,c\}$	$\emptyset$
G_3	$\{d,e\}$	$\{\{d,e\}\}$
G_4	$\{f,g,h\}$	$\emptyset$
G_5	$\{i,j,k\}$	$\{\{i,j\}\}$
G_6	$\{l,m,n\}$	$\{\{l,m\},\{m,n\}\}$
G_7	$\{o,p,q\}$	$\{\{o,p\},\{p,q\},\{o,q\}\}$

Adjacency

> If $e = \{u,v\}$ is an edge, then u and v are called **adjacent** vertices and u and e (also v and e) are said to be **incident**.

Thus in graph G_7 the vertices o and p are adjacent, as are the vertices p and q and the vertices o and q. In graph G_6, l and m are adjacent and m and n are adjacent, but l and n are not adjacent. In graph G_7, the vertex p and the edge $\{p,q\}$ are incident, while p and $\{o,q\}$ are not incident.

> If e_1 and e_2 are distinct edges incident with a common vertex, then e_1 and e_2 are called **adjacent** edges.

For example, in G_7 the edges $\{p,q\}$ and $\{q,o\}$ are adjacent. In fact, any two edges of G_6 or G_7 are adjacent. An edge $\{u,v\}$ is usually denoted by uv or vu.

Consider the graph $G = (V,E)$ diagramed in Figure 12.18. The vertex set for this graph is $V = \{u,v,w,x,y\}$, and the edge set is

$$E = \{\{u,y\},\{v,w\},\{w,x\}\}$$

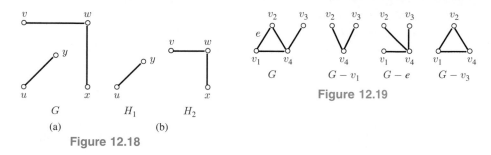

Figure 12.18

Figure 12.19

In this graph, v is adjacent to w and w is adjacent to x, but x is not adjacent to v. Also u and y are adjacent to each other but are not adjacent to any other vertex. The vertex w is incident to the edges vw and xw, and these edges are adjacent. The edge uy is not adjacent to edges vw or wx.

Subgraphs and Supergraphs

> If G is a graph, the vertex set of G is frequently denoted $V(G)$ and the edge set $E(G)$. A graph H is a **subgraph** of a graph G if $V(H) \subseteq V(G)$ and $E(H) \subseteq E(G)$; in such a case G is also called a **supergraph** of H.

Thus a subgraph of a graph G is a graph H whose vertex set is a subset of $V(G)$ and whose edge set is a subset of $E(G)$. For example, let G be the graph diagramed in Figure 12.18. Then the graph $H_1 = (V_1, E_1)$, where $V_1 = \{u, y\}$ and $E_1 = \{\{u, y\}\}$, is a subgraph of G. Also the graph $H_2 = (V_2, E_2)$, where $V_2 = \{v, w, x\}$ and $E_2 = \{\{v, w\}, \{w, x\}\}$, is a subgraph of G. There are many other subgraphs of G. Can you describe them?

The simplest type of subgraph of a graph G is that obtained by deletion of a vertex or edge. If G is a graph with at least two vertices and $v \in V(G)$, then $G - v$ denotes the subgraph of G with vertex set $V(G) - \{v\}$, that is, the vertices of $V(G)$ except v, and whose edges are all those of G not incident with v. If $e \in E(G)$, then $G - e$ is the subgraph of G with vertex set $V(G)$ and edge set $E(G) - \{e\}$. Examples of such subgraphs are given in Figure 12.19.

Multigraphs

For certain applications a generalization of the concept of a graph is useful. In some situations a vertex may be adjacent to itself, or there may be more than one edge joining two vertices, such as in the Königsberg Bridge problem. An edge from a vertex to itself is called a **loop**.

> In a **multigraph** loops and multiple edges are allowed.

More precisely, a multigraph is a finite nonempty set V together with a collection E of one- and two-element subsets some of which may be repeated. (In a graph, only two-element subsets are allowed as edges, and none are repeated.) For example, if $V = \{a, b, c, d, e, f\}$ and

$$E = \{\{a, b\}, \{a, b\}, \{b, d\}, \{c, d\}, \{c, d\}, \{c\}, \{c\}, \{b, e\}\}$$

then $G = (V, E)$ is a multigraph. The one-element subsets are the loops, and the number of times a subset is repeated is the **multiplicity** of that edge. The multiplicity, in this case, of edges $\{a, b\}$, $\{c, d\}$, and $\{c\}$ is 2, and the multiplicity of the other edges is 1. This multigraph is diagramed in Figure 12.20.

Definitions of adjacency, submultigraphs, etc., are analogous to the corresponding definitions for graphs. A (p, q) multigraph is a multigraph with p vertices and q edges. For example, the multigraph of Figure 12.20 is a $(6, 8)$ multigraph. The graph G of Figure 12.19 is a $(4, 4)$ graph.

Degree of a Vertex

The **degree** of a vertex v in a multigraph G is the number of edges of G incident with v (a loop at v counts as two edges incident with v). The degree of v is denoted deg v. A vertex is **odd** or **even** depending on whether its degree is odd or even. A 0-degree vertex is an **isolated vertex**, and a 1-degree vertex is an **end vertex**.

For example, in the Figure 12.20 multigraph, deg $a = 2$, deg $b = 4$, deg $c = 6$, deg $d = 3$, deg $e = 1$, deg $f = 0$. Thus a, b, c, f are even vertices; d, e are odd; e is an end vertex; and f is isolated. Notice that the sum of the degrees is 16, which is twice the number of edges. The next theorem states that this is always the case.

Theorem 12.2   Let G be a (p, q) multigraph with $V(G) = \{v_1, v_2, \ldots, v_p\}$. Then
$$\deg v_1 + \deg v_2 + \cdots + \deg v_p = 2q$$

Proof: Since each edge is incident with two vertices, when the degrees of the vertices are summed, each edge is counted twice.   □

Notice that in the above multigraph there are an even number of odd vertices. The next theorem states that this always holds. The proof of this theorem is worked out in Exercises 25 through 28.

Theorem 12.3   In any multigraph G there are an even number of odd vertices.

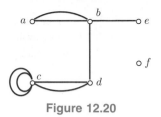

Figure 12.20

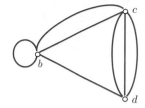

Figure 12.21

EXERCISES

1. For the graph of Figure 12.15, which vertices are adjacent, which edges are adjacent, and which vertices and edges are incident?
2. Answer Exercise 1 for the graph of Figure 12.13.
3. Diagram the graph $G = (V, E)$, where

$$V = \{a, b, c, d, e\}$$
$$E = \{\{a, b\}\{a, d\}, \{c, d\}, \{b, e\}, \{d, e\}\}$$

Answer the questions in Exercise 1 for this graph.

4. Diagram all the different graphs with four vertices.
5. Let G be the multigraph of Figure 12.20. Diagram $G - f$, $G - c$, $G - a$, $G - a - b$, $G - ab$.
6. Give an example of a graph G and a vertex $v \in V(G)$ in which $G - v$ has five fewer edges than G.
7. If G is a (p, q) graph and $e \in E(G)$, prove that $G - e$ is a $(p, q - 1)$ graph.
8. If G is a (p, q) graph and $v \in V(G)$, must $G - v$ be a $(p - 1, q)$ graph? Why?
9. A graph is **regular** if all its vertices have the same degree. Diagram all the different regular graphs with four or five vertices.
10. Explain why a graph is a multigraph but a multigraph need not be a graph.
11. Diagram the following multigraph and give the degree of each vertex. Which vertices are odd, even, isolated, and end vertices?

$$V = \{a, b, c, d, e, h\}$$
$$E = \{\{a\}, \{a\}, \{a\}, \{a, b\}, \{a, b\}, \{a, b\}, \{b, c\}, \{b, c\}, \{b, d\}, \{d, e\}\}$$

12. What is the multigraph diagramed in Figure 12.21? Give the degree of each vertex of this multigraph.
13. Show that Theorems 12.2 and 12.3 hold for the multigraph of Exercise 11.
14. Show that Theorems 12.2 and 12.3 hold for the multigraph of Figure 12.21.
15. If $a_1, \ldots, a_n$ are odd numbers and $a_1 + \cdots + a_n$ is even, prove that n is even.
16. Diagram all the different $(4, 2)$ graphs.
17. Diagram all the different $(4, 3)$ graphs.
18. Diagram all the different $(5, 3)$ graphs.
19. Let G be a (p, q) graph each of whose vertices is odd. Prove that p is even.
*20. Let G be a (p, q) graph all of whose vertices have degree 3 and $q = 2p - 3$. What can be said about G?

21. Which of the following statements are true for every multigraph? For those statements that are not true for every multigraph, give an example to show they are not.
    (a) There are an odd number of even vertices.
    (b) There are an even number of even vertices.
    (c) There are an odd number of odd vertices.
    (d) There are an even number of odd vertices.

*22. Let S be a set with p elements. Prove that the number of two-element subsets S is $p(p-1)/2$. For a (p, q) graph, prove that $q \leq p(p-1)/2$.

*23. If G is a (p, q) graph, how many edges does $\overline{G}$ have?

*24. Let $S = \{1, 2, \ldots, p\}$. Write a BASIC program that prints all the two-element subsets of S.

(Optional Exercises)

In Exercises 25–28, we shall prove Theorem 12.3.

Let G be a (p, q) multigraph with $V(G) = \{v_1, v_2, \ldots, v_p\}$. Since zero is an even number, if G has no odd vertices we are finished. Otherwise, let $v_1, \ldots, v_k$ be the odd vertices and $v_{k+1}, \ldots, v_p$ the even vertices.

*25. Why is $\deg v_1 + \deg v_2 + \cdots + \deg v_p$ even?

*26. Why is $\deg v_{k+1} + \deg v_{k+2} + \cdots + \deg v_p$ even?

*27. Use Exercises 23 and 24 to show that $\deg v_1 + \cdots + \deg v_k$ is even.

*28. Use Exercise 15 to prove that k is even.

12.3 Instant Insanity

What is the difference between method and device? A method is a device which you can use twice.
— GEORGE POLYA, Twentieth-Century Mathematician

In this section multigraphs will be used to find a solution to the popular colored cube-stacking game called *Instant Insanity*. After seeing this solution a student of the author stated:

> *I can see it now. The Instant Insanity puzzle is brought out. I feign frustrated inability for fifteen minutes or so, then mumble something about being able to do it if I really wanted to. Screech—the trap opens. I'm goaded into a wager with all present. Clank! The jaws snap shut. I whip out my pencil and paper and graph-theorize myself into financial independence.*

This puzzle consists of four cubes, each face of which is colored with one of four colors so that every color appears on at least one face of each cube. The object is to make a straight stack of the four cubes one on top of the other so that each of the four colors appears exactly once on each side of the stack.

Solving by Chance

There is little likelihood of solving this puzzle purely by chance since there are 41,472 possible combinations of the four cubes. Why? You can put the bottom cube down in

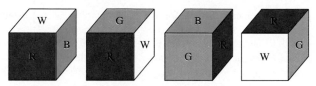

Instant Insanity Puzzle

many different ways, but only three of them are essentially different since there are only three pairs of faces that can be covered (one on bottom, one on top), leaving the other four faces exposed. The second cube, however, may be put on top of cube 1 on any of cube 2's 6 faces, and then cube 2 may be rotated in 4 possible ways, giving $(6)(4) = 24$ possibilities. The third and fourth cubes also have these 24 possibilities; so there are $(3)(24^3) = 41,472$ different possibilities. Unless you are incredibly lucky (in which case you should be in Las Vegas anyway), you need a system.

Alternative methods of solution have been devised using computers, but they are inefficient and expensive.

Suppose the four colors are red, white, green, and blue, abbreviated R, W, G, B. Label the cubes 1, 2, 3, 4. Unfortunately a solution to the game need not exist, as the "unfolded" cubes in Figure 12.22 show. Here you see that no matter how you stack these cubes the sides of the stack will show eight blue squares when a solution demands only four. Fortunately, the manufacturer of the puzzle does not produce such a diabolical set.

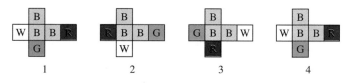

Figure 12.22

Sample Game

Before looking at the following solution, why don't you play a game for a while? You can either use a commercial *Instant Insanity* puzzle or make one yourself. A simple way to make a puzzle is to use four sugar cubes. You can color the sides with food coloring or merely label the sides arbitrarily with the letters R, W, G, B. After you have played with your set for a while, color or label a set as in Figure 12.23 and follow the steps in the solution. If you get tired or hungry, you can always close this book and eat the sugar cubes.

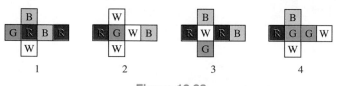

Figure 12.23

Stack Multigraph

Suppose the set consists of the cubes in Figure 12.23. Construct a multigraph (called the **stack** multigraph) as follows. There are four vertices, each representing a different color, and two vertices are adjacent if and only if the colors are *opposite each other* in some cube. Label each edge 1, 2, 3, 4 according to which cube it describes. Figure 12.24 depicts the stack multigraph for the set of cubes in Figure 12.23.

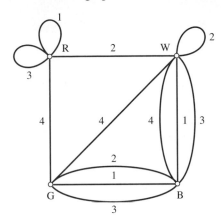

Figure 12.24

The stack multigraph gives all the possible combinations of opposites for this particular set of cubes. From the stack multigraph a particular combination of opposites must be extracted that will solve the problem.

Oppositeness Multigraphs

In order to see what combinations of opposites give solutions, consider an *Instant Insanity* puzzle that has been solved (Figure 12.25).

Now form an f-b (front-back) and l-r (left-right) oppositeness multigraph for this stack. This is given in Figure 12.26.

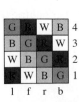

Figure 12.25

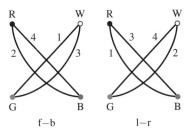

Figure 12.26

What characterizes f-b and l-r oppositeness multigraphs that give solutions? Since each color appears exactly once on the front, say, of the stack and exactly once on

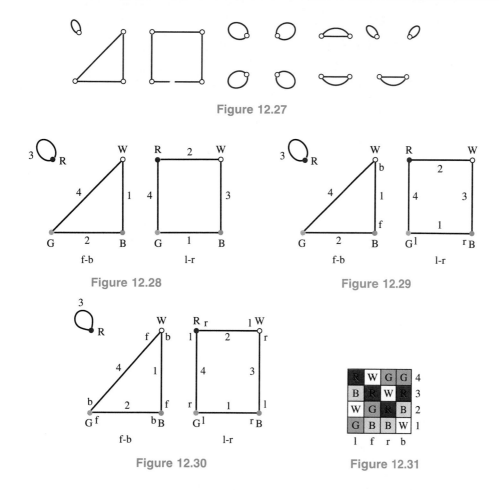

Figure 12.27

Figure 12.28

Figure 12.29

Figure 12.30

Figure 12.31

the back, they are precisely those multigraphs with four vertices all of whose vertices have degree 2. There are only five different multigraphs of this type. These are given in Figure 12.27.

Now return to the stack multigraph in Figure 12.24. In order to find a solution, find two submultigraphs with no edges in common of the type in Figure 12.27 (with each edge labeled differently) to serve as f-b and l-r oppositeness multigraphs. These are shown in Figure 12.28 (the f-b and l-r designations are arbitrary).

To realize the actual arrangements of the cubes, it only remains to determine which of the two, say, cube 1 colors is to be on the front of the stack and which on the back. Similarly, which cube 1 color will be on the left and which on the right. The designations are arbitrary for the **first** cube. Let B be on the front, W the back, G the left, and B the right of cube 1. These are designated in Figure 12.29.

The requirement that each color appear exactly once then forces the other faces to be in the reverse orders, as the continuation of the labeling shows in Figure 12.30. From the solution arrangement of Figure 12.30, the solution is given in Figure 12.31.

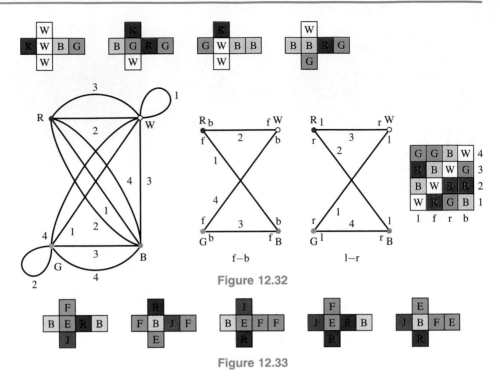

Figure 12.32

Figure 12.33

Solution Algorithm

In summary, the following is an algorithm (procedure) for solving any *Instant Insanity* stacking problem:

> (1) Form the stack multigraph.
> (2) Extract two edge-disjoint oppositeness submultigraphs of the type in Figure 12.27.
> (3) Label the colors in the multigraphs of Step 2 for front-back and left-right and read off the solution.

To illustrate the method, another stacking problem is solved in Figure 12.32.

In the Museum of Childhood in Edinburgh there is a five-cube puzzle similar to *Instant Insanity*. In this puzzle each face of the five cubes is covered by a flag of Belgium, France, Japan, Russia, or England. The object is to stack the five cubes so that each vertical wall contains each of the five flags. Using an analysis similar to that of the *Instant Insanity* puzzle, there are $(3)(24^4) = 995,328$ possible positions. If you try one possibility every 10 seconds, it will take over 2764 hours or over 115 days straight work to try them all. Or if you work 8 hours a day, 7 days a week, it will take almost a year to try every possibility. It is thus virtually impossible to solve this puzzle by trial and error. The systematic method given here takes less than 10 minutes. The unfolded puzzle at Edinburgh is shown in Figure 12.33.

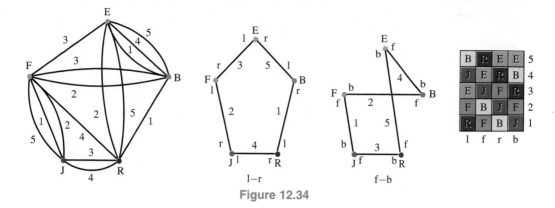

Figure 12.34

To solve this puzzle, form the stack multigraph and find two edge-disjoint sub-multigraphs each of whose vertices has degree 2. The solution is given in Figure 12.34.

EXERCISES

1. If you tried a different position every 10 seconds, how many days would it take to try every possibility in the *Instant Insanity* puzzle?

2. Show that the only different multigraphs on four vertices all of whose vertices have degree 2 are those in Figure 12.27.

3. Find the multigraphs on five vertices all of whose vertices have degree 2.

4. Solve the *Instant Insanity* problem for the cubes in Figure 12.35.

5. Solve the *Instant Insanity* problem for the cubes in Figure 12.36.

6. Make up your own *Instant Insanity* problem and solve it.

7. Construct the f-b and l-r multigraphs for the stack in Figure 12.37.

8. Solve the five-flags puzzle in Figure 12.38.

9. How would the *Instant Insanity* puzzle and its solution change if there were three cubes? Six cubes?

*10. What can you say about the uniqueness of solutions to the *Instant Insanity* puzzle?

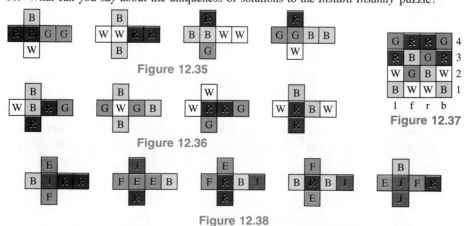

Figure 12.35

Figure 12.36

Figure 12.37

Figure 12.38

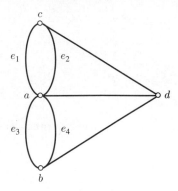

Figure 12.39

12.4 Königsberg Bridge Problem

So many paths that wind and wind,
When just the art of being kind
Is all this sad world needs.
— ELLA WHEELER WILCOX, Twentieth-Century Poet

As discussed in Section 12.1, Euler used the multigraph depicted in Figure 12.39 to solve the Königsberg Bridge problem. Euler wanted to determine whether it was possible to begin at one of the vertices, traverse each edge exactly once, and end at the original vertex.

Trails

In discussing problems like the Königsberg Bridge problem it is convenient to consider different ways in which a traverse from one vertex to another can be made. For example, in Figure 12.39, starting at vertex d a traverse can be made across edge dc to vertex c and then across edge e_1 to vertex a. This is an example of a d-a trail. We denote this trail by d, dc, c, e_1, a. Another example of a d-a trail is d, da, a. In an arbitrary multigraph, a u_1-u_n **trail** is an alternating sequence of vertices and edges of the form

$$u_1, u_1u_2, u_2, u_2u_3, \dots, u_{n-1}, u_{n-1}u_n, u_n$$

where all the edges are distinct. We call u_1 and u_n the **initial** and **final** vertices, respectively. Thus a u_1-u_n trail is an alternating sequence of vertices and edges that has three properties:

(1) u_1 is the first vertex, and u_n is the last vertex.
(2) The edges are distinct.
(3) Each edge is immediately preceded and succeeded by the vertices it contains.

Examples of b-c trails in Figure 12.39 are b, e_3, a, e_1, c and b, bd, d, dc, c.

A trail whose initial and final vertices are the same is called a **circuit**. Thus a u-u trail is a circuit. In Figure 12.39, an example of a circuit is a, e_3, b, e_4, a. This circuit has a as its initial and final vertex. Another example of a circuit with initial and final vertex a is a, e_3, b, bd, d, da, a.

Eulerian Circuit

Now return to the Königsberg Bridge problem. This problem is equivalent to determining whether the multigraph in Figure 12.39 has a circuit containing all its edges. Such a circuit is called an **Eulerian circuit**. Euler proved the following theorem.

Theorem 12.4   If a multigraph G has an Eulerian circuit, then every vertex of G is even.

Proof: Let C be an Eulerian circuit, and let $v \in V(G)$. If v is isolated, then deg $v = 0$ is even. If v is not isolated, it must occur in C. If v is not the initial vertex (and hence not the final vertex either), then each time v is encountered in C, it is entered and left via distinct edges. Hence each occurrence of v in C represents a contribution of 2 to the degree of v; so deg v is even. If v is the initial vertex of C, then C begins and ends with v, contributing 2 to the degree, and every other occurrence of v gives an additional 2 to its degree; so deg v is even.   □

Not every vertex in the multigraph of Figure 12.39 is even; so, applying the above theorem, this multigraph has no Eulerian circuit. This solves the Königsberg Bridge problem.

Eulerian circuits have been put to practical application by determining the most efficient routes for such things as garbage collection and mail delivery. For example, a letter carrier wants to start at the post office, deliver the mail, and return in the most efficient manner. If we represent the streets by edges and the intersections by vertices of a multigraph, then an Eulerian circuit allows the letter carrier to go down every street precisely once and return. Such a circuit eliminates the need to cover a street more than once. Theorem 12.4 states that if such a route is possible, then an even number of streets to be covered meet at each intersection. A letter carrier on foot would certainly appreciate this kind of route.

For an example of an Eulerian circuit, remove vertex d in Figure 12.39. The resulting graph has vertices a, b, and c and edges e_1, e_2, e_3, and e_4. Now each vertex is even, and an Eulerian circuit is given by $c, e_1, a, e_3, b, e_4, a, e_2, c$.

EXERCISES

1. In Figure 12.39, what are the initial and final vertices of the trail b, e_3, a, e_2, c?
2. Find a c-b trail in Figure 12.39.
3. Find a circuit in Figure 12.39 with d as its initial and final vertex.
4. In Figure 12.39, is b, e_4, a, e_1, c, cd, d a trail?

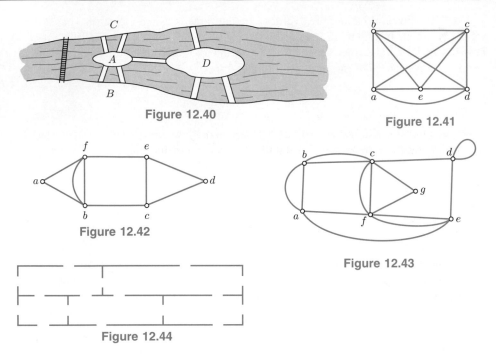

Figure 12.40

Figure 12.41

Figure 12.42

Figure 12.43

Figure 12.44

5. In Figure 12.39, is $d, dc, c, e_1, a, e_2, c, cd, d$ a trail?

6. In Figure 12.39, is c, e_1, a, e_3, b, ad, d a trail?

7. A u_1-u_n **walk** is an alternating sequence of vertices and edges that has Properties 1 and 3 of this section. In Figure 12.39, give an example of a walk that is not a trail.

8. A u_1-u_n **path** is an alternating sequence of vertices and edges that has Properties 1 and 3; also the vertices are distinct. Show that a path is a trail. In Figure 12.39, give an example of a trail that is not a path.

9. There are now nine bridges in Königsberg; one is a railroad bridge (see Figure 12.40). Is there now an Eulerian circuit with the railroad bridge excluded? With the railroad bridge included?

10. Suppose that in addition to the seven bridges of Königsberg a new bridge is constructed anywhere. Does there now exist an Eulerian circuit?

11. An **Eulerian trail** is a trail that includes all the edges. Prove that if a multigraph G contains an Eulerian trail with different initial and final vertices, then there were precisely two odd vertices. Was there an Eulerian trail for the seven bridges of Königsberg? What if another bridge is added anywhere?

12. Find an Eulerian circuit for the multigraph in Figure 12.41.

13. Does the multigraph in Figure 12.42 have an Eulerian circuit? If not, can you add one edge so that it does?

14. Find an Eulerian circuit in the multigraph in Figure 12.43.

15. In Figure 12.42 find a walk that is not a trail. Find a trail that is not a path. Find a path (see Exercises 7 and 8).

16. Figure 12.44 is a floor plan of a house. Is it possible to walk through each door exactly once?

$\boxed{12.5}$ Trees

A fool sees not the same tree that a wise man sees.
— WILLIAM BLAKE (1757–1827), British Poet and Artist

It was pointed out in Section 12.1 that Kirchhoff and Cayley encountered special types of graphs called trees in their investigations of electric networks and organic chemistry. In this section some of the basic properties of trees will be developed.

Connectedness

Two distinct vertices u, v of a multigraph G are **connected** if there is a u-v trail; G is **connected** if every two of its vertices are connected; if G is not connected, it is **disconnected**. A **component** of G is a maximal connected submultigraph of G; that is, a component of G is a connected submultigraph of G that is not contained in a larger connected submultigraph of G.

For example, the multigraph in Figure 12.45 has three components. The vertices v and w are connected, while u is not connected to v or w. Intuitively, a component is a part of a multigraph that is linked together in one piece. Different components have no links between them. If a multigraph is connected, then it is a component of itself. The number of components of a multigraph G is denoted $c(G)$. Notice that a multigraph G is connected if and only if $c(G) = 1$.

This theory is useful for studying communication, transportation, assembly line, and production systems. For example, in a communication system the vertices represent communication stations, and the edges represent communication channels such as telephone lines or microwave beams. If u and v are adjacent, they represent stations that can communicate directly with each other. If u and v are connected, they can communicate with each other but may have to go through intermediate stations to do so. It is desirable to communicate along trails with distinct vertices since then there is no redundancy in stations or channels. Also, if the needed lines are not tied up, communication along the shortest possible trail is more efficient and economical. A component is the largest possible subsystem in which stations can communicate. Communication cannot take place between different components.

Definition of a Tree

A connected graph with no circuits is called a **tree**.

For example, in Figure 12.17, G_2, G_4, and G_5 are not trees since they are not connected; G_7 is not a tree since it has a circuit; while G_1, G_3, and G_6 are trees.

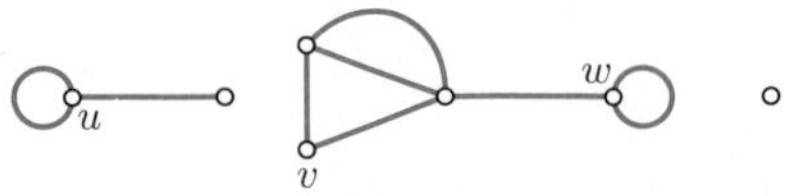

Figure 12.45

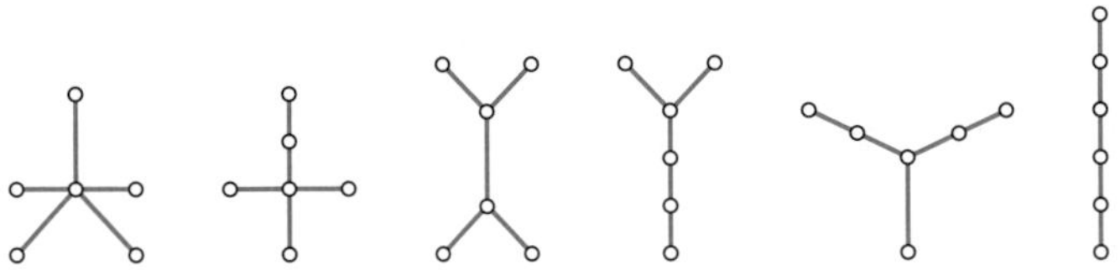

Figure 12.46

Figures 12.3 and 12.4 give other examples of trees. In Figure 12.46 all the trees with six vertices are diagramed.

The graph G_1 that has only one vertex is called the **trivial graph**. Notice that all the trees considered, except for G_1, have at least two end vertices. This is true for any nontrivial tree. The fact that any nontrivial tree has at least two end vertices is intuitively quite reasonable. Since a tree has no circuits, it cannot wrap around to itself and it must begin and end somewhere, that is, at end vertices. For example, in Figure 12.46 the trees have five, four, four, three, three, and two end vertices, respectively.

Theorem 12.5   Every nontrivial tree has at least two end vertices.

Proof: Let T be a nontrivial tree, and suppose that T has only one end vertex v_1. Since T is nontrivial and connected, v_1 must be adjacent to some vertex v_2. Since v_2 is not an end vertex, v_2 must be adjacent to another vertex $v_3 \neq v_1$. Again since v_3 is not an end vertex, v_3 must be adjacent to some vertex $v_4 \neq v_2$. Also $v_4 \neq v_1$ since otherwise $v_1, v_1v_2, v_2, v_2v_3, v_3, v_3v_1, v_1$ would be a circuit, which contradicts the fact that T has no circuits. Now find a $v_5, v_6, \ldots, v_n$. This process must stop, say at v_n, since there can be only a finite number of vertices. Now v_n cannot be one of the previous v_i's or be adjacent to one of the previous v_i's since then there would be a circuit. But since v_n is not an end vertex, it must be adjacent to some new vertex v_{n+1}. This gives a contradiction. Hence T cannot have only one end vertex. Now suppose T has no end vertex. Starting with any vertex of T and proceeding as above, a contradiction results. $\square$

Notice that for every tree that has been considered so far there is one fewer edge than there are vertices. For example, the trees in Figure 12.46 all have 5 edges and 6 vertices, the trees in Figure 12.4 have 13 edges and 14 vertices. In Figure 12.3, the first tree has 3 edges and 4 vertices, the second tree has 6 edges and 7 vertices, the third 4 edges and 5 vertices, and the fourth has 7 edges and 8 vertices. The next theorem states that this is always true. The proof of this theorem is worked out in Exercises 28 through 33.

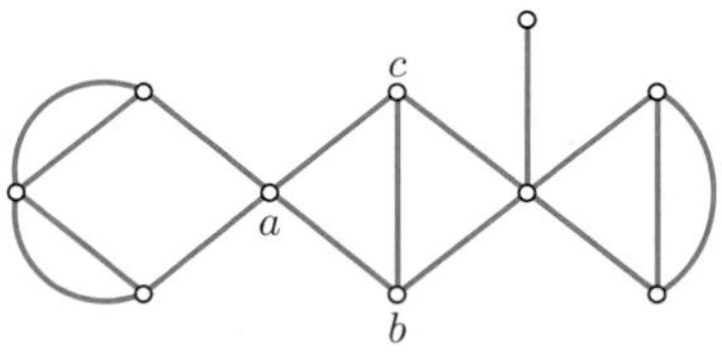

Figure 12.47

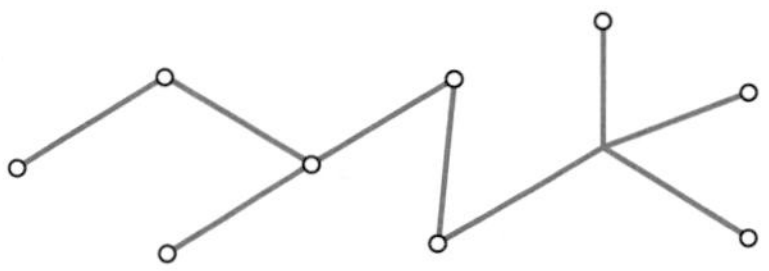

Figure 12.48

Theorem 12.6   If T is a (p, q) tree, then $q = p - 1$.

EXAMPLE 1   Consider the multigraph in Figure 12.47. Suppose this multigraph represents a communication network in which the vertices correspond to stations and the edges correspond to communication links. What is the largest number of edges that can be deleted while still allowing the stations to communicate with each other?

Solution.   First, since multiple edges are not necessary, all but one of a set of multiple edges can be deleted. Second, circuits are not necessary, since circuits give two ways in which stations can communicate and all that is needed is one way. For example, the circuit a, ac, c, cb, b, ba, a gives two ways for a and c to communicate, namely along a, ac, c or along a, ab, b, bc, c. If the edge ab is deleted, then a and c can still communicate. Thus one edge of each circuit can be deleted. The edges that are now left are the fewest necessary to maintain communication between all stations. One way of accomplishing this (there are many other ways) is shown in Figure 12.48.

Since communication is maintained between the stations, the graph in Figure 12.48 is still connected. Also, since one edge from each circuit has been deleted, there are no circuits left. Thus the graph in Figure 12.48 is a tree. This tree is an example of a spanning tree for the multigraph in Figure 12.47.   □

Spanning Tree

A spanning tree for a communication system gives a minimum set of links that need to be maintained to retain communication between all stations. The general definition of a spanning tree is the following.

> If G is a multigraph, a **spanning tree** T for G is a subgraph of G that is a tree that contains all the vertices of G.

Theorem 12.7   Every connected multigraph has a spanning tree.

Proof: Let G be a connected multigraph. Delete from $E(G)$ all the loops and all but one of each set of multiple edges. From the edges that remain, delete one edge from each circuit. Since the circuits have been broken, the remaining graph T has no circuits. The graph T is still connected because a circuit gives two trails connecting vertices in the circuit and deleting one edge still leaves one trail. Thus T is a tree containing all the vertices of G and hence is a spanning tree for G.   □

Corollary 12.1 If G is a connected (p, q) multigraph, then $q \geq p - 1$.

Proof: By Theorem 12.7, G has a spanning tree T with p vertices. Applying Theorem 12.6, T has $p - 1$ edges. But G has at least as many edges as T, and so $q \geq p - 1$. $\square$

It follows from Corollary 12.1 that a tree has the fewest number of edges possible to remain connected. Thus the removal of any edge from a tree will disconnect it.

EXERCISES

1. Draw a graph that has one component.
2. Draw a graph that has two components.
3. Draw a graph that has one circuit.
4. If a nontrivial graph has an isolated vertex, can it be a tree?
5. Give an example of a multigraph (a) with 10 components, (b) with n components.
6. Give examples other than those considered in the text of the following.
   (a) A disconnected graph with no circuits
   (b) A connected graph with circuits
   (c) A tree
   (d) A graph with no end vertices
7. Give an example of a graph (a) with three end vertices, (b) with n end vertices.
8. Prove that a vertex v is isolated if and only if $c(G - v) < c(G)$.
9. Find a spanning tree for the multigraph of Figure 12.47.
10. Find spanning trees for the multigraphs of Figures 12.41 to 12.43.
11. (a) Give an example of a disconnected multigraph in which Corollary 12.1 does not hold.
    (b) Give an example of a disconnected multigraph in which Corollary 12.1 does hold.
12. Can a disconnected multigraph have a spanning tree? Why?
13. Give an example of a graph with no circuits in which the conclusion of Theorem 12.6 does not hold. Give an example of a connected graph in which the conclusion of Theorem 12.6 does not hold.
14. Diagram all the different trees with four vertices.
15. Diagram all the different trees with five vertices.
16. Show that a graph G is a tree if and only if every two distinct vertices of G are connected by a unique trail of G.
17. Show that the components of a graph without circuits are trees. (A graph without circuits is called a **forest**.)
*18. Show that a circuitless (p, q) graph with $q = p - 1$ is a tree.
*19. Show that a connected (p, q) graph with $q = p - 1$ is a tree.
20. A vertex v of a multigraph G is a **cut vertex** if $c(G - v) > c(G)$. What are the cut vertices in the multigraph of Figure 12.47? What are the cut vertices in the multigraph of Figure 12.46? What is the significance of a cut vertex in a communication network?
*21. Show that w is a cut vertex if and only if there are vertices $u \neq v$ distinct from w such that every u-v trail contains w. (See Exercise 20 for the definition of a cut vertex.)

22. Show that no cut vertex is an end vertex and that no end vertex is a cut vertex. (See Exercise 20 for the definition of a cut vertex.)

23. A **complete** graph is a graph in which every pair of vertices are adjacent. The complete graph with p vertices is denoted K_p. Diagram K_1, K_2, K_3, K_4, and K_5.

24. Show that K_p is a subgraph of K_{p+1} (see Exercise 23).

*25. How many edges does K_p have? (See Exercise 23.)

26. Show that K_p is connected. Which K_p's are trees? (See Exercise 23.)

27. Show that K_p has no cut vertices (see Exercises 20 and 23).

(Optional Exercises)

In Exercises 28–33 we shall prove Theorem 12.6. Let T be a (p, q) tree. We must show that $q = p - 1$.

28. If $p = 1$, show that the result holds.

29. If $p = 2$, show that the result holds.

30. If $p = 3$, show that the result holds.

31. Suppose $p = 4$ and let v be an end vertex. Prove that $T - v$ is a tree with three vertices.

32. Use Exercises 30 and 31 to show that the result holds for $p = 4$.

33. If $p = 5$, proceed as above to show that the result holds. Continuing in this way it follows that the theorem holds for any finite number of vertices.

12.6 Planar Graphs

Mathematicians assume the right to choose, within the limits of logical contradiction, what path they please in reaching their results.
— HENRY ADAMS (1838–1918), American Historian

Suppose you have the job of designing a safe and economical highway system. Certain areas must be linked by highways, and you are to decide where these highways are to be placed. Since intersections are inconvenient and unsafe for high-speed driving and underpasses and overpasses are costly to construct, the best highway system will keep the number of crossing roads to a minimum. If you represent the different areas by vertices and the highways by edges, you will have a graph. What you need is a graph with edges that do not cross.

There are many important uses of graphs with edges that do not cross. Such graphs have recently become important in the design of miniature electronic circuits. The conductors in these circuits are printed on a thin wafer as a fine silver or copper line. These lines cannot cross or there will be a short circuit. Thus these electronic circuits must be represented by graphs with noncrossing edges. Information about such graphs can be very useful to this technology. An important question is: How can you tell if a graph can be drawn so that edges do not cross? This question will be answered in this section.

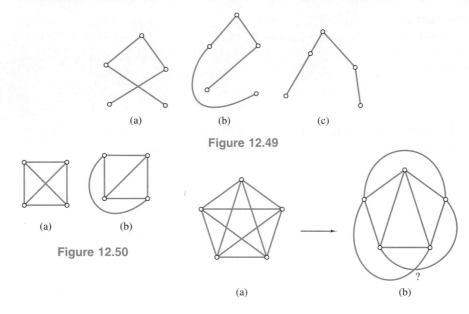

(a)   (b)   (c)

Figure 12.49

(a)   (b)

Figure 12.50

(a)   (b)

Figure 12.51

Definition of a Planar Graph

A graph is **planar** if it can be diagramed on a plane so that distinct edges do not cross.

For example, the graph in Figure 12.49a is planar since it can be diagramed as in Figure 12.49b or 12.49c. Although the diagrams locate the vertices and edges of the graph in Figure 12.49a in different ways, the underlying graph is still the same. The graph in Figure 12.50a is planar since it can be diagramed as in Figure 12.50b.

A Nonplanar Graph

An example of a nonplanar graph is K_5, diagramed in Figure 12.51a. To gain some insight into why K_5 is not planar, try to diagram K_5 so that distinct edges do not cross. Figure 12.51b shows such an attempt. No matter how hard you try there is always one edge that must cross some other edge. A rigorous proof that K_5 is nonplanar will be given later.

A graph is **complete** if every pair of its vertices are adjacent.

The complete graph with p vertices is denoted K_p. Thus K_5 is the complete graph with five vertices. Figure 12.50a is a diagram of K_4, G_1 in Figure 12.17 is K_1, G_3 in Figure 12.17 is K_2, and G_7 in Figure 12.17 is K_3. K_5 is sometimes called the

pentagram and has appeared in many places in history, for example Solomon's seal. It was used in the magic symbol of the ancient Pythagoreans. The pentagram appeared on the inner face of Sir Gawain's shield to remind him of the power of his five virtues and the strong interconnections between them.

Another Nonplanar Graph

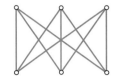

Figure 12.52

Another nonplanar graph is given by the story of the three cabins and the three wells. There were paths from each cabin to each of the wells, but since the inhabitants of the different cabins did not like each other, they decided to make paths from their cabins to each of the wells so that the paths did not cross. In this way they would not have to see their neighbors. But no matter how hard they tried, they could not construct the paths. This graph is diagramed in Figure 12.52 and is denoted $K(3, 3)$. It will be proved later that $K(3, 3)$ is nonplanar.

Regions

When planar graphs are considered, it will always be assumed that they are diagramed on the plane so that distinct edges do not cross.

> Given a planar graph G, a **region** is a maximal portion of the plane that is bounded by the edges of the graph so that any two points of the region can be joined by an arc that does not cross a vertex or edge.

Intuitively, if the planar graph is thought of as a map, the edges and vertices serve as boundaries and the regions may be thought of as the countries or states. There is always one unbounded region called the **external** region. For the graph in Figure 12.53 there are five regions, $R_1, \ldots, R_5$, and R_5 is the external region. Notice that a tree has only one region.

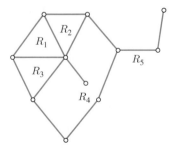

Figure 12.53

Euler's Formula

Let G be a connected planar graph with p vertices, q edges, and r regions. A formula relating these three numbers was discovered by Euler over 200 years ago. See if you

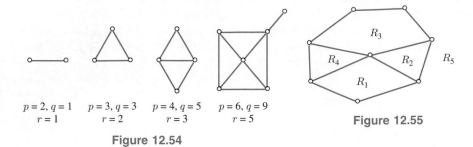

$$p = 2, q = 1 \qquad p = 3, q = 3 \qquad p = 4, q = 5 \qquad p = 6, q = 9$$
$$r = 1 \qquad\qquad r = 2 \qquad\qquad r = 3 \qquad\qquad r = 5$$

Figure 12.54

Figure 12.55

can guess this formula. In Figure 12.54, p, q, and r have been found for four graphs. Can you make a conjecture from these examples?

Theorem 12.8   **(Euler's formula)** If G is a connected planar graph with p vertices, q edges, and r regions, then $p - q + r = 2$.

The proof of Theorem 12.8 is worked out in Exercises 30 through 32. As an illustration of Euler's formula, consider the planar graphs in Figure 12.54. In the first graph $p - q + r = 2 - 1 + 1 = 2$, in the second graph $p - q + r = 3 - 3 + 2 = 2$, in the third graph $p - q + r = 4 - 5 + 3 = 2$, and in the last graph $p - q + r = 6 - 9 + 5 = 2$.

Let P be a planar graph with q edges and r regions $R_1, R_2, \ldots, R_r$. Suppose each edge of P is in at least one circuit and that q_i edges bound region R_i, $i = 1, 2, \ldots, r$. For example, in Figure 12.55, $q = 11$, and there are five regions, for which $q_1 = 4$, $q_2 = 3$, $q_3 = 5$, $q_4 = 3$, and $q_5 = 7$. If the q_i's are added, then each edge is counted twice so that $q_1 + q_2 + \cdots + q_r = 2q$. For example, in Figure 12.55,

$$q_1 + q_2 + q_3 + q_4 + q_5 = 22 = 2q$$

This result does not hold if there are edges that are not in a circuit. For example, in Figure 12.53,

$$q_1 + q_2 + q_3 + q_4 + q_5 = 3 + 3 + 3 + 7 + 9 = 25 \quad \text{while} \quad 2q = 38$$

The observation made in the previous paragraph together with Euler's formula can be used to show that $K(3, 3)$ is nonplanar. Indeed, suppose $K(3, 3)$ is planar and let P be a plane representation of it in which distinct edges do not cross. Since each edge of $K(3, 3)$ is in at least one circuit, this also holds for the edges of P. Also, since $K(3, 3)$ has no triangles, there can be no triangles in P; so each region in P is bounded by at least four edges. Suppose there are r regions and that q_i edges bound the ith region, $i = 1, 2, \ldots, r$. Then since $q_i \geq 4$ for $i = 1, 2, \ldots, r$, $2q = q_1 + q_2 + \cdots + q_r \geq 4r$. For $K(3, 3)$, $q = 9$ and $p = 6$; so by Euler's formula $r = 2 - p + q = 5$. But the above inequality gives $r \leq (1/2)q = 9/2$. Since this is a contradiction, $K(3, 3)$ cannot be planar.

Triangulated Graphs

> A planar graph is **triangulated** if each of its regions is bounded by a triangle, that is, three edges.

In Figure 12.56, examples of some triangulated graphs are shown. In Figure 12.56a, $p = 3$, $q = 3$; in Figure 12.56b, $p = 4$, $q = 6$; and in Figure 12.56c, $p = 5$, $q = 9$. There is a formula that relates p and q for triangulated graphs. Can you guess it? Let us see if we can derive such a formula using the information we have at hand. Suppose there are r regions. Each region is bounded by three edges. If the number of edges bounding a region are summed over all regions, the result is $3r$. But such a sum counts each edge twice (try this on the graphs in Figure 12.56). Hence $3r = 2q$. Thus $r = (2/3)q$, and so Euler's formula becomes

$$2 = p - q + r = p - q + \frac{2}{3}q = p - \frac{1}{3}q$$

Hence $6 = 3p - q$ or $q = 3p - 6$. We have thus proved the following theorem.

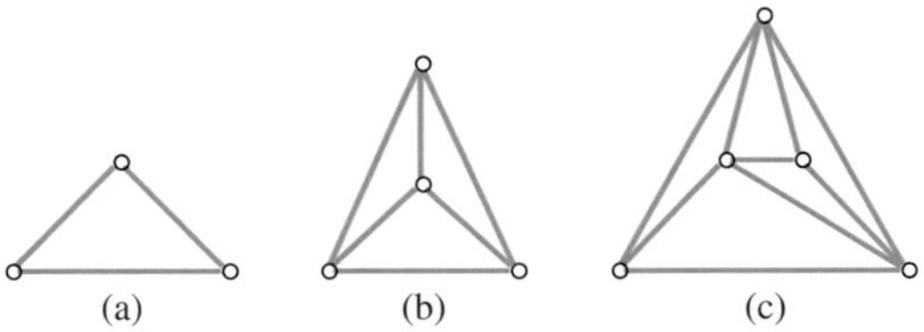

(a)   (b)   (c)

Figure 12.56

Theorem 12.9   If G is a triangulated (p, q) graph, then $q = 3p - 6$.

As an illustration of this theorem, consider the triangulated graphs in Figure 12.56.

In Figure 12.56a, $3p - 6 = 9 - 6 = 3 = q$
in Figure 12.56b, $3p - 6 = 12 - 6 = 6 = q$
and in Figure 12.56c, $3p - 6 = 15 - 6 = 9 = q$

Corollary 12.2   If G is a planar (p, q) graph with $p \geq 3$, then $q \leq 3p - 6$.

Proof: Add to G sufficiently many edges for the resulting (p', q') graph G' to be triangulated. Then $p' = p$ and $q' \geq q$. Hence $q \leq q' = 3p - 6$.   □

Corollary 12.3 K_5 is nonplanar.

Proof: Suppose K_5 were planar. Since $p = 5$, $q = 10$, applying Corollary 12.2 gives $10 = q \leq 3p - 6 = 9$, which is a contradiction.   □

Corollary 12.4 Every planar graph contains a vertex v with deg $v \leq 5$.

Proof: Let G be a planar (p, q) graph with $V(G) = \{v_1, v_2, \ldots, v_p\}$. If all the vertices have degree greater than 5, then by Corollary 12.2 and Theorem 12.2,

$$6p \leq \text{deg } v_1 + \cdots + \text{deg } v_p = 2q \leq 6p - 12$$

which is impossible. Hence there must be a vertex of degree less than or equal to 5. $\square$

Subdivisions

It has been shown that K_5 and $K(3, 3)$ are nonplanar. A deep and surprising theorem due to Kuratowski shows that the converse essentially holds. To state this theorem a definition is needed. An **elementary subdivision** of a graph G with at least one edge is obtained from G by the removal of some edge uv and the addition of a new vertex w and the edges uw and vw. For example, in Figure 12.57, G_2 and G_3 are elementary subdivisions of G_1. A **subdivision** of G is a graph H that is the same as G or is obtained from G by a succession of elementary subdivisions. For example, G_4 is a subdivision of G_1 in Figure 12.57.

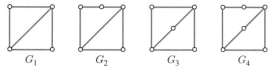

Figure 12.57

Kuratowski's theorem states that a graph is nonplanar if and only if it contains a subdivision of K_5 or $K(3, 3)$ as a subgraph. The proof is long and will not be given here.

EXERCISES

1. Why does a tree have only one region?
2. How many regions do each of the graphs in Figure 12.54 have?
3. How many regions do each of the graphs in Figure 12.56 have?
4. Draw a connected planar graph with five vertices.
5. Draw a connected planar graph with six vertices.
6. Is there a connected planar graph with n vertices for any natural number n?
7. Is there a nonplanar graph with n vertices for any natural number $n \geq 5$?
8. Prove that any graph with four or fewer vertices is planar.
9. If, instead of Figure 12.52, one of the cabins had a path to only two of the wells and the other two had a path to all three wells, would the graph be planar? Can you draw it?
10. Does Theorem 12.9 hold if $p < 3$?
11. Using Kuratowski's theorem, show that K_n is nonplanar for $n \geq 5$.
12. Show that K_n is nonplanar for $n \geq 5$ using Corollaries 12.2 and 12.4.
*13. Give an example of a planar graph that contains no vertex of degree less than 5.

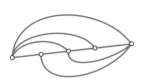

Figure 12.58

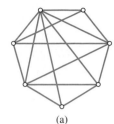
(a)

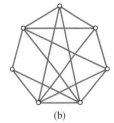
(b)

Figure 12.59

14. Draw the graph in Figure 12.58 in the plane so that each edge is a straight-line segment. Is this graph triangulated?

15. Show that the graph in Figure 12.59a is planar.

16. Show that the graph in Figure 12.59b is nonplanar.

17. If a connected planar graph has the same number of vertices as edges, how many regions does it have?

18. Is there a connected planar graph with 8 vertices, 6 regions, and 14 edges?

19. A certain connected planar graph has five regions and six vertices. How many edges does it have? Sketch such a graph.

20. Prove that if a connected planar graph has an even number of edges and regions, it must have an even number of vertices.

21. Give an example of a triangulated planar graph that is not complete.

22. Give an example of a planar graph for which Euler's formula does not hold.

23. Test Euler's formula to see if it holds for the graphs in Figures 12.53 and 12.55.

24. Test Euler's formula to see if it holds for the graph in Figure 12.5.

25. Give an example of a connected planar graph for which $p \geq 3$ but $q \neq 3p - 6$.

26. Prove that a triangulated graph is connected.

27. Give an example of a (p, q) graph with $p \geq 3$ for which the conclusion of Corollary 12.2 does not hold.

28. Prove that a subdivision of K_5 or $K(3, 3)$ is nonplanar.

29. Give an example of a graph all of whose vertices have degree greater than 5.

(Optional Exercises)

In Exercises 30–32 we shall prove Euler's formula.

Let G be a connected planar graph with p vertices, q edges, and r regions. We must show that $p - q + r = 2$.

30. If G is a tree, show that the result holds.

31. If G is not a tree, then by Theorem 12.7, G has a spanning tree T. By reversing the process of finding a spanning tree, G can be obtained from T by adding edges one by one that complete certain circuits. If one edge is added to complete one additional circuit, show that q and r both increase by 1.

32. Use Exercises 30 and 31 to complete the proof.

*33. Prove that if a planar graph has two components, p vertices, q edges, and r regions, then $p - q + r = 3$. State and prove the corresponding formula for the case of n components.

*34. Which of the results of this section hold for multigraphs? For those that do not hold, give a counterexample.

Photo 12.2
The pyramids at Giza are examples of polyhedra.

12.7 Platonic Solids

Regular polyhedra have an absolutely non-human character. They are not inventions of the human mind, for they existed as crystals in the earth's crust long before mankind appeared on the scene.
— M. C. ESCHER, Twentieth-Century Dutch Artist

The ancient Greeks made a very careful study of the platonic solids, or the **regular polyhedra**, which will be precisely defined shortly. The platonic solids were an important part of the Greek scheme of the universe. Plato believed that all matter in the universe was composed of mixtures of four basic elements. He believed that each of these four elements had the shape of one of the five platonic solids and that the universe itself had the shape of the fifth. Thus the Greek scheme was based upon the fact that there are only five regular polyhedra. This startling fact has amazed people for thousands of years. Why are there only five platonic solids? Read on and you will see.

Polyhedra

There are important applications of planar graphs to the study of polyhedra. Recall that a **polygon** is a closed plane figure bounded by a finite number of straight-line segments. Some of the most common polygons are given in Figure 12.60.

A **polyhedron** is a three-dimensional solid bounded by finitely many polygons.

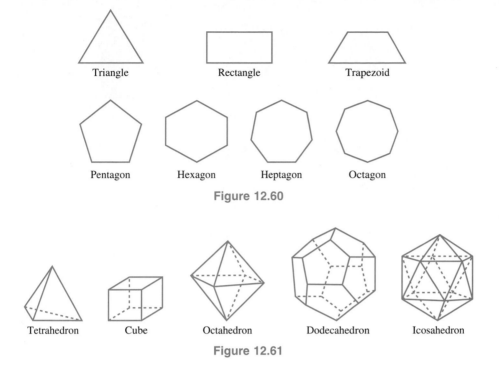

Figure 12.60

Figure 12.61

The most common polyhedra are illustrated in Figure 12.61.

A polyhedron P is **convex** if for any two points $x, y \in P$ the line segment from x to y is contained in P.

A convex polyhedron is intuitively one with no indentations, holes, or cavities. The polyhedra of Figure 12.61 are all convex. In Figure 12.62 are examples of polyhedra that are not convex.

Denote the number of vertices, edges, and faces of a polyhedron by V, E, and F, respectively. Find these numbers for the polyhedra of Figure 12.61 and see if they agree with Table 12.2.

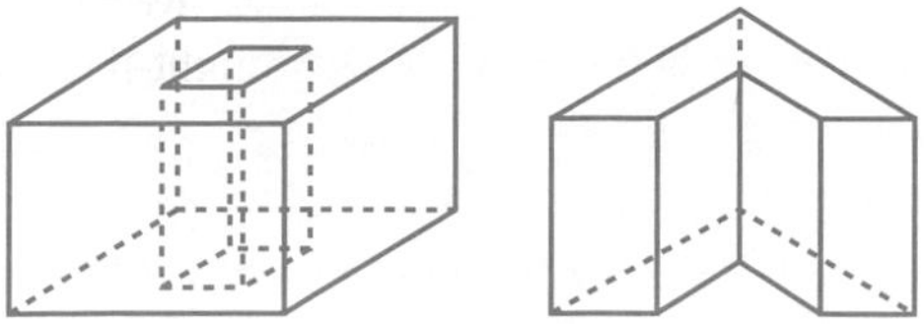

Figure 12.62

Table 12.2

Polyhedron	V	E	F
Tetrahedron	4	6	4
Cube	8	12	6
Octahedron	6	12	8
Dodecahedron	20	30	12
Icosahedron	12	30	20

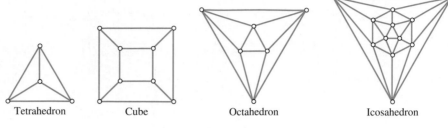

| Tetrahedron | Cube | Octahedron | Icosahedron |

Figure 12.63

Compute V, E, and F for some other convex polyhedra and see what you get. Do you see a formula relating these three numbers? You will soon find that $V - E + F = 2$. It is no coincidence that this formula is the same as Euler's formula for connected planar graphs. This is because every convex polyhedron P is associated with a connected planar graph $G(P)$ that has the same number of edges and vertices as P and for which the number of regions equals the number of faces of P. Intuitively, $G(P)$ is found by removing one of the faces of P and then "squashing" P flat like a tin can. The graph corresponding to the dodecahedron is given in Figure 12.5; the graphs corresponding to the other polyhedra in Figure 12.61 are given in Figure 12.63. In this way the formula $V - E + F = 2$ follows from Euler's formula. It is also concluded from Corollary 12.4 that every convex polyhedron has a vertex of degree 3, 4, or 5.

If P is a polyhedron, denote the number of vertices of degree n by V_n and the number of faces bounded by n edges by F_n. For example, for the tetrahedron,

$$V_3 = 4, \quad V_4 = V_5 = \cdots = 0$$
$$F_3 = 4, \quad F_4 = F_5 = \cdots = 0$$

For the cube,
$$V_3 = 8, \quad V_4 = V_5 = \cdots = 0$$
$$F_3 = 0, \quad F_4 = 6, \quad F_5 = F_6 = \cdots = 0$$

In order for P to be a three-dimensional solid, $V_n = F_n = 0$ for $n \le 2$. Just as for graphs, the sum of the degrees of the vertices is $2E$. In a similar way, the sum of the number of edges bounding every face counts each edge twice, and so again $2E$ is obtained. Hence

$$2E = 3V_3 + 4V_4 + \cdots = 3F_3 + 4F_4 + \cdots \tag{12.1}$$

For example, for the tetrahedron, $3V_3 = 12$ and $3F_3 = 12$, which is twice the number of edges. Also, for the cube, $3V_3 = 24$ and $4F_4 = 24$ is twice the number of edges. These formulas will be useful in proving the following theorem.

Theorem 12.10   At least one face of every convex polyhedron is a triangle, quadrilateral, or pentagon.

Proof: If $F_3 = F_4 = F_5 = 0$, then $2E = 6F_6 + 7F_7 + \cdots \geq 6(F_6 + F_7 + \cdots) = 6F$ and so $E \geq 3F$. Also, $2E = 3V_3 + 4V_4 + \cdots \geq 3(V_3 + V_4 + \cdots) = 3V$. But then Euler's formula gives

$$E = V + F - 2 \leq \frac{2}{3} E + \frac{1}{3} E - 2 = E - 2$$

which is a contradiction.   □

Regular Polyhedra

> A polygon is **regular** if the lengths of its sides are equal and its interior angles are all equal.

For example, equilateral triangles and squares are regular polygons. Two polygons are **congruent** if by a rotation or a translation in space (or both) one can be made coincident with the other (see Figure 12.64).

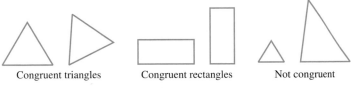

Congruent triangles   Congruent rectangles   Not congruent

Figure 12.64

> A **regular** polyhedron is a polyhedron whose faces are congruent regular polygons and each vertex of which is incident with the same number of edges.

The second requirement does not follow from the first, as can be seen by gluing two tetrahedra along a common face. Five examples of regular polyhedra are the tetrahedron, the cube, the octahedron, the dodecahedron, and the icosahedron. For a regular polyhedron note that $V = V_k$ for some $k \geq 3$ and $F = F_h$ for some $h \geq 3$. For example, for the tetrahedron, the vertices are all of degree 3 (three edges are incident with each vertex) and each face is bounded by three edges, so $V = V_3$, $F = F_3$, and for the cube, $V = V_3$, $F = F_4$.

These five regular polyhedra have a great amount of symmetry. They were known to the ancient Greeks. They were studied by Pythagoras and his school and were important in the philosophy of Plato. Plato claimed that the atoms of the elements had these shapes. The Greeks believed there were four elements (fire, earth, air, and water) and that all matter was constructed from them. Plato asserted that the atoms of fire had the shape of tetrahedra; atoms of earth, the shape of cubes; atoms of air, octahedra; and atoms of water, icosahedra. The universe itself had the shape of a dodecahedron. Although chemists have long known that atoms do not have the shape of regular polyhedra, many crystals do. For example, sodium chlorate crystals are

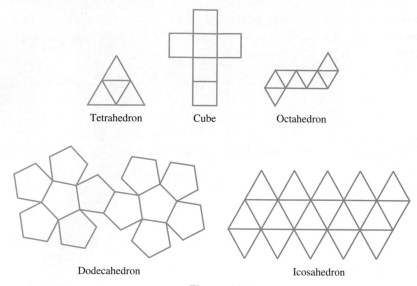

Figure 12.65

frequently cubes or tetrahedra, and chrome alum crystals are octahedra. Although the ancients were familiar with the five convex regular polyhedra, they could not find any more and eventually proved that no others exist. This beautiful and amazing result was proved later by Euler using graph theory. (The proof is worked out in Exercises 15 through 19.)

Theorem 12.11 There are exactly five convex regular polyhedra.

EXERCISES

1. Why are the polyhedra in Figure 12.62 not convex?

2. Draw a polygon that is not convex.

3. If two polygons are congruent, do they have the same number of sides? The same area?

4. If two polygons are congruent, must they be regular?

5. Does Euler's formula hold for the polyhedra in Figure 12.62?

6. Compute V, E, and F for five polyhedra that have not been considered in the text. Check to see if Euler's formula holds.

7. (a) Draw a polyhedron that is not convex for which Euler's formula does hold.
   (b) Draw a polyhedron that is not convex for which Euler's formula does not hold.

*8. Make a model of each of the five regular polyhedra from paper or light cardboard (see Figure 12.65).

9. Show that a polygon need not be regular if the lengths of its sides are equal. Show that a polygon need not be regular if all its interior angles are equal.

10. For each of the graphs of the five regular polyhedra, find a circuit that includes every vertex (such circuits are called **Hamiltonian** circuits).

11. Prove that a convex polyhedron cannot have the same number of vertices as it has edges.

12. Is there a convex polyhedron with 8 vertices, 6 faces, and 14 edges?

13. A certain convex polyhedron has five faces and six vertices. How many edges does it have? Sketch such a polyhedron.

14. Prove that if a convex polyhedron has an even number of edges and faces, it must have an even number of vertices.

(Optional Exercises)

In Exercises 15–19 we shall prove Theorem 12.11.

Let P be a convex regular polyhedron. As previously noted, $F = F_h$ and $V = V_k$ for some $h \geq 3$ and $k \geq 3$.

*15. Use equation (12.1) to show that $2E = kV_k = hF_h$.

*16. Use Euler's formula to show that $(4 - h)F_h + (4 - k)V_k = 8$.

*17. Apply Theorem 12.10 and Corollary 12.4 to show that $3 \leq h \leq 5$ and $3 \leq k \leq 5$. Show that this gives nine possible cases for h and k.

*18. Prove the following.
   (a) If $h = 3$ and $k = 3$, P is a tetrahedron.
   (b) If $h = 3$ and $k = 4$, P is an octahedron.
   (c) If $h = 3$ and $k = 5$, P is an icosahedron.
   (d) If $h = 4$ and $k = 3$, P is a cube.
   (e) If $h = 5$ and $k = 3$, P is a dodecahedron.

*19. Prove that the following cases are impossible.
   (a) $h = 4$ and $k = 4$ (b) $h = 4$ and $k = 5$
   (c) $h = 5$ and $k = 4$ (d) $h = 5$ and $k = 5$

Chapter 12 Summary of Terms

complement $\overline{G}$ of a graph G

adjacent vertices and edges

incident edge

subgraph of a graph

degree of a vertex

degree formula $\deg v_1 + \deg v_2 + \cdots + \deg v_p = 2q$

Instant Insanity

circuits and Eulerian circuits

trees

tree formula   For a (p, q) tree, $q = p - 1$

spanning tree

planar graphs and regions

Euler's formula for connected planar graphs $p - q + r = 2$

regular polyhedra and platonic solids

Chapter 12 Test

1. Draw a $(6, 4)$ graph. How many edges does the complement of this graph have?

2. (a) Draw two different regular graphs with six vertices in which each vertex has degree 2.
   (b) Draw a regular graph with six vertices in which each vertex has degree 3.

3. Let G be a regular graph with p vertices in which each vertex has degree d. Show that either p or d is even. How many edges does G have?

4. (a) Is there a graph with 5 vertices and 11 edges? Why?
   (b) Is there a multigraph with 5 vertices and 11 edges? Why?

5. If a tree has 10 edges, how many vertices does it have?

6. Diagram the following multigraph

$$V = \{a, b, c, d, e\}$$

$$E = \{\{a\}, \{a\}, \{a, b\}, \{b, c\}, \{b, c\}, \{a, c\}, \{b, d\}, \{c\}, \{a, d\}, \{c, d\}\}$$

Give the degree of each vertex. Which vertices are odd, even, and isolated?

7. Solve the following *Instant Insanity* problem.

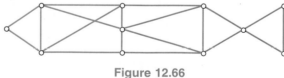

8. Let T be a tree with p vertices. If v is an end vertex, prove that $T - v$ is a tree with $p - 1$ vertices. How many edges does $T - v$ have?

9. If a connected planar graph has one region, show that it is a tree.

10. If a connected planar (p, q) graph has an even number of regions, show that p and q are both even or both odd.

11. Draw K_3 and all of its subgraphs.

12. What is the maximal number of edges for graphs with five vertices?

13. Are the following statements true or false?
    (a) A multigraph with a loop is not a graph.
    (b) The graph K_8 has 28 edges.
    (c) A (p, q) graph has p edges and q vertices.
    (d) A subgraph of G can have more edges than G.
    (e) The complement of G has the same number of vertices as G.
    (f) The complement of G has the same number of edges as G.
    (g) Every multigraph has an odd number of even vertices.

14. For a connected (p, q) multigraph, how many edges are there for a spanning tree?

15. Find a spanning tree for the highway system shown in Figure 12.66.

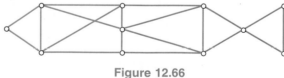

Figure 12.66

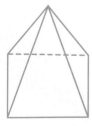

Figure 12.67

16. Figure 12.67 illustrates a pyramid. Find the number of vertices, edges, and faces of this solid and verify Euler's formula. Is this pyramid a convex regular polyhedron?

EPILOGUE

Does the pursuit of truth give you as much pleasure as before? Surely it is not the knowing but the learning, not the possessing but the acquiring, not the being-there but the getting-there, that afford the greatest satisfaction. If I have clarified and exhausted something, I leave it in order to go again into the dark. Thus is that insatiable man so strange: when he has completed a structure it is not in order to dwell in it comfortably, but to start another.
— KARL FRIEDRICH GAUSS (1777–1855), German Mathematician

This mathematical journey is now completed. As with most journeys, it is sad that it is over, but it is good to be home. You can now savor your memories and recall your experiences at your leisure. In the future you will have flashbacks of your experiences on the trip and the knowledge you have gained will add to your perspective. You may be glad the journey is over and never plan to take another, or you may be weary but happy that you went, or you may have come home enthusiastic and want to set off immediately on another adventure. Whatever your feelings, you must admit that your mathematical journey has helped you to grow and mature a little. It is now a little bit easier to face the world, to make decisions.

Now that the journey is over, what have you come away with, what have you learned? You need not concern yourself with details now. What are your overall impressions? Although it still may be difficult to define precisely what mathematics is, you should now have some general ideas as to what mathematics is about, what some of its properties are, what its concerns are, and how it can be applied to problems of the real world.

You have seen in every chapter of this book that mathematics is not only useful but also has an existence in its own right. Mathematics is not, as is commonly thought, computation and manipulation of numbers; mathematics is concerned with pattern and structure. These patterns and structures are discovered by inductive reasoning, and results describing their properties are proved using deductive reasoning. Two concepts that have appeared in every chapter of this book are sets and functions. These concepts are important tools that aid in investigating pattern and structure. Think back over the chapters you have studied and see if you can illustrate these statements with specific examples. Are there other general statements you can make about mathematics?

It is hoped that this will not be your last journey into the mathematical realm. If you have seen some of the enjoyment, excitement, and challenge of mathematics, then your trip has been a success. If this is the case, then you have just begun, and your mental travels within this intellectual realm will take you on many more journeys throughout your life.

ANNOTATED BIBLIOGRAPHY

CHAPTER 1 Orientation

There are many good books containing mathematical games and puzzles.

Franco Agostini, *Math and Logic Games*, Facts on File, New York, 1983.

Walter W. Ball and H. S. Coxeter, *Mathematical Recreations and Essays*, University of Toronto Press, Toronto, 1974.

Martin Gardner, *Wheels, Life and Other Mathematical Amusements*, W. H. Freeman, New York, 1983.

Martin Gardner, *Scientific American Book of Mathematical Puzzles and Diversions*, Simon and Schuster, New York, 1963, 1965.

Boris Kordemsky, *The Moscow Puzzles*, Charles Scribner, New York, 1972.

Theoni Pappas, *The Joy of Mathematics*, Wide World Publishing, San Carlos, California, 1989.

Hugo Steinhaus, *Mathematical Snapshots*, Oxford University Press, New York, 1983.

The following books contain further reading on sets and functions.

P. R. Halmos, *Naive Set Theory*, Springer-Verlag, New York, 1974.

Ernst Sandheimer and Allan Rogerson, *Numbers and Infinity: A Historical Account of Mathematical Concepts*, Cambridge University Press, New York, 1981.

Robert R. Stoll, *Set Theory and Logic*, Dover, New York, 1979.

CHAPTER 2 Mathematics and Art

Books and articles on mathematics and art:

William Camfield, "Juan Gris and the *Golden Section*," *Art Bulletin* **47**, 128–134 (1965).

Lynda Dalrymple, "A new facet of Cubism: The 'fourth dimension' and 'non-Euclidean' geometry reinterpreted," *Art Quarterly* **34**, 411–433 (1971).

Michael Holt, *Mathematics in Art*, Van Nostrand Reinhold, New York, 1971.

P. Lodopoulas, "Fine arts and geometry," *Journal of Aesthetics and Art Criticism* **28**, 535–540 (1970).

Kenneth May, "Mathematics and art," *The Mathematics Teacher*, October 1967, pp. 568–572.

Joseph Schillinger, *The Mathematical Basis of the Arts*, Philosophical Library, New York, 1948.

James Shaw, "Mathematics as a fine art," *The Mathematics Teacher*, November 1967, pp. 738–747.

A periodical devoted entirely to interactions between art and science:

Leonardo, International Journal of the Contemporary Artist, Pergamon Press, Oxford, quarterly.

Books on art that contain mathematical ideas:

Faber Birren, *Color, Form and Space*, Reinhold, New York, 1961.

Edmund Feldman, *Varieties of Visual Experience*, Prentice-Hall, Englewood Cliffs, New Jersey, 1972.

Rene Parola, *Optical Art: Theory and Practice*, Reinhold, New York, 1969.

Books containing computer art:

Clifford Pickover, *Computers, Pattern, Chaos, and Beauty*, St. Martin's Press, New York, 1990.

Jasia Reichardt, *Cybernetics, Art and Ideas*, New York Graphic Society, Greenwich, Connecticut, 1971.

Jasia Reichardt, *The Computer in Art*, Van Nostrand Reinhold, New York, 1971.

Books on Gestalt psychology:

Kurt Koffka, *Principles of Gestalt Psychology*, Harcourt, Brace, New York, 1935.

Wolfgang Köhler, *Gestalt Psychology*, Liveright, New York, 1947.

Books and articles on symmetry and proportion:

Donald Crowe, "The geometry of African art; I: Bakuba art," *Journal of Geometry* **12**, 169–182 (1971).

H. E. Huntley, *The Divine Proportion: A Study in Mathematical Beauty*, Dover, New York, 1970.

Caroline Macgillavry, *Symmetry Aspects of M. C. Escher's Periodic Drawings*, A. Oostolk's Uitgeversmaatschappy NV, Utrecht, 1965.

Arthur Razzel and K. Watts, *Symmetry*, Doubleday, Garden City, New York, 1968.

Doris Schattschneider, *Visions of Symmetry: Notebooks, Periodic Drawings, and Related Work of M. C. Escher*, W. H. Freeman, New York, 1990.

Hermann Weyl, *Symmetry*, Princeton University Press, Princeton, 1952.

M. C. Escher and tessellations:

M. C. Escher, *The Graphic Work of M. C. Escher*, trans. John Brigham, Ballantine Books, New York, 1967.

M. C. Escher, *The Graphic Work of M. C. Escher*, Hawthorn, New York, 1970.

J. L. Locher, ed., *The World of M. C. Escher*, Harry N. Abrams, New York, 1971.

Evan Maletsky, "Designs with tessellations," *The Mathematics Teacher*, April 1974, pp. 335–360.

Ernest Ranucci, "Master of tessellations: M. C. Escher, 1898–1972," *The Mathematics Teacher*, April 1974, pp. 299–306.

Joseph Teeters, "How to draw tessellations of the Escher type," *The Mathematics Teacher*, April 1974, pp. 307–310.

Books on art and illusions:

E. H. Gombrich, *Art and Illusion*, Pantheon Books, London, 1968.

M. Luckiesh, *Visual Illusions*, Dover, New York, 1965.

Residue designs, Poinsot stars, and spirolaterals:

Phillip Davis and William Chinn, *3.1416 and All That*, Simon and Schuster, New York, 1969.

Sonia Forseth and Andria Troutman, "Using mathematical structures to generate artistic designs," *The Mathematics Teacher*, May 1974, pp. 393–398.

Phil Locke, "Residue designs," *The Mathematics Teacher*, March 1972, pp. 260–263.

Books on artists considered in this chapter:

Alberto Busignani, *Mondrian*, Grosset and Dunlap, New York, 1968.

David Larkin, ed., *Magritte*, Ballantine Books, New York, 1972.

Maurice de Sauismarez, *Bridget Riley*, New York Graphic Society, New York, 1970.

Victor Vasarely, *Victor Vasarely*, New York Graphic Society Ltd., Greenwich, Connecticut, 1973.

Some journals on computer graphics:

Computer Graphics Forum, European Association for Computer Graphics, Oxford, England.
Computer Graphics World, PennWell, Westford, Massachusetts.
Computers and Graphics, Pergamon Press, Tarrytown, New York.
Visual Computer, International Computer Graphics Society, Springer-Verlag, Berlin.

CHAPTER 3 Logic

Elementary books on logic:

Lewis Carroll, *Symbolic Logic and the Game of Logic*, Dover, New York, 1973.
J. N. Crossley, *What is Mathematical Logic?*, Oxford University Press, New York, 1972.
Martin Gardner, *Logic Machines and Diagrams*, Chicago University Press, Chicago, 1982.

More advanced books on logic:

George Boole, *The Mathematical Analysis of Logic*, Barnes and Noble, New York, 1965.
Elliott Mendelson, *Boolean Algebra and Switching Circuits,* Schaum's Outline Series, McGraw-Hill, New York, 1970.
W. V. Quine, *Methods of Logic*, Harvard University Press, Cambridge, Massachusetts, 1982.

CHAPTER 4 Number Theory

A delightful little book about numbers:

C. Reid, *From Zero to Infinity*, Thomas Crowell, New York, 1964.

Some elementary books on number theory:

Albert Beiler, *Recreations in the Theory of Numbers*, Dover, New York, 1964.
Tobias Dantzig, *Number: The Language of Science*, Free Press, New York, 1967.
J. Maxfield and M. Maxfield, *Discovering Number Theory*, W. B. Saunders, Philadelphia, 1972.
Oystein Ore, *Number Theory and Its History*, McGraw-Hill, New York, 1948.

Two texts with interesting chapters on number theory:

A. Beck, M. D. Bleicher, and D. Crowe, *Excursions into Mathematics*, Worth, New York, 1969.
S. Stein, *Mathematics: The Man-Made Universe*, W. H. Freeman, San Francisco, 1969.

CHAPTER 5 Geometry

Elementary and easy reading:

Edwin Abbott, *Flatland, A Romance of Many Dimensions*, Barnes and Noble, New York, 1963.
Dionys Burger, *Sphereland*, Thomas Crowell, New York, 1965.
Marvin Greenberg, *Euclidean and Non-Euclidean Geometries: Development and History*, W. H. Freeman, San Francisco, 1980.

More specialized and a little more difficult:

Leonard Blumenthal, *A Modern View of Geometry*, W. H. Freeman, San Francisco, 1961.
Thomas Heath, *A Manual of Greek Mathematics*, Dover, New York, 1963.

Some interesting books on fractals:

M. F. Barnsley, *Fractals Everywhere*, Academic Press, Boston, 1988.
M. F. Barnsley et al., *The Science of Fractal Images*, Springer-Verlag, New York, 1988.
Hans Lauwerier, *Fractals*, Princeton University Press, Princeton, 1991.
B. Mandelbrot, *The Fractal Geometry of Nature*, W. H. Freeman, San Francisco, 1983.

CHAPTER 6 Consumer Mathematics

Elementary books with chapters on consumer mathematics and finance:

David Johnson and Thomas Mowry, *Mathematics: A Practical Odyssey*, Wadsworth, Belmont, California, 1992.

Charles Miller, Vern Heeren, and John Hornsby, *Mathematical Ideas*, Scott, Foresman, Glenville, Illinois, 1990.

Karl Smith, *The Nature of Mathematics*, Brooks/Cole, Pacific Grove, California, 1991.

A more advanced book on mathematical economics:

Ann Hughes, *Applied Mathematics: For Business, Economics, and the Social Sciences*, Richard D. Irwin, Homewood, Illinois, 1983.

CHAPTER 7 Computer Science

Computer history and how computers work:

Herman Goldstine, *The Computer from Pascal to von Neumann*, Princeton University Press, Princeton, 1972.

Peter Laurie, *The Joy of Computers*, Little, Brown, Boston, 1983.

Readings from Scientific American: Computers and Computation, W. H. Freeman, San Francisco, 1971.

Two books on BASIC with a minimum of mathematics:

Robert Albrecht, LeRoy Finkel, and Jerald Brown, *BASIC*, John Wiley, New York, 1984.

*My Computer Likes Me * when i speak in BASIC*, Dymax, Menlo Park, California, 1972.

Some books on BASIC requiring more mathematics:

James Coan, *Basic BASIC*, Hayden, New York, 1986.

John Kemeny and Thomas Kurtz, *BASIC Programming,* 4th ed., John Wiley, New York, 1988.

John Skelton, *An Introduction to the BASIC Language*, Holt, Rinehart and Winston, New York, 1983.

CHAPTER 8 Statistics

Elementary statistics texts:

Marek Fisz, *Probability Theory and Mathematical Statistics*, Krieger, Melbourne, Florida, 1980.

David Freedman, Robert Pisani, and Roger Purves, *Statistics*, Norton, New York, 1978.

Books on statistical sampling:

Leslie Kish, *Survey Sampling*, John Wiley, New York, 1965.

F. F. Stephan and P. J. McCarthy, *Sampling Opinions*, John Wiley, New York, 1958.

Books on correlation analysis:

C. Daniel and F. S. Wood, *Fitting Equations to Data*, John Wiley, New York, 1971.

N. R. Draper and H. Smith, *Applied Regression Analysis*, John Wiley, New York, 1966.

CHAPTER 9 Probability Theory

Standard books on probability theory, at a little higher level than the present text:

William Feller, *An Introduction to Probability Theory and Its Applications,* Vol. 1, John Wiley, New York, 1968.

Seymour Lipschutz, *Probability,* Schaum's Outline Series, McGraw-Hill, New York, 1968.

An easy-to-read account of probability theory:

David Bergamini and the Editors of *Life*, *Mathematics*, Life Science Library, Time, Inc., New York, 1963.

Popularized articles on probability theory:

A. J. Ayers, "Chance," *Scientific American* **213**, 44–54 (1965).

Mark Kac, "Probability," *Scientific American* **211**, 92–108 (1965).

Warren Weaver, "Probability," *Scientific American* **183**, 44–47 (1950).

Interesting problems in probability theory:

Frederick Mosteller, *Fifty Challenging Problems in Probability with Solutions*, Addison-Wesley, Reading, Massachusetts, 1965.

CHAPTER 10 Gamble if You Must

Many of the ideas in Section 10.2 can be found in the following article, where a different approach is taken:

Andrew Sterrett, "Gambling doesn't pay!," *The Mathematics Teacher*, March 1967, pp. 210–214.

An interesting account of a gambling mathematician:

O. Ore, *Cardano: The Gambling Scholar*, Princeton University Press, Princeton, 1953.

Books on the mathematics of gambling:

L. Dubins and L. Savage, *How to Gamble if You Must*, McGraw-Hill, New York, 1965.

W. Weaver, *Lady Luck*, Anchor Books, Garden City, New York, 1963.

Interesting nonmathematical books on gambling:

Major Ridde, *The Weekend Gambler's Handbook*, Random House, New York, 1963.

J. D. Williams, *The Compleat Strategyst*, McGraw-Hill, New York, 1954.

CHAPTER 11 Matrices

Applications of matrices:

Ann Hughes, *Applied Mathematics: For Business, Economics, and the Social Sciences*, Richard D. Irwin, Homewood, Illinois, 1983.

Matrices are an important part of linear algebra. Some standard texts on linear algebra:

Stephen Friedberg, Arnold Insel, and Lawrence Spence, *Linear Algebra*, Prentice-Hall, Englewood Cliffs, New Jersey, 1989.

Bill Jacob, *Linear Algebra*, W. H. Freeman, New York, 1990.

Dennis Schneider, *Linear Algebra*, Macmillan, New York, 1987.

CHAPTER 12 Graph Theory

An excellent book on graph theory on the level of the present text:

Oystein Ore, *Graphs and Their Uses*, L. W. Singer, New York, 1963.

Textbooks on graph theory on a higher level than the present text:

M. Behzad and G. Chartrand, *Introduction to the Theory of Graphs*, Allyn and Bacon, Boston, 1971.

F. Harary, *Graph Theory*, Addison-Wesley, Reading, Massachusetts, 1969.

R. Wilson, *Introduction to Graph Theory*, Academic Press, New York, 1972.

A collection of articles on graph theory, some of which are at the research level:

F. Harary, ed., *The Many Facets of Graph Theory*, Springer-Verlag, New York, 1969.

ANSWERS TO SELECTED EXERCISES

CHAPTER 1

Section 1.1

1. Divide S into sixteen 1-inch squares showing that S has area 16 sq in.

3. Divide S into sixteen 1-inch squares, eight 1 by 1/2-inch rectangles, and one 1/2-inch square. Put the rectangles together to get 1 square inch for each pair of rectangles and use Exercise 2 to get an area of 1/4 square inch for the 1/2-inch square. This gives a total of $20\frac{1}{4}$ square inches. **5.** The area is doubled.

7. In this case the curve starting from the lower left in Figure 1.1 is steadily rising, and the curve starting from the upper left is steadily falling. Hence these two curves must meet at precisely one point.

9. $n(n+1)/2$ **11.** $(3^{11} - 1)/2$ **13.** 24; 720; 100; 380 **15.** Figure A1.1

17. The tie costs $2 and the shirt costs $7.50. **19.** $3/2$ **21.** 43 **23.** TET

27. In this case it does not make any difference whether you switch or not. You have only one chance out of three of winning. **29.** Seven hamburgers, 33 hotdogs, and 60 mints. Yes.

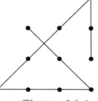

Figure A1.1

Section 1.2

1. 8, 9, 16, 125. **3.** $a^4 = a \times a \times a \times a$. Also, $a^2 \times a^2 = a \times a \times a \times a$.

5. $a^{n+m} = a \times a \times \cdots \times a$ ($n + m$ times). Also, $a^n a^m = a \times a \times \cdots \times a$ ($n + m$ times).

7. Letting $m = -n$ in the formula $a^{n+m} = a^n a^m$ gives $a^0 = a^n a^{-n}$. Since $a^0 = 1$ it follows that $a^{-n} = 1/a^n$.

11. Let a, b, c, d be natural numbers and suppose $ab \le cd$. Dividing both sides by bd gives $a/d \le c/b$. **13.** 800 miles **23.** There are three girls and four boys in the family.

29. See Section 12.1. **33.** 0

Section 1.3

1. $A \cup B = \{1, 2, 3, 4\}$, $A \cap B = \{2, 3\}$

3. (a) $\{2, 4, 6, 8\}$ (b) $\{1, 3, 5, 7\}$ (c) $\{1, 2, 7, 8\}$ (d) $\{2, 3, 4, 5, 6, 7, 8\}$ (e) $\emptyset$ (f) U

5. Figure A1.2 **7.** $(A - B) \cup (B - A)$ **9.** $A \cap B' \cap C$ **11.** $A \cap B \cap C$

13. If $a \in A \cup A$, then $a \in A$. If $a \in A$, then $a \in A \cup B$. The other statement is similar.

15. Both $A \cup B$ and $B \cup A$ consist of the members of A or B. Both $A \cap B$ and $B \cap A$ consist of the members of A and B.

17. If $A \subseteq B'$, then $A \cap B \subseteq B' \cap B = \emptyset$ so $A \cap B = \emptyset$. Assume $A \cap B = \emptyset$. If $a \in A$, then $a \notin B$ so $a \in B'$. Hence $A \subseteq B'$.

19. Using Figure A1.3, $(A \cap B) \cup C = (2 \cup 3) \cup (3 \cup 4 \cup 6 \cup 7) = 2 \cup 3 \cup 4 \cup 6 \cup 7$ and $(A \cup C) \cap (B \cup C) = (1 \cup 2 \cup 3 \cup 4 \cup 6 \cup 7) \cap (2 \cup 3 \cup 5 \cup 6 \cup 4 \cup 7) = 2 \cup 3 \cup 4 \cup 6 \cup 7$.

21. Using Figure A1.3, $(A \cup B) \cup C$ and $A \cup (B \cup C)$ both equal $1 \cup 2 \cup 3 \cup 4 \cup 5 \cup 6 \cup 7$ while $(A \cap B) \cap C$ and $A \cap (B \cap C)$ both equal 3.

23. (a) If $a \in A \cap B$, then $a \in A$ and hence $A \cap B \subseteq A$. The other parts are similar.

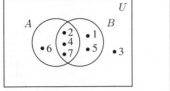

Figure A1.2

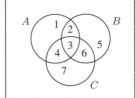

Figure A1.3

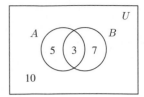

Figure A1.4

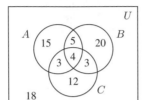

Figure A1.5

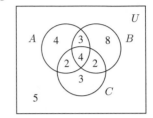

Figure A1.6

25. By Exercise 23, $A \cap B$ is contained in both A and B. If C is a subset of both A and B, then $C \subseteq A \cap B$ and hence $A \cap B$ is the largest subset of both A and B. The other part is similar.

27. In Figure 1.15b, $A - B = 1$ and $A - A \cap B = 1$. Also, $A \cup B = 1 \cup 2 \cup 3$ while $A - B = 1$, $A \cap B = 2$, and $B - A = 3$. Finally, $A - B = 1$ and $A \cap B = 2$, so $(A - B) \cap (A \cap B) = \emptyset$.

Section 1.4

1. 0   **3.** (a) 8   (b) 53   (c) 22   (d) 16   **5.** $|B| = 5$ **7.** $\emptyset$ **9.** 2   **11.** 12

13. $A = B$ **15.** 10   **17.** (a) 18   (b) 22   (c) 4   (d) 6

19. Figure A1.4   **21.** Figure A1.5   **23.** Figure A1.6

25. Since $(A \cap C) \cap (B \cap C) = A \cap B \cap C$, we have by (1.1) that $|A \cap C| + |B \cap C| = |(A \cap C) \cup (B \cap C)| + |A \cap B \cap C|$. Now applying (1.1) gives $|A| + |B| + |C| = |A \cup B| + |C| + |A \cap B| = |A \cup B \cup C| + |(A \cup B) \cap C| + |A \cap B| = |A \cup B \cup C| + |(A \cap C) \cup (B \cap C)| + |A \cap B| = |A \cup B \cup C| + |A \cap B| + |A \cap C| + |B \cap C| - |A \cap B \cap C|$

Section 1.5

1. $f(1) = f(2) = f(3) = 1$; $g(1) = g(2) = g(3) = 2$; $h(1) = h(2) = h(3) = 3$. There are 24 other functions. Can you describe some of them?

5. The only function $f \colon S \to T$ is the function given by $f(s) = b$ for all $s \in S$.

7. $f(x) = 1$ works for each case. Can you give other examples?

9. Figure A1.7   **11.** (a) 1   (b) 3   (c) 1   (d) 19   (e) -9

13. (a) 3   (b) 2   (c) 1   (d) 1/2   (e) 1/3   **15.** Figure A1.8   **17.** Figure A1.9

19. $-b/a$ **21.** -1 **23.** 0, 1   **25.** (a) $C = 15n + 0.1x$ (b) \$110

27. \$1.65. The graph is similar to Figure 1.30.   **29.** Figure A1.10

31. (a) $(f + g)(x) = 4x + 1$ (b) $(f + g)(x) = 4x^2 + 2x + 3$ (c) $(f + g)(x) = x^3 + x^2 + x + 1$

33. If $f(x) = a_1 x^2 + b_1 x + c_1$ and $g(x) = a_2 x^2 + b_2 x + c_2$ are quadratic functions, then $(f + g)(x) = (a_1 + a_2)x^2 + (b_1 + b_2)x + (c_1 + c_2)$ is a quadratic function if $a_1 \neq -a_2$. Otherwise it is a linear function.

35. If $f(x) = ax + b$ and $g(x) = cx + d$ are linear functions, then $fg(x) = (ac)x^2 + (ad + bc)x + bd$ is a quadratic function if $a \neq 0$, $c \neq 0$. Otherwise, it is a linear function.   **37.** No   **39.** 4

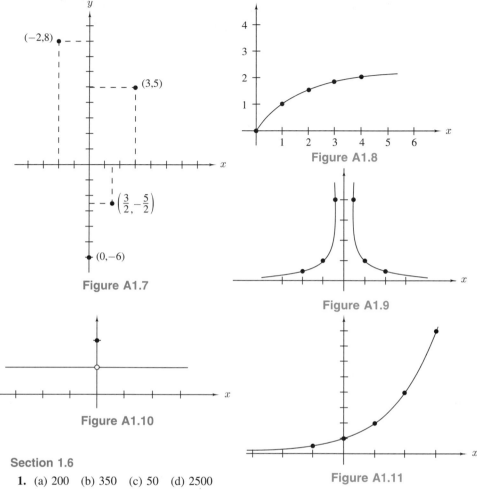

Figure A1.7

Figure A1.8

Figure A1.9

Figure A1.10

Figure A1.11

Section 1.6

1. (a) 200   (b) 350   (c) 50   (d) 2500
3. (a) 31.62   (b) 199.53   (c) 158,489.32   (d) 0.0630957   (e) 0.3162278
5. (a) 3.5   (b) 7   (c) −0.4   **7.** Figure A1.11
9. Many buildings in Cairo were poorly constructed.   **11.** 1.16   **13.** 3.16   **19.** 0.60
21. 12.95 yr   **23.** 4.5   **25.** 2.828   **33.** 13;   17;   yes;   no
35. f is one-to-one, g is not one-to-one. f is onto, g is not onto.   **39.** Yes, it is onto.

Section 1.7

1. 250 billion sq ft   **3.** 53.5 billion people   **5.** $112.36   **7.** 32
9. About 585 billion yr   **11.** $2^{64} - 1$ grains   **13.** 34.31 days   **15.** 68.97
17. 100;   105;   110;   115   **19.** $f_n = 3^{n-1}f_1 + (3^{n-1} - 1)/2$ **21.** 7.27 yr   **23.** $f_n = n + 1$
25. $f_8 = 1280$; $g_8/f_8 \approx 0.35$, $f_9 = 2560$; $g_9/f_9 \approx 0.20$; $f_{10} = 5120$; $g_{10}/f_{10} \approx 0.11$
27. $f_n = a(r^n - 1)/(r - 1)$ **29.** $f_n = n(n + 1)/2$

Section 1.8

1. The supersets of winning coalitions are either themselves or A, so Axiom 1 holds. The complement of a winning coalition is either $\emptyset$ or a one-element set. Since these are not winning coalitions, Axiom 2 holds.

3. Check to see that the supersets of $\{a, b\}$ $\{a, c\}$, $\{b, c\}$ are winning coalitions. The supersets of the three-element winning coalitions are either themselves or A. Hence, Axiom 1 holds. The complements of the three-element winning coalitions are one-element sets and $A' = \emptyset$. Also $\{a, b\}' = \{c, d\}$, $\{a, c\}' = \{b, d\}$, and $\{b, c\}' = \{a, d\}$. Since none of these complements are winning coalitions, Axiom 2 holds.

5. Since A is the only winning coalition, Axiom 1 holds. Since $A' = \emptyset$, Axiom 2 holds.

7. In Example 1, any two winning coalitions have at least two members in common. Hence the intersection of two winning coalitions is nonempty.

11. The minimal winning coalitions are $\{a, b\}$, $\{a, c\}$, and $\{b, c, d\}$, and all winning coalitions are supersets of these.

13. Examples 2, 3, and 5 have no blocking coalitions. The blocking coalitions in Example 4 are $\{c\}$, $\{b\}$, $\{a\}$, $\{a, b\}$, $\{a, c\}$, $\{a, d\}$, and their complements.

15. The dictator in Example 5 is a. The other examples have no dictators.

17. $n/2 + 1$; $(n + 1)/2$

23. If $\{x\}$ is a blocking coalition, then $\{x\}'$ is also a blocking coalition. Suppose there is a winning coalition K that does not contain x. Then $K \subseteq \{x\}'$. But then $\{x\}'$ is a winning coalition, which is a contradiction.

25. If L_1 and L_2 are losing coalitions, then L_1' and L_2' are winning coalitions. By Theorem 1.2, $L_1' \cap L_2' \neq \emptyset$. Hence, $(L_1' \cap L_2')' \neq A$. But $(L_1' \cap L_2')' = L_1 \cup L_2$.

27. Let a, b, c, d have 2, 1, 1, and 1 vote, respectively.

29. No, because the complement of the winning coalition $\{a\}$ is the winning coalition $\{b, c, d\}$, for example.

Chapter 1 Test

1. 120; 90; 9900   **3.** 20,100   **5.** 5/6   **7.** $\{1, 2, 3, 4\}$; $\{4\}$; $\{1, 3\}$; $\{1, 3\}$

9. U; A **11.** Figure A1.12   **13.** 13

15. According to the data $|C \cap A| = 9$ and $|C \cap B \cap A'| = 1$ so $|C| \geq 10$. But $|C| = 9$, which is inconsistent.

17. $\log \dfrac{1}{a} = \log 1 - \log a = -\log a$ **19.** $\dfrac{1}{2}\left(\dfrac{1}{\log 3} - 1\right) = 0.54795$ **21.** 10.409%

23. (a) T   (b) F   (c) T   (d) F   (e) F   (f) T   (g) T   (h) F   (i) T   (j) T   (k) F   (l) T
(m) T   (n) F   (o) T   (p) T   (q) T

CHAPTER 2

Section 2.1

5. 10   **7.** $1 + 2 + \cdots + n = n(n + 1)/2$ **9.** There are n rows and n columns.

11. Figure A2.1   **13.** $5(1 + 2 + \cdots + n) = 5n(n + 1)/2$ **15.** 36   **17.** 16   **19.** $4n(n + 1)$

21. $\sqrt{2}$

Section 2.2

5. 2, 3, 4   **7.** a

9. $1/\left(\sqrt{2} + 1\right) = \left(\sqrt{2} - 1\right) / \left(\sqrt{2} + 1\right)\left(\sqrt{2} - 1\right) = \sqrt{2} - 1$ and
$1/\left(\sqrt{5} + 1\right) = \left(\sqrt{5} - 1\right) / \left(\sqrt{5} + 1\right)\left(\sqrt{5} - 1\right) = \left(\sqrt{5} - 1\right)/4$ **13.** $2(w + \ell)$

Section 2.4

1. If you move the figure a certain distance to the right or left, it will coincide with the original figure.

3. Two types of rotational symmetry and three types of bilateral symmetry.

5. $180°$ rotational symmetry; bilateral symmetry about a vertical and horizontal axis.

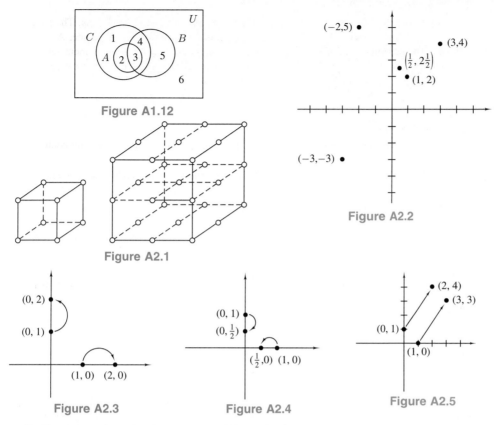

Figure A1.12

Figure A2.2

Figure A2.1

Figure A2.3   Figure A2.4   Figure A2.5

7. Two types of rotational symmetry (120° and 240°).

9. Figure 2.23b has bilateral symmetry about the two diagonals; Figure 2.23c does not.

11. Figure 2.23b

Section 2.5

1. Figure A2.2

3. $(1,2) \rightarrow (2,3/2)$; $(3,4) \rightarrow (6,7/2)$; $(-2,5) \rightarrow (-4,3/2)$; $(-3,-3) \rightarrow (-6,-3)$; $(1/2,5/2) \rightarrow (1,3/2)$

5. The transformations of this section are functions from the plane into the plane.

9. They are inverses of each other.   **11.** An expansion of the square by a factor of 2.

15. Figures A2.3–A2.5 illustrate   (a) the dilation $(x,y) \rightarrow (2x,2y)$, (b) the contraction $(x,y) \rightarrow (x/2,y/2)$,   (c) the translation $(x,y) \rightarrow (x+2,y+3)$.

17. Figure A2.6   **19.** Figure A2.7   **21.** Figure A2.8   **23.** Figure A2.9

25. The half-plane to the right of the y-axis.   **27.** The upper right quadrant.

29. The straight line through the origin at a 45° angle to the x-axis (Figure A2.10).

31. Figure A2.11

Section 2.6

1. The upper figure is maluma, and the lower figure is takete.   **3.** A, C, E

5. Let x and y be points in the intersection. Then x and y are in all the convex sets. Since the sets are convex, the line segment from x to y is in all the convex sets. Hence this segment is in the intersection.

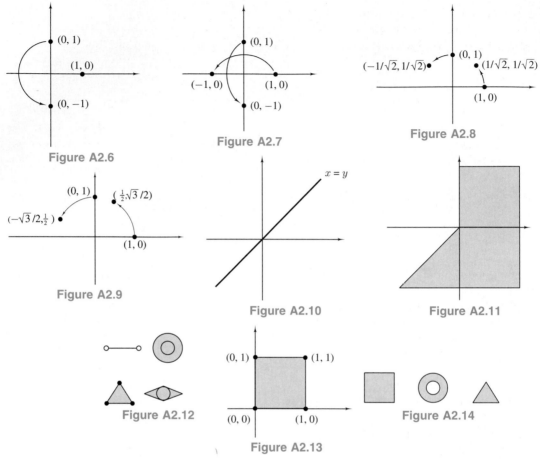

Figure A2.6

Figure A2.7

Figure A2.8

Figure A2.9

Figure A2.10

Figure A2.11

Figure A2.12

Figure A2.13

Figure A2.14

7. Let $x, y \in A_1 \cup A_2 \cup \cdots$. Then $x \in A_m$ and $y \in A_n$ for some m and n. Suppose $A_m \subseteq A_n$. Then $x, y \in A_n$. Hence the line segment from x to y is in A_n. Thus this line segment is in $A_1 \cup A_2 \cup \cdots$. If $A_n \subseteq A_m$, the proof is similar. **9.** Figure A2.12 **11.** Figure A2.13

Section 2.7

1. (a), (b), and (d) are closed. (c) is open. (e) and (f) are neither.

3. Let $A_1, A_2, \ldots, A_n$ be open sets, and let $x \in A_1 \cap A_2 \cap \cdots \cap A_n$. Then there are disks $D_1, D_2, \ldots, D_n$ centered at x such that $D_i \subseteq A_i$, $i = 1, 2, \ldots, n$. One of these disks, say D_j, has minimal radius. Then $D_j \subseteq A_1 \cap A_2 \cap \cdots \cap A_n$. Hence $A_1 \cap A_2 \cap \cdots \cap A_n$ is open.

5. Let A_1 and A_2 be open, and let $x \in A_1 \cup A_2$. Then there is a disk D centered at x such that $D \subseteq A_1$ or $D \subseteq A_2$. But then $D \subseteq A_1 \cup A_2$. Hence $A_1 \cup A_2$ is open.

7. The union of a finite number of closed sets is closed, and the intersection of any collection of closed sets is closed. **9.** Figure A2.14

11. A set S in three-dimensional space is open if for any $x \in S$ there exists a ball B centered at x such that $B \subseteq S$. A set S in one-dimensional space is open if for any $x \in S$ there exists an interval I centered at x such that $I \subseteq S$.

Section 2.8

5. (a) 0 (b) 1 (c) 0 (d) ∞ (infinity)

Section 2.9

1. 11; 12   **3.** 11; 4

5. $12 = 1 + 11 = 2 + 10 = 3 + 9 = 4 + 8 = 5 + 7 = 6 + 6$ and $13 = 1 + 12 = 2 + 11 = 3 + 10 = 4 + 9 = 5 + 8 = 6 + 7$

7. No; yes   **9.** 2, 3, 5, 7, 11, 13, 17, 19, 23, 29

Section 2.10

1. (a) 0 (b) 1 (c) 0 (d) 4   **3.** 11, 18, 25   **5.**

+	0	1	2
0	0	1	2
1	1	2	0
2	2	0	1

·	0	1	2
0	0	0	0
1	0	1	2
2	0	2	1

Chapter 2 Test

3. 28   **5.** 6   **7.** $\sqrt{a^2 + 2ab + b^2} = a + b$ **9.** A plane

11. (a) $(0, 0)$ (b) $(3, -1)$ (c) $(-2, 0)$ (d) $(1, -5)$

13. The points $(1, 1)$ and $(-1, -1)$ are in the set, but the straight line between them contains $(0, 0)$, which is not in the set.

15. 2, 3/2, 4/3, 5/4, 6/5. The sequence converges to 1.   **19.** 3

CHAPTER 3

Section 3.1

1. No   **3.** Yes   **5.** Yes   **7.** Yes   **9.** Conjunction   **11.** Conditional

13. Disjunction   **15.** $2 \geq 3$ **17.** $a \notin A$ **19.** She is tall or she has blue eyes.

21. She is not tall.   **23.** She is tall or she does not have blue eyes.

25. She is not tall and she does not have blue eyes.

27. It is not the case that she is not tall and she has blue eyes.

29. $p =$ His name is George; $q =$ His name is Jack; $p \vee q$

31. $p = x + 2 = 3$; $q = x = 1$; $p \rightarrow q$

33. My age is not less than 65 and my gross income is at least \$6,400.

35. I am the head of a household. Moreover, my age is less than 65 and my gross income is at least \$7150 or my age is not less than 65 and my gross income is at least \$8000.

39. $p \vee (q \wedge r) = (p \vee q) \wedge (p \vee r)$ **41.** $p \vee (q \vee r) = (p \vee q) \vee r$ **43.** $(p \rightarrow q) \rightarrow (\sim q \rightarrow \sim p)$

Section 3.2

1. F   **3.** T   **5.** T   **7.** T   **9.** T   **11.** T   **13.** T   **15.** F   **17.** T   **19.** F

21. FTTT   **23.** FFTF   **25.** TTTTTTFF   **27.** FTFTTTFT

Section 3.3

1. If it rains, then I wash my car.   **3.** If it does not rain, then I do not wash my car.

5. It rains if and only if I wash my car.   **7.** It rains if and only if I do not wash my car.

9. If it rains, then I wash my car or it does not rain.

11. If it rains or I wash my car, then it rains and I wash my car

13. T   **15.** T   **17.** T   **19.** F   **21.** F

23. Converse: If I quit my job, then I win the lottery.
Inverse: If I don't win the lottery, then I won't quit my job.
Contrapositive: If I don't quit my job, then I don't win the lottery.

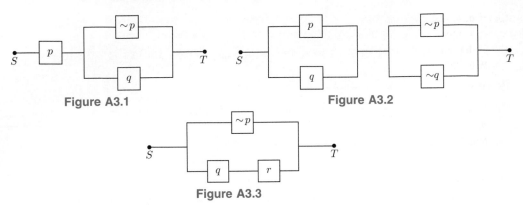

Figure A3.1

Figure A3.2

Figure A3.3

25. Converse: If I will not get the flu, then I get a flu shot.

Inverse: If I don't get a flu shot, then I will get the flu.

Contrapositive: If I will get the flu, then I don't get a flu shot.

27. TTFT **29.** TFTT **31.** TFFT **33.** TTTFTTTF

35. T **37.** T **39.** F **41.** T **43.** F

Section 3.4

1. This bed is not too hard or this bed is not too soft.

3. You had told me and I would not have known.

5. I cannot take it and I cannot leave it. **7.** We are here and we are all there.

9. TTTT **11.** TTTT **13.** TTTT **15.** TTFT **17.** FFTF

19. TFTTFFFF **21.** TTTF **23.** TTFF **25.** TFFT **27.** TTFF

29. $q \to p \equiv \sim q \lor p \equiv \sim(\sim p) \lor \sim q \equiv \sim p \to \sim q$ **31.** $\sim(\sim p \lor \sim q) \equiv \sim(\sim p) \land \sim(\sim q) \equiv p \land q$

33. $(p \land q) \lor (p \land \sim q) \equiv p \land (q \lor \sim q) \equiv p \land \mathrm{T} \equiv p$ **35.** $p \lor \sim p \equiv \mathrm{T} \equiv q \lor \sim q$

37. If $x = 2$, then $x^2 = 2(2) = 4$. If $x^2 = 4$, then x could be -2.

39. If $x = 1$, then since $1 < 2$ we have $x < 2$. If $x < 2$, then x could be 0, for example, in which case $x \neq 1$.

Section 3.5

3. $p \land \sim p \equiv 0$ **5.** All of them. **7.** t **9.** Figure A3.1 **11.** Figure A3.2

13. Figure A3.3 **15.** $p \land (q \lor r)$ **17.** I **19.** $p \land (q \lor r \lor s)$

Section 3.6

7. Valid **9.** Valid **11.** Invalid **13.** Valid

17. Valid; *modus ponens* **19.** Valid; *disjunctive syllogism* **21.** Valid; *modus tollens*

23. Valid; reasoning by transitivity

Section 3.7

3. 121 **5.** Suppose N is the largest even number. Then $N + 2$ is even and $N < N + 2$. This is a contradiction, so there is no largest even number.

7. Suppose $ab = 0$ and $a \neq 0$, $b \neq 0$. Since $a \neq 0$, we can divide both sides of the equation $ab = 0$ by a to obtain $b = 0$. This is a contradiction, so a or b must be zero.

9. Suppose $a + b < 100$ and $a \geq 50$, $b \geq 50$. Then $a + b \geq 50 + 50 = 100$. This is a contradiction, so $a < 50$ or $b < 50$.

11. Inverse: If the sum of two numbers is greater than or equal to 100, then the two numbers are each greater than or equal to 50.

Converse: If one of two numbers is less than 50, then the sum of the two numbers is less than 100. No.

13. If a, b, and c are the lengths of the sides of a triangle and $a^2 + b^2 = c^2$, then the triangle is a right triangle. We shall see in the next chapter that the converse is true.

17. If the first digit on the left of n is 1, then $10 \leq n \leq 19$. Hence, $50 \leq 5n \leq 95$, so $5n$ has two digits.

19. $1^2 + 2^2 + \cdots + n^2 = \dfrac{n(n+1)(2n+1)}{6}$

23. $5^2 + 2(6) = 6^2 + 1$ and $n^2 + 2(n+1) = (n+1)^2 + 1$. To prove this, we have $(n+1)^2 = n^2 + 2n + 1 = n^2 + 2(n+1) - 1$.

25. No; for example $3 + 5 = 8$ is even but 3 and 5 are odd.

Chapter 3 Test

 1. FTFTFFFT

 3. (a) It snows. I ski.   (b) If I ski, then it snows. If it does not snow, then I do not ski. If I do not ski, then it does not snow.   (c) Contrapositive   **5.** False

 7. $p \wedge (q \vee {\sim}q) \equiv p \wedge \mathrm{T} \equiv p$ **9.** $r = {\sim}p \wedge {\sim}q$ **11.** An absurdity; a tautology

13. $(p \wedge q) \wedge [(p \wedge r) \vee (q \wedge {\sim}r)]$ **15.** Yes;   the argument is equivalent to
$$p \to q$$
$$q \to s$$
$$s \to {\sim}r$$
$$p \to {\sim}r$$

17. A natural number m is odd if and only if $m = 2n - 1$ for some $n \in \mathbb{N}$. But $2n - 1 = 2(n - 1) + 1$ where $n - 1 \in \mathbb{N} \cup \{0\}$.

19. (a) F   (b) F   (c) F   (d) F   (e) T   (f) F   (g) F   (h) F   (i) F   (j) T   (k) T

23. (a) T   (b) F   (c) T

CHAPTER 4

Section 4.1

 1. 12   **3.** 11   **5.** $n = m^3$ for some $m \in \mathbb{N}$;   1, 8, 27, 64

 7. 3, 6, 9, 12, 15   **9.** 21, 28   **13.** 1, 1, 2, 3, 5, 8, 13, 21, 34, 55, 89, 144, 233, 377, 610

17. The number of pairs is given by the Fibonacci sequence.   **19.** 5

21. The $(n+1)$th triangular number adds $n+1$ pebbles to the bottom row of the nth triangular number. Hence, $t_{n+1} = t_n + n + 1$.

23. If a_i is the ith Fibonacci number, then $a_1 + a_2 + \cdots + a_n = a_{n+2} - 1$.

25. 1, 3, 4, 7, 11, 18, 29, 47, 76, 123   **27.** $123 = (7)(5) + 11(8)$, $199 = (7)(8) + 11(13)$

29. $2 + 2 = (2)(2)$ **31.** $1 + 1 + 2 + 4 = (1)(1)(2)(4)$ **33.** $1 + 1 + 1 + 1 + 2 + 6 = (1)(1)(1)(1)(2)(6)$

Section 4.2

 1. $1 + 6$, $2 + 5$, $3 + 4$, $1 + 1 + 5$, $1 + 2 + 4$, $1 + 3 + 3$, $2 + 2 + 3$, $1 + 1 + 1 + 4$, $1 + 1 + 2 + 3$, $1 + 2 + 2 + 2$, $1 + 1 + 1 + 1 + 3$, $1 + 1 + 1 + 2 + 2$, $1 + 1 + 1 + 1 + 1 + 2$, $1 + 1 + 1 + 1 + 1 + 1 + 1$

 3. $n/2$ if n is even, $(n-1)/2$ if n is odd.   **5.** One; $1 + 1 + \cdots + 1 + 2$ $((n-2)$ ones)

 9. The number of ways of slicing a pie into pieces whose angles are a multiple of $360°/n$ equals the number of partitions of n.   **13.** $25^2 = 24^2 + 7^2$

17. Any odd number can be written as a sum of two consecutive natural numbers.

19. $24 = 11 + 13$, $26 = 13 + 13$, $28 = 11 + 17$, $30 = 13 + 17$ **21.** 11, 17, 23, 27, 29

23. If there were a finite number of squares, then there would be a largest square n^2. But $(n+1)^2$ is a square and $(n+1)^2 > n^2$, which is a contradiction. Another way to prove this is that the squares are the infinite set of numbers $\{n \in \mathbb{N}: n = k^2, \quad k = 1, 2, 3, \ldots\}$.

Table A4.1

29	30	31	32	33	34	35	36	37	38	39	40
p	c	p	c	c	c	c	c	p	c	c	c

Table A4.2

n	Divisors of n	$d(n)$
13	1, 13	2
14	1, 2, 7, 14	4
15	1, 3, 5, 15	4
16	1, 2, 4, 8, 16	5
17	1, 17	2
18	1, 2, 3, 6, 9, 18	6
19	1, 19	2
20	1, 2, 4, 5, 10, 20	6
21	1, 3, 7, 21	4
22	1, 2, 11, 22	4
23	1, 23	2

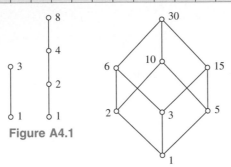

Figure A4.1

Figure A4.2

Section 4.3

1. Tiles of size 4, 6, 8, and 12 in will work.   3. (b), (d)   5. $n/1 = n$, $n/n = 1$

7. Table A4.1   9. Table A4.2

11. The divisors of 36 are 1, 2, 3, 4, 6, 9, 12, 18, and 36. There are 9 divisors. Since $36 = (2^2)(3^2)$, Theorem 4.2 states that 36 has $(2 + 1)(2 + 1) = 9$ divisors.

15. 210; $d(210) = 16$

17. Primes greater than 2 are odd and the product of two odd numbers is odd.

19. Figure A4.1   21. Figure A4.2

23. Since $k \mid m$, $m = kt$ for some $t \in \mathbb{N}$, and since $m \mid n$, $n = ms$ for some $s \in \mathbb{N}$. Hence $n = kts$, so $k \mid n$.

25. $k \mid n \leftrightarrow n = kt$ for some $k \in \mathbb{N} \leftrightarrow n/k = t \in \mathbb{N}$.

27. Not necessarily. For example, let $m = 2$, $n = 3$, and $k = 6$.

Section 4.4

1. 1, 2, 4, 5, 10, 20   3. See Figure 4.16;   6;   36;   3, 36   5. m 7. 2; 87,316

9. $m \mid n$ 13. If $p \mid n$, $\gcd(p, n) = p$. If $p \nmid n$, $\gcd(p, n) = 1$.

15. Both gcd's equal 3 and both lcm's equal 36.

Section 4.5

1. No   3. No; $H(2) = 8$ is not prime.

5. 46;   62;   78;   95;   109;   125;   139;   154

7. Let $n = 1,000,000,001$. Then $n! + 2, n! + 3, \ldots, n! + n$ gives a billion consecutive composite numbers.   9. The odd numbers form the infinite set $\{2k - 1 \colon k \in \mathbb{N}\}$.

11. If $n = 2$, then n is prime and $n + 2 = 4$ is not prime. If $n = 3$, then n is prime and so is $n + 2 = 5$.

13. 1, 2, 4, 6, 10   15. For example, all even numbers greater than 2 are composite.

17. $F(n) = 2n$, $n \neq 1$ 19. Not necessarily

21. (2, 4), (3, 5), (5, 7), (7, 9), (11, 13), (13, 15), (17, 19), (19, 21), (23, 25), (29, 31), (31, 33), (37, 39), (41, 43), (47, 49)

23. Not necessarily; the natural number could have the form $n = p^3$, where p is a prime number.

Section 4.6

1. $1 + 2 + 3 + 4 + 6 = 16 \neq 12$ **3.** $1 + 2 + 4 + 8 + 16 + 31 + 62 + 124 + 248 = 496$

5. 4; 6; 10   **7.** $2p$ **9.** 1; 3; 7; 6; 12; 8; 15; 13; 18   **11.** Yes; no

13. If n is prime, then its only divisors are 1 and n. Hence, $s(n) = n + 1$.

15. $s(n) = 2n$ if and only if the sum of the divisors of n other than itself is n. But this is the definition of perfect.   **17.** 12; 18; 20

21. (a) $s(6 \cdot 35) = s(6)s(35)$ (b) $s(3 \cdot 6) \neq s(3)s(6)$ (c) $s(3 \cdot 5) = s(3)s(5)$
(d) $s(4 \cdot 6) \neq s(4)s(6)$ **25.** 1; 10; 100; 1000; ...

27. If $n = pq$, where p and q are distinct primes, then p and q are the only divisors of n besides 1 and n.   **29.** Yes; use either Exercise 27 or 28 to show this.

Section 4.7

1. $47 = 4 \times 10 + 7 \times 10^0$ **3.** $1239 = 1 \times 10^3 + 2 \times 10^2 + 3 \times 10 + 9 \times 10^0$ **5.** 10^9

7. n **9.** (a) 10 (b) 100 (c) 1000 (d) 10000   **11.** 1000000000000000

13. (a) 2 (b) 4 (c) 8 (d) 16

15. (a) $25 = 16 + 8 + 1 = 2^4 + 2^3 + 0 \times 2^2 + 0 \times 2 + 2^0 = 11001$ (Figure A4.3).

17. (a) 10111   23   **19.** The ternary number 12 has the base 10 representation 5.
     +11010   +26
     110001   49

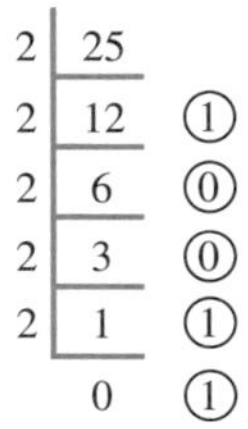

Figure A4.3

Section 4.8

1. There will be no 1s in the first row; whenever a 1 appears in the second row, there is also a 1 in the third row.   **3.** None of these positions are strategic.

5. Take nine sticks from the first, second, or third pile.

Chapter 4 Test

1. If $n = 2n'$ and $m = 2m'$ are even, then $nm = 4n'm'$ is even.

3. If $n = 2n' - 1$ and $m = 2m' - 1$ are odd, then $nm = 4n'm' - 2(n' + m') + 1$ is odd.

5. $7+1, 6+2, 5+3, 4+4, 1+1+6, 1+2+5, 1+3+4, 2+2+4, 2+3+3, 1+1+1+5, 1+1+2+4,$
$1+1+3+3, 1+2+2+3, 2+2+2+2, 1+1+1+1+4, 1+1+1+2+3, 1+1+2+2+2,$
$1+1+1+1+1+3, 1+1+1+1+2+2, 1+1+1+1+1+1+2, 1+1+1+1+1+1+1+1$

7. 24   **9.** 2; 174,632   **11.** $4200 = (2)(2)(2)(3)(5)(5)(7)$ **13.** 8   **15.** 11000101

17. (a) T (b) F (c) F (d) F (e) F (f) F (g) T (h) T (i) F (j) T

19. $1 + 1 + 2 + 2 + 2 = 8$, $1 + 1 + 1 + 2 + 5 = 10$.

CHAPTER 5

Section 5.1

9. Let $Ax + By = C$ be the standard form for the line. If the line is horizontal, then $A = 0$, in which case $y = C/B$. Hence, the slope is zero. Conversely, if the slope is zero, then the slope-intercept form is $y = b$. In this case $A = 0$, so the line is horizontal.

11. $-mx + y = y_1 - mx_1$ **13.** Figure A5.1   **15.** Figure A5.2   **17.** Figure A5.3

19. x-intercept is 5, no y-intercept   **21.** x-intercept is 2, y-intercept is $3/2$

23. x-intercept is $5/2$, y-intercept is $-5/4$ **25.** 0, $-3/4$ **27.** $-1/3$

29. $(0, 5/3), (-5/2, 0), (-1, 1)$ **31.** $y = x$ **33.** $V(t) = 5000 - 1000t, V(3) = \$2000$

Section 5.2

1. $x = 0$ **3.** $y = 2$ **5.** $y - 3 = 5(x - 2)/3$ **7.** The point $(1000, 1997)$ is on the line.

9. $(-2, 3)$ **11.** Both lines are horizontal.   **13.** Both lines have slope -1.

15. $x + y = 4$ **17.** The slopes are $m_1 = -1$ and $m_2 = 1$ and $m_1 m_2 = -1$.

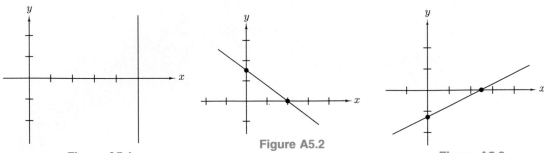

Figure A5.1   **Figure A5.2**   **Figure A5.3**

19. $y = -x/3$ **21.** $y = 2$ **23.** $-3/4$

25. $f(\ell) = \{(x - 1, y + 1): (x, y) \in \ell\} = \{(x - 1, -(2/3)x + 8/3): x \in \mathbb{R}\}$
$= \{(x, -(2/3)x + 2): x \in \mathbb{R}\}$
Hence, the equation for $f(\ell)$ is $y = -(2/3)x + 2$ or $2x + 3y = 6$.

Section 5.3

1. 8   **3.** $\sqrt{2}$ **5.** Yes   **7.** Yes   **9.** Yes   **11.** 24, 74   **13.** $\sqrt{52}$

15. $d(A, B) = \sqrt{121 + 4} = 5\sqrt{5}$, $d(A, C) = \sqrt{9 + 16} = 5$, $d(B, C) = \sqrt{64 + 36} = 10$, and $[d(A, B)]^2 = [d(A, C)]^2 + [d(B, C)]^2$ **19.** $(x + 5)^2 + (y - 2)^2 = 25$

21. The radius of the circle is $r = \sqrt{25 + 16} = \sqrt{41}$.   **23.** Figure A5.4

Section 5.4

1. 12   **3.** 40   **5.** 5π **7.** 49   **9.** 15   **11.** 70

13. 24   **15.** 16π **17.** π **19.** $A = \frac{1}{2}(b + x)h - \frac{1}{2}xh = \frac{1}{2}bh$

21. 24   **23.** 100π **25.** $\pi(R^2 - r^2)$ **27.** $\frac{1}{3}\pi r^2 h$ **29.** They are equal.

Section 5.5

5. 3   **7.** $\frac{65}{4}$ **9.** $55\frac{1}{3}$ **11.** $\frac{49}{20}$

13. The area under $g(x)$ from a to b is at least as large as the area under $f(x)$ from a to b.

15. $4\frac{2}{3}$ **17.** $\frac{1}{6}$ **19.** $\int_0^1 f_n(x) = \int_0^1 x^n = \frac{1}{n + 1}$ **21.** $\frac{256}{5}$ **23.** $\frac{2}{3}(2\sqrt{2} - 1)$ **25.** $\frac{2}{7}$

Section 5.6

1. The longitudes (or meridians of longitude) are great circles. The latitudes are not (except the equator).

3. $12 \times 30 = 360$

5. An icosahedron has 12 vertices, 30 edges, and 20 faces. Each face accounts for 10 dimples, each edge accounts for 5 dimples, and each vertex accounts for 1 dimple. This gives a total of $20(10) + 30(5) + 12 = 362$.

Section 5.7

1. 4   **3.** 2   **7.** $\log 2/\log 2.5 = 0.7565$ **9.** $\log 8/\log 4 = 1.5000$

11. $\log 4/\log 2 = 2.0000$

Chapter 5 Test

1. x-intercept 9/5; y-intercept $-9/7$;   slope 5/7   **3.** $3x - 4y = -6$

5. $(x - 1)^2 + (y - 1)^2 = 2$ **7.** $A = 13$; $P = 16$ **9.** 19;   20;   21   **11.** 64   **13.** 9

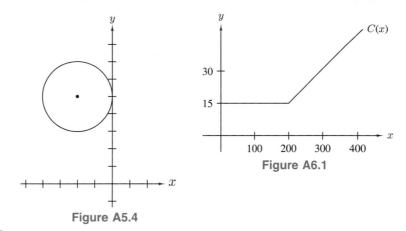

Figure A5.4

Figure A6.1

CHAPTER 6

Section 6.1

1. Since the interest is $I = Prt$, the total amount is $A = P + I = P + Prt = P(1 + rt)$.
3. $900 **5.** $78.75 **7.** 11.67 yr **9.** $11,040.20 **11.** Company Y **13.** 8.24%
15. 14.29 **17.** 7.87 **19.** 11.58 **21.** 5.165%; 5.095% **25.** 20.15 **27.** 10.24

Section 6.2

1. $2838.14 **3.** $7389.06 **5.** $16,444.65 **7.** 5.13% **9.** 6.72%
11. e^2 **13.** $499.60 **15.** $5186.48 **17.** $1730.85 **19.** $1/e^{at} = e^{-at}$
21. 13.86 **23.** 9.24 **29.** $8986.58
31. The certificates of deposit will be worth $1,006.88 in ten years. **33.** $1094.16

Section 6.3

1. (a) $8212.46 **3.** $2153.84 **5.** $33,701.28 **7.** $30,246.85 **9.** $391.93
11. $1193.84 **13.** $43,081.98; $25,081.98 **15.** (a) $1937.06 (b) $33,776.60

Section 6.4

1. $33,003.87 **5.** Multiply numerator and denominator by $(1 + r)^n$.
7. $222.44; $3346.67 **9.** $952.32; $242,836.42 **11.** $392.22; $1766.46
13. $268.27; $2877.20 **15.** $13,534.02

Section 6.5

1. (i), (iii), (v) are fixed costs and (ii), (iv) are variable costs. **3.** $P(x) = 4.10x - 390$
5. $1570; $2000; $430 **7.** (a) $75 (b) $15 + 0.15x$ **9.** $C_F = \$5000; C_V(x) = 50x$
11. Figure A6.1
13. 21 **15.** $C(x) = 300x + 2100$ **17.** 10 **19.** 1.19, -4.19 **21.** $-1, -3/2$

Section 6.6

1. (a) $8300 (b) $380; $10,200 (c) 100; $9000 (d) 2.55; 130.78
3. $20 **5.** 220
7. (a) $C(x) = 1000 + 40x$ (b) $R(x) = 55x - 0.01x^2$ (c) $P(x) = 15x - 0.01x^2 - 1000$
  (d) $P(100) = \$400; P(500) = \$4000; P(1000) = \$4000$ **9.** $1.62; 73.33
11. $1.44; 56.59 **13.** $(0, 1)$ **15.** $(5.12, 17.25), (-3.12, 0.75)$ **17.** $(0.75, 1.56)$

Chapter 6 Test

1. $3054.60 **3.** $11,127.71 **5.** $33,680.63 **7.** $447.93; $23,751.87
9. 2000; $65 **11.** $x = 20$ **13.** There are no real roots.

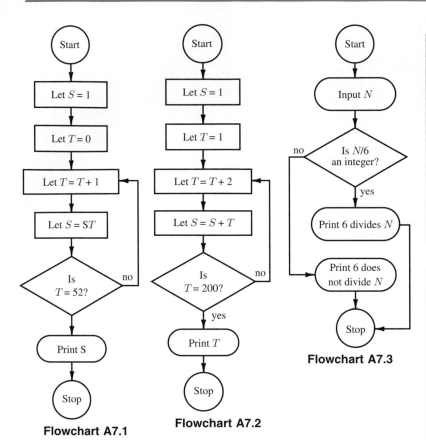

Flowchart A7.1

Flowchart A7.2

Flowchart A7.3

Table A7.1

Step	What computer does
1	Start
2	Let $N = 2$
4	Let $N = 4$
3	Let $N = 16$
5	Print 16
6	Yes
4	Let $N = 18$
3	Let $N = 324$
5	Print 324
6	No
7	Stop

Table A7.2

Step	What computer does
1	Start
2	Let $N = 2$
3	Let $N = 4$
5	Print 4
4	Let $N = 6$
6	Yes
3	Let $N = 36$
5	Print 36
4	Let $N = 38$
6	Yes
3	Let $N = 1444$
5	Print 1444
4	Let $N = 1446$
6	No
7	Stop

CHAPTER 7

Section 7.2

9. 1010   **11.** 1100   **13.** 10000   **15.** 20   **17.** 15   **19.** 32

21. 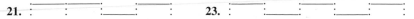   **23.**

Section 7.4

7. Flowchart A7.1   **9.** Flowchart A7.2   **11.** Flowchart A7.3

13. Flowchart 8 instructs the computer to print the first 100 natural numbers.

15. Flowchart 10 instructs the computer to find 50! and print the answer.

17. Flowchart 12 instructs the computer to print 6, 38, and 1446.

19. The computer will now print the numbers 16 and 324.

21. (a) Table A7.1   (b) Table A7.2

Table A7.3

Step	What computer does
1	Start
2	Input $N = 11$
3	Let $D = 1$
4	Let $D = 2$
5	Is $\frac{11}{2}$ an integer? No
7	Is $2 \leq \sqrt{11}$? Yes
4	Let $D = 3$
5	Is $\frac{11}{3}$ an integer? No
7	Is $3 \leq \sqrt{11}$? Yes
4	Let $D = 4$
5	Is $\frac{11}{4}$ an integer? No
7	Is $4 \leq \sqrt{11}$? No
8	Print 11 is prime
9	Stop

Table A7.4

Step	What computer does
1	Start
2	Input $A = 10$, $B = 15$
3	Let $N = 10$
4	Is $\frac{15}{10}$ an integer? No
7	Let $N = 9$
4	Is $\frac{15}{9}$ an integer? No
7	Let $N = 8$
4	Is $\frac{15}{8}$ an integer? No
7	Let $N = 7$
4	Is $\frac{15}{7}$ an integer? No
7	Let $N = 6$
4	Is $\frac{15}{6}$ an integer? No
7	Let $N = 5$
4	Is $\frac{15}{5}$ an integer? Yes
5	Is $\frac{10}{5}$ an integer? Yes
6	Print $5 = \gcd(10, 15)$
8	Stop

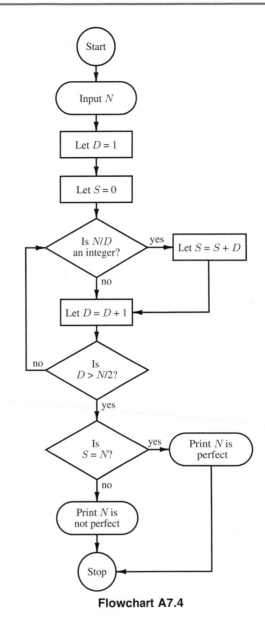

Flowchart A7.4

23. Flowchart 14 instructs the computer to print the numbers 3, 5, 7, 9, 11, ... This causes an infinite loop that never stops.

25. Table A7.3   **27.** (a) Table A7.4   **31.** Flowchart A7.4

Section 7.5

1. (a) $7 * X \uparrow 4 + 3 * X \uparrow 2 - 2 * X + 4$

3. The expression could have the values 67 or 22 or others.

5. The computer will print   12   35   25   49

7. The computer will print   13   25   169

9. 10 LET N = 1
20 LET S = 1
30 LET N = N + 1
40 LET S = S + N ↑ 2
50 If N < 100 THEN 30
60 PRINT S
70 END

11. 5 INPUT N
10 INPUT M
15 IF N * M = 0 THEN 35
20 IF N * M > 0 THEN 45
25 PRINT N * M; "< 0"
30 GO TO 50
35 PRINT N * M; "= 0"
40 GO TO 50
45 PRINT N * M; "> 0"
50 END

13. 50 GO TO 100

15. 10 LET A = 2 * (B + C)
20 GO TO 30
30 LET V = L * W * H
40 LET C = A + B
50 LET B = (A + C)/5
60 IF A < B THEN 90
70 IF A = B THEN 90
80 IF M >= N THEN 90
90 IF M >= N THEN 120
100 LET A = 3
110 INPUT A, B, C
120 PRINT N
130 END
RUN
140 LET A = 2
150 IF N = 10 THEN 20

17. 2 IS A FACTOR OF 100
4 IS A FACTOR OF 100
5 IS A FACTOR OF 100
10 IS A FACTOR OF 100

19. 10 INPUT A, B
20 IF INT(SQRT(A↑ 2 + B ↑ 2)) = SQRT(A ↑ 2 + B ↑ 2) THEN 60
30 PRINT A; "AND"; B; "ARE NOT THE FIRST TWO";
40 PRINT "TERMS OF A PYTHAGOREAN TRIPLE"
50 GO TO 80
60 PRINT A; "AND"; B; "ARE THE FIRST TWO";
70 PRINT "TERMS OF A PYTHAGOREAN TRIPLE"
80 END

21. 10 INPUT A, B, C
20 IF A >= B THEN 60
30 IF B >= C THEN 100
40 PRINT C
50 GO TO 110
60 IF A >= C THEN 80
70 GO TO 40
80 PRINT A
90 GO TO 110
100 PRINT B
110 END

23. 10 INPUT N
20 PRINT 2
30 LET A = 1
40 LET A = A + 2
50 LET B = 3
60 IF A > N THEN 130
70 IF B <= SQRT(A) THEN 100
80 PRINT A
90 GO TO 40
100 IF A /B = INT(A/B) THEN 40
110 LET B = B + 2
120 GO TO 70
130 END

25. 10 INPUT N
20 LET D = 1
30 LET S = 0
40 IF N /D > INT(N/D) THEN 60
50 LET S = S + D
60 LET D = D + 1
70 IF D <= N/2 THEN 40
80 IF S = N THEN 110
90 PRINT N; "IS NOT PERFECT"
100 GO TO 120
110 PRINT N; "IS PERFECT"
120 END

Section 7.6

1. 8
48

3. 5
13

5. 55

7. 10 READ X
20 PRINT X ↑ 5
30 DATA 2, 3, 4, 5, 6, 7, 8
40 IF X = 8 THEN 60
50 GO TO 10
60 END

9. The numbers 1 through 10 are printed in a vertical column.

11. The program instructs the computer to find the sum $2 + 2^2 + 2^3 + \cdots + 2^{10}$ and print 2046.

13. 10 PRINT "WHAT IS N";
20 INPUT N
30 LET S = 0
40 FOR I = 1 TO 20
50 LET S = S + N ↑ I TO 10
60 NEXT I
70 PRINT S
80 END

15. 10 READ A, B, C, D, E, F, G, H, I, J
20 PRINT 2 ↑ A + 2 ↑ B + 2 ↑ C + $\cdots$ + 2 ↑ J
30 DATA 1, 2, 3, 4, 5, 6, 7, 8, 9, 10
40 END

Note: Line 20 must be entered in full.

17. 1   4   9   16   25   36   49   64   81   100   121

19. 10 FOR I = 1 TO 10
20 FOR J = 1 TO 10
30 PRINT I; "+"; J; "="; I + J,
40 NEXT J
50 NEXT I
60 END

21. The computer will compute $1 + \cfrac{2}{1 + \cfrac{2}{1 + \cfrac{2}{1 + \cfrac{2}{1 + 2}}}}$ and print 2.047619048.

23. 10 INPUT M
20 LET N = 1
30 FOR I = 1 TO M
40 LET N = 1 + 1/N
50 NEXT I
60 PRINT N −(SQRT(5) + 1)/2
70 END
Run for $M = 1, 5, 10, 20, 25, 26,$ and 27.

25. GOSUB instructs the computer to return to the GOSUB command after the subroutine is completed.

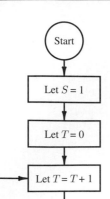

Flowchart A7.5

Table A8.1

Value	Frequency
2	1
1	2
0	1

Frequency

Figure A8.1

Table A8.2

Value	Frequency
0	1
1	4
2	6
3	4
4	1

Table A8.3

Class interval	Frequency
(58, 64]	2
(64, 70]	10
(70, 76]	4
(76, 82]	2
(82, 88]	4
(88, 94]	2
(94, 100]	2

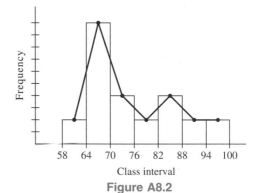

Figure A8.2

Chapter 7 Test

5. $(X \uparrow 2 + 2 * X + 1)/(2 * X + 1) + (5 * X \uparrow 2 + 1) \uparrow 2/(3 * X)$ **7.** Flowchart A7.5

9. 5 LET N = 0
10 LET N = N + 1
15 PRINT N
20 IF N < 100 THEN 10
25 END

11. 10 READ A, B, C, D, E, F
20 DATA 94, 88, 85, 90, 86, 87
30 LET X = (A + B + C + D)/4
40 PRINT (0.6) * X + (0.3) * E + (0.1) * F
50 END

13. 1 1
2 8
3 27
4 64
5 125

CHAPTER 8

Section 8.1

3. 75.81   **5.** Let A = 4, B = 3, C = 2, D = 1, F = 0, and compute the mean

$$\frac{2 \cdot 0 + 4 \cdot 1 + 12 \cdot 2 + 4 \cdot 3 + 4 \cdot 4}{26} = 2.15$$

This is the grade point average. The mean is slightly higher than a C.

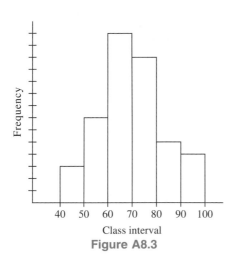
Figure A8.3

Table A8.4

X value	Frequency
0	3
1	2
2	2
3	1

Table A8.5

Y value	Frequency
0	3
1	2
2	1
3	2

Table A8.6

Z value	Frequency
0	5
2	1
4	1
9	1

Table A8.7

n	S_n
1	0.50000
5	0.83333
10	0.90909
15	0.93750
20	0.95238

7. Table 8.3 and Figure 8.2 are the same except that we replace $(55, 60], (60, 65], \ldots$ by $[56, 61), [61, 66), \ldots$. **9.** Table A8.1 and Figure A8.1 **11.** Table A8.2

13. Table A8.3 and Figure A8.2 **15.** Figure A8.3

17. Table A8.4, Table A8.5, and Table A8.6

Section 8.2

1. $\displaystyle\sum_{i=1}^{1} a_i = a_1, \qquad \sum_{i=1}^{2} a_i = a_1 + a_2, \qquad \sum_{i=1}^{3} a_i = a_1 + a_2 + a_3$

$\displaystyle\sum_{i=1}^{4} a_i = a_1 + a_2 + a_3 + a_4, \qquad \sum_{i=1}^{5} a_i = a_1 + a_2 + a_3 + a_4 + a_5$

3. (b) median = 74.5, mode = 65 **5.** (a) $m = 4.33$, median = 3 **7.** $m = 10.72$

11. The new median, mean, and mode are the old ones plus c.

13. If n is odd there are $(n-1)/2$ values below $x_{(n+1)/2}$ and $(n-1)/2$ values above. If n is even, the two "middle" values are $x_{n/2}$ and $x_{n/2+1}$, since there are $n/2 - 1$ values below and above these two values.

15. 55

17. Since the mean and median coincide,

$$x_2 = \frac{x_1 + x_2 + x_3}{3} \tag{A8.1}$$

Hence, $2x_2 = x_1 + x_3$ and $x_3 - x_2 = x_2 - x_1$. Conversely, if (8.5) holds, then $2x_2 = x_1 + x_3$, so (A8.1) holds.

19. There are many examples. One is a data set with one data value. Another is a data set in which all the data values coincide. Another is $\{0, 2, 2, 4\}$. Can you find more?

21. Let $\displaystyle S_n = \sum_{i=1}^{n} \frac{1}{i(i+1)}$. We then have Table A8.7.

25. (a) $\displaystyle\sum_{i=1}^{n} ca_i = ca_1 + ca_2 + \cdots + ca_n = c(a_1 + a_2 + \cdots + a_n) = c\sum_{i=1}^{n} a_i$

(b) $\displaystyle\sum_{i=1}^{n}(a_i + b_i) = a_1 + b_1 + a_2 + b_2 + \cdots + a_n + b_n$

$$= a_1 + a_2 + \cdots + a_n + b_1 + b_2 + \cdots + b_n = \sum_{i=1}^{n} a_i + \sum_{i=1}^{n} b_i$$

(c) No. For $n = 2$, the left side is $a_1 b_1 + a_2 b_2$ while the right side is $(a_1 + a_2)(b_1 + b_2) = a_1 b_1 + a_2 b_2 + a_1 b_2 + a_2 b_1$. In general, these do not coincide. For example, let $a_1 = a_2 = b_1 = b_2 = 1$.

27. Using a common denominator gives $\dfrac{1}{i} - \dfrac{1}{i+1} = \dfrac{i+1-i}{i(i+1)} = \dfrac{1}{i(i+1)}$. Hence, $\displaystyle\sum_{i=1}^{n} \dfrac{1}{i(i+1)} =$

$$\sum_{i=1}^{n} \frac{1}{i} - \sum_{i=1}^{n} \frac{1}{i+1} = 1 + \frac{1}{2} + \frac{1}{3} + \cdots + \frac{1}{n} - \left(\frac{1}{2} + \frac{1}{3} + \cdots + \frac{1}{n} + \frac{1}{n+1}\right) = 1 - \frac{1}{n+1}$$

Section 8.3

1. Substitute 0 for m in (8.12).   **3.** (a) 1.3   (b) 11.31   (e) 2.55   (h) 2.36   (j) 7.64

5. (a) $-1.35, -0.58, 0.19, 0.96, 1.73$ **7.** 1.08

9. The variance is multiplied by c^2 and σ is multiplied by $|c|$.

11. $x_1 = 0$, $x_2 = 2$ or $x_1 = 2$, $x_2 = 0$ **13.** Both have the value $\dfrac{1}{n}\displaystyle\sum_{i=1}^{n} x_i^2$.

15. By Schwarz's inequality, we have mad $= \dfrac{\sum |x_i - m| f_i}{\sum f_i} = \dfrac{\sum |x_i - m| f_i^{1/2} f_i^{1/2}}{\sum f_i}$

$$\leq \frac{\left[\sum (x_i - m)^2 f_i\right]^{1/2} \left[\sum f_i\right]^{1/2}}{\sum f_i} = \left[\frac{\sum (x_i - m)^2 f_i}{\sum f_i}\right]^{1/2} = \sigma$$

Section 8.4

1. (a) 0.23   (b) 0.42   (c) 0.73   **3.** 0.30   **5.** 0.64   **7.** $P[a, b] = (b^2 - a^2)/c^2$

9. The area A is the sum of the areas of the rectangle and triangle $A = ca + (c/2)(b - a)$
$= (c/2)(a + b)$

11. The first equality is $P(a \leq X \leq b) = P(X = a) + P(a < X \leq b) = P(a < X \leq b)$.

13. The area under the density function from a to c is the area from a to b plus the area from b to c.

15. Applying (8.23) gives $P(-\sigma \leq X \leq \sigma) = \sigma(2 - \sigma) = 0.65$ and $P(-2\sigma \leq X \leq 2\sigma)$
$= 2\sigma(2 - 2\sigma) = 0.97$

17. Applying (8.21) gives

$$P(a \leq X \leq b) = P(0 \leq X \leq 1) + P(0 \leq X \leq -a) = \frac{1}{2} - \frac{a}{2}(2 + a) = \frac{1}{2}[1 - a(2 + a)]$$

19. Applying (8.21) gives

$$P(a \leq X \leq b) = P(-1 \leq X \leq 0) + P(0 \leq X \leq b) = \frac{1}{2} + \frac{b}{2}(2 - b) = \frac{1}{2}[1 + b(2 - b)]$$

21. The area under the graph of g from $-1/4$ to $1/4$ is $1/2$ and the areas from -1 to $-1/4$ and $1/4$ to 1 are each $1/4$.   **23.** (a) $1/4$ (b) $1/8$

25. $P\left(-\dfrac{1}{2} \leq X \leq \dfrac{1}{2}\right) = P\left(-\dfrac{1}{2} \leq X \leq 0\right) + P\left(0 \leq X \leq \dfrac{1}{2}\right) = \dfrac{1}{2} \cdot \dfrac{1}{3} + \dfrac{1}{2} \cdot \dfrac{2}{3} = \dfrac{1}{2}$

27. For Case 1 we have $P(a \leq X \leq b) = 2(1 - a)/3$.
For Case 3 we have $P(a \leq X \leq b) = -a/3 + 2b/3 = (2b - a)/3$.

y
1
1/2
g(x)
x
−1
1

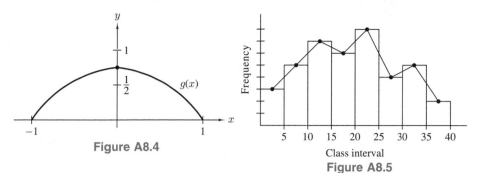

Figure A8.4

Frequency
5 10 15 20 25 30 35 40
Class interval

Figure A8.5

Section 8.5

1. Applying (8.25) gives $m = \int_{-\infty}^{0} x\, g(x) + \int_{0}^{\infty} x\, g(x)$
Replacing x by $-y$, since $g(-y) = g(y)$ we have

$$m = \int_{0}^{\infty} (-y) g(-y) + \int_{0}^{\infty} x\, g(x) = -\int_{0}^{\infty} y\, g(y) + \int_{0}^{\infty} x\, g(x) = 0$$

3. Applying (8.27) we have

$$\text{var} = \int_{-\infty}^{\infty} (x-m)^2\, g(x) = \int_{-\infty}^{\infty} (x^2 - 2mx + m^2)\, g(x)$$

$$= \int_{-\infty}^{\infty} x^2\, g(x) - 2m \int_{-\infty}^{\infty} x\, g(x) + m^2 \int_{-\infty}^{\infty} g(x) = m_2 - 2m^2 + m^2 = m_2 - m^2$$

5. $m = \int_{-\infty}^{\infty} x\, g(x) = \int_{-1}^{0} x(1+x) + \int_{0}^{1} x(1-x) = \int_{-1}^{0} x + \int_{-1}^{0} x^2 + \int_{0}^{1} x - \int_{0}^{1} x^2 = 0$

7. $P(a \le X \le b) = \int_{a}^{b} g_c(x) = \int_{a}^{0} g_c(x) + \int_{0}^{b} g_c(x) = \frac{1}{c^2}\int_{a}^{0} (c+x) + \frac{1}{c^2}\int_{0}^{b} (c-x)$

$$= \frac{1}{c}\int_{a}^{0} 1 + \frac{1}{c^2}\int_{a}^{0} x + \frac{1}{c}\int_{0}^{b} 1 - \frac{1}{c^2}\int_{0}^{b} x = -\frac{a}{c} - \frac{a^2}{2c^2} + \frac{b}{c} - \frac{b^2}{2c^2}$$

$$= \frac{1}{2c}\left(2b - 2a - \frac{a^2}{c} - \frac{b^2}{c}\right) = \frac{1}{2c}\left[b\left(2 - \frac{b}{c}\right) - a\left(2 + \frac{a}{c}\right)\right]$$

9. $m = 0$. By Exercise 2 we have $\text{var} = m_2 = 2\int_{0}^{1} x^2\, g(x) = 2\int_{0}^{1/4} x^2 + \frac{2}{3}\int_{1/4}^{1} x^2$

$$= \frac{2}{3}\left(\frac{1}{4^3}\right) + \frac{2}{9}\left(1 - \frac{1}{4^3}\right) = \frac{2}{9} + \frac{4}{9}\left(\frac{1}{4^3}\right) = \frac{33}{9 \cdot 16} \quad \text{and} \quad \sigma = \sqrt{\text{var}} = \frac{\sqrt{33}}{12} = 0.48.$$

11. Figure A8.4

13. $m = \int_{-1}^{1} x\, g(x) = \frac{3}{4}\int_{-1}^{1} x(1-x^2) = \frac{3}{4}\int_{-1}^{1} x - \frac{3}{4}\int_{-1}^{1} x^3 = \frac{3}{4}\left(\frac{1}{2} - \frac{1}{2}\right) - \frac{3}{4}\left(\frac{1}{4} - \frac{1}{4}\right) = 0$

15. $m_2 = \int_{-1}^{1} x^2\, g(x) = \frac{3}{4} \int_{-1}^{1} x^2 - \frac{3}{4} \int_{-1}^{1} x^4 = \frac{3}{4}\left(\frac{1}{3}+\frac{1}{3}\right) - \frac{3}{4}\left(\frac{1}{5}+\frac{1}{5}\right) = \frac{1}{2} - \frac{3}{10} = \frac{1}{5}$

$\sigma = \frac{1}{\sqrt{5}} = 0.45$

17. Applying Exercise 14 gives $P(-2\sigma \leq X \leq 2\sigma) = 3(4\sigma)/4 - (16\sigma^3)/4 = 0.99$.

19. $\int_{0}^{1} g(x) = 6 \int_{0}^{1} x - 6 \int_{0}^{1} x^2 = 6\left(\frac{1}{2}\right) - 6\left(\frac{1}{3}\right) = 1$

21. $P(a \leq X \leq b) = \int_{a}^{b} g(x) = 6 \int_{a}^{b} x - 6 \int_{a}^{b} x^2 = 3(b^2 - a^2) - 2(b^3 - a^3)$

23. $m_2 = \int_{0}^{1} x^2\, g(x) = 6 \int_{0}^{1} x^3 - 6 \int_{0}^{1} x^4 = 6\left(\frac{1}{4}\right) - 6\left(\frac{1}{5}\right) = \frac{3}{10}$

$\text{var} = m_2 - m^2 = \frac{3}{10} - \frac{1}{4} = \frac{1}{20}$　and　$\sigma = \sqrt{\text{var}} = \frac{1}{2\sqrt{5}} = 0.22$

Section 8.6

1. (a) 0.9554　(b) 0.2743　(d) 0.0107　**3.** (a) 0.3707　**5.** 98.4%　**7.** 86.6%　**9.** 1.656
11. $\sigma = 1/\sqrt{2\pi} = 0.3989$

Section 8.7

3. $r = 1$　**5.** $r = -1$. The σ line is the line $y = -x$.
7. $m_X = 69$, $m_Y = 162.43$, $m_{2X} = 4765$, $m_{2Y} = 26{,}463.29$, $m_{XY} = 11{,}221.43$, $\text{var}_X = 4$,
$\text{var}_Y = 79.79$, $\text{cov}(X, Y) = 13.76$, $\sigma_X = 2$, $\sigma_Y = 8.93$, $r = 0.77$, $s = 4.47$
9. If $c > 0$, $r = 1$, and the σ line is the line $y = cx$. If $c < 0$, $r = -1$, and the σ line is the
line $y = cx$.
11. If $a > 0$, $r = 1$, and the σ line is the line $y = ax + b$. If $a < 0$, $r = -1$, and the σ line is
the line $y = ax + b$.
13. $m_X = 2.67$, $m_{2X} = 11$, $\sigma_X = 1.97$, $m_Y = 32{,}000$, $m_{2Y} = 1.0713 \times 10^9$, $\sigma_Y = 6879.92$,
$m_{XY} = 98{,}000$, $\text{cov}(X, Y) = 12{,}560$, $r = 0.93$

Chapter 8 Test

1. Figure A8.5　**3.** (15, 20]　**5.** No; let $n = 2$, $a_1 = a_2 = 1$　**7.** 3.17, 2.5, 2
9. The measurement is very accurate.

11. (a) $\int_{0}^{1} 4x^3 = 4\left(\frac{1}{4}\right) = 1$　(b) $m = \int_{0}^{1} x\, g(x) = 4 \int_{0}^{1} x^4 = 4\left(\frac{1}{5}\right) = 0.800$

(c) $m_2 = \int_{0}^{1} x^2\, g(x) = 4 \int_{0}^{1} x^5 = 4\left(\frac{1}{6}\right) = 0.667$
$\text{var} = m_2 - m^2 = 0.02667$, $\sigma = \sqrt{\text{var}} = 0.1633$

(d) $P\left(\frac{1}{2} \leq X \leq \frac{3}{4}\right) = \int_{1/2}^{3/4} g(x) = \int_{1/2}^{3/4} 4x^3 = \left(\frac{3}{4}\right)^4 - \left(\frac{1}{2}\right)^4 = 0.254$

13. $m_X = 2.50$, $m_{2X} = 9.17$, $\sigma_X = 1.71$, $m_Y = 5$, $m_{2Y} = 30.67$, $\sigma_Y = 2.38$, $m_{XY} = 16.33$,
$\text{cov}(X, Y) = 3.83$, $r = 0.94$

15. (a) T　(b) T　(c) F　(d) F　(e) T　(f) T　(g) F　(h) F　(i) F　(j) T

CHAPTER 9

Section 9.2

1. {TH, TT}

3. {HH, HT}. This event equals A. This event is the complement of the event in Exercise 1.

5. 1/2, 1/4   **7.** {THH, THT, TTH, TTT}; 2/3   **9.** Ω; 1

11. $C = \{$HH, HT, TT$\}$, $\{$HH, HT$\} \subseteq C$, $\{$HT$\} \subseteq C$; $C' = \{$TH$\}$, $P(C) = 3/4$, $P(C') = 1/4$

13. $1 - 25^3/17{,}576$ **15.** 1/18; 5/12   **17.** Harry. Why?   **19.** 1/(26)(26)   **21.** 25/52

25. (a) 18/100   (b) 19/100   (c) 81/100

27. (a) 1/3   (b) 7/15   (c) 8/15   (d) 1/5   **29.** 3/5   **31.** 1/6

Section 9.3

1. 2/3   **3.** 7/12   **5.** 3/5, 2/5   **7.** $P(\emptyset) = P(\Omega') = 1 - P(\Omega) = 0$

9. If $A \cap B = \emptyset$, then $P(A) + P(B) = P(A \cup B) \le 1$.   **11.** Use (9.7)   **13.** Use (9.7)

15. Letting $B = A'$ in (9.4) gives $1 = P(\Omega) = P(A \cup A') = P(A) + P(A')$.

17. (a) 0.35   (b) 0.40   (c) 0.70   (d) 0.30   (e) 0.35   **19.** 7/216; 7/209   **21.** 3/5

23. No   **25.**   (a) $P(A \cup B) = P(A) + P(B) - P(A \cap B) \le P(A) + P(B)$

(b) $P(B) = P[A \cup (B - A)] = P(A) + P(B - A) \ge P(A)$

27. Consider the case in which each sample point has the same probability.

29. 1/1   **31.** 2/5   **33.** \$10   **35.** \$30

Section 9.4

1. 0   **3.** 0   **5.** 1/6   **7.** 1/6   **9.** $((0.25)(0.05) + (0.35)(0.04))/(0.25 + 0.35)$

11. The prisoner should put one white ball in three of the urns and the remaining 97 white balls and the 100 black balls in the fourth urn. Then $P(W) = 1/4 + 1/4 + 1/4 + (1/4)(97/197)$ $\approx 7/8$.

13. $P(B \mid B) = \dfrac{P(B \cap B)}{P(B)} = \dfrac{P(B)}{P(B)} = 1$

15. First $P(B) = P[(A \cup A') \cap B] = P[(A \cap B) \cup (A' \cap B)] = P(A \cap B) + P(A' \cap B)$. Hence

$$P(A' \mid B) = \frac{P(A' \cap B)}{P(B)} = \frac{P(B) - P(A \cap B)}{P(B)} = 1 - \frac{P(A \cap B)}{P(B)} = 1 - P(A \mid B)$$

17. $P(A \cup B \mid C) = \dfrac{P[(A \cup B) \cap C]}{P(C)} = \dfrac{P[(A \cap C) \cup (B \cap C)]}{P(C)} = \dfrac{P(A \cap C)}{P(C)} + \dfrac{P(B \cap C)}{P(C)}$

$= P(A \mid C) + P(B \mid C)$

19. 1/2   **21.** 500/525

23. (a) There are 216 sample points, and each has probability 1/216.

(b) $A \subseteq C$, $B \subseteq C$, $A \cap B = \emptyset$ (c) $P(A) = 1/36$, $P(B) = 1/72$, $P(C) = 4/9$

(d) $P(A \mid B) = 0$, $P(A \mid C) = P(A)/P(C) = 1/16$, $P(B \mid A) = 0$

$P(B \mid C) = P(B)/P(C) = 1/32$, $P(C \mid A) = 1$, $P(C \mid B) = 1$

25. (a) All ordered pairs of different cards.   (b) (52)(51)   (c) $A \subseteq C$, $A \subseteq B$

(d) $P(A) = (13)(12)/(52)(51)$, $P(B) = 4P(A)$, $P(C) = 13/52$

(e) $P(A \cap B) = P(A)$, $P(A \cap C) = P(A)$, $P(B \cap C) = P(A)$

(f) $P(A \mid B) = 1/4$, $P(B \mid A) = 1$, $P(A \mid C) = 12/51$, $P(C \mid A) = 1$, $P(B \mid C) = 12/51$,

$P(C \mid B) = 1/4$

27. 2/3   **29.** 1/2, 7/8   **31.** 3/25   **33.** 0.324

35. (a) 126   (b) 1/8   (c) $\left[2 - (1/2^5)\right]/12$

37. (a) $r/(b+r)$ (b) $(c+b)/(b+r+c+d)$ (c) $b(c+b)/(b+r)(b+r+c+d) + r(b+d)/(b+r)(b+r+c+d)$

(d) $b(c + b)/(b + r)(b + r + c + d)$ (e) $b(c + b)(2c + b)/(b + r)(b + r + c + d)(b + r + 2c + 2d)$

Section 9.5

1. $3/11$ **5.** $1/3$; $1/4$; $2/3$ **9.** $4/19$; $3/17$; $9/19$ **13.** $23/45$

Section 9.6

1. Yes   **3.** $1/3^4$ **5.** $8/27$ **7.** Yes   **9.** (a) $18/125$ (b) $103/250$
11. (a) 0 or 1   (b) 0 or 1   (c) 1   **13.** $7/8$
15. $P(A \cap B) = P(A) + P(B) - P(A \cup B) = P(A) - P(A)P(B') = P(A)\left[1 - P(B')\right]$
$= P(A)P(B)$
17. (a) $1/8$, $5/8$ (b) We are assuming female is independent of blue eyes.
19. $11/12$ **21.** $1/169$; $3/(13)(51)$ **23.** $6.98   **25.** (a) 0.81   (b) 0.01   (c) 0.1

Section 9.7

1. $1/4$ **3.** $3/16$ **5.** TS   **7.** 23.75% **9.** (a) 0.3; 0.7   (b) 0.09; 0.49; 0.42
11. $P_1(YY) = 0.09$, $P_1(GG) = 0.49$, $P_1(GY) = 0.42$

Chapter 9 Test

1. $1/2$ **3.** $P(A \cap B) = P(A) + P(B) - P(A \cup B) = 1 + 1 - 1 = 1$ **5.** 0.035; $1/7$
7. $P(B) = 1 - \dfrac{P(A \cup B) - P(B)}{P(A)} = \dfrac{P(A) + P(B) - P(A \cup B)}{P(A)} = \dfrac{P(A \cap B)}{P(A)}$
9. 0.36; 0.16; 0.48   **11.** $5/9$ **13.** 799; $144/799$; $7/799$; $648/799$
15. 100,000;   0.32805;   0.00045;   0.59049;   0.40951
17. (a) $2/3$ (b) $1/3$ (c) $4/9$ (d) $2/9$ **19.** $2 \times 10^{-6} - 10^{-12}$

CHAPTER 10

Section 10.2

1. 0   **3.** $15.1666\ldots$ **5.** 5.5   **7.** 19.5¢   **9.** 2
11. $E(f \cdot g) = 4/9$; $E(f)E(g) = 8/9$
13. $E(f \cdot g) = E(f)E(g) = 1/4$ **15.** 7   **17.** 3   **19.** $4/9$

Section 10.3

1. (a) -2.7¢   **3.** (a) -20¢   (b) -10¢   (c) 0¢   **5.** -50¢
7. -4.6¢   **11.** -5.3¢ per play
13. 0.040386   **15.** 2.86¢ per play;   11.66 hr   **17.** After eight consecutive losses.

Section 10.4

1. 20; 35; 120; 840   **5.** 50; 2450; 117,600   **7.** $(16)(15)(14)(13) = 43{,}680$ **9.** 60
11. 120   **13.** 120   **15.** (a) $(26)^3$ (b) $_{26}P_3$ (c) $(3)(26)(25)$ (d) $1/26$ **17.** $n = 8$
19. The combinations taken 2 at a time are $\{1,2\}$, $\{1,3\}$, $\{1,4\}$, $\{2,3\}$, $\{2,4\}$, $\{3,4\}$
21. $(a + b)^3 = a^3 + 3a^2b + 3ab^2 + b^3 = {}_3C_0a^3 + {}_3C_1a^2b + {}_3C_2ab^2 + {}_3C_3b^3$

Section 10.5

3. 4,496,388   **7.** -14.6¢
9. (a) $1/3838380$ (b) 0.000053147   (c) 0.0021923   (d) 0.0311798   (e) -39.7¢
11. (a) $1/169$;   (b) $1/13$ **19.** (a) $(4)(13/52)(12/51)(11/50)(10/49)$
(b) $(13)(4/52)(3/51)(2/50)(1/49)$
(c) $(52)(4/52)(3/51)(2/50)$
21. (a) $\dfrac{4}{{}_{52}C_{13}}$ (b) $\dfrac{(52)(39)}{{}_{52}C_{13}}$ (c) $\dfrac{{}_{16}C_{13}}{{}_{52}C_{13}}$ **23.** $5/52$

25. (a) $(5/13)(48/51)(47/50)(46/49)(45/48)$ (b) $\dfrac{(_4C_2)(48)(47)(46)}{(3!)(_{52}C_5)}$

(c) $\dfrac{(_4C_3)(_{48}C_2)}{_{52}C_5}$ (d) $\dfrac{48}{_{52}C_5}$ (e) 0 **27.** $\dfrac{(12)(_{40}C_4)}{_{52}C_5}$; $\dfrac{(_{12}C_5)(_{40}C_3)}{_{52}C_5}$; $\dfrac{_{12}C_5}{_{52}C_5}$

Section 10.6

1. Given a_4, the player wins if a 4 is tossed and loses if a 7 is tossed. Since there are three ways of tossing a 4 and six ways of tossing a 7, there are three chances in nine of winning. Hence $P(w \mid a_4) = 1/3$.

3. $(5/11)(5/36)/(0.4929)$ **5.** 0.1255 **7.** $(5/36)^2$

9. The complete probability tree is the same as Figure 10.4 except that the w and ℓ probabilities are reversed in all but the first two paths. The probability of winning is $306/495$.

Section 10.7

1. 0.6 **3.** $\$300$ **5.** $\$1500$ **7.** $2/3$

9. $\dfrac{1}{2}(cf_{n+1}+dg_{n+1})+\dfrac{1}{2}(cf_{n-1}+dg_{n-1}) = c\left[\dfrac{1}{2}f_{n+1} + \dfrac{1}{2}f_{n-1}\right]+d\left[\dfrac{1}{2}g_{n+1} + \dfrac{1}{2}g_{n-1}\right] = cf_n+dg_n$

11. Use the equation in the text with $P(L_1) = p$ and $P(W_1) = (1 - p)$.

Chapter 10 Test

1. $1/2$ **3.** $-15¢$

5. $_{n+1}C_r = \dfrac{(n + 1)!}{(n + 1 - r)!r!} = \left(\dfrac{n + 1}{n + 1 - r}\right)\dfrac{n!}{(n - r)!r!} = \left(\dfrac{n + 1}{n + 1 - r}\right)\,_nC_r$. Hence,

$_nC_r = \left(\dfrac{n + 1 - r}{n + 1}\right)\,_{n+1}C_r = \left(1 - \dfrac{r}{n + 1}\right)\,_{n+1}C_r.$

7. $(_4C_2)(_4C_2)(_{13}C_2)/(_{52}C_4)$ **9.** 0.909 **11.** 4845

13. Since $f(a_i) \geq 0$ and $P\{a_i\} \geq 0$, $i = 1,\ldots,n$, we have $E(f) = f(a_1)P\{a_1\} + \cdots + f(a_n)P\{a_n\} \geq 0$.   **15.** 0.0238476

CHAPTER 11

Section 11.1

1. 3×2; 2×3; 3×3; 3×3; 1×3; 3×1 **3.** (a) 0 (b) -3 (c) 6

5. $A^t = \begin{bmatrix} 5 & 2/3 & 7 \\ 0 & 1 & 3 \\ -1 & 6 & 3 \\ 2 & -2 & 8 \end{bmatrix}$

7. If A is a $1 \times m$ row vector, then A^t has order $m \times 1$, which is a column vector.

9. Denote the entries of D by d_{ij} and the entries of D^t by d_{ij}^t. Then $d_{ii}^t = d_{ii}$ and $d_{ij}^t = d_{ij} = 0$ for $i \neq j$.

11. $\begin{bmatrix} 1 & 2 & 3 & 4 \\ 1/2 & 1 & 3/2 & 2 \\ 1/3 & 2/3 & 1 & 4/3 \\ 1/4 & 1/2 & 3/4 & 1 \end{bmatrix}$ **13.** $\begin{bmatrix} 0 & 2 & 3 & 4 \\ 2 & 0 & 3 & 4 \\ 3 & 3 & 0 & 4 \\ 4 & 4 & 4 & 0 \end{bmatrix}$

15. $0; 6; 8$ **17.** $1; 1; 1; 1$ **21.** $m \cdot n$ **23.** $\begin{bmatrix} 3.00 & 2.60 \\ 2.75 & 2.80 \\ 3.10 & 2.40 \end{bmatrix}, C^t$

Section 11.2

1. $\begin{bmatrix} 1 & 5 & 4 \\ 10 & 0 & -1 \\ -2 & 3 & 4 \end{bmatrix}$ **3.** $\begin{bmatrix} 2 & 7 & 6 \\ 14 & -3 & 3 \\ -4 & 7 & 7 \end{bmatrix}$ **5.** $\begin{bmatrix} -3 & 5 & 10 \\ 2 & -6 & -3 \\ 2 & -1 & 4 \end{bmatrix}$ **7.** $\begin{bmatrix} 1 & 2 & 2 \\ 4 & -3 & 4 \\ -2 & 4 & 3 \end{bmatrix}$

9. $\begin{bmatrix} -1/2 & 5/2 & 7/2 \\ 3 & -3/2 & -1 \\ 0 & 1/2 & 2 \end{bmatrix}$ **11.** $\begin{bmatrix} 0 & -7 & -9 \\ -10 & 6 & -2 \\ 2 & -5 & -7 \end{bmatrix}$ **13.** $\begin{bmatrix} 14 & -6 & -25 \\ 16 & 15 & 17 \\ -14 & 16 & -2 \end{bmatrix}$

15. $\dfrac{1}{18}\begin{bmatrix} 12 & -7 \\ 10 & -6 \\ 15 & -2 \end{bmatrix}$ **17.** $\dfrac{1}{15}\begin{bmatrix} 0 & 10 & -12 \\ 5 & -45 & 2 \end{bmatrix}$ **19.** $X = \dfrac{1}{3}A = \begin{bmatrix} 0 & 1 \\ 4/3 & -2/3 \end{bmatrix}$

21. $X = B = \begin{bmatrix} 1 & -1/2 \\ 3 & 0 \end{bmatrix}$ **23.** $X = A + C - B = \begin{bmatrix} 0 & 7/2 \\ 2 & -4 \end{bmatrix}$

27. $(A + B) + C = \left[a_{ij} + b_{ij} \right] + \left[c_{ij} \right] = \left[(a_{ij} + b_{ij}) + c_{ij} \right]$

$\quad = \left[a_{ij} + (b_{ij} + c_{ij}) \right] = \left[a_{ij} \right] + \left[b_{ij} + c_{ij} \right] = A + (B + C)$

29. $c(A + B) = c\left[a_{ij} + b_{ij} \right] = \left[ca_{ij} + cb_{ij} \right] = \left[ca_{ij} \right] + \left[cb_{ij} \right]$

$\quad = c\left[a_{ij} \right] + c\left[b_{ij} \right] = cA + cB$

31. $(kA)^t = \left[ka_{ij} \right]^t = \left[ka_{ji} \right] = k\left[a_{ji} \right] = k\left[a_{ij} \right]^t = kA^t$

Section 11.3

1. Not defined **3.** 4×5 **5.** 7×3 **7.** 12 **9.** $25/12$ **11.** $\begin{bmatrix} 9 & -14 & 21 \\ -5 & 12 & -12 \\ 5 & 9 & 9 \end{bmatrix}$

13. $\begin{bmatrix} 18 & 0 & 0 \\ 0 & -5 & 0 \\ 0 & 0 & -7 \end{bmatrix}$ **15.** $\begin{bmatrix} 2.5 & 3 & 2 & 1 \end{bmatrix}\begin{bmatrix} 20 \\ 12 \\ 15 \\ 25 \end{bmatrix} = \begin{bmatrix} 141 \end{bmatrix}$ **17.** 32

19. Each row of O consists entirely of zeros, so the vector multiplication of any row of O with any column of A equals zero.

21. If $C = AB$ then $c_{ij} = \begin{cases} 0 & \text{if } i \neq j \\ a_i b_i & \text{if } i = j \end{cases}$

Section 11.4

1. $\begin{bmatrix} 1 \\ 18 \\ -1 \end{bmatrix}$ **3.** $\begin{bmatrix} 6 \\ -1 \\ -2 \end{bmatrix}$ **5.** $\begin{bmatrix} 9 & -2 & 5 & 6 \end{bmatrix}$ **7.** $\begin{bmatrix} 2 & 0 & 1 \\ 36 & 0 & 18 \\ -2 & 0 & -1 \end{bmatrix}$

9. $\begin{bmatrix} 6 & 0 & 24 \\ -1 & 0 & -4 \\ -2 & 0 & -8 \end{bmatrix}$ **11.** \$4115

13. They both equal $\begin{bmatrix} -3 \\ 13 \\ 11 \end{bmatrix}$ **15.** They both equal $\begin{bmatrix} 1 & 18 & -1 \end{bmatrix}$

Section 11.5

1. They both equal $\begin{bmatrix} -5 & 5 \\ -3 & 13 \end{bmatrix}$ **3.** They both equal $\begin{bmatrix} 1 & 9 \\ 7 & 19 \end{bmatrix}$

5. They both equal $\begin{bmatrix} 8 & 7 \\ -1 & 11 \end{bmatrix}$　**7.** No　**9.** $\begin{bmatrix} 1 & 0 & -4 & 4 \\ 2 & 1 & -2 & 3 \\ 3 & 5 & -1 & 2 \end{bmatrix}$

11. $\begin{bmatrix} 6 & 3 & -6 & 9 \\ 1 & 0 & -4 & 4 \\ 3 & 5 & -1 & 2 \end{bmatrix}$　**13.** $\begin{bmatrix} 2 & 1 & -2 & 3 \\ 1 & 0 & -4 & 4 \\ -3 & -5 & 1 & -2 \end{bmatrix}$　**15.** $\begin{bmatrix} 2 & 1 & -2 & 3 \\ -1 & -1 & -2 & 1 \\ 3 & 5 & -1 & 2 \end{bmatrix}$

17. $\begin{bmatrix} 7/2 & 7/2 & -5/2 & 4 \\ 1 & 0 & -4 & 4 \\ 3 & 5 & -1 & 2 \end{bmatrix}$　**21.** $\begin{bmatrix} 7 & -2 \\ -3 & 1 \end{bmatrix}$　**23.** $\dfrac{1}{2}\begin{bmatrix} 3 & 2 \\ 1 & 0 \end{bmatrix}$

25. Singular　**27.** $\dfrac{1}{13}\begin{bmatrix} 39 & -26 & -26 & 26 \\ 29 & -21 & -29 & 26 \\ -13 & 13 & 13 & -13 \\ -2 & 1 & 2 & 0 \end{bmatrix}$

29. For AB to be defined, B must be an $n \times p$ matrix for some $p \in \mathbb{N}$. For BA to be defined, p must equal m so B is an $n \times m$ matrix.

31. $\begin{bmatrix} 1/3 & 0 \\ 0 & 1 \end{bmatrix}$　**33.** I　**35.** Multiply both sides of $AB = 0$ on the right by B^{-1}.

37. If the diagonal entries of D are $d_1, d_2, \ldots, d_n$, then D^{-1} is a diagonal matrix with diagonal entries $d_1^{-1}, d_2^{-1}, \ldots, d_n^{-1}$.　**39.** Since $AA^{-1} = I$, we have $(A^{-1})^{-1} = A$.

Section 11.6

1. $x = 3/2$, $y = 1/2$　**3.** $x = 0$, $y = 2$, $z = -1$

5. Multiplying both sides of the second equation by $-1/3$ gives $2x - 3y = -2/3$. This contradicts the first equation.

7. $x = -1$, $y = -1$, $z = 3$　**11.** 61, 113, 151, 51, 111, 144, 24, 70, 89, 41, 86, 109

13. $x = 63$, $y = -22$　**15.** $x = 0$, $y = 10$　**17.** $x = 2.057$, $y = 2.533$, $z = 1.311$

19. 225, 56, 191, 90, 21, 88, 213, 44, 207, 312, 81, 250　**21.** $\dfrac{1}{12}\begin{bmatrix} 4 & -12 & 0 \\ -10 & 24 & 6 \\ 5 & -6 & -3 \end{bmatrix}$

23. $w = -5/7$, $x = -3/14$, $y = 4/7$, $z = 45/14$

Chapter 11 Test

1. (a) 4　(b) 0　(c) 2×3　(d) $\begin{bmatrix} -1 & 3 & -4 \\ -5 & 0 & -1 \end{bmatrix}$　(e) $a_{21} = 5$　(f) 3×2

3. $\begin{bmatrix} 1 & 1 & 12 \\ 5 & -3 & 4 \\ 4 & 9 & -12 \end{bmatrix}$, $\begin{bmatrix} 3 & 1 & -4 \\ -9 & 3 & 0 \\ -2 & 1 & -2 \end{bmatrix}$, $\begin{bmatrix} -3 & 1 & 4 \\ 9 & -3 & 0 \\ 2 & -1 & 2 \end{bmatrix}$, $\begin{bmatrix} 2 & -2 & 1 \\ 1 & 0 & 5 \\ 4 & 2 & -7 \end{bmatrix}$,

$\begin{bmatrix} -3 & 0 & 24 \\ 21 & -9 & 6 \\ 9 & 12 & -15 \end{bmatrix}$　**5.** $\begin{bmatrix} 17 & 13 & -2 \\ 8 & 8 & -26 \\ 13 & -43 & 53 \end{bmatrix}$, $\begin{bmatrix} 6 & 39 & -60 \\ 22 & 17 & 8 \\ -7 & -22 & 55 \end{bmatrix}$

7. Let $V = \begin{bmatrix} v_1 & v_2 & \cdots & v_n \end{bmatrix}$. Then $VV^t = v_1^2 + v_2^2 + \cdots + v_n^2 \geq 0$. If $VV^t = 0$, then $v_1 = v_2 = \cdots = v_n = 0$, so $V = 0$.

9. (a) The first row is unchanged and the other rows are all zeros.
(b) The first column is unchanged and the other columns are all zeros.

11. Show that the steps for finding the inverse cannot be carried out.　**13.** -5; -9

15. (a) T　(b) F　(c) F　(d) T　(e) F　(f) T　(g) T　(h) T　(i) F　(j) F　(k) F

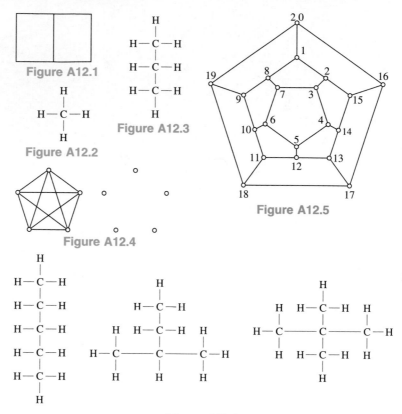

Figure A12.1

Figure A12.2

Figure A12.3

Figure A12.4

Figure A12.5

Figure A12.6

CHAPTER 12

Section 12.1

1. Figure A12.1   **7.** Figure A12.2   **9.** Figure A12.3   **13.** Figure A12.4

17. 3, 3, 2, 3   **19.** Figure A12.5   **21.** Figure A12.6

23. If the number of edges incident to v in G is m, then the number of edges incident to v in $\overline{G}$ is $n - 1 - m$.

Section 12.2

1. The following vertices are adjacent: c and b, a and b, a and d, b and d. The following edges are adjacent: cb and ab, ba and ad, ad and bd, cb and bd, ab and bd. Finally, c is incident with cb; b is incident with bc, ba, and bd; a is incident with ab and ad; d is incident with ad and db.

3. Figure A12.7   **5.** Figure A12.8

7. $G - e$ has the same number of vertices as G and one fewer edge.   **9.** Figure A12.9

11. See Figure A12.10. $d(a) = 9$, $d(b) = 6$, $d(c) = 2$, $d(d) = 2$, $d(e) = 1$, $d(h) = 0$. Vertices a and e are odd, and the others are even. Vertex h is isolated, and e is an end vertex.

13. The sum of the degrees is 20, and there are 10 edges. There are two odd vertices.

15. Suppose n is odd. Since each of the a_i's is an even number minus 1, the sum of the a_i's gives an even number minus the odd number n. But an even number minus an odd number is odd. This is a contradiction.   **17.** Figure A12.11

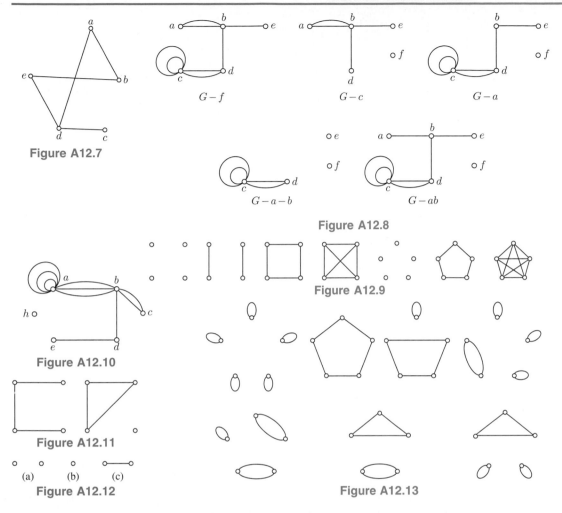

Figure A12.7

$G-f$ $G-c$ $G-a$

Figure A12.8

$G-a-b$ $G-ab$

Figure A12.9

Figure A12.10

Figure A12.11

(a)   (b)   (c)
Figure A12.12

Figure A12.13

19. The sum of the degrees is an even number since this sum equals twice the number of edges. But the sum of p odd numbers is even only if p is even (see Exercise 15).

21. Only part (d) is always true. See Figure A12.12.   **23.** $p(p-1)/2 - 8$

25. This follows from Theorem 12.2.   **27.** The difference of even numbers is even.

Section 12.3

1. 4.8 days   **3.** Figure A12.13   **5.** Figure A12.14   **7.** Figure A12.15

Section 12.4

1. The initial vertex is b, and the final vertex is c.   **3.** d, dc, c, e_2, a, ad, d

5. No, the edge dc is repeated.   **7.** Exercise 5 gives such an example.   **9.** No; no

11. The proof is the same as that of Theorem 12.4 except that since the initial and final vertices are different, they must have odd degrees. No; yes

13. No; yes   **15.** b, bc, c, cb, b; a, ab, b, bf, f, fb, b; a, ab, b

Section 12.5

1. Figure A12.16   **3.** Figure A12.17

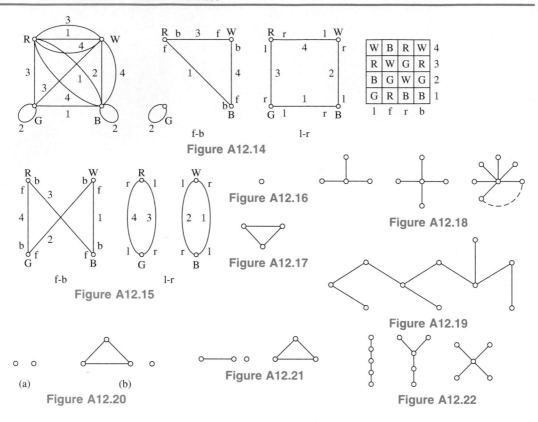

Figure A12.14

Figure A12.15

Figure A12.16

Figure A12.17

Figure A12.18

Figure A12.19

(a)   (b)

Figure A12.20

Figure A12.21

Figure A12.22

5. (a) 10 isolated vertices   (b) n isolated vertices   **7.** Figure A12.18

9. Figure A12.19   **11.** Figure A12.20   **13.** Figure A12.21   **15.** Figure A12.22

17. The components of a graph without circuits are connected graphs without circuits.

19. Let G be a connected (p, q) graph with $q = p - 1$. Suppose G has a circuit C. Let e be an edge of C. Then $G - e$ is a connected graph with p vertices and $p - 2$ edges. This contradicts Corollary 12.1. Hence G has no circuits and is therefore a tree.

21. If w is a cut vertex, then its removal must disconnect two vertices. Hence there must exist vertices u and v such that every $u - v$ trail contains w. Conversely, if every $u - v$ trail contains w, then the removal of w disconnects u and v. This increases the number of components; so w is a cut vertex.   **23.** Figure A12.23   **25.** $p(p - 1)/2$

27. If a vertex of K_p is removed, all the remaining vertices are still connected.

29. The only tree with two vertices is G_3 in Figure 12.17 and this tree has one edge.

31. The graph $T - v$ is still connected since v is an end vertex. We cannot produce a circuit by deleting v, so $T - v$ has no circuits. Hence, $T - v$ is a tree with three vertices.

33. Delete an end vertex and proceed as before.

Section 12.6

1. Since a tree has no circuits, it cannot enclose a region. Hence, the only region of a tree is the external region.   **3.** 2, 4, 6   **5.** Any of the graphs in Figure 12.46 would do.

7. Yes; for example, supergraphs of K_5 can be constructed with any number of vertices greater than 5.

9. Yes; see Figure A12.24.   **11.** If $n \geq 5$, then K_n contains K_5 as a subgraph.

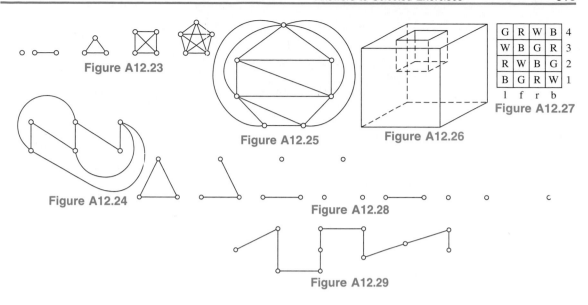

Figure A12.23

Figure A12.24

Figure A12.25

Figure A12.26

G	R	W	B	4
W	B	G	R	3
R	W	B	G	2
B	G	R	W	1

l f r b

Figure A12.27

Figure A12.28

Figure A12.29

13. The graph of the icosahedron in Figure 12.63.   **15.** Figure A12.25   **17.** Two

19. Nine   **21.** Figure 12.56c   **25.** G_1 in Figure 12.57   **27.** K_5 **29.** K_7

31. Each new circuit encloses an addition region, so q and r both increase by 1.

Section 12.7

1. These polyhedra each contain a pair of points that are the endpoints of line segments that do not lie entirely in the polyhedron.   **3.** Yes; yes

5. Yes, in the first polyhedron $V = 16$, $E = 24$, and $F = 10$; in the second $V = 12$, $E = 18$, and $F = 8$.

7. (a) Euler's formula holds for the polyhedra in Figure 12.62.
(b) Euler's formula does not hold for the polyhedron in Figure A12.26. For the polyhedron in Figure A12.26, $V = 16$, $E = 24$, and $F = 11$.

9. A parallelogram with equal sides;   a rectangle.

11. If there is such a convex polyhedron, then $V = E$, and so $V - E + F = 2$ implies $F = 2$. But this is impossible.   **13.** Nine

15. Since all the V_k's are the same and all the V_h's are the same, this follows from (12.1).

17. Since one face is a triangle, quadrilateral, or pentagon, they all must be one of these. Similarly, since one vertex has degree 3, 4, or 5, they all must have one of these degrees. Hence, $3 \leq h \leq 5$ and $3 \leq k \leq 5$. Since h and k each have three possible values, there are nine possible cases for h and k.

Chapter 12 Test

1. 11   **3.** Since $pd = 2q$ is even, p or d must be even. There are $q = pd/2$ edges.

5. 11   **7.** Figure A12.27

9. The graph cannot have any circuits, since then it would have more than one region.

11. Figure A12.28   **13.** (a) T   (b) T   (c) F   (d) F   (e) T   (f) F   (g) F

15. Figure A12.29

Photo Credits

Chapter 1 **Opener:** Harvey Lloyd/The Stock Market. **Photo 1.2:** Lester Sloan/Woodfin Camp and Associates. **Photo 1.3:** Steve Goldberg/Monkmeyer Press. **Photo 1.4:** Courtesy of S & S Worldwide. **Photo 1.5:** Mark Reinstein/The Image Works.

Chapter 3 **Opener:** Scott Applewhite/AP/Wide World Photos. **Photo 3.1:** Stanley Rowin/The Picture Cube. **Photo 3.2:** R. Michael Stuckey/Comstock.

Chapter 4 **Opener:** Hanley and Savage/The Stock Market. **Photo 4.1:** R. B. Sanchez/The Stock Market. **Photo 4.2:** Jack Elness/Comstock. **Photo 4.3:** UPI/Bettmann Newsphotos. **Photo 4.4:** Jim Pickerell/The Image Works.

Chapter 5 **Opener:** Peter Vadnai/The Stock Market. **Photo 5.1:** Berenholz/The Stock Market. **Photo 5.2:** Tony Craddock/Science Photo Library/Photo Researchers. **Photo 5.3:** Gregory Sams/Science Photo Library/Photo Researchers. **Photo 5.4:** Gregory Sams/Science Photo Library/Photo Researchers. **Photo 5.5:** Gregory Sams/Science Photo Library/Photo Researchers.

Chapter 6 **Opener:** Gabe Palmer/The Stock Market. **Photo 6.1:** Paul Barton/The Stock Market. **Photo 6.2:** Craig Hammell/The Stock Market. **Photo 6.3:** Craig Hammell/The Stock Market. **Photo 6.4:** S. Smith/The Image Works.

Chapter 7 **Opener:** Gabe Palmer/The Stock Market. **Photo 7.1:** Culver Pictures. **Photo 7.2:** Blair Seitz/Photo Researchers. **Photo 7.3:** Historical Pictures/Stock Montage. **Photo 7.4:** Culver Pictures. **Photo 7.5:** NASA/The Image Works. **Photo 7.6:** Scott Berner/The Picture Cube. **Photo 7.7:** Jim Pickerell/The Image Works.

Chapter 8 **Opener:** Brownie Harris/The Stock Market. **Photo 8.1:** Hugh Rogers/Monkmeyer Press. **Photo 8.2:** R. Michael Stuckey/Comstock.

Chapter 9 **Opener:** Ken Wood/Photo Researchers. **Photo 9.1:** Joseph Nettis/Photo Researchers. **Photo 9.2:** Dick Luria/Photo Researchers. **Photo 9.3:** Comstock.

Chapter 10 **Opener:** Tom Campbell/FPG International. **Photo 10.1:** George Goodwin/Monkmeyer Press. **Photo 10.2:** Culver Pictures. **Photo 10.3:** Culver Pictures. **Photo 10.4:** Culver Pictures. **Photo 10.5:** Culver Pictures. **Photo 10.6:** Culver Pictures. **Photo 10.7:** Mimi Forsyth/Monkmeyer Press. **Photo 10.8:** Mimi Forsyth/Monkmeyer Press. **Photo 10.9:** Mimi Forsyth/Monkmeyer Press. **Photo 10.10:** James Steinberg/Photo Researchers. **Photo 10.11:** Spencer Grant/Monkmeyer Press.

Chapter 11 **Opener:** Michael Abramson/Woodfin Camp and Associates. **Photo 11.1:** Mimi Forsyth/Monkmeyer Press.

Chapter 12 **Opener:** Tom McHugh/Photo Researchers. **Photo 12.1:** Tom Grill/Comstock. **Photo 12.2:** Dick Durrance/Woodfin Camp and Associates.

Cartoon Credits

Page 42: *B.C.* reprinted by permission of Johnny Hart and Creators Syndicate, Inc. **Page 164:** *HERMAN* © Jim Unger. Reprinted by permission of Universal Press Syndicate. All rights reserved. **Page 198:** *BLOOM COUNTY* © 1984, Washington Post Writers Group. Reprinted by permission. **Page 233:** *FOR BETTER OR FOR WORSE* © Lynn Johnston Prod. Reprinted by permission of Universal Press Syndicate. All rights reserved. **Page 317:** *ZIGGY* © Ziggy & Friends. Reprinted by permission of Universal Press Syndicate. All rights reserved. **Page 417:** *POOR ARNOLD'S ALMANAC* © 1989 Creators Syndicate. Reprinted by permission of Arnold Roth. **Page 513:** *ROBOTMAN* reprinted by permission of NEA, Inc.

INDEX